Structural Analysis

Principles, Methods
and Modelling

Structural Analysis

Principles, Methods and Modelling

Gianluca Ranzi and Raymond Ian Gilbert

CRC Press
Taylor & Francis Group
Boca Raton   London   New York

CRC Press is an imprint of the
Taylor & Francis Group, an **informa** business

A SPON PRESS BOOK

CRC Press
Taylor & Francis Group
6000 Broken Sound Parkway NW, Suite 300
Boca Raton, FL 33487-2742

© 2015 by Taylor & Francis Group, LLC
CRC Press is an imprint of Taylor & Francis Group, an Informa business

No claim to original U.S. Government works

Printed on acid-free paper
Version Date: 20140617

International Standard Book Number-13: 978-0-415-52644-9 (Paperback)

Visit the Taylor & Francis Web site at
http://www.taylorandfrancis.com

and the CRC Press Web site at
http://www.crcpress.com

Contents

Preface

This book is intended as a text for undergraduate students of Civil or Structural Engineering about to embark on the adventure of learning how to analyse engineering structures. It provides a unique in-depth treatment of structural analysis where fundamental aspects and derivations of the analytical and numerical formulations are outlined and illustrated by numerous simple, yet informative, worked examples.

The book is divided into four parts. The first part comprises Chapters 1 to 4 and covers the analysis of statically determinate structures. Although it is assumed that the student has already completed courses in statics and mechanics of solids, some of the material revises concepts and procedures that have been covered previously. The second part of the book includes Chapters 5 to 9 and deals with the classical methods for the analysis of statically indeterminate structures. These methods are suitable for hand calculation, where the deformation characteristics and the geometry of the structure, as well as considerations of equilibrium, are used to establish the internal actions and structural deformations. Although practising structural engineers usually use computer software packages to analyse structures, these classical methods provide the background knowledge that is essential for the preparation of appropriate input for structural analysis software and the correct interpretation of the output. The third part (Chapters 10 to 12) covers the stiffness method of analysis that underpins most computer applications and commercially available structural analysis software, while the fourth part (Chapters 13 to 15) deals with more advanced topics, including the finite element method, structural stability and problems involving material nonlinearity. Finally, three appendices are included that provide additional background material that is of use throughout the book.

Every topic is illustrated with *numerous worked examples* that lead the student step by step through the solution process. Sections entitled *Reflection Activities* invite students to reflect on the material covered by questioning some of the details of the procedures or extending their applicability to a broader range of problems. The detailed sequence of steps required by different methods of analysis are described in particular sections entitled *Summary of Steps*. At the end of most chapters, a wide range of tutorial problems are set to assist the student to practise the various analysis techniques and to build critical thinking.

The book is complemented by a comprehensive set of educational support material for both instructors and students as described below. We hope that the book will prove useful to both students and instructors.

ADDITIONAL RESOURCES AVAILABLE FOR STUDENTS AND INSTRUCTORS

Resources available for download at http://www.crcpress.com/product/isbn/9780415526449

- *MATLAB scripts* of the worked examples included in Chapters 10 to 13 and 15 are available and will enable students to gain a clear understanding of all steps involved in the structural analysis solution process when implemented in a computer program.

Resources for instructors who adopt the book are available from CRC Press upon request. Please send an email to orders@crcpress.com or contact your sales representative.

- *Solutions Manual* presents detailed solutions for every tutorial problem included in the book.
- *PowerPoint presentations* files available for face-to-face lectures.
- *A PowerPoint presentation* to introduce MATLAB to students with no prior knowledge of it.
- *Videos* supported by a voice narration and are available to enable students to review selected material covered in the book at their own pace. These videos are viewable with computers and smart devices.

MATLAB® is a registered trademark of The MathWorks, Inc. For product information, please contact:

The MathWorks, Inc.
3 Apple Hill Drive
Natick, MA 01760-2098 USA
Tel: 508-647-7000
Fax: 508-647-7001
E-mail: info@mathworks.com
Web: www.mathworks.com

FEEDBACK

As this is the first edition of the book, we would welcome your feedback. Please feel free to send us any comments, criticisms or suggestions regarding any aspect of the book. We will greatly appreciate your input. Please send your feedback to gianluca.ranzi@sydney.edu.au.

Acknowledgments

Thanks are extended to the colleagues and students who provided valuable assistance in the preparation of the textbook and of the solution manual, in particular, Peter Ansourian, Massimiliano Bocciarelli, Graziano Leoni, and Alessandro Zona, who reviewed parts of the manuscript; Lingzhu Chen, Glen Clifton, Anthony Joseph, Charles K. S. Loo Chin Moy, and Osvaldo Vallati, who assisted in the preparation of the material for the solution manual and in the formatting of the figures.

The authors acknowledge the support given by their respective institutions, the University of Sydney and the University of New South Wales.

Chapter 1

Introduction

1.1 STRUCTURAL ANALYSIS AND DESIGN

Structural engineering involves the *analysis* and *design* of structures and is one of the core sub-disciplines of *civil engineering*. Civil engineering structures take a variety of forms and include buildings, bridges, towers, marine structures, dams, tunnels, retaining walls and other infrastructure. The most common materials used for the construction of these structures are concrete, steel and timber, although a variety of other materials are used including stone, aluminium, polymers, carbon fibre, glass and many more.

Structural engineering underpins and sustains the built environment, where bridges, buildings and other structures must be safe, serviceable, durable, aesthetically pleasing and economical. It is concerned primarily in developing structural solutions to resist loads and other forces, and in devising ways to provide safe load paths for these forces. It is an applied science, founded on mathematical laws and physical concepts applied to engineering materials, both traditional and advanced, for the provision of infrastructure and technological innovation. The demands of new and existing structures imposed by society and by economics and the use of new or advanced materials require solutions that challenge and unite creativity and scientific rigour.

Structural design involves the determination of the type of structure that is suitable for a particular purpose, the materials from which the structure is to be constructed, the loads and other actions that the structure must sustain and the arrangement, layout and dimensions of the various components of the structure. This involves detailed calculations to ensure that the structure is stable and that every structural member, and every connection between members, has adequate strength to resist the design loads. It also involves determination of the deformation of each part of the structure to ensure that the structure remains serviceable throughout its design life and is able to perform its intended function. Structural design involves careful detailing of every part of the structure, including the preparation of detailed structural drawings that effectively communicate the engineering design to the contractors who are engaged to build the structure.

Structural analysis is an integral part of structural design. It involves the calculation of the response of the structure to the design loads and imposed deformations that it will be required to resist during its lifetime. This involves the determination of the internal forces within the various components of the structure and the deformation of these components. Calculation of the internal forces in a structure will allow the structural designer to select materials and member sizes that provide the structure with adequate strength and ensure that the chances of collapse are acceptably small. Calculation of the deformation of the structure will permit the assessment of serviceability. Whether or not a structure is acceptable for a particular purpose depends on its deformation, as well as its strength.

1

The mathematical algorithms used for structural analysis range from classical methods, suitable for manual calculation (often assuming linear elastic material behaviour), to more complex non-linear numerical analysis, using modern matrix methods and high-performance computers. The choice depends on the type and complexity of the structure and the computational power available to the structural analyst. All methods involve the application of structural mechanics to an idealised structure, where approximations are made concerning the geometry of the structure and its support conditions, the applied loads and deformations, and the material modelling laws. The interpretation of the results of the analysis requires both experience and engineering judgment.

1.2 STRUCTURAL IDEALISATION

It is not possible to undertake an exact analysis of a structure. Real structures have complex geometries that are never known exactly at the time of the analysis. Even structural members that are supposed to be straight are never exactly straight; cross-sectional dimensions that are supposed to be uniform along a member are never exactly uniform; and the dimensions and rotational capacity of connections between structural members and structural supports are never known precisely. Real materials have properties that vary from point to point in a structure and the actual variation and distribution of material properties is never known with a great deal of precision. In addition, the magnitude and distribution of the loads imposed on a structure are rarely known accurately. Structural analysis is therefore undertaken on an *idealised structure*, where simplifying assumptions are made concerning the geometry of the structure and its supports, the material properties and the applied loads so that the conditions approximate those of the real structure. These simplifying approximations introduce errors, some small and some not so small. However, the aim of the idealisation is to simplify the analysis, so that the calculated loads, internal actions, reactions, stresses and deformations are not too different from those in the real structure and adequately describe the behaviour of the structure.

In this process, *loads* applied to any structure can take a variety of forms and may be *idealised* as either concentrated loads or distributed loads, including line loads and surface loads. The structure is *idealised* as a combination of various *components* and *members*, adequately connected to each other and capable of transferring the applied actions through the structure to the supporting foundations. The magnitudes and directions of the forces exerted on a structure by its supports depend on the *types of support*. These structural idealisations (members, loads and supports) are briefly discussed in the following sections.

1.3 STRUCTURAL MEMBERS AND ELEMENTS

Structures are composed of various components and members, that can be categorised according to their dimensions and the way they carry loads.

Many common structural members can be adequately described as *one-dimensional elements*. This is an appropriate classification for elements whose lengths are larger than their cross-sectional dimensions. This is illustrated in Figure 1.1 for a steel member and a reinforced concrete member, whose lengths L are greater than the size of their cross-sections depicted by B and D. For the purpose of structural idealisation, it is possible in these cases to replace the member by a one-dimensional line element.

Line elements are further classified according to the internal actions they resist. For example, a line element that carries only axial forces (either compression or tension) is

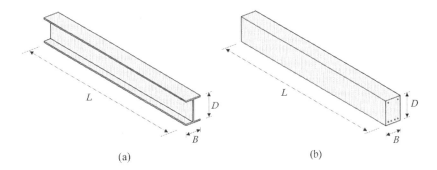

Figure 1.1 One-dimensional (line) elements. (a) Structural steel. (b) Reinforced concrete.

usually referred to as a *bar* or a *truss element*. Figure 1.2 illustrates a truss element before and after deformation. The element elongates when resisting a tensile force and shortens when resisting a compressive force. A bar carrying axial tensile loads applied at each end, with each cross-sections subjected only to axial tension, is called a *tie* (Figure 1.2a), while a bar carrying axial compression applied at each end, with each cross-sections subjected only to axial compression, is called a *strut* (Figure 1.2b).

Whether an axially loaded element in a structure is classified as a tie or a strut often depends on its position in the structure and the way the loads are transferred to it. Let us consider a bookshelf attached to a wall as shown in Figure 1.3a. The weight of the shelf is resisted by compressive forces induced in the diagonal members below the shelf, and the diagonal elements are classified as struts. Let us now consider the arrangement shown in Figure 1.3b. To resist the weight of the shelf, the diagonals above the shelf are in tension and are therefore classified as ties. Sometimes, flexible wires, chains or ropes are used to carry axial tension and are referred to as *cables*. Because of their flexibility, cables are unable to resist compressive loads.

A one-dimensional line element that carries transverse loads in bending and shear (with or without torsion) is called a *beam* or a *girder* and these are very common in structures. Their main features can be illustrated with the help of a simple ruler. If you hold the ruler at its ends and bend it, as depicted in Figure 1.4a, you will apply a couple at each end and a constant curvature will exist along the length of the ruler. The ruler is subjected to a constant bending moment along its length and there are no transverse loads applied to the ruler and

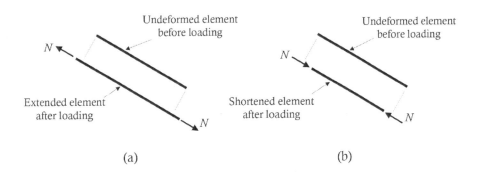

Figure 1.2 One-dimensional (line) elements — ties and struts. (a) Bar in axial tension: a tie. (b) Bar in axial compression: a strut.

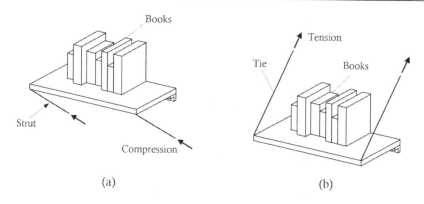

Figure 1.3 Example of struts and ties. (a) Struts supporting shelf from below. (b) Ties supporting shelf from above.

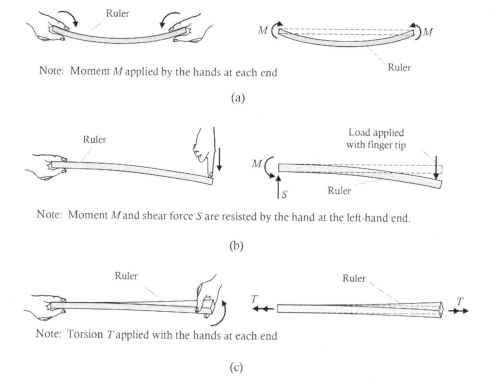

Figure 1.4 Example of line elements acting as beams or girders. (a) Constant moment. (b) Shear force and varying. (c) Constant torsion.

consequently no shear forces. If you are unfamiliar with the concepts of bending moment and shear force, do not worry as they are introduced and explained in Chapters 2 and 3.

If you now hold the same ruler at one end only and apply a downward force with your fingertip at the other end, as shown in Figure 1.4b, each cross-sections of the ruler will be subjected to a constant shear force and the bending moment increases linearly with distance from the applied downward load. A beam can also be subjected to torsion or twisting about its longitudinal axis. Torsion will be induced in the ruler, if you hold it at its ends and twist one side relative to the other, as illustrated in Figure 1.4c.

Columns are one-dimensional elements primarily loaded in axial compression but may also carry bending moments, shear forces and torsion.

Members with one dimension (thickness t) smaller than the other two (length L and width B) can be represented by two-dimensional elements, also referred to as planar elements. It is common to refer to two-dimensional flat elements as *plates* (Figure 1.5a) and curved elements as *shells* (Figure 1.5b). Plates can be subdivided into *slabs* and *walls*, where *slabs* are usually in a horizontal plane and withstand transverse loads by a combination of axial force, bending moments, torsion and shear at their cross-sections. *Walls* consist of planar elements, usually in a vertical plane, and resist both in-plane and transverse forces. Slabs are commonly found in the floor systems of buildings and form part of most bridge decks, while walls also form part of most buildings and are often used to retain soils and water. *Shells* usually have curved surfaces capable of resisting axial forces, bending, torsion and shear.

Membranes are two-dimensional elements that support applied loads by means of a biaxial tensile state. A simple example of this kind of structure is the jumping bed depicted in Figure 1.6, which is capable of supporting the transverse loads induced by the self-weight of a person by stretching its material as shown. The tension induced in the membrane material of the jumping bed by virtue of its deformation is resisted by a compressive force in the ring strut.

Arches are curved structural members that transfer the applied loads by means of compressive forces along the arch. Depending on the actual geometry of the arch and on the nature of the applied loads, the compressive axial force might be combined with different

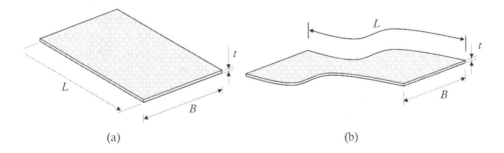

(a)   (b)

Figure 1.5 Two-dimensional (planar) elements. (a) Plate. (b) Shell.

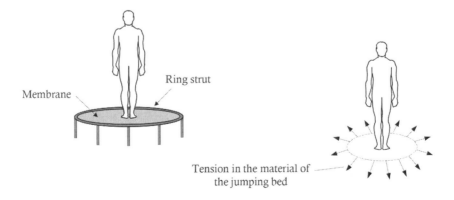

Ring strut

Membrane

Tension in the material of
the jumping bed

Figure 1.6 A membrane structure.

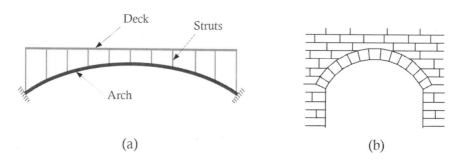

Figure 1.7 Common arches. (a) Arch supporting a bridge deck. (b) Arch in a masonry wall.

levels of moment and shear forces that need to be carefully evaluated. Arches are popular in bridge applications. A typical arch supporting a bridge deck is shown in Figure 1.7a. Arches are also commonly used around openings in masonry construction, as depicted in Figure 1.7b.

1.4 STRUCTURAL SYSTEMS

Structural systems consist of combinations of structural members and are classified according to the type of elements and the way these elements are connected to each other. A real structure is usually represented by a combination of one-, two- and three-dimensional elements. Only some of the most common structural systems are considered in this section and these include *trusses, frames, arch and cable structures* and *surface structures*.

Trusses are two- or three-dimensional systems of struts and ties connected at their ends by simple pinned connections. This implies that the ends of the members connected together at a node can rotate relative to each other. It is common to assume that loads are mainly applied at the nodes and, under these conditions, the truss elements are subjected to either axial tensile or compressive forces. In reality, truss members may be loaded, between the nodes and must also resist their self-weight. Depending on the members' dimensions, these *member loads* may need careful consideration for an accurate prediction of the structural response. The main components of a simple plane truss are shown in Figure 1.8. The horizontal top and bottom members are usually referred to as *top and bottom chords*, respectively. The distance between the top and bottom chord greatly affects the magnitude of the axial forces in the chords and the structural performance of the truss. Under the same

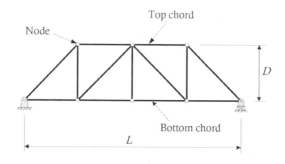

Figure 1.8 Typical components of a plane (two-dimensional) truss.

applied loading, a reduction in the truss depth (i.e. a reduction in the dimension D in Figure 1.8) will lead to higher forces in the top and bottom chords, and the deflection of the truss will also increase. The *diagonal members* and *vertical members* are either struts or ties depending on the directions and magnitudes of the applied loads.

Frames are two- or three-dimensional systems formed by beams and columns connected by rigid, semi-rigid or pinned connections. In the case of rigid connections, no relative rotations can occur between the ends of the members framing into the connection. In these systems, the structural elements are usually subjected to a combination of axial force, bending moment and shear force. Low- to medium-height buildings often employ systems of frames, consisting of horizontal beams and slabs, and vertical columns and walls to resist both gravity and lateral loads. High-rise buildings often use frames to support gravity loads and include stiff shear walls to assist the frames to carry the lateral loads caused by wind or earthquakes. Skeletons of some simple two-dimensional frames are illustrated in Figure 1.9. In particular, Figure 1.9a shows a typical plane frame that may be adopted in a low-rise building, while frames shown in Figure 1.9b represent two possible layouts of *portal frames* commonly used for warehouses or factory buildings.

Cable structures are those in which at least one of the main load-carrying elements is a flexible cable. Flexible cables can resist only axial tension and are unable to resist bending moment, shear force or axial compression. A cable can resist lateral loads if the ends of the cable are securely attached to firm supports. It does this by taking up a particular shape depending on the loading such that the bending moment at every point along the cable is zero and only axial tension exists at every point along the cable. Many of the longest span bridges are cable structures, as are many sports stadium roofs. Some examples of cable structures are illustrated in Figure 1.10.

Surface structures, such as plates and shells, are three-dimensional structures formed by elements whose thickness is usually much smaller than its other dimensions. These can be arranged to produce a wide range of structures, such as flat floor slabs, bridge decks, tanks and silos, cooling towers, dam walls, tunnels, roofs and coverings in a variety of forms, including domes and hyperbolic paraboloids. The response of this type of structure is highly three-dimensional and needs careful attention at the analysis stage. Examples of a shell structure and a dome structure are shown in Figure 1.11.

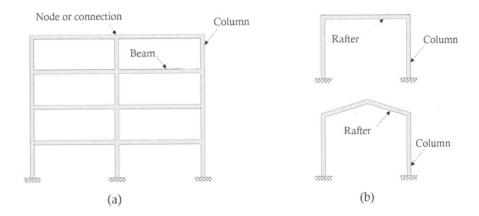

Figure 1.9 Examples of plane frames. (a) Typical rigid frame for buildings. (b) Typical portal frames for warehouses.

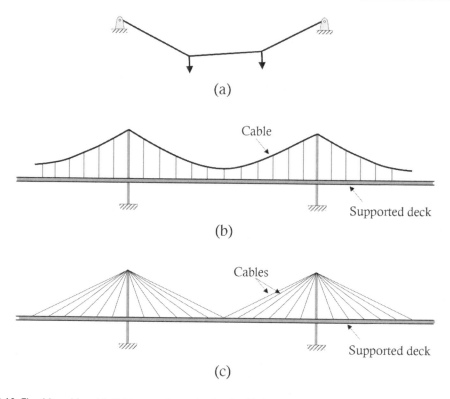

Figure 1.10 Flexible cables. (a) Cable carrying point loads. (b) Suspension bridge. (c) Cable-stayed bridge.

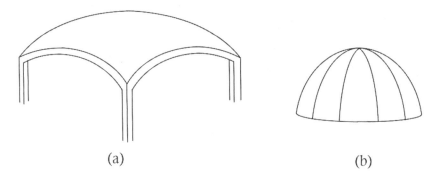

Figure 1.11 Surface structures. (a) Shell structure. (b) Dome.

1.5 TYPES OF LOADS

The loads to be used in the design of a structure depend on the type of structure, its location and its purpose. They include dead loads, live loads, wind loads, earthquake loads, earth pressure, liquid pressure, rain and snow loads and prestress. Loads may also develop owing to restraint of deformation caused by a wide range of actions, including, for example, temperature changes, support settlement or shrinkage of concrete or timber.

Dead loads, also referred to as *permanent actions*, are generally defined as those loads imposed by both the structural and non-structural components. Dead loads include the self-weight of the structure and the forces imposed by all walls, floors, roofs, ceilings, permanent partitions, service machinery and other permanent construction. Dead loads are usually permanent, are fixed in position and can be estimated with reasonable accuracy from the density of the relevant material or type of construction.

Live loads, also called *imposed actions*, are loads that are attributed to the intended use or purpose of the structure and are generally specified by regional or national codes and specifications. The *specified* live load depends on the expected use or occupancy of the structure and usually includes allowances for impact and inertia loads (where applicable) and for possible overload. Both uniformly distributed and concentrated live loads are usually specified. The magnitude and distribution of the actual live load are never known exactly at the design stage, and it is by no means certain that the specified live load will not be exceeded at some stage during the life of the structure. Live loads may or may not be present at any particular time; they are not constant and their position can vary. Although part of the live load is transient, some portion may be permanently applied to the structure and will have effects similar to dead loads. Live loads also arise during construction owing to stacking of building materials, the use of equipment or the construction procedure (such as the loads induced by floor-to-floor propping in multi-storey construction).

Wind, earthquake, snow and *temperature loads* (or *actions*) depend on the geographical location. The magnitudes of the loads adopted for use in structural design also depend on the relative importance or design life of the structure (and are expressed as a function of the mean return period). Wind loads also depend on the surrounding terrain as well as the height and shape of the structure above the ground. Earthquake loads are also highly dependent on the soil conditions present below the foundation and by the type of foundation adopted in the design. These *environmental* loads are specified in national codes.

Loads on structures take a variety of forms and may be idealised as either *concentrated loads* or *distributed loads*, including line loads and surface loads. A *concentrated load* is a force that acts at a point on a structure. Of course, real loads act over finite areas, but it is not unreasonable to treat many loads as if they were located at a point. The wheel loads from a truck on a bridge deck, for example, are often treated as concentrated loads, as illustrated in Figure 1.12. A *line load* is a load distributed along a line in a structure. The transverse load arising from the weight of a partition wall on a floor slab is an example of a line load. Surface loads are distributed over an area of a structure. Examples of surface loads are the self-weight of a floor slab, the wind load on the sides of a building, the pressure exerted by water on the sides of a dam and the snow load on a roof.

The magnitude of the distributed load on a structural member at a particular location (per unit length or per unit area) is called the *load* (or *force*) *intensity*. If the load intensity is

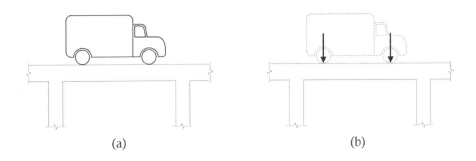

(a)   (b)

Figure 1.12 Example of wheel loads replaced by concentrated forces. (a) Wheel loads. (b) Concentrated forces replacing wheel loads.

constant over a length of beam or an area of floor, the load is said to be *uniformly distributed*. Sometimes the force intensity and the nature of its variation are well defined, in which case the determination of the magnitude and position of the resultant presents no problem. In other cases, it may be necessary to introduce approximating assumptions. For example, it is usually assumed that the loads imposed on the floor by the occupants of a building structure are uniformly distributed, even though the imposed loads on different parts of a floor are likely to be quite different at any time. The structural designer must often arrange the design loads on a floor to produce the most adverse effect.

1.6 SUPPORTS FOR STRUCTURES

A *structure* is usually supported on a foundation or on another structure. The magnitude and direction of the forces exerted on the structure by its supports depend on the type of support. Common support conditions include *fixed supports*, *pinned supports* and *roller supports*, and these are briefly illustrated below considering some very common objects that we encounter in our day-to-day life.

For example, let us consider the sign shown in Figure 1.13a. It needs to be able to resist its own self-weight as well as environmental loads, such as wind, rain or snow. In order for such a structure to keep its stability, it needs to be fixed to the ground to prevent any movement or rotation at the support location; that is, both translation and rotation must be prevented. This particular support condition is referred to as a *fixed support* (or *built-in* support). A member fixed at one end and free at the other is called a *cantilever*. Examples include balconies (Figure 1.13b), awnings and walls that are fixed at the base to a footing and free at the top as illustrated in Figure 1.13c. Figure 1.13d illustrates the symbol that we will use to describe a fixed support.

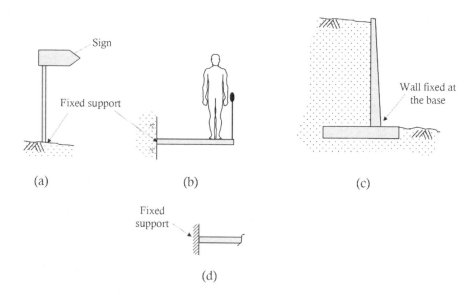

(a) (b) (c)

(d)

Figure 1.13 Examples of structures with fixed supports. (a) Sign post. (b) Cantilevered balcony. (c) Cantilevered retaining wall. (d) Symbol.

Let us now consider a typical see-saw as depicted in Figure 1.14a. The support enables the board to rotate freely (as one person goes up and the other goes down) while preventing translations in any direction. This is an example of a *pinned support*. Any support that holds the structure in position but allows relatively free rotation is considered to be a pinned support. The supports illustrated in Figure 1.14b may all be considered as pinned supports, even though at real supports some restraint against rotation is inevitable. The validity of these assumptions depends on the detailing of the connection and relative rigidities of the connecting members. In this book, we will adopt the symbols shown in Figure 1.14c to indicate a pinned support.

A *roller support* is a support that allows both rotation and translation parallel to the surface on which the roller moves, but it does not permit translation perpendicular to the surface on which the roller is located. Roller supports are commonly used in bridges, as illustrated in Figure 1.15a, to enable the bridge to expand or contract freely by permitting movement at one end. This static configuration, with one end of the beam pinned and the other one on a roller support, is very common in practice and such a structure is known as a *simply-supported beam*. In this case, the beam is free to rotate at each end and also free to move horizontally at the right-hand support, as shown. It is not free to move vertically (upwards or downwards) at either end. The common symbols that we will use for a roller support are shown in Figure 1.15b.

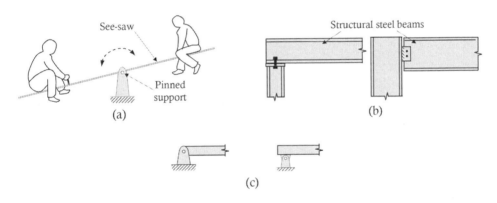

Figure 1.14 Pinned supports. (a) Example of see-saw. (b) Steel brain supports assumed to be pinned. (c) Symbols.

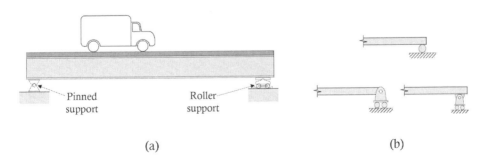

Figure 1.15 Roller supports. (a) Simply-supported bridge beam. (b) Symbols.

Chapter 2

Statics of structures
Equilibrium and support reactions

2.1 INTRODUCTION

This chapter provides a review of a number of fundamental concepts relevant to the statics of structures. After introducing the coordinate systems that we will use throughout the book, we revisit the concepts of forces and moments, and how these are specified in a two-dimensional plane and a three-dimensional space. We then describe the typical supports used to restrain the movements of structures and present the use of free-body diagrams to identify all forces acting on a structure. On the basis of equilibrium considerations, we then show the procedures required to calculate the reactions. This chapter deals with structural systems whose reactions (and internal actions as outlined in the following chapters) are determined using only the principles of statics. This class of structures is referred to as *statically determinate* and the analysis of such structures is independent of the deformation of the structure and the properties of the materials from which the structure is made. This will become clearer in subsequent chapters when we will be dealing with structures that require consideration of the structural deformations and material properties, as well as statics, to evaluate the reactions.

2.2 COORDINATE SYSTEMS

Throughout the book, we will assume coordinate systems to be *orthogonal* and to satisfy the *right-hand rule*. The *orthogonal* condition implies that all axes are perpendicular to each other and, with the *right-hand rule*, the positive directions of the x-, y- and z-axes are oriented in the same way as axes located along the thumb, forefinger, and middle finger, respectively, of the right hand, as illustrated in Figure 2.1.

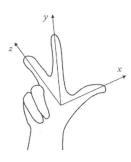

Figure 2.1 Use of the right-hand rule.

A large number of structures can be idealised as two-dimensional and analysed in a two-dimensional plane. In these cases, it is convenient to assign the x- and y-axes to the plane of interest and to have the positive direction of the z-axis pointing out of the plane. For example, the x–y coordinate system of Figure 2.2a defines the coordinates in the plane of the page. Applying the right-hand rule to Figure 2.2a, the thumb of the right hand should point toward the positive direction of x and the forefinger in the positive direction of y (Figure 2.2b). The positive direction of the z-axis should then point out of the page, as depicted by the middle finger in Figure 2.2b.

The positive orientation of rotations around the different axes can be evaluated by the right-hand rule as illustrated in Figure 2.3. This approach requires that you grasp the relevant axis with your right hand, making sure that the thumb is pointing in the positive direction of the axis. The rotation produced by your fingers when closing the hand (curling) around the axis identifies positive rotation about the axis (see Figure 2.3).

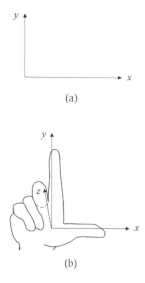

(a)

(b)

Figure 2.2 Coordinate systems assigned on a plane. (a) Coordinate system in the plane of the page defined by the x- and y-axes. (b) Use of right-hand rule.

Figure 2.3 Positive rotations with respect to an axis.

2.3 FORCE

A *force* is a vector quantity defined by its magnitude (expressing the intensity of the force), its line of action (depicting the direction of the force), and its point of application (where the force is applied).

Let us consider a force of magnitude F applied in the two-dimensional plane at a point O, as shown in Figure 2.4. Adopting an orthogonal reference system x–y, the components of the force parallel to the x- and y-axes are referred to as F_x and F_y, respectively. If θ is the angle formed between the force and the x-axis (Figure 2.4a), components F_x and F_y can be calculated from trigonometry as:

$$F_x = F \cos \theta \tag{2.1a}$$

$$F_y = F \sin \theta \tag{2.1b}$$

Alternatively, it is possible to express components F_x and F_y in terms of the angles between the force and the x- and y-axes, θ_x and θ_y, respectively, as shown in Figure 2.4b. That is:

$$F_x = F \cos \theta_x = Fl \quad \text{where} \quad l = \cos \theta_x \tag{2.2a}$$

$$F_y = F \cos \theta_y = Fm \quad \text{where} \quad m = \cos \theta_y \tag{2.2b}$$

where l and m are referred to as the *direction cosines* of the force, θ_x defines the angle between the force and the x-axis, and θ_y represents the angle between the force and the y-axis. In two-dimensional problems, the Fl and Fm notation has little advantage over the notation $F \cos \theta$ and $F \sin \theta$, but in three-dimensional problems, there are often significant advantages, as we will see later.

From Figure 2.4, it can be observed that the magnitude and direction of F can be expressed in terms of its components F_x and F_y as:

$$F = \sqrt{F_x^2 + F_y^2} \qquad \theta = \theta_x = \arctan\left(\frac{F_y}{F_x}\right) \tag{2.3a,b}$$

where the trigonometric function $\arctan(F_y/F_x)$ can also be written as $\tan^{-1}(F_y/F_x)$ or $\text{atan}(F_y/F_x)$. From Equation 2.3a, it can be readily shown that $l^2 + m^2 = 1$ (as $F = \sqrt{F_x^2 + F_y^2} = F\sqrt{l^2 + m^2} = F \times 1$).

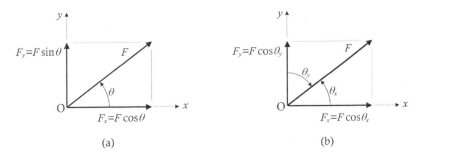

(a)   (b)

Figure 2.4 Components of a force in a (two-dimensional) plane. (a) Components of a force F based on θ. (b) Components of a force F based on θ_x and θ_y.

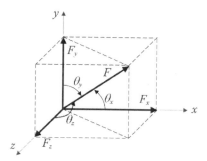

Figure 2.5 Components of a force in a three-dimensional space.

When considering a three-dimensional space, the components of a force of magnitude F can be described on the basis of the direction cosines that it forms with the x-, y- and z-axes, as depicted in Figure 2.5. The force components parallel to the x-, y- and z-axes are denoted as F_x, F_y and F_z, respectively, and are determined from:

$$F_x = F \cos \theta_x = l\,F \qquad \text{where} \qquad l = \cos \theta_x \tag{2.4a}$$

$$F_y = F \cos \theta_y = m\,F \qquad \text{where} \qquad m = \cos \theta_y \tag{2.4b}$$

$$F_z = F \cos \theta_z = n\,F \qquad \text{where} \qquad n = \cos \theta_z \tag{2.4c}$$

where l, m and n are the *direction cosines* of the force, while θ_x, θ_y and θ_z are the angles of the force with the x-, y- and z-axes, respectively, as shown in Figure 2.5. The magnitude of F can be calculated from its components as:

$$F = \sqrt{F_x^2 + F_y^2 + F_z^2} \tag{2.5}$$

from which it can be observed that $l^2 + m^2 + n^2 = 1$.

2.4 MOMENT OF A FORCE

The *moment of a force* about any point is the product of the force and the perpendicular distance of the point from the line of action of the force. For example, if we consider force F applied at point A in the x–y plane, as shown in Figure 2.6a, the moment of F about the point B is determined as:

$$M_B = F\,d \tag{2.6}$$

where d is the perpendicular distance from point B to the line of action of F.

When dealing with two-dimensional problems, it is implicit that the moment of a force about a point in the x–y plane is actually the moment of the force about the axis coming out of the page at the point in the z direction. The sign convention adopted in Equation 2.6 assumes anticlockwise moments to be positive (in accordance with the right-hand rule of

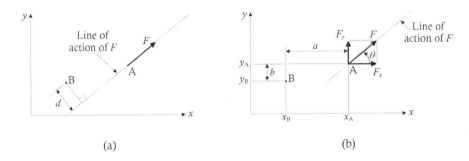

Figure 2.6 Moment of a force in a (two-dimensional) plane. (a) Moment calculated as $M_B = F\,d$. (b) Moment calculated as $M_B = F_y a - F_x b$.

Figure 2.3 applied to rotations with respect to the z-axis) and the subscript of M_B denotes that the moment is evaluated about point B.

The moment M_B may be thought of as a measure of the tendency of the force F to cause rotation about the z-axis through point B. If an object is pivoted at B, the force F acting on the object, in the absence of any other forces, will cause rotation about B.

It is often convenient to determine the moment of a force as the sum of the moments of its components. Reconsidering the force F of Figure 2.6a, the moment previously calculated in Equation 2.6 can be rewritten in terms of the components F_x and F_y (Figure 2.6b):

$$M_B = F_y a - F_x b \tag{2.7}$$

where a and b are the perpendicular distances from point B to F_y and F_x, respectively. This calculation can also be expressed in terms of the coordinates of points A (x_A, y_A) and B (x_B, y_B). Noting that $a = x_A - x_B$ and $b = y_A - y_B$, we have:

$$M_B = F_y(x_A - x_B) - F_x(y_A - y_B) \tag{2.8}$$

When calculating the moment of a force applied in a three-dimensional space, it is usually convenient to sum the moments of the force components parallel to the coordinate axes. For example, referring to Figures 2.7a and b, the moment about the x-axis at B produced by force F is denoted $M_{x(B)}$ and can be evaluated as:

$$M_{x(B)} = F_z b - F_y c \tag{2.9a}$$

This can be verified by considering the components of F in the y–z plane, as shown in Figure 2.7b. The subscript of $M_{x(B)}$ depicts the fact that the moment is calculated at point B with respect to the x-axis.

Adopting a similar notation, the moments at B determined with respect to the y- and z-axes are denoted as $M_{y(B)}$ and $M_{z(B)}$, respectively, and can be determined as (Figures 2.7c and d)

$$M_{y(B)} = F_x c - F_z a \qquad M_{z(B)} = F_y a - F_x b \tag{2.9b,c}$$

where a, b and c depict the perpendicular distances between point B and the lines of actions of the force components (F_x, F_y and F_z). Considering the coordinates of points A(x_A, y_A, z_A)

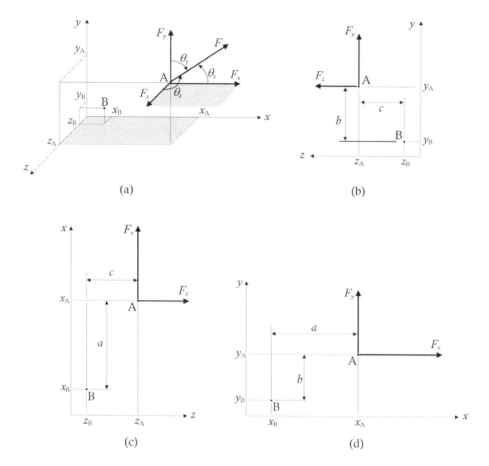

Figure 2.7 Moment of a force in a three-dimensional space. (a) Force *F* applied at point A. (b) Moment about the *x*-axis at B. (c) Moment about the *y*-axis at B. (d) Moment about the *z*-axis at B.

and $B(x_B, y_B, z_B)$, the expressions for the moments of Equation 2.9 can be simplified by substituting a, b and c with $a = x_A - x_B$, $b = y_A - y_B$ and $c = z_A - z_B$, to give:

$$M_{x(B)} = F_z(y_A - y_B) - F_y(z_A - z_B) \tag{2.10a}$$

$$M_{y(B)} = F_x(z_A - z_B) - F_z(x_A - x_B) \tag{2.10b}$$

$$M_{z(B)} = F_y(x_A - x_B) - F_x(y_A - y_B) \tag{2.10c}$$

which describes the moment calculated with respect to B for a force applied at point A.

The magnitude of the moment vector at B (based on Equations 2.9 or 2.10, and expressed in terms of $M_{x(B)}$, $M_{y(B)}$ and $M_{z(B)}$) and its direction cosines are given by:

$$M = \sqrt{M_{x(B)}^2 + M_{y(B)}^2 + M_{z(B)}^2} \tag{2.11a}$$

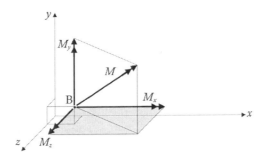

Figure 2.8 Representation of a moment in a three-dimensional space.

$$l = \frac{M_{x(B)}}{M} \qquad m = \frac{M_{y(B)}}{M} \qquad n = \frac{M_{z(B)}}{M} \qquad \text{(2.11b–d)}$$

The representation of the moment in three dimensions is outlined in Figure 2.8, which shows the moment resultant and its components. Moment vectors are usually represented by a double arrow, to distinguish them from force vectors.

2.5 RESULTANT FORCE AND MOMENT

A group of forces applied to a rigid body can be replaced by a *resultant force and moment* applied at a particular point. It is usually convenient for the resultant force and moment to be calculated at the centre of gravity of the body. The two resultants produce an equivalent effect to that induced by the group of forces. Because a single large force may cause local damage that would not be produced by a large number of smaller forces, the resultants are often said to be *statically equivalent* to the group of forces, that is, equivalent only as far as statics is concerned. The component of the resultant force in any direction is the same as the sum of the components of all the forces in the group in that direction. In a similar manner, the resultant moment calculated with respect to any axis is equal to the sum of the moments caused by all the forces in the group about that axis.

For a two-dimensional group of forces acting in the x–y plane, the components of the resultant force R_x and R_y can be expressed as the sum of the components of each force in the group in the x and y directions, respectively:

$$R_x = \Sigma F_x \qquad \text{(2.12a)}$$

$$R_y = \Sigma F_y \qquad \text{(2.12b)}$$

and the moment that the resultant produces about the z-axis through any point O, here referred to as $M_{R(O)}$, is equal to the sum of the moments produced by the forces in the group about that point:

$$M_{R(O)} = \Sigma M_{F(O)} \qquad \text{(2.12c)}$$

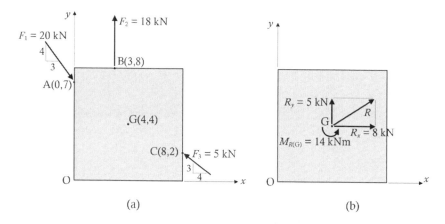

Figure 2.9 Resultant force and moment produced by a group of forces in a two-dimensional plane (note: all node coordinates are in metres). (a) Group of forces. (b) Resultant force and moment.

Let us consider the set of forces applied to the rigid square plate shown in Figure 2.9a and the calculation of the resultant force and moment at the centre of gravity of the plate located at its geometric centre G. From Equations 2.12a and b:

$$R_x = \Sigma F_x = F_{x1} + F_{x2} + F_{x3} = 12 + 0 - 4 = 8 \text{ kN} \tag{2.13a}$$

$$R_y = \Sigma F_y = F_{y1} + F_{y2} + F_{y3} = -16 + 18 + 3 = 5 \text{ kN} \tag{2.13b}$$

The moment resultant is the sum of the moments $M_{F(G)}$ of each force about the point G(4,4) calculated using Equation 2.8 as:

$$M_{F1(G)} = F_{y1}(x_A - x_G) - F_{x1}(y_A - y_G) = (-16)(0-4) - 12 \times (7-4) = 28 \text{ kNm} \tag{2.14a}$$

$$M_{F2(G)} = F_{y2}(x_B - x_G) - F_{x2}(y_B - y_G) = 18 \times (3-4) - 0 \times (8-4) = -18 \text{ kNm} \tag{2.14b}$$

$$M_{F3(G)} = F_{y3}(x_C - x_G) - F_{x3}(y_C - y_G) = 3 \times (8-4) - (-4)(2-4) = 4 \text{ kNm} \tag{2.14c}$$

and, from Equation 2.12c, the moment resultant at G is:

$$M_{R(G)} = M_{F1(G)} + M_{F2(G)} + M_{F3(G)} = 28 + (-18) + 4 = 14 \text{ kNm} \tag{2.14d}$$

The resultant force and moment are illustrated in Figure 2.9b.

The procedure required to calculate the resultants is revisited in Worked Example 2.1.

Let us now consider a set of parallel forces. Since the directions of all the forces are the same, the resultant is parallel to the forces and its magnitude may be found by algebraic addition.

WORKED EXAMPLE 2.1

Five forces $(F_1, F_2, F_3, F_4, F_5)$ are applied in the x–y plane to the rigid rectangular plate at the points and in the directions shown in Figure 2.10. Determine the resultants, expressed in terms of a force and a moment located at (i) the centre of gravity G(8,5) and (ii) the point H(16,10).

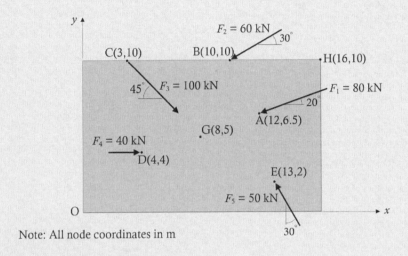

Note: All node coordinates in m

Figure 2.10 Rectangular plate for Worked Example 2.1.

The calculation of the components of the five forces is carried out using the direction cosines as specified in Equations 2.2. For clarity, all angles θ_x and θ_y related to each force are plotted in Figure 2.11 and tabulated below, together with the values obtained for the x and y components $(F_x$ and $F_y)$, and the coordinates of the point of application. The angles θ_x and θ_y have been measured on the basis of the sign convention illustrated in Figure 2.4b (from the positive x- and y-axes to the force vector).

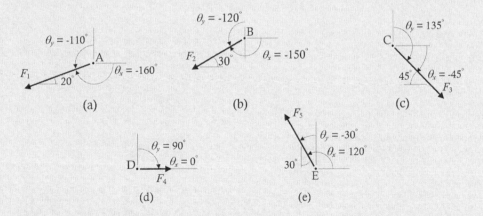

Figure 2.11 Angles formed by the five forces with the x- and y-axes. (a) F_1. (b) F_2. (c) F_3. (d) F_4. (e) F_5.

Force ID	Force (kN)	θ_x	θ_y	$F_x = F\cos\theta_x$ (kN)	$F_y = F\cos\theta_y$ (kN)	Point of Application of the Force	
						x (m)	y (m)
F_1	80	$-160°$	$-110°$	-75.18	-27.36	12	6.5
F_2	60	$-150°$	$-120°$	-51.96	-30	10	10
F_3	100	$-45°$	$135°$	70.71	-70.71	3	10
F_4	40	$0°$	$90°$	40	0	4	4
F_5	50	$120°$	$-30°$	-25	43.30	13	2

The resultant force is calculated in terms of R_x and R_y based on Equations 2.12a and b as:

$$R_x = \Sigma F_x = -41.43 \text{ kN} \quad \text{and} \quad R_y = \Sigma F_y = -84.77 \text{ kN}$$

and therefore $R = \sqrt{(-41.43)^2 + (-84.77)^2} = 94.35 \text{ kN}$.

The inclination of the line of action of R is obtained from Equation 2.3b and considering the signs of the resultant components:

$$\theta = \tan^{-1}\frac{R_y}{R_x} = 243.96°$$

Clearly, with negative x and y components, R is located in the third quadrant. Unlike the resultant force, the value of the resultant moment varies according to the point about which it is calculated. For this reason, the moments evaluated at points G and H are considered separately in the following. In the calculations, all lengths are in metres and forces are in kilonewtons.

(i) Resultant moment about point G(8,5):
The moment of each force about G is obtained from Equation 2.8:
For F_1: $M_{F_1(G)} = F_{y1}(x_A - x_G) - F_{x1}(y_A - y_G) = (-27.36)(12 - 8) - (-75.17)(6.5 - 5) = 3.32 \text{ kNm}$
For F_2: $M_{F_2(G)} = F_{y2}(x_B - x_G) - F_{x2}(y_B - y_G) = (-30)(10 - 8) - (-51.96)(10 - 5) = 199.80 \text{ kNm}$
For F_3: $M_{F_3(G)} = F_{y3}(x_C - x_G) - F_{x3}(y_C - y_G) = (-70.71)(3 - 8) - (70.71)(10 - 5) = 0 \text{ kNm}$
For F_4: $M_{F_4(G)} = F_{y4}(x_D - x_G) - F_{x4}(y_D - y_G) = (0)(4 - 8) - (40)(4 - 5) = 40 \text{ kNm}$
For F_5: $M_{F_5(G)} = F_{y5}(x_E - x_G) - F_{x5}(y_E - y_G) = (43.30)(13 - 8) - (-25)(2 - 5) = 141.51 \text{ kNm}$
and from Equation 2.12c: $M_{R(G)} = \Sigma M_{F(G)} = 384.63 \text{ kNm}$.

The resultants located at point G are shown in Figure 2.12a.

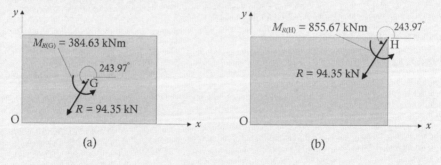

Figure 2.12 Resultants for Worked Example 2.1. (a) Resultants located at point G. (b) Resultants located at point H.

(ii) Resultant moment about point H(16,10):

$M_{F_1(H)} = F_{y1}(x_A - x_H) - F_{x1}(y_A - y_H) = (-27.36)(12 - 16) - (-75.17)(6.5 - 10) = -153.67 \text{ kNm}$

$M_{F_2(H)} = F_{y2}(x_B - x_H) - F_{x2}(y_B - y_H) = (-30)(10 - 16) - (-51.96)(10 - 10) = 180 \text{ kNm}$

$M_{F_3(H)} = F_{y3}(x_C - x_H) - F_{x3}(y_C - y_H) = (-70.71)(3 - 16) - (70.71)(10 - 10) = 919.24 \text{ kNm}$

$M_{F_4(H)} = F_{y4}(x_D - x_H) - F_{x4}(y_D - y_H) = (0)(4 - 16) - (40)(4 - 10) = 240 \text{ kNm}$

$M_{F_5(H)} = F_{y5}(x_E - x_H) - F_{x5}(y_E - y_H) = (43.30)(13 - 16) - (-25)(2 - 10) = -329.90 \text{ kNm}$

The resultant moment is therefore (Equation 2.12c): $M_{R(H)} = \Sigma M_{F(H)} = 855.67 \text{ kNm}$, as shown in Figure 2.12b.

In the case of two parallel forces of equal magnitude but opposite in direction (i.e. opposite sign), such as the two horizontal forces in Figure 2.13a, the resultant force is zero (i.e. $F - F$), but the moment of the two forces about any point is clearly not zero and equals $F\ell$. Such a pair of forces is called *a couple* and $F\ell$ is called the moment of the couple, represented by the moment resultant M_R:

$$M_R = F \times \ell \tag{2.15}$$

A couple tends to cause rotation without translation. It cannot be replaced by a single force, which must necessarily tend to cause translation as well as rotation. A system of forces is equivalent to a couple if the forces have a resultant force of zero magnitude but have a non-zero moment about any point. The couple is often represented by a single symbol (as shown in Figure 2.13b), showing its sense (clockwise or anticlockwise), with its magnitude specified. The couple has no component force in any direction and its moment is the same about all points.

For a three-dimensional space, the force and moment resultants are calculated following a procedure similar to the one adopted for two-dimensional planes. The components of R are given by:

$$R_x = \Sigma F_x \qquad R_y = \Sigma F_y \qquad R_z = \Sigma F_z \tag{2.16a-c}$$

and the moment components about any point O are:

$$M_{xR(O)} = \Sigma M_{xF(O)} \qquad M_{yR(O)} = \Sigma M_{yF(O)} \qquad M_{zR(O)} = \Sigma M_{zF(O)} \tag{2.17a-c}$$

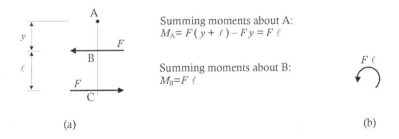

Summing moments about A:
$M_A = F(y + \ell) - Fy = F\ell$

Summing moments about B:
$M_B = F\ell$

(a)

(b)

Figure 2.13 Moment of a couple.

WORKED EXAMPLE 2.2

Consider the rigid body shown in Figure 2.14, which is subjected to two forces (F_1, F_2) whose components are as follows:

- For F_1 applied at A(20,8,0): $F_{1x} = 40$ kN, $F_{1y} = 30$ kN, $F_{1z} = 20$ kN
- For F_2 applied at B(0,0,12): $F_{2x} = 15$ kN, $F_{2y} = 10$ kN, $F_{2z} = 30$ kN

Calculate the resultant force and moment applied at the centre of gravity G(10,4,6).

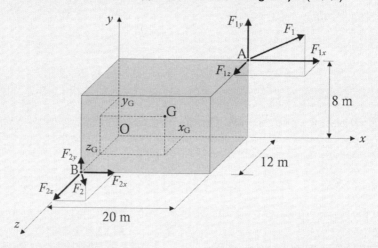

Figure 2.14 Rigid body for Worked Example 2.2.

The components of the resultant force are calculated by summing the components of all forces in the same direction:

$$R_x = \Sigma F_x = 55 \text{ kN}; \qquad R_y = \Sigma F_y = 40 \text{ kN}; \qquad R_z = \Sigma F_z = 50 \text{ kN}$$

and the resultant force is:

$$R = \sqrt{55^2 + 40^2 + 50^2} = 84.41 \text{ kN}$$

The inclination of the line of action of R is defined by the direction cosines, evaluated from R and its components using Equations 2.4:

$$I = \frac{R_x}{R} = 0.652; \qquad m = \frac{R_y}{R} = 0.474; \qquad n = \frac{R_z}{R} = 0.592$$

The moment calculated about each coordinate axis through G(10,4,6) is determined using Equations 2.10 for each of the applied forces:

$$M_{xR(G)} = F_{z1}(y_A - y_G) - F_{y1}(z_A - z_G) + F_{z2}(y_B - y_G) - F_{y2}(z_B - z_G)$$
$$= 20(8 - 4) - 30(0 - 6) + 30(0 - 4) - 10(12 - 6) = 80 \text{ kNm}$$
$$M_{yR(G)} = 40(0 - 6) - 20(20 - 10) + 15(12 - 6) - 30(0 - 10) = -50 \text{ kNm}$$
$$M_{zR(G)} = 30(20 - 10) - 40(8 - 4) + 10(0 - 10) - 15(0 - 4) = 100 \text{ kNm}$$

The resultant moment M_R at point G is therefore:

$$M_{R(G)} = \sqrt{80^2 + (-50)^2 + 100^2} = 137.5 \text{ kNm}$$

2.6 REACTIONS

Loads applied to a structure are transmitted to the foundations or to other parts of the structure by means of supports. The supports prevent movement of the structure in different directions, and when the structure is loaded, restraining forces (called *reaction forces* or simply *reactions*) develop at the supports. The specific reactions depend on the type of support.

Typical supports and their role in the behaviour of two-dimensional structures are discussed below and in subsequent sections. The two-dimensional representation (usually in the *x–y* plane) is very useful for practical applications as many structural systems, such as beams, columns, and frames, can often be conveniently idealised and analysed as two-dimensional structures. In these cases, it is usually assumed that the depth (or thickness) of each member is small when compared to its length and, because of this, the member can be modelled as a line element and its depth can be ignored. The case of reactions in three-dimensional structures is dealt with separately in Section 2.11.

A support provides a restraining action at a particular point in a structure and this can be represented by one or more reactions depending on the number of movements that are prevented. A number of common support conditions are illustrated in Table 2.1.

A *roller support*, for example, prevents movement in the direction perpendicular to the supporting plane. Such restraint is provided by means of a single reaction force applied at the support point *in that direction*.

A *pinned support* (or *hinged support*) allows rotation but does not permit translation in any direction. The reaction consists of a single force (*R*) whose line of action passes through the pinned support and whose magnitude and direction depend on the magnitude and direction of the applied loads. The reaction *R* can therefore act in any direction and, for convenience, is often replaced by its vertical and horizontal components (*V* and *H*, respectively). The reaction at a pinned support cannot include a couple (as rotation is permitted to occur freely).

The pinned support and the roller support are often called *simple supports*, and a single span beam, with a pinned support at one end and a roller support at the other, is called a *simply-supported beam*.

A *fixed support* (or a *built-in support*) does not permit either translation or rotation, and the reaction at a fixed support consists of a force acting through the support (with vertical and horizontal components *V* and *H*, respectively) and a couple *M*.

2.7 FREE-BODY DIAGRAM

A *free-body diagram* (FBD) is a very useful sketch of the structure showing all forces (including couples) applied to it and having all supports replaced with their corresponding reactions.

Let us now go through the steps involved in drawing the free-body diagram of the structure shown in Figure 2.15a. We first need to identify all the supports of the structure. In this case, they consist of a pinned support at A and a roller support at B. We then need to replace each support with the possible reactions that can develop at that support, as shown in Figure 2.15b (refer to Table 2.1).

The free-body diagram is then obtained by redrawing the structure with all external loads and all reactions as illustrated in Figure 2.15c. The pinned support at A has been replaced by the horizontal and vertical components of the reaction force at A (H_A and V_A, respectively) and the roller support at B is replaced by the vertical reaction force at B (V_B). The *x*-axis is

Table 2.1 Types of supports for two-dimensional structures

Support type	Symbol	Possible reactions	Restrained and unrestrained movements
Fixed support		M H V 3 reactions: H, V, M	Restrained: all movements, including rotation and translations Unrestrained: none
Pinned support		H V H V 2 reactions: H, V	Restrained: translations of the node Unrestrained: rotation
Hinge		V H H V 2 reactions: H, V	Restrained: relative translations between connecting members Unrestrained: relative rotation between connecting members
Guided support		M H 2 reactions: H, M	Restrained: rotation and translation perpendicular to supporting plane Unrestrained: translation parallel to supporting plane
Roller support		V V 1 reaction: V	Restrained: translation perpendicular to supporting plane Unrestrained: rotation and translation parallel to supporting plane

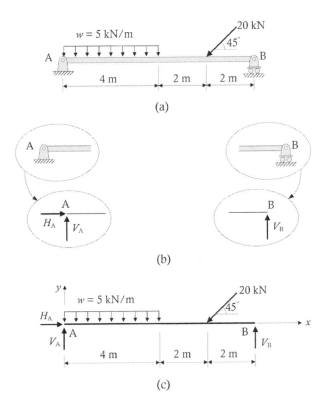

Figure 2.15 Example of a free-body diagram. (a) Idealized structure. (b) Replacement of support with equivalent reactions. (c) Free-body diagram.

taken as the longitudinal axis of the beam and, according to the right-hand rule (Figure 2.1), the y-axis is vertical (upwards) and the z-axis is at right angles to the x–y plane coming out of the page (Figure 2.2).

We can specify an arbitrary positive direction for the reactions when drawing them on the free-body diagram. For example, for the roller support isolated in Figure 2.16a, the vertical reaction force can be either positive upward (option 1) or downward (option 2). It does not matter which option is selected. If option 1 is selected and the reaction is taken to be upward, when the reaction is calculated, it will have a positive value if it is in fact

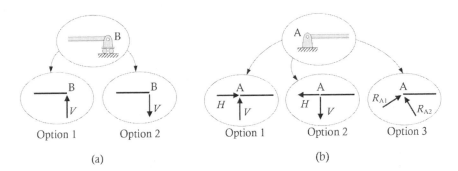

Figure 2.16 Arbitrary positive directions for the support reactions. (a) Roller support. (b) Pinned support.

acting in the assumed upward direction, and it will have a negative value if it is acting downward.

Similar considerations apply to the pinned support with the two reaction components shown in various options in Figure 2.16b. The pinned support can be replaced by any set of two reaction forces including the inclined ones (shown as option 3 in Figure 2.16b), as long as they are not parallel to each other. For example, all three options of Figure 2.16b are valid and the selection of a particular set of forces will only influence the terms included in the equilibrium equations used for the determination of the reactions, as outlined in the next section.

SUMMARY OF STEPS 2.1: Drawing a free-body diagram

1. Inspect the structure and, for each support, identify the corresponding set of reactions (see Table 2.1).

2. Draw the free-body diagram of the structure specifying all external loads applied and replacing all supports with the corresponding set of reactions identified in step 1. (Remember: positive directions adopted for the reactions are arbitrary.)

2.8 EQUILIBRIUM EQUATIONS FOR PLANAR STRUCTURES

A structure is in equilibrium only if the forces acting upon it have zero resultant force and zero resultant moment. In two-dimensional or planar structures, for the resultants to be zero, the sum of the components of the forces in any two (non-coincident) directions in the plane of the structure must be zero:

$$\Sigma F_x = 0 \tag{2.18a}$$

$$\Sigma F_y = 0 \tag{2.18b}$$

In the analysis of a two-dimensional structure, it is usual to select the x and y directions at right angle to each other (as shown in Figure 2.2), with the x-axis running along the longitudinal axis of the member. When Equations 2.18a and b are satisfied, the system is either in equilibrium or it is acted on by a couple (Figure 2.13) and is therefore rotating. Rotational or moment equilibrium is satisfied (i.e. rotation will not occur) if the sum of the moments about any point is also zero. That is:

$$\Sigma M_z = 0 \tag{2.18c}$$

These three equations (Equations 2.18a, b and c) are the conditions for equilibrium of two-dimensional structures and are used extensively in structural engineering.

In summary, a two-dimensional or planar structure is in equilibrium if the sum of the force components acting on the structure in each of two independent directions is equal to zero and the sum of the moments of the forces acting on the structure about any axis perpendicular to the plane of the structure is also zero as specified in Equations 2.18.

Alternatively, a structure will be in equilibrium if two or three equations of rotational equilibrium are satisfied, such as:

$$\Sigma F_x = 0 \qquad \Sigma M_{z(A)} = 0 \qquad \Sigma M_{z(B)} = 0 \qquad\qquad (2.19a\text{–}c)$$

or

$$\Sigma M_{z(A)} = 0 \qquad \Sigma M_{z(B)} = 0 \qquad \Sigma M_{z(C)} = 0 \qquad\qquad (2.20a\text{–}c)$$

where the points A, B and C are in the plane of the structure and are not co-linear. Equations 2.19 can only be used to verify equilibrium if the line connecting the two points (A and B) is not perpendicular to the direction of the force equilibrium equation (i.e. Equation 2.19a).

2.9 EXTERNAL STATICAL DETERMINACY AND STABILITY

The equations of equilibrium may be employed to express relationships between all forces (including moments or couples) acting on a structure. For two-dimensional structures, the three equilibrium equations (as expressed in Equations 2.18, 2.19 and 2.20) may be used to evaluate three unknowns.

Before proceeding with further considerations related to the use of the equilibrium equations, it is useful to examine the stability of the structure. A structure is said to be *internally stable* if, when all its reactions are removed, it does not undergo any change of shape. For example, reconsider the beam of Figure 2.15a. It is clear that, if we remove the supports, the beam will not change its shape (i.e. it will still remain a straight beam) even though it is no longer in equilibrium and will be moving in the direction of the resultant of the applied loads on the beam. This is not the case, however, for the frame shown in Figure 2.17. After removing the two pinned supports, the two rigid parts of the structure ABC and CDE can move relative to each other by rotating with respect to point C (owing to the presence of the hinge at C). A structure that changes its shape after the removal of the supports is classified as *internally unstable*. On the basis of this definition, the structure of Figure 2.17 is internally unstable. This is not to say the structure is unstable as, with an appropriate arrangement of supports (such as the one shown in Figure 2.17), the previously observed changes in shape are prevented.

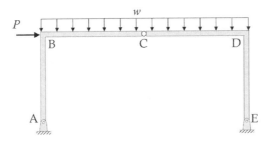

Figure 2.17 Pin-supported frames with internal hinge.

The external statical determinacy and stability of structures will now be dealt with separately considering first internally stable systems followed by internally unstable systems.

2.9.1 Internally stable structures

When dealing with internally stable structures, there are three equilibrium equations available from statics for the calculation of three unknown reactions. For example, the beam of Figure 2.18a is supported by a pinned support and a roller support that can be replaced by two reactions at A and one reaction at B. For this case, the number of unknown reactions n_r is 3 and these can be determined with the three equilibrium equations. These types of structures are usually referred to as *statically determinate externally*, to highlight the fact that all reactions can be determined from statics alone. The term *externally* is included to clarify that the member is statically determinate with regard to the calculation of the

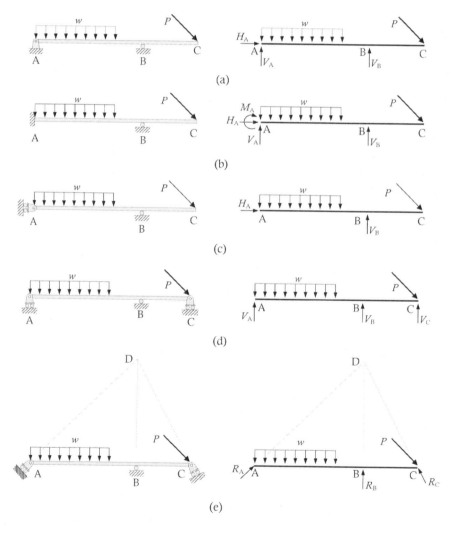

Figure 2.18 Layouts and free-body diagrams of structures. (a) Statically determinate externally. (b) Statically indeterminate externally. (c) Statically unstable externally. (d) Geometrically unstable externally: parallel reactions. (e) Geometrically unstable externally: concurrent reactions.

reactions. Some members that are statically determinate *externally* may not necessarily be statically determinate *internally*. That is, the internal actions in the member may not be able to be determined from statics alone. Such members will be discussed further in Section 3.4.

If the support at A was changed to a fixed support, as shown in Figure 2.18b, there are now four unknown reactions ($n_r = 4$) and the three equilibrium equations are not sufficient to determine them. Such a structure is classified as *statically indeterminate* and is said to be *n*-fold indeterminate (where $n = n_r - 3$). In the case of Figure 2.18b, the structure is 1-fold indeterminate (i.e. $n = 1$ since $n_r = 4$). The analysis of statically indeterminate structures requires consideration of the stiffness (or deformability) of the structure and this depends on the size and material properties of the various parts of the structures in addition to equilibrium. This will be considered in subsequent chapters.

If the beam is supported by the two roller supports outlined in Figure 2.18c, there are only two unknown reactions. When the number of reactions n_r is less than 3 (i.e. number of equilibrium equations available from statics), the structure is said to be *statically unstable*.

There are cases where the structure is still unstable despite possessing a number of reactions n_r equal to or greater than 3. These situations can occur when the reactions are not arranged to provide geometric stability and are either parallel (see Figure 2.18d) or concurrent (see Figure 2.18e). Structures with these inadequate support arrangements are denoted as *geometrically unstable externally*.

The determination of the reactions using the equilibrium equations alone is only possible when dealing with structures that are statically determinate.

2.9.2 Internally unstable structures

A structure that is internally unstable requires an appropriate arrangement of the supports in order to make it stable. Although there are only three equations of equilibrium for a planar structure, it is possible to have a statically determinate structure with more than three reaction components. For example, reconsider the frame shown in Figure 2.17, with its four unknown reaction forces: two at A and two at E (shown in the free-body diagram of Figure 2.19a). In addition to the three equations of equilibrium applied to the entire structure, it is possible to write an equation to recognise that the bending moment at the internal hinge at C is known to be zero, i.e. $M_C = 0$. This can be carried out by cutting the structure at C and enforcing moment equilibrium with respect to the hinge point on either the left or the right free-body diagram of the structure, illustrated in Figures 2.19b and c, respectively. This additional equation (i.e. additional to the three equilibrium equations available from statics) is called an *equation of condition*, because it expresses the condition imposed by the presence of the hinge. For a particular structure, each equation of condition increases by one the number of statically determinate reaction components

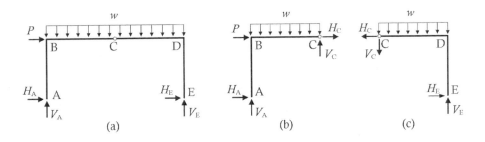

Figure 2.19 Free-body diagrams of a frame with an internal hinge.

that can be determined by statics. If n_c is the number of available equations of condition, the minimum number of reaction components required for a planar structure to be stable is $n_c + 3$.

If the actual number of reaction components for a structure is n_r, for a structure to be externally statically determinate, then $n_r = n_c + 3$.

For example, the structure in Figure 2.20a possesses four unknown reactions that can be calculated from the three equations of equilibrium and the one equation of condition provided by the presence of the hinge at B. Such a structure is classified as statically determinate externally (as well as internally unstable), because $n_r = n_c + 3 = 1 + 3 = 4$.

If $n_r < n_c + 3$, the structure as a whole is always unstable, and is denoted *statically unstable* or a *mechanism*, in which case the structure can undergo a change of shape without any deformation of individual members. Examples of unstable structures or mechanisms are shown in Figure 2.21. Note that for the frame shown in Figure 2.21d, it is the substructure CEF (with a hinge at C and E and a roller support at F) that is unstable, even though the number of reaction components for the whole structure exceeds $n_c + 3$.

If a structure has more reactions than can be determined by statics alone, it is said to be statically indeterminate and the deformational characteristics and material behaviour of the structural members, as well as the equations of equilibrium, are required to determine the reactions. A stable indeterminate structure is said to be n-fold indeterminate (with $n = n_r - n_c - 3$). Examples of idealised indeterminate beams are shown in Figure 2.22. The structure in Figure 2.22a is 1-fold indeterminate, while the one in Figure 2.22b is 2-fold indeterminate. The analysis of statically indeterminate structures is covered in subsequent chapters.

Having the minimum number of $n_c + 3$ reactions does not necessarily ensure stability. As we have seen for internally stable structures, the reactions must be so arranged that the structure is externally stable (i.e. systems of reactions must not form sets of parallel or concurrent forces).

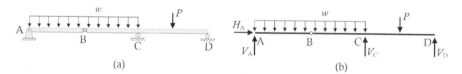

Figure 2.20 Example of a structure statically determinate externally (and internally unstable).

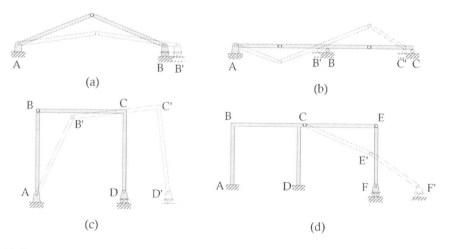

Figure 2.21 Structures statically unstable externally.

Figure 2.22 Structures statically indeterminate externally (and internally unstable).

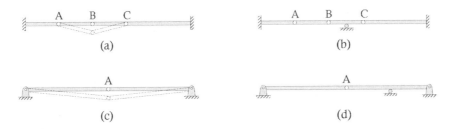

Figure 2.23 Examples of structures with three hinges placed on the same line.

There are also other cases where, despite $n_r \geq n_c + 3$, the structure is unstable. For example, this could occur when three or more hinges are placed on the same line, as shown in Figure 2.23a, for which $n_r = 3 + 3 = 6$. In this case, the structure is unstable because it is not able to carry any vertical load applied at B and the structure will undergo a change in shape as shown. This mechanism can be avoided by adding adequate restraints to the beam within the length defined by the three hinges (i.e. within the length ABC), as shown, for example, in Figure 2.23b where a roller support has been added between points B and C. In this case, the structure is now 1-fold statically indeterminate externally. A careful selection of the supports is needed when three or more hinges are placed on the same line. Another similar unstable case is illustrated in Figure 2.23c, which is shown to highlight the fact that this problem can also occur when an internal hinge is located between two adjacent pinned supports on the same line. The mechanism of Figure 2.23c is modified to a 1-fold externally statically indeterminate structure in Figure 2.23d by adding a roller support between the internal hinge and the right support.

SUMMARY OF STEPS 2.2: Evaluation of whether a structure is externally statically determinate, indeterminate, or unstable

1. Draw a free-body diagram of the structure (see Summary of Steps 2.1).

2. Verify that the system of reactions do not form a set of parallel or concurrent forces when considering either the entire structure or a part of it. Such a structure is classified as geometrically unstable externally. In the case where 3 or more hinges are placed on the same line, verify that the support arrangement provided is adequate and does not lead to a mechanism.

3. Evaluate whether the structure is internally stable or unstable. A structure is internally stable if the structure does not change its shape after all supports are removed. It is internally unstable if the structure, or a part of it, can change its shape after the reactions are removed.

4. Count the number of unknown reactions n_r and the number of equations of condition n_c (if any) and classify the structure as follows:

a. Statically determinate externally when $n_r = n_c + 3$.

b. Statically indeterminate externally when $n_r > n_c + 3$. The structure is said to be n-fold indeterminate externally, where $n = n_r - (n_c + 3)$.

c. Statically unstable when $n_r < n_c + 3$ or when part of the structure possesses a mechanism (as shown, for example, in Figure 2.21).

The number of equations of condition n_c corresponds to the number of internal hinges that cause internal instability.

WORKED EXAMPLE 2.3

Specify for each of the structures shown in Figure 2.24 whether they are statically determinate, indeterminate, or unstable externally.

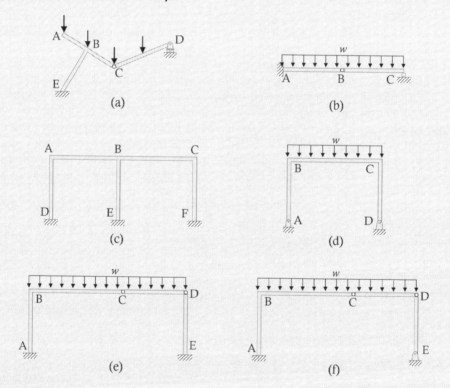

Figure 2.24 Structures for Worked Example 2.3.

The solution is provided below following the procedure detailed in Summary of Steps 2.2.

(1) The free-body diagrams of all structures are provided in Figure 2.25.

(2) The reactions of all structures do not form a system of concurrent or parallel forces.

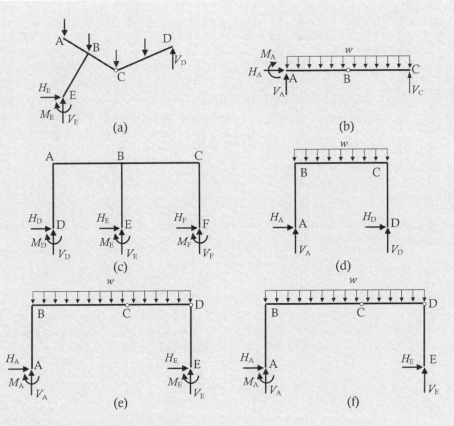

Figure 2.25 Free-body diagrams for Worked Example 2.3.

(3) The internal stability of the structures varies from case to case as follows:

a. Internally unstable because if the supports are removed, the structure can change its shape by rotating with respect to the hinge located at point C

b. Internally unstable

c. Internally stable because if the supports are removed, the structure does not change its shape

d. Internally stable

e. Internally unstable

f. Internally unstable

(4) The classification of external statical determinacy is provided below:

a. The structure possesses four unknown reactions and there is one equation of condition available because of the presence of the hinge at C. Therefore, $n_r = 4$ and $n_c = 1$ from which $n = 0 (= n_r - (n_c + 3) = 4 - (1 + 3) = 0)$. Based on this, the structure is classified as statically determinate.

b. $n_r = 4$ and $n_c = 1$: $n = n_r - (n_c + 3) = 0$. The structure is statically determinate.

c. $n_r = 9$: $n = n_r - 3 = 6$. The structure is 6-fold statically indeterminate.

d. $n_r = 4$: $n = n_r - 3 = 1$. The structure is 1-fold statically indeterminate.

e. $n_r = 6$ and $n_c = 2$: $n = n_r - (n_c + 3) = 1$. The structure is 1-fold statically indeterminate.

f. $n_r = 5$ and $n_c = 2$: $n = n_r - (n_c + 3) = 0$. The structure is statically determinate.

2.10 DETERMINATION OF REACTIONS

The reaction forces of a statically determinate two-dimensional structure are calculated by applying the equilibrium equations (Equations 2.18, 2.19 and 2.20) to a free-body diagram of the entire structure or a free-body diagram of a part of the structure. For example, reconsidering the simply-supported beam of Figure 2.15, the three unknown reactions, H_A, V_A and V_B, can be calculated by applying the three equilibrium equations to the free-body diagram of the beam. Although not always possible, a careful choice of the sequence in which the equilibrium equations are considered may allow the reactions to be determined without the need to solve simultaneous equations. For the free-body of Figure 2.15c, applying the equation of horizontal force equilibrium $\left(\Sigma F_x = 0 \right)$ gives:

$$\xrightarrow{+} \Sigma F_H = 0: \quad H_A - 20\cos 45 = 0 \quad \therefore H_A = 14.14 \text{ kN}$$

By next applying the equation of moment equilibrium about the z-axis passing through the support at A $\left(\Sigma M_A = 0 \right)$, the vertical reaction at B is obtained:

$$\left(+ \Sigma M_A = 0: \quad -5 \times 4 \times 2 - 20\sin 45 \times 6 + V_B \times 8 = 0 \quad \therefore V_B = 15.61 \text{ kN} \right.$$

The choice of applying rotational equilibrium about A is convenient because it limits the number of unknowns in the equation to one. This is because the components of the reactions at A pass through A and therefore do not cause any moment about that point.

Likewise, the vertical component of the reaction at A can be found by applying the equation of moment equilibrium about the z-axis passing through the support at B:

$$\left(+ \Sigma M_B = 0: \quad -V_A \times 8 + 5 \times 4 \times 6 + 20\sin 45 \times 2 = 0 \quad \therefore V_A = 18.53 \text{ kN} \right.$$

It is noted that the line of action of the horizontal component of the reaction at A passes through B and, therefore, does not cause any moment about B.

Moment equilibrium could have been taken about any two other points resulting in two equations in two unknowns (V_A and V_B), or we could have used horizontal and vertical force equilibrium equations together with one moment equilibrium equation. Despite the increase in complexity of the solution, the results would still be identical.

These results may be verified by checking that equilibrium is satisfied in the vertical direction:

$$\uparrow + \Sigma F_V = 0: \quad V_A - 5 \times 4 - 20\sin 45 + V_B = 0 \Rightarrow 18.53 - 5 \times 4 - 14.14 + 15.61 = 0 \quad \therefore \text{OK}$$

It is good practice to routinely apply such a check to verify the accuracy of the calculations.

The main steps required for the determination of the reactions are outlined below followed by two worked examples.

SUMMARY OF STEPS 2.3: Determination of the reactions

1. Draw the free-body diagram of the structure (see Summary of Steps 2.1).

2. Evaluate whether the structure is statically determinate externally, indeterminate externally, or unstable (see Summary of Steps 2.2).

3. If the structure is statically determinate, apply the equations of equilibrium and, if required, the equations of conditions to determine the unknown reactions. The calculation of the reactions in statically indeterminate structures will be dealt with in coming chapters.

WORKED EXAMPLE 2.4

The bent beam ABCD shown in Figure 2.26 is supported by a pin at A and a roller at D. Determine the reactions at A and D.

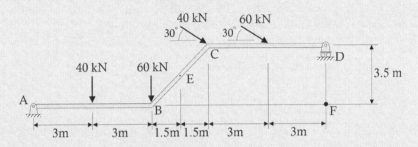

Figure 2.26 Structure for Worked Example 2.4.

(1) Draw the free-body diagram

The free-body diagram of the whole structure is presented in Figure 2.27, with the x and y directions assigned as shown and the three reaction components, H_A, V_A and V_D, included. For later calculations, it is convenient to replace the inclined forces with their horizontal and vertical components. In the solution process, we have assumed that the depth (or thickness) of the beam is small compared to its length and can be ignored.

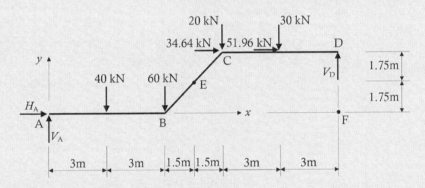

Figure 2.27 Free-body diagram for Worked Example 2.4.

(2) Determine the statical determinacy

The bent beam is internally stable as it does not change its shape if the supports are removed. As there are three reaction components that do not form a system of concurrent or parallel forces and there are three equilibrium equations, the structure is statically determinate.

(3) Determine the reactions

When finding reactions by hand calculations, it is desirable to determine each reaction independently as far as possible to minimise the complexity of the calculation. It is also recommended to check the reactions by an alternative calculation before proceeding with the analysis. To obtain an equation involving only V_D, we take moments about A:

$$\left(\underset{+}{\curvearrowleft}\right)\Sigma M_A = 0: \quad -(40 \times 3) - (60 \times 6) - (34.64 \times 3.5) - (20 \times 9) - (51.96 \times 3.5) - (30 \times 12)$$

$$+ (V_D \times 15) = 0 \quad \therefore V_D = 88.21\,\text{kN}$$

To obtain an equation involving only V_A, we can take moments about point F (which is the point of intersection of the lines of actions of H_A and V_D, as shown in Figure 2.27):

$$\left(\underset{+}{\curvearrowleft}\right)\Sigma M_F = 0: \quad -(V_A \times 15) + (40 \times 12) + (60 \times 9) - (34.64 \times 3.5) + (20 \times 6) - (51.96 \times 3.5)$$

$$+ (30 \times 3) = 0 \quad \therefore V_A = 61.79\,\text{kN}$$

Summing the forces vertically confirms that the above values are correct (or have compensating errors which is unlikely):

$$\uparrow + \Sigma F_y = 0: \quad V_A - 40 - 60 - 20 - 30 + V_D = 61.79 - 40 - 60 - 20 - 30 + 88.21 = 0 \quad \therefore \text{OK}$$

By equating the forces in the x direction to zero, we obtain an equilibrium equation with H_A as the only unknown:

$$\underset{+}{\pm}\Sigma F_x = 0: \quad H_A + 34.64 + 51.96 = 0 \quad \therefore H_A = -86.60\,\text{kN}$$

The reaction H_A is therefore acting in the opposite direction to that shown in Figure 2.27. The calculated reactions are summarised in Figure 2.28.

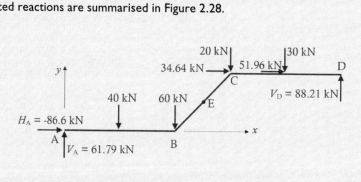

Figure 2.28 Summary of loads and reactions for Worked Example 2.4.

A further check by equating to zero the sum of the moments of all forces about another point (such as either C or D) will confirm that the calculated reactions are correct.

WORKED EXAMPLE 2.5

Calculate the reactions at the pinned supports at A and E of the frame ABCDE shown in Figure 2.29a. There is a frictionless hinge at C, so that the single equation of condition is $M_C = 0$. The free-body diagram of the entire frame is shown in Figure 2.29b and the free-body diagrams of the substructures CDE and ABC are shown in Figures 2.29c and d, respectively.

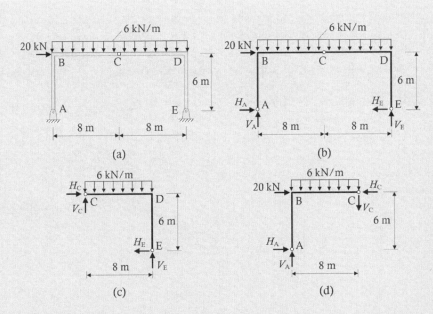

Figure 2.29 Structure and free-body diagrams for Worked Example 2.5. (a) Idealized frame. (b) Free-body diagram of whole frame. (c) Free-body diagram of CDE. (d) Free-body diagram of ABC.

With a total of four unknown reaction components ($n_r = 4$), three equilibrium equations and one equation of condition ($n_c = 1$), the number of equations equals the number of unknowns and the frame is statically determinate. The direction assumed for each of the reaction components is arbitrary, but cannot be changed once the solution process has started.

For the free-body diagram of the whole frame (Figure 2.29b), with three of the four reaction components passing through A, moment equilibrium about the hinge at A gives:

$$V_E \times 16 - 6 \times 16 \times 8 - 20 \times 6 = 0 \implies V_E = 55.5 \text{ kN}$$

Considering the free-body diagram CDE (Figure 2.29c), knowing that the moment at C is zero and with V_E having been determined, moment equilibrium about C gives the horizontal reaction component at E:

$$55.5 \times 8 - H_E \times 6 - 6 \times 8 \times 4 = 0 \implies H_E = 42.0 \text{ kN}$$

Returning to the free-body diagram of the entire frame (Figure 2.29b), with H_E now known, summing the horizontal forces gives:

$$H_A + 20 - 42.0 = 0 \implies H_A = 22.0 \text{ kN}$$

Summing the vertical forces on the frame allows the determination of the vertical reaction component at A:

$$V_A - 6 \times 16 + 55.5 = 0 \implies V_A = 40.5 \text{ kN}$$

In this example, all four reaction components are positive, indicating that all reactions are acting in the directions selected in the free-body diagram (Figure 2.29b).

To check these results, consider the free-body diagram ABC shown in Figure 2.29d. Moment equilibrium about the hinge at C shows that the calculated reaction components at A are correct:

$$\curvearrowright \quad H_A \times 6 + 6 \times 8 \times 4 - V_A \times 8 = 22.0 \times 6 + 6 \times 8 \times 4 - 40.5 \times 8 = 0 \quad \therefore OK$$

Although the components of the force passing through the hinge at C (H_C and V_C) are readily determined by considering force equilibrium on either Figure 2.29c or d, they are not required to determine the reactions at supports A and E.

The calculated reactions are summarised in Figure 2.30.

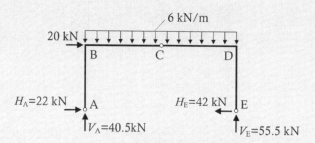

Figure 2.30 Summary of loads and reactions for Worked Example 2.5.

2.11 EQUILIBRIUM AND REACTIONS IN THREE-DIMENSIONAL STRUCTURES

The procedure to be used for the calculation of the reactions of three-dimensional structures is similar to that presented for plane structures, except that now there are six equilibrium equations available from statics. The resultant force in each of the three mutually perpendicular directions (i.e. the sum of all force components in each direction) must be zero:

$$\Sigma F_x = 0 \qquad \Sigma F_y = 0 \qquad \Sigma F_z = 0 \tag{2.21a–c}$$

and the resultant couple about each of three mutually perpendicular axes (i.e. the sum of the moments of all forces about each axis) must also be zero:

$$\Sigma M_x = 0 \qquad \Sigma M_y = 0 \qquad \Sigma M_z = 0 \tag{2.21d–f}$$

The six equations (Equations 2.21a through f) are the general conditions of equilibrium of forces in space.

Different support conditions are available in three-dimensional structures that can combine different combinations of restraints to the displacements or rotations at the supports.

In some three-dimensional problems, when determining unknown forces by hand, a judicious choice of axes about which to take moments will often shorten the solution. For instance, any particular force is eliminated from the moment calculation if moments are taken about an axis intersecting this force or an axis parallel to it.

When determining the reactions of a three-dimensional beam or frame, it is often convenient to consider what motion will be permitted if a certain reaction is removed. This provides an indication of what particular equilibrium equation can be used to evaluate this reaction. For instance, if removal of a given reaction would leave the body free to rotate about the y-axis, then an equation of moments about the y-axis will enable that reaction to be calculated directly.

Similar to a two-dimensional structure, the determination of the reactions for a statically determinate three-dimensional structure can be carried out using only the equilibrium equations. The evaluation of the statical determinacy of a three-dimensional structure is now dealt with in Reflection Activity 2.1, which is then followed by Worked Example 2.6.

REFLECTION ACTIVITY 2.1

Reconsider the Summary of Steps 2.2 outlining the procedure to be performed to evaluate whether a structure is statically determinate, indeterminate, or externally unstable, and suggest how the steps should be modified to be applicable to three-dimensional structures.

The first three points (points 1–3) included in Summary of Steps 2.2 are applicable in their current form to three-dimensional structures and, hence, do not need to be changed.

The conditions to be used for the determination of the statical determinacy need to be modified to account for the fact that the number of equilibrium equations available in three-dimensional is 6 (and not 3 as for plane structures).

Point 4 is therefore revised as follows:

(4) count the number of unknown reactions n_r and the number of equations of condition n_c (if any) and classify the structure as follows:

a. Statically determinate externally when $n_r = n_c + 6$.

b. Statically indeterminate externally when $n_r > n_c + 6$. The structure is said to be n-fold indeterminate externally, where $n = n_r - (n_c + 6)$.

c. Statically unstable when $n_r < n_c + 6$ or when part of the structure possesses a mechanism.

WORKED EXAMPLE 2.6

The beam ABC shown in Figure 2.31 has a right angle bend at B and lies in the horizontal x–z plane. The segments AB and BC are both 6 m long. The support at C can exert a reaction in any direction (with components in the x, y and z directions shown as R_4, R_5 and R_6, respectively). The support at B does not exert a reaction in the direction BC and so the reaction at B has components in the y and z directions, R_2 and R_3, respectively. The support at A can exert a reaction in the vertical y direction only, R_1.

Loads are applied at the midpoints of AB and BC, with the 8 kN and 12 kN loads both horizontal and normal to the beam, and the 20 kN load applied in the y direction.

Calculate the six unknown reaction components.

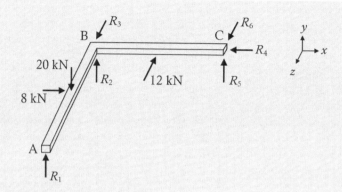

Figure 2.31 Free-body diagram for Worked Example 2.6.

(1–3) The FBD of the structure is depicted in Figure 2.31 from which it can be observed that:
a. The reactions do not form a system of concurrent or parallel forces.
b. The structure is internally stable.

(4) Reaction forces R_2, R_3, R_5 and R_6 and the 12 kN applied load all intersect the x-axis along BC. In addition, the 8 kN applied load and the reaction R_4 are parallel to the x-axis. By taking moment about the x-axis through B and C, the reaction R_1 is readily determined:

$$\Sigma(M_x)_{BC} = 0: \qquad 20 \times 3 - R_1 \times 6 = 0 \qquad\qquad \therefore R_1 = +10\,\text{kN}$$

Since all forces except R_5 either intersect the z-axis along AB or are parallel to it, taking moment about the z-axis gives:

$$\Sigma(M_z)_{AB} = 0: \qquad R_5 \times 6 = 0 \qquad\qquad \therefore R_5 = 0$$

Summing the forces in the y direction gives the remaining vertical reaction:

$$\Sigma F_y = 0: \qquad R_1 + R_2 + R_5 - 20 = 0 \qquad\qquad \therefore R_2 = 10\,\text{kN}$$

Summing the forces in the x direction, we get:

$$\Sigma F_x = 0: \qquad 8 - R_4 = 0 \qquad\qquad \therefore R_4 = 8\,\text{kN}$$

Only three forces cause a moment about the y-axis through B, namely, the 8 kN and 12 kN applied loads and the reaction R_6. Therefore:

$$\Sigma(M_y)_B = 0: \qquad 8 \times 3 + 12 \times 3 - R_6 \times 6 = 0 \qquad \therefore R_6 = +10\,\text{kN}$$

Summing the forces in the z direction gives:

$$\Sigma F_z = 0: \qquad R_3 + R_6 - 12 = 0 \qquad\qquad \therefore R_3 = 2\,\text{kN}$$

As a check on the above calculations, take moment about the y axis through C:

$$\Sigma(M_y)_C = 0: \qquad 8 \times 3 + R_3 \times 6 - 12 \times 3 = 0 \qquad \therefore \text{OK}$$

The calculated reactions and applied loads are summarised in Figure 2.32.

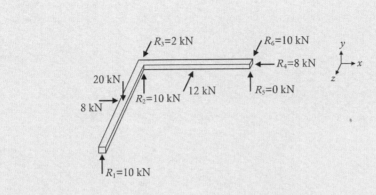

Figure 2.32 Summary of loads and reactions for Worked Example 2.6.

PROBLEMS

2.1   Determine the components F_x and F_y parallel to the x- and y-axes for the forces F shown.

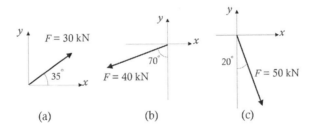

(a)   (b)   (c)

2.2   Calculate the magnitude and direction of the force F described by its components F_x and F_y illustrated below. Determine the direction cosines of F with respect to the x- and y-axes.

$F_x = 10$ kN	$F_x = -12$ kN	$F_x = 10$ kN
$F_y = 15$ kN	$F_y = -15$ kN	$F_y = -10$ kN
(a)	(b)	(c)

2.3 Consider the forces F shown and determine their components F_x, F_y and F_z parallel to the x-, y- and z-axes, respectively.

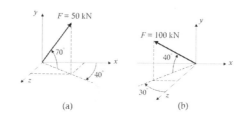

(a)   (b)

2.4 Determine the magnitude of F and the direction cosines related to the x-, y- and z-axes for the forces associated with the components shown.

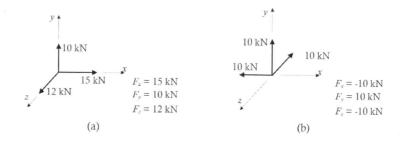

(a)   (b)

2.5 Calculate the moments of the forces F with respect to points A and B shown (dimensions of coordinates are in metres).

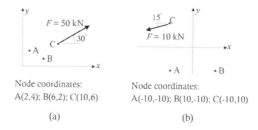

Node coordinates:
A(2,4); B(6,2); C(10,6)

(a)

Node coordinates:
A(-10,-10); B(10,-10); C(-10,10)

(b)

2.6   Consider the forces described by the components F_x, F_y and F_z parallel to the x-, y- and z-axes provided in the figure below (dimensions of coordinates are in metres). Determine the magnitude and direction cosines of the moments they induce with respect to point A.

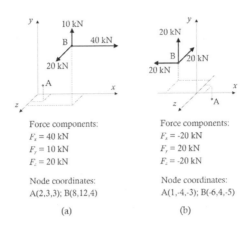

Force components:
F_x = 40 kN
F_y = 10 kN
F_z = 20 kN

Node coordinates:
A(2,3,3); B(8,12,4)

(a)

Force components:
F_x = -20 kN
F_y = 20 kN
F_z = -20 kN

Node coordinates:
A(1,-4,-3); B(-6,4,-5)

(b)

2.7   Find the magnitude and direction of the force resultant of the system of concurrent forces shown.

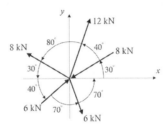

2.8   For the force systems shown, determine the magnitude of force F_1 and the direction θ of force F_2, if F_2 is the resultant of the remaining forces.

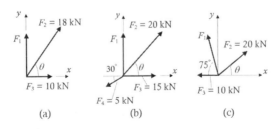

(a)   (b)   (c)

2.9   For the force systems considered in Problem 2.8, determine the magnitude of force F_1 and the direction θ of force F_2, if F_1 is the resultant of the remaining forces.

2.10 If the angle between the lines of action of two concurrent forces F_1 and F_2 is θ, show that the magnitude of the resultant of the two forces is:

$$R = \sqrt{F_1^2 + F_2^2 + 2F_1F_2 \cos\theta}$$

and the angle α between the lines of action of the resultant and the force F_1 is:

$$\alpha = \tan^{-1}\left(\frac{F_2 \sin\theta}{F_1 + F_2 \cos\theta}\right)$$

2.11 The resultant of two concurrent forces $F_1 = 10$ kN and $F_2 = 15$ kN is $R = 5$ kN. Find the angle between F_1 and F_2.

2.12 Find the resultants, expressed in terms of a force and a moment located at (i) point O(0,0) and (ii) point H(9,4), for the system of forces shown.

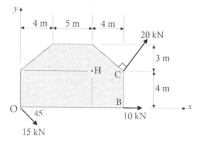

2.13 The three forces shown act on the sides of a triangular plate ABC.
   (i) Find the magnitude of the resultant force.
   (ii) Replace the three forces by a statically equivalent force system consisting of a single force and couple applied at A.

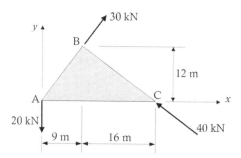

2.14 The 2 m square plate shown is acted on by two forces, each of 80 kN, and a clockwise couple of 100 kNm as shown. Determine the resultants, expressed in terms of a force and a moment applied at points (i) G(1,1), (ii) B(2,2) and (iii) C(2,0).

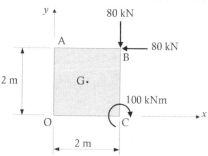

2.15 Determine the magnitude of the resultants expressed in terms of a force and a moment applied at A, produced by the three forces acting on the bar ABCD.

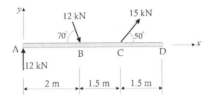

2.16 The cranked bar shown is in equilibrium. Calculate the forces R_1, R_2 and R_3.

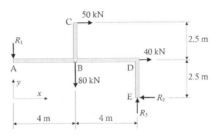

2.17 Calculate the components of the resultant force and moment applied at (i) point O and (ii) point H for the system of forces shown.

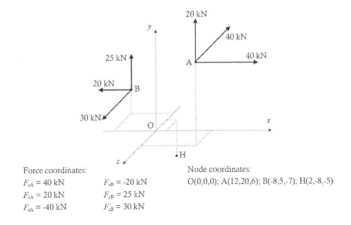

Force coordinates:

F_{xA} = 40 kN F_{xB} = -20 kN
F_{yA} = 20 kN F_{yB} = 25 kN
F_{zA} = -40 kN F_{zB} = 30 kN

Node coordinates:

O(0,0,0); A(12,20,6); B(-8,5,-7); H(2,-8,-5)

2.18   Determine whether the beams are externally statically determinate, indeterminate or unstable.

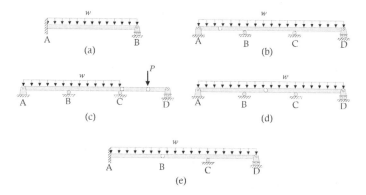

(a)   (b)

(c)   (d)

(e)

2.19   Consider the frames shown below and evaluate whether they are externally statically determinate, indeterminate or unstable.

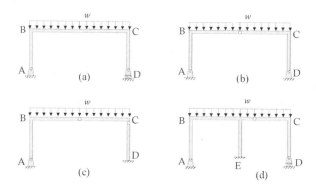

(a)   (b)

(c)   (d)

2.20   Draw the free-body diagram of the simply-supported beam shown below and calculate the reactions at each support.

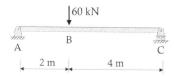

2.21   Consider the simply-supported beam and sketch its free-body diagram. Determine the reactions at each support.

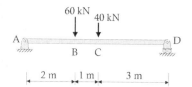

2.22 Sketch the free-body diagram for the simply-supported beam shown below and evaluate the reactions at the supports at A and C.

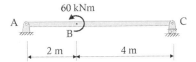

2.23 Draw the free-body diagram for the simply-supported beam and determine the reactions at each support.

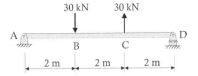

2.24 Consider the frame illustrated below and sketch its free-body diagram. Calculate the reactions provided by the pinned support at A and by the roller support at D.

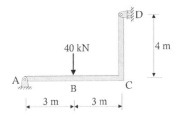

2.25 For the beam outlined below, determine the reactions at the pinned and roller supports.

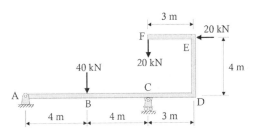

2.26 The frame has a pinned support at A and a roller at C. Draw its free-body diagram and calculate the reactions at the two supports.

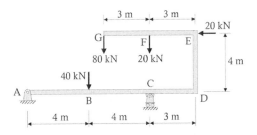

2.27 For the beam shown below, draw the free-body diagram and determine the reactions at the pinned support at A and at the roller at E.

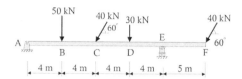

2.28 The beam ABCDEF shown is pinned at A and supported on rollers at C, D and F, and has two internal hinges as shown. Calculate the reactions at each support.

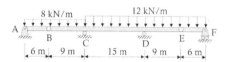

2.29 For the beams shown, calculate the reactions at A and D.

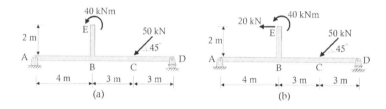

2.30 Reconsider the frames of Problem 2.29 and evaluate the effect on the reactions in each case if the 40 kNm couple is moved from E to C.

2.31 The horizontal beam shown is supported on rollers at A and E, with the roller planes at 60° and 30°, respectively, to the horizontal. The beam is also supported by a roller at C (on a horizontal roller plane). For the loading shown, find the reactions at the three supports.

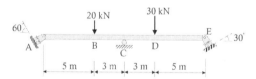

2.32 The simply-supported beam is acted on by a linearly varying distributed load, as shown. Determine the reactions at A and B.

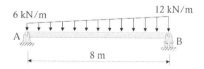

2.33 The beam is pinned at A and has a roller at D (with the reaction at D perpendicular to the beam). Determine the reactions at the two supports.

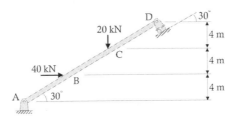

2.34 The simply-supported beam shown is subjected to two couples applied at B and C. Find the reactions at A and D.

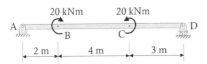

2.35 The frame ABCD is fixed at B. For the loading shown, evaluate the reactions at A.

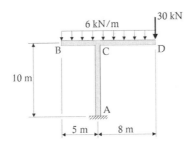

2.36 The bent beam is pinned at F and supported by a vertical roller at A (i.e. the reaction at A is horizontal). Calculate the reactions at the supports at A and F.

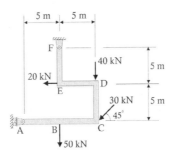

2.37 Consider the beam outlined below. Determine the reactions at the pinned support at A and at the roller at G.

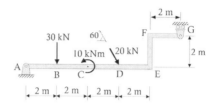

2.38 The structure shown has a pinned support at A and a roller at D. Calculate the reactions at A and D.

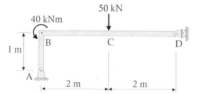

2.39 The beam is subjected to a uniformly distributed load over the segment BC. Calculate the reactions at the pinned support at A and at the roller at B.

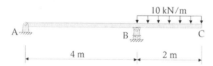

2.40 For the structure shown, determine the reactions at A and D.

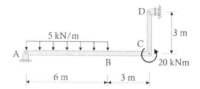

2.41 For the portal frame ABCDE, determine the reactions at the pinned supports located at A and E.

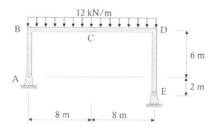

2.42   For the structure shown, calculate the vertical and horizontal reaction components at the supports at A, B and C.

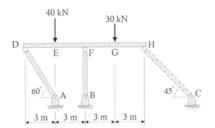

2.43   Consider the portal frame ABCDE and determine the reactions at A and E.

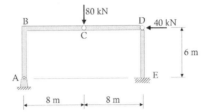

2.44   Reconsider the portal frame ABCDE of Problem 2.43 and assume the support at A is fixed and the one at E is pinned. Determine the effects of this change in support conditions on the reactions calculated in the previous problem.

2.45   Consider the three pinned arch shown and calculate the reactions at the pinned supports at A and G.

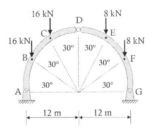

2.46   The cantilever beam ABCD is bent at right angles in the $x–z$ plane at B and at right angles in the $y–z$ plane at C. It is fixed in position in each direction at A so that six reaction components can develop at that support. The lengths of the three legs of the beam are as follows: AB = 6 m, BC = 5 m and CD = 4 m.
For the loading shown, determine the six reactions at A.

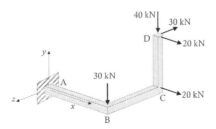

2.47 The structure shown has three legs, AB in the z direction, BC in the x direction, and BD in the y direction. AB and BC are 6 m long and BD is 5 m long. It is supported at A, C and D. The support at A restrains movement in the z direction only (R_1) and rotation about the y-axis only (R_2). The support at C restrains movement in the y direction only (R_3) and rotation about the x-axis only (R_4). The support at D restrains movement in the x and y directions only (R_5 and R_6, respectively).

The 6 kN and 8 kN loads act at the midpoint of AB, while the 12 kN and 14 kN loads act at the midpoint of BC. The 6 kN load acts in the x direction, the 8 kN and 14 kN loads act in the y direction, and the 10 kN and 12 kN loads act in the z direction.

Find the six reactions.

Note: the double-headed arrows in the figure denote that the reactions R_2 and R_4 are couples about an axis in the direction of the arrow.

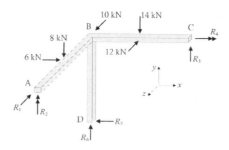

Chapter 3

Internal actions of beams and frames

3.1 INTRODUCTION

This chapter deals with the calculation of internal actions in statically determinate beams and frames. We have seen that statically determinate structures are those for which the reactions at the supports and the internal actions on any cross-section may be determined using only the equations of equilibrium. We will see subsequently that for statically determinate structures, a small deformation of one part of the structure does not induce reactions at the supports and does not cause a change in deformation of any other part of the structure.

Only two-dimensional beams and frames will be considered in this chapter, i.e. beams and frames that lie in a single plane (the x–y plane) with applied forces and reactions in the same plane and with moments about an axis perpendicular to that plane. In addition, we will be concerned here only with beams and frames whose deformations are small compared to the dimensions of the structure, so that the change in geometry of the structure when the loads are applied is small enough to be ignored.

3.2 INTERNAL ACTIONS AT A CROSS-SECTION

Most structures are built up of several components connected together. Such components exert forces upon one another at their connections, and it is necessary to evaluate these forces. Such forces are *internal* to the structure as a whole, and no information can be obtained about them by considering the equilibrium of the complete structure. According to Newton's third law, the force exerted by component X upon component Y is equal and opposite to the force exerted by Y upon X. So even if an attempt is made to include these forces in the equilibrium equations, they will cancel each other out. The only way information can be obtained about such forces is to consider the equilibrium of a part of the structure. The part is selected such that the internal force in question becomes *external* to that part.

In Section 2.7, the concept of the *free-body diagram* was introduced when considering the equilibrium of a complete structure, isolated from its supports and acted upon by the external applied loads, as well as the reactions exerted by the supports. The concept of the free-body diagram can be extended to apply to any part of a structure. The corresponding free-body diagram is a diagram showing the particular part or component of the structure together with all the forces that are external to that component. These forces consist of the external forces applied directly to the component, as well as those internal forces exerted by the removed part of the structure upon the part being considered. The free-body diagram of part of a structure was briefly introduced in Chapter 2 when considering the equations of condition for the calculation of the reactions of a structure.

For example, a single beam may be arbitrarily divided into two parts and either part considered as a free-body diagram. By considering equilibrium of either of these free-body diagrams, the force transmitted at the interface between the two parts may be determined. The internal force in question may be treated as an external force acting on the partial free-body diagram and may be readily calculated using the equations of statics. Let us consider the straight beam ABCD shown in Figure 3.1a. The beam reactions may be determined by enforcing equilibrium and are shown on the free-body diagram of Figure 3.1b. Suppose that we need to determine the force transmitted across the section C. For this purpose, we cut the beam at C and consider equilibrium of either segment AC or CD. Figure 3.1c shows the free-body diagram of the portion AC that is acted upon by the given forces at A and B and by an unknown force Q at C acting at the centroid of the cross-section and an unknown couple M acting about the centroidal axis (refer to Appendix A). Since the free-body AC is in equilibrium, the unknowns Q and M are the *equilibrants* of the other two forces acting on AC and may be determined by simple statics.

These equilibrants Q and M are the *internal actions* exerted on the length of beam AC by the length of beam CD. Equal and opposite actions Q' and M' must be exerted upon CD by AC, and these actions may be found if the equilibrium of free-body diagram CD is considered (Figure 3.1d). It is convenient in practice to resolve the force Q into its two

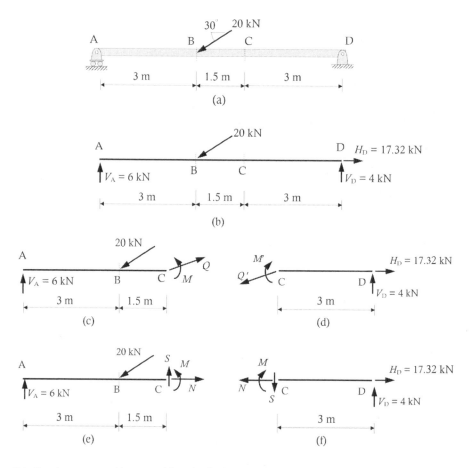

Figure 3.1 Simply-supported beam and free-body diagrams.

components, one parallel to the axis of the beam and one perpendicular to it (Figure 3.1e and f). Thus, the equilibrants are replaced by a statically equivalent system comprising two forces and a couple, all acting at the centroid of the cross-section C. The component force parallel to the axis of the member is called the *axial force*, denoted here by N. The component force perpendicular to the axis is called the *shear force*, denoted by S (or sometimes V). The couple is called the *bending moment*, denoted by M.

These definitions are applicable also to curved members, in which case the component forces of the internal action at any section are taken parallel and perpendicular to the tangent to the curve at that section. Based on this, the definition of the internal actions at any cross-section can then be generalised to the following.

- The *axial force* is the component of the internal action in a direction parallel to the longitudinal axis of the member at the section.
- The *shear force* is the component of the internal action in a direction normal to the longitudinal axis of the member at the section.
- The *bending moment* is the moment of the internal action about the point where the longitudinal axis of the member intersects the given cross-section, i.e. about the centroid of the cross-section (or more precisely about an axis through the centroid of the cross-section and perpendicular to the plane of the structure, i.e. *centroidal axis*).

Not only is it convenient to express the internal action at a cross-section of a structural member by the axial force, shear force and bending moment, but engineers find that this procedure facilitates the design of the member.

3.3 SIGN CONVENTION OF INTERNAL ACTIONS

It is important to discuss the sign convention of internal actions before engaging in their calculation. Figures 3.1e and f show the left- and right-hand free-body diagrams on either side of a particular cross-section (at C) of the simply-supported beam ABCD. The internal actions on the cross-section (N, S and M) are shown in the positive sense on each free-body diagram. In this illustration, if we determine the bending moment by considering free-body diagram AC (Figure 3.1e), we find that M is anticlockwise. On the other hand, if we consider the equilibrium of CD (Figure 3.1f), we find that M is clockwise, being the opposite to the couple exerted on AC. Thus, if the terms *clockwise* and *anticlockwise* are adopted as criteria of positive and negative, the sign of the bending moment would differ according to whether AC or CD is considered. Besides being inconvenient, this does not reflect the physical behaviour of the term *bending moment*. Similar remarks apply to the axial force N and shear force S. A more satisfactory sign convention is obtained by considering the effect that these actions have on the deformation of a small portion of the beam.

Suppose that at the part of the beam under consideration two cuts C_1 and C_2 are made very close together, thus isolating a small element of the beam as indicated by the side elevation shown in Figure 3.2. At the cut C_1, the three internal actions are shown acting on each cut surface. They are similarly shown at the cut C_2. The actions at C_1 and C_2 are nearly identical, since C_1 and C_2 are close together.

The small element is thus subjected to three *pairs* of actions that tend to distort its shape. It is most important not to confuse these internal actions with applied forces and couples. An applied *force* would act in a particular direction. For a horizontal member (as in Figure 3.2), the axial force is represented by a *pair* of horizontal forces, one of which acts to the left and one to the right. Similarly, a shear force is depicted by a *pair* of vertical forces, one up and one down, and a bending moment by a *pair* of couples, one clockwise and one anti-clockwise.

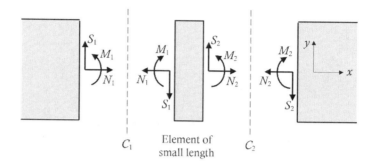

Figure 3.2 Free-body diagram of beam element of small length.

The small element in Figure 3.2 is subjected to a pair of forces N_1 and N_2 that in this instance tend to increase its length, that is, the element is in tension. The axial force N is said to be positive if it stretches the element and puts it into tension, and negative if it causes compression and shortens its length (see Figure 3.3a). The element is subjected to a pair of forces S_1 and S_2 that tend to cause a shearing type of deformation. If the forces S_1 and S_2 are in the directions shown in Figure 3.2, the shear force S is said to be positive (see Figure 3.3b). The element is subjected to a pair of couples M_1 and M_2 that tend to bend it. The bending moment M is said to be positive if the element bends concave upward (as shown in Figure 3.3c) or concave towards a specified positive direction if the bar is not horizontal. These adopted sign conventions are summarized in Figure 3.3.

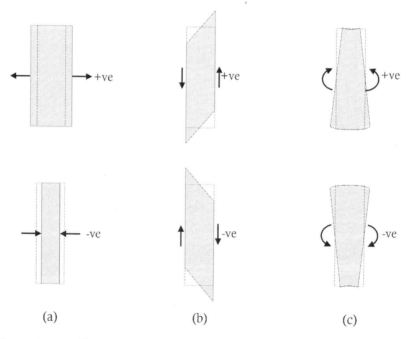

Note: +ve = positive;   - ve= negative

Figure 3.3 Sign convention for axial force, shear force and bending moment. (a) Axial force. (b) Shear force. (c) Bending moment.

The illustrations of Figure 3.3 outline the physical significance of the above definition of the sign conventions. In many practical problems, the signs may be determined by imagining the nature of the deformation. For instance, a simply-supported beam carrying downward loads will bend so that it becomes concave on the top (Figure 3.4a), inducing a shortening in the top fibres of the member and an elongation in the bottom fibres. Such bending is denoted as positive bending, sometimes called *sagging* bending. The same simply-supported beam carrying upward loads will bend so that it becomes concave on the bottom (Figure 3.4b), and the bottom fibres of the member shorten while the top fibres elongate. This is negative bending, sometimes called *hogging* bending. These definitions of positive (sagging) and negative (hogging) moments are useful for beam arrangements, but they lose their meaning for vertical members, for which it is not possible to distinguish between the top and bottom fibres.

In more general problems, the physical determination of signs is less simple and analytical rules are more convenient. The x-axis is generally defined as running along the beam, as shown in Figure 3.5. If the beam is cut at a particular cross-section, the x-axis is directed outward on one cut face, that is, the left-hand face in Figure 3.5, and this is the face of *positive incidence*. If the forces N, S and M on this face agree with the direction of the x- and y-axes, then they are defined as positive. On the other cut face, the x-axis is directed inward as on the right-hand side of Figure 3.5. This face is called the face of *negative incidence* and, on this face, N, S and M are positive if they *disagree* with the x- and y-axes.

In Figure 3.1e and f, the left- and right-hand free-body diagrams were shown for the beam of Figure 3.1a cut at C. On each diagram, the internal actions were depicted acting in their positive sense as defined above. On the left-hand free-body diagram ABC, where the cut face at C is the face of positive incidence, positive N and S are shown in the directions of x and y, and positive M is anticlockwise. On the right-hand free-body diagram CD, the cut face at C is a face of negative incidence, and the positive directions of N and S are opposed to x and y, and positive M is clockwise. Applying the equations of equilibrium to either free-body diagram will give the values of N, S and M with the correct signs.

Figure 3.4 Example of positive and negative moments. (a) Positive (sagging) moment. (b) Negative (hogging) moment.

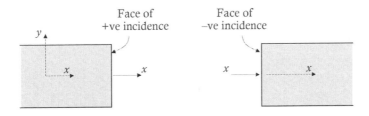

Figure 3.5 Faces of positive and negative incidence.

3.4 DETERMINATION OF INTERNAL ACTIONS AND STATICAL DETERMINACY

For the evaluation of internal actions at a particular location, the member is first cut at that location and the internal actions N, S and M are then determined by applying the equations of equilibrium to one of the two free-body diagrams produced by the cut.

We will now determine the internal actions at the cross-section at C for the beam shown in Figure 3.1a (and reproduced in Figure 3.6a). The free-body diagram for the entire beam with the reactions determined from statics is shown in Figure 3.6b. The first step in the solution is to cut the structure at C to produce the two free-body diagrams ABC (Figure 3.6c) and CD (Figure 3.6d). On each free-body diagram, there are three unknown internal actions, that is, N, S and M, and these can be determined using the three equations available from statics. Applying the equilibrium equations (Equations 2.18) to the free-body ABC (Figure 3.6c) gives:

$$\Sigma F_x = 0: \quad -20\cos 30 + N_C = 0 \quad \therefore N_C = +17.32 \text{ kN}$$

$$\Sigma F_y = 0: \quad +6 - 20\sin 30 + S_C = 0 \quad \therefore S_C = +4.0 \text{ kN}$$

and taking moments about C:

$$\Sigma M_C = 0: \quad -6 \times 4.5 + 20\sin 30 \times 1.5 + M_C = 0 \quad \therefore M_C = +12.0 \text{ kNm}$$

Similar results can be obtained by applying the equilibrium equations to the free-body diagram CD in Figure 3.6d:

$$\Sigma F_x = 0: \quad -N_C + 17.32 = 0 \quad \therefore N_C = +17.32 \text{ kN}$$

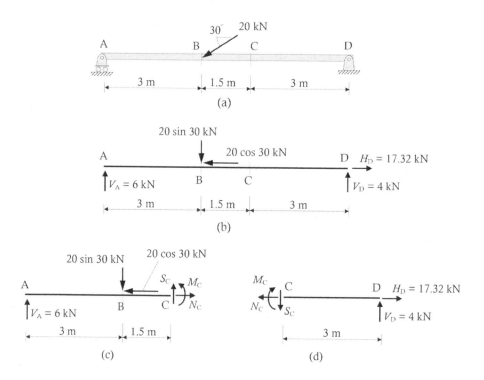

Figure 3.6 Simply-supported beam and free-body diagrams.

$$\Sigma F_y = 0: \quad -S_C + 4 = 0 \quad \therefore S_C = +4.0 \text{ kN}$$

and taking moments about C:

$$\Sigma M_{(C)} = 0: \quad -M_C + 3 \times 4 = 0 \quad \therefore M_C = +12.0 \text{ kNm}$$

When writing these equilibrium equations, the positive directions for the various forces F_x, F_y and the couples M are taken to the right, upwards and anticlockwise, respectively. This is arbitrary, as it makes no difference to the answers if the opposite directions are selected.

Let us now consider the simply-supported beam of Figure 3.7a with the aim to determine the internal actions at G. The free-body diagram of the whole structure is shown in Figure 3.7b. To determine N_G, S_G and M_G, we must perform a cut through the structure at G, and the presence of the closed loop BCDE means that any cut through G will always intersect the structure at two locations, as highlighted in the cut shown in Figures 3.7c and d. We now have six unknown internal actions, that is, N_G, S_G and M_G at G and N_H, S_H and M_H at H. Clearly, the three equilibrium equations of statics are not sufficient to determine the unknown internal actions in this structure.

Structures for which the internal actions cannot be calculated from statics are denoted as *internally statically indeterminate* and one needs to also consider the geometry of the cross-sections and the material properties of the structure in order to calculate the internal actions.

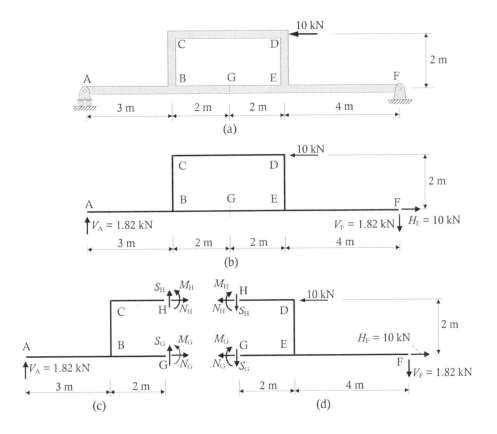

Figure 3.7 Simply-supported beam with a closed loop and free-body diagrams.

In general, when there is a closed loop in a beam or frame, its internal actions cannot be evaluated from statics alone.

The steps involved in the determination of the internal actions in a statically determinate structures are presented in Summary of Steps 3.1.

SUMMARY OF STEPS 3.1: Determination of internal actions

1. Draw the free-body diagram of the entire structure (see Summary of Steps 2.1) highlighting all reactions and applied loads.

2. Identify locations along the beam or frame where the internal actions need to be calculated.

3. Perform a cut through the structure at each location of interest and calculate the internal actions. This is carried out by applying the three equations of equilibrium to the free-body diagram containing the three unknown internal actions created by the cut. In the case of closed loops, the methods of analysis presented in subsequent chapters need to be used for the evaluation of the internal actions, as the number of unknowns is higher than the number of available equations of equilibrium (the structure is said to be statically indeterminate internally). As already observed for the calculation of the reactions in Section 2.9, the presence of hinges in the closed loops might add equations of conditions in the calculation of the internal actions.

Note: The above steps assume that the three internal actions at the cut section are the only unknowns on the free-body diagram and that all the reactions at each of the supports have previously been calculated.

SUMMARY OF STEPS 3.2: Evaluation of whether a structure is statically determinate, indeterminate and unstable

For the case of plane structures, the procedure involved in the classification of external statical determinacy has been presented in Summary of Steps 2.2, while the aspects related to the possibility of calculating the internal actions from equilibrium considerations have been outlined in Summary of Steps 3.1.

A structure is defined as *statically determinate* only if the following two conditions are satisfied: (i) it is statically determinate externally and the reactions at each support can be determined from statics, and (ii) its internal actions along each member can be calculated from statics.

A structure is defined as *statically indeterminate* if at least one of the following two conditions is satisfied: (i) it is statically indeterminate externally and the reactions cannot be determined from statics alone, or (ii) its internal actions at any point cannot be calculated from statics.

For a statically indeterminate structure, the degree of statical indeterminacy depends on both the number of external redundants (i.e. redundant support reactions) and the number of redundant internal actions n_{ir}. For example, in Figure 3.7c: $n_{ir} = 3$. If n_r is the number of external support reactions and n_c is the number of equations of condition (as defined in Summary of Steps 2.2), the degree of statical indeterminacy n is calculated from:

$$n = n_r + n_{ir} - (n_c + 3)$$

and the structure is said to be n-fold indeterminate.

A structure is *unstable* if n calculated above is negative or when part of the structure possesses a mechanism (as previously discussed in Summary of Steps 2.2).

A stable structure is statically determinate if n calculated above is equal to zero.

WORKED EXAMPLE 3.1

The reactions at the supports of the cranked beam shown in Figure 3.8 were calculated in Worked Example 2.4. Determine the axial force, shear force and bending moment at the cross-section at E (at the mid-point of the length of beam BC).

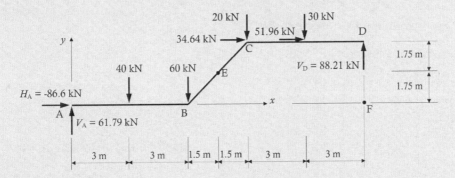

Figure 3.8 Free-body diagram from Worked Example 2.4.

First, we cut the structure at E, thereby dividing the structure into two parts, ABE and ECD. The free-body diagrams of ABE and ECD are shown in Figure 3.9a and b, respectively. The internal actions at E will be calculated by considering equilibrium of both free-body diagrams to demonstrate that they both lead to the same values. We start from the free-body diagram of ABE (Figure 3.9a), which includes the reaction forces at A. The inclination of the x-axis at E to the horizontal is:

$$\theta = \tan^{-1}(1.75/1.5) = 49.4°.$$

Equilibrium requires that the sum of the components of all forces in the x'-direction is zero. Therefore:

$+\nearrow:$ $N_E - 60 \sin 49.4 - 40 \sin 49.4 + 61.79 \sin 49.4 - 86.6 \cos 49.4 = 0$

$\therefore N_E = +85.37$ kN

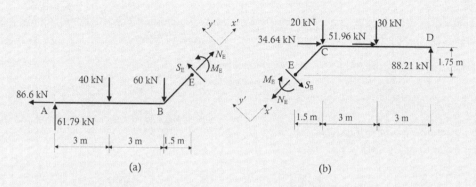

Figure 3.9 Free-body diagrams ABE and ECD for Worked Example 3.1.

Similarly, in the y'-direction:

$+\nwarrow$: $\quad S_E - 60 \cos 49.4 - 40 \cos 49.4 + 61.79 \cos 49.4 + 86.6 \sin 49.4 = 0$

$\therefore S_E = -40.88$ kN

Taking moment about E gives:

$\curvearrowright$: $\quad M_E + 60 \times 1.5 + 40 \times 4.5 - 61.79 \times 7.5 - 86.6 \times 1.75 = 0$

$\therefore M_E = +345.0$ kNm.

Alternatively, the internal actions at E may be determined from the right-hand free-body diagram of ECD (shown in Figure 3.9b). Summing the forces in the x'-direction in Figure 3.9b gives:

$+\nearrow$: $\quad -N_E + 34.64 \cos 49.4 + 51.96 \cos 49.4 - 20 \sin 49.4 - 30 \sin 49.4 + 88.21 \sin 49.4 = 0$

$\therefore N_E = +85.37$ kN

Similarly in the y'-direction:

$+\nwarrow$: $\quad -S_E - 34.64 \sin 49.4 - 51.96 \sin 49.4 - 20 \cos 49.4 - 30 \cos 49.4 + 88.21 \cos 49.4 = 0$

$\therefore S_E = -40.88$ kN

Taking moment about E gives:

$\curvearrowright$: $\quad -M_E - 34.64 \times 1.75 - 51.96 \times 1.75 - 20 \times 1.5 - 30 \times 4.5 + 88.21 \times 7.5 = 0$

$\therefore M_E = +345.0$ kNm.

3.5 AXIAL FORCE, SHEAR FORCE AND BENDING MOMENT DIAGRAMS

The internal actions vary from point to point along a beam or frame and it is often convenient to express this variation in algebraic terms. If the bending moment, for example, is computed for a typical point at a distance x from a chosen origin, an expression is obtained for M in terms of x. Similar expressions may also be derived for the axial force N and shear force S. These expressions are usually valid only over a limited part of the structure. This is the case if the loading is discontinuous, i.e. if concentrated loads are applied to the structure or if the magnitude of a distributed load suddenly changes. It is also the case if the beam or frame suddenly changes direction or if the beam contains an interior support.

It is frequently convenient to illustrate the variation of axial force, shear force and bending moment by plotting their graphs against distance along the beam or frame. Such graphs are called *axial force, shear force* and *bending moment diagrams* and may be obtained either by plotting the algebraic functions mentioned above or by simply calculating N, S or M at a number of isolated points along the structure and plotting these values as ordinates on the member longitudinal axis, i.e. the x-axis.

Throughout the book, the bending moment diagrams will be drawn on the tension side of the beam or frame, meaning that the line showing the variation of the moment will be located on the side of the cross-section subjected to tensile deformations. For example, for the simply-supported beam of Figure 3.10a, tension occurs at the bottom and the bending moment diagram is therefore plotted as shown in Figure 3.10b.

SUMMARY OF STEPS 3.3: Drawing of axial force diagram (AFD), shear force diagram (SFD) and bending moment diagram (BMD)

1. Identify segments of the structures along which the expressions for the external axial force, shear force and moment are continuous and locate the positions where discontinuities of these expressions occur.

2. Perform a cut in each of the segments identified at point 1 at some arbitrary location placed at a distance x along the structure and derive expressions for N, S and M (as a function of x) by applying the equations of equilibrium one at a time.

3. Plot the expressions determined at step 2 on the member axis to produce the axial force diagram (AFD), shear force diagram (SFD) and bending moment diagram (BMD).

The sign convention adopted in this book for the plotting of the BMD is to plot the curve on the tension side of the members.

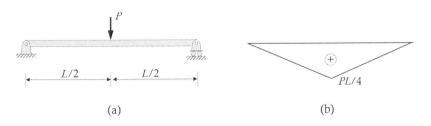

(a)   (b)

Figure 3.10 Sign convention for plotting of bending moment diagrams. (a) Simply-supported beam. (b) Bending moment diagram (plotted on tension side).

WORKED EXAMPLE 3.2

The free-body diagram of a simply-supported beam ABCD is shown in Figure 3.11. Express the axial force, shear force and bending moment for each of the beam segments AB, BC and CD as functions of the distance x from end A and draw the axial force, shear force and bending moment diagrams for the whole beam.

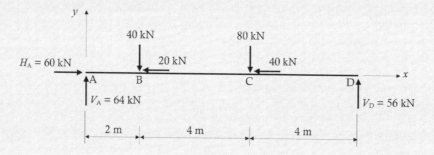

Figure 3.11 Free-body diagram of a simply-supported beam ABCD.

(1) The segments along which the internal actions remain continuous are AB, BC and CD, because of the presence of the applied forces at B and C.

(2) We perform a cut at an arbitrary point within each segment and consider equilibrium of either the left- or the right-hand free-body diagrams to determine the expressions for N, S and M.

<u>Segment AB</u>: If we cut the beam at any cross-section X between points A and B (at x from end A, as shown in the left-hand free-body diagram in Figure 3.12a), the internal actions on the cut cross-section can be readily determined from the equations of equilibrium:

$$\Sigma M_x = 0: \quad M = 64x \text{ kNm} \tag{3.1a}$$

$$\Sigma F_y = 0: \quad S = -64 \text{ kN} \tag{3.1b}$$

$$\Sigma F_x = 0: \quad N = -60 \text{ kN} \tag{3.1c}$$

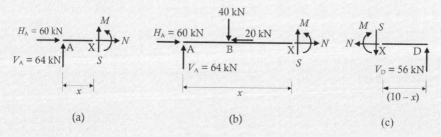

Figure 3.12 Free-body diagrams for Worked Example 3.2. (a) In segment AB. (b) In segment BC. (c) In segment CD.

These expressions apply to every cross-section between A and B (i.e. from $x = 0$ to 2 m).

Segment BC: To determine the internal actions at any cross-section between B and C, consider the free-body diagram of Figure 3.12b. The beam has been cut at a cross-section x from A between points B and C. The equations of equilibrium give:

$$\Sigma M_X = 0: \quad M = 64x - 40(x - 2) = 24x + 80 \text{ kNm} \tag{3.2a}$$

$$\Sigma F_y = 0: \quad S = -64 + 40 = -24 \text{ kN} \tag{3.2b}$$

$$\Sigma F_x = 0: \quad N = -60 + 20 = -40 \text{ kN} \tag{3.2c}$$

These expressions apply to every cross-section between B and C (i.e. from $x = 2$ to 6 m).

Segment CD: We now consider cross-sections between C and D. It is easier to consider the free-body to the right of the cut section as shown in Figure 3.12c. In this case, the cut cross-section is x from A or $(10 - x)$ from D. The equations of equilibrium give:

$$\Sigma M_X = 0: \quad M = 56(10 - x) = 560 - 56x \text{ kNm} \tag{3.3a}$$

$$\Sigma F_y = 0: \quad S = +56 \text{ kN} \tag{3.3b}$$

$$\Sigma F_x = 0: \quad N = 0 \text{ kN} \tag{3.3c}$$

These expressions apply to every cross-section between C and D (i.e. from $x = 6$ to 10 m).

(3) In Figure 3.13, the three expressions for bending moment, shear and axial force are plotted against x. It must be remembered that each expression is valid only for a particular range of x.

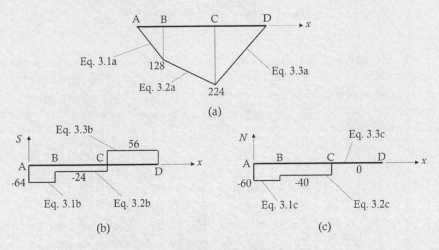

(a)

(b)   (c)

Figure 3.13 Internal action diagrams for Worked Example 3.2. (a) BMD (in kNm) (drawn on tension side). (b) SFD (in kN). (c) AFD (in kN).

With regard to bending moments, each of Equations 3.1a, 3.2a and 3.3a are linear and the bending moment diagram is drawn as a series of straight lines in the regions between the points where the transverse concentrated loads are applied (i.e. A, B, C and D). It is also noted that between the points where the transverse loads are applied, the shear force is constant and, between the points where longitudinal loads are applied, the axial force is constant.

WORKED EXAMPLE 3.3

The statically determinate two-span beam shown in Figure 3.14a is pinned at A and supported on rollers at C and D. It contains a frictionless hinge at B. The free-body diagram of the beam is shown in Figure 3.14b. Determine the reactions at each support and the expressions for M and S in segments AB, BC and CD, and then sketch the bending moment and shear force diagrams for the beam. Note that because there are no longitudinal (horizontal) loads applied to this beam, the axial force is everywhere zero.

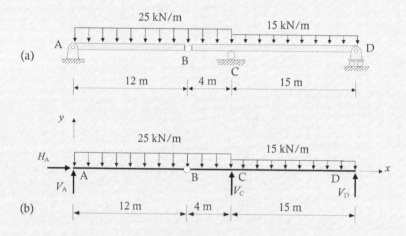

Figure 3.14 Beam and free-body diagram for Worked Example 3.3.

With a total of four unknown reaction components ($n_r = 4$) and with three equilibrium equations and one equation of condition ($n_c + 3 = 4$), the beam is statically determinate. The direction assumed for each of the reaction components is arbitrary. A positive answer will indicate the selected direction is correct. A negative result will indicate that the correct direction is opposite to that assumed.

Determine reactions

Consider the free-body in Figure 3.14b. Equating the sum of the forces in the x-direction to zero $\left(\Sigma F_x = 0\right)$ gives $H_A = 0$ kN.

Next, consider the free-body diagram of segment AB (shown in Figure 3.15). Since we know that the moment at the hinge at B is zero, taking moments about B gives:

$$\curvearrowright \quad -V_A \times 12 + 25 \times 12 \times 6 = 0 \quad \Rightarrow \quad V_A = 150.0 \text{ kN}$$

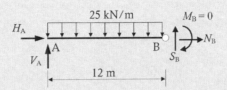

25 kN/m $M_B = 0$

H_A

A   B N_B

V_A S_B

12 m

Figure 3.15 Free-body diagram of segment AB.

Returning to the free-body diagram of the whole beam (Figure 3.14b), with V_A now known, taking moment about the support at C gives:

$$\text{↻} \quad -150 \times 16 + 25 \times 16 \times 8 - 15 \times 15 \times 7.5 + V_D \times 15 = 0 \quad \Rightarrow \quad V_D = 59.17 \text{ kN}$$

Similarly, taking moments about D gives:

$$\text{↻} \quad -150 \times 31 + 25 \times 16 \times 23 - V_C \times 15 + 15 \times 15 \times 7.5 = 0 \quad \Rightarrow \quad V_C = 415.83 \text{ kN}$$

The equation $\Sigma F_y = 0$ can now be applied to check that the vertical reactions calculated above are correct:

$$+\uparrow \quad V_A + V_C + V_D - 25 \times 16 - 15 \times 15 = 0$$

$$150.0 + 415.83 + 59.17 - 25 \times 16 - 15 \times 15 = 0 \quad \therefore \text{ OK}$$

Determine internal actions

<u>Segment AC</u>: For any cross-section x m from A within the length of beam AC, expressions for M and S can be obtained from the free-body diagram of Figure 3.16a.

Taking moments about the cross-section x m from A gives:

$$M = 150x - 25 \times x \times x/2 = 150x - 12.5x^2 \text{ (kNm)} \tag{3.4a}$$

$$S = -150 + 25x \text{ (kN)} \tag{3.4b}$$

These expressions apply to every cross-section between A and C (i.e. from $x = 0$ to 16 m).

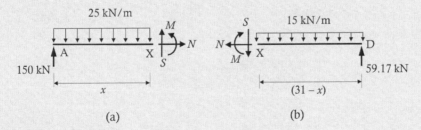

25 kN/m S 15 kN/m

M

A   X N N X   D

150 kN S M

x $(31 - x)$ 59.17 kN

(a)   (b)

Figure 3.16 Determination of the internal actions. (a) Free-body of segment AB (and BC). (b) Free-body of segment CD.

<u>Segment CD</u>: For any cross-section between C and D, consider the free-body to the right of the cut section as shown in Figure 3.16b. In this case, the cut cross-section is x m from A or $(31 - x)$ m from D. The equations of equilibrium give:

$$M = 59.17 \times (31 - x) - 15 \times (31 - x)(31 - x)/2$$

$$= 405.83x - 7.5x^2 - 5373.23 \text{ kNm} \tag{3.5a}$$

$$S = 59.17 - 15 \times (31 - x) = 15x - 405.83 \text{ kN} \tag{3.5b}$$

Equations 3.5 apply to every cross-section between C and D (i.e. from $x = 16$ to 31 m). Equations 3.4 and 3.5 are plotted in Figure 3.17 to form the bending moment and shear force diagrams.

Alternatively, for the free-body in Figure 3.16b, if the length of the beam segment had been defined as x (so that x varied from 0 at D to 15 m at C), the equations for M and S would become:

$$M = 59.17x - 7.5x^2 \text{ (kNm)} \quad \text{and} \quad S = 59.17 - 15x \text{ (kN)}.$$

These give the same numerical value for M and S at any point along the span CD.

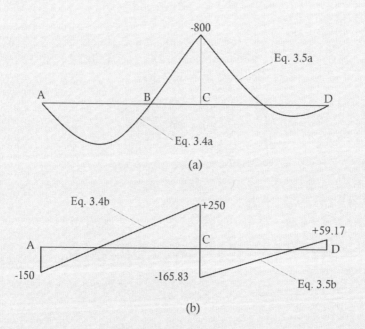

(a)

(b)

Figure 3.17 Bending moment and shear force diagrams for Worked Example 3.3. (a) Bending moment diagram (kNm). (b) Shear force diagram.

WORKED EXAMPLE 3.4

The free-body diagram of the frame of Figures 2.29 and 2.30 is shown in Figure 3.18 and includes the four reactions determined in Worked Example 2.5. Draw the axial force, shear force and bending moment diagrams for the frame.

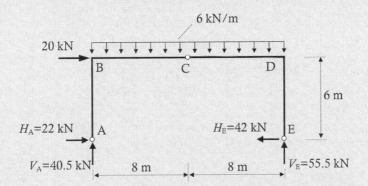

Figure 3.18 Free-body diagram of the frame for Worked Example 3.4.

(1) The axial force, shear force and bending moment at any point on the frame can be determined by considering the free-body diagram on either the left or right of the point considered. Appropriate free-body diagrams for each segment of the frame are shown in Figure 3.19, which also highlight the positive direction of the x-axis adopted along each member. The segments along which the internal actions remain continuous are AB, BD and DE.

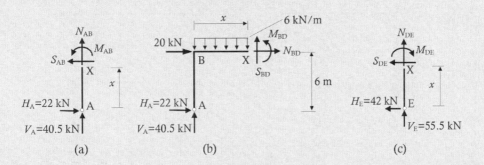

Figure 3.19 Free-body diagrams for Worked Example 3.4. (a) Segment AB. (b) Segment BD. (c) Segment DE.

(2) We now derive the expressions for the internal actions by performing cuts along segments AB, BD and DE.

<u>Segment AB</u>: In Figure 3.15a, x varies from 0 at A to 6 m at B:

$$N_{AB} = -40.5 \text{ kN}; \quad S_{AB} = +22.0 \text{ kN}; \quad \text{and } M_{AB} = -22.0x \text{ kNm}$$

At A where $x = 0$: $N_A = -40.5$ kN; $S_A = +22.0$ kN;   and $M_A = 0$.
At B where $x = 6$ m: $N_B = -40.5$ kN; $S_B = +22.0$ kN;   and $M_B = -132$ kNm.

Segment BD: In Figure 3.19b, x varies from 0 at B to 16 m at D (note that the presence of the hinge does not modify the expressions of the internal actions):

$$N_{BD} = -42.0 \text{ kN};   S_{BD} = 6x - 40.5 \text{ kN; and}$$

$$M_{BD} = 40.5x - 22.0 \times 6 - 6x^2/2 = -3x^2 + 40.5x - 132.0 \text{ kNm}$$

At B where $x = 0$: $N_B = -42.0$ kN; $S_B = -40.5$ kN;   and $M_B = -132.0$ kNm.
At C where $x = 8$ m: $N_C = -42.0$ kN; $S_C = +7.5$ kN;   and $M_C = 0$.
At D where $x = 16$ m: $N_D = -42.0$ kN; $S_D = +55.5$ kN;   and $M_D = -252.0$ kNm.
The correctness of the expression obtained for the bending moment is verified by double-checking that the moment present at the hinge at C is zero.

Segment DE: In Figure 3.19c, x varies from 6 m at D to 0 at E:

$$N_{DE} = -55.5 \text{ kN};   S_{DE} = -42.0 \text{ kN};   and M_{DE} = -42.0x \text{ kNm}$$

At D where $x = 6$ m: $N_D = -55.5$ kN; $S_D = -42.0$ kN;   and $M_D = -252.0$ kNm.
At E where $x = 0$: $N_E = -55.5$ kN; $S_B = -42.0$ kN;   and $M_B = 0$.

(3) The expressions for the axial force, shear force and bending moment diagrams are plotted in Figure 3.20.

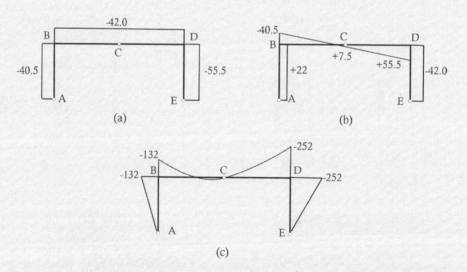

Figure 3.20 Internal actions for Worked Example 3.4. (a) Axial force diagram (kN). (b) Shear force diagram (kN). (c) Bending moment diagram (kNm).

WORKED EXAMPLE 3.5

A simply-supported beam of span 12 m is subjected to a linearly varying load (varying from 0 kN/m at the left support A to 15 kN/m at the right support B). A free-body diagram of the whole beam is shown in Figure 3.21a. Determine and plot expressions for M, S and N.

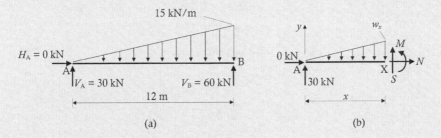

(a)   (b)

Figure 3.21 Free-body diagrams for Worked Example 3.5.

(1) The internal actions are continuous over the entire beam length.

(2) The free-body diagram of the portion of the beam to the left of a cross-section x from the support A is shown in Figure 3.21b. The load intensity w_x at the cut cross-section is obtained from geometry as:

$$\frac{w_x}{x} = \frac{-15}{12} \quad \therefore w_x = -\frac{5}{4}x \text{ (kNm)} \tag{3.6}$$

where the negative sign in the expression of w_x identifies that the distributed load is pointing in the opposite direction to the positive of the y-axis.

The resultant of the triangular distributed load acting on the free-body of Figure 3.21b is equal to $\frac{1}{2}w_x x = -\frac{5}{8}x^2$ and it acts at a distance $x/3$ from the cut cross-section.

Summing the forces vertically on Figure 3.21b, we get:

$$S + 30 - \frac{5}{8}x^2 = 0 \quad \therefore S = \frac{5}{8}x^2 - 30 \text{ (kN)} \tag{3.7}$$

and taking moment about the cut section gives:

$$M - 30x + \frac{5}{8}x^2\frac{x}{3} = 0 \quad \therefore M = 30x - \frac{5}{24}x^3 \text{ (kNm)} \tag{3.8}$$

From horizontal equilibrium: $N = 0$. $\tag{3.9}$

(3) The diagrams for M, S and N are obtained by plotting Equations 3.7, 3.8 and 3.9 and are shown in Figure 3.22.

For the bending moment and shear force diagrams shown in Figures 3.17 and 3.22, it would appear that the point of maximum positive moment corresponds to that point on the span where the shear force is zero. In Chapter 5, we will demonstrate that this is in fact always the case.

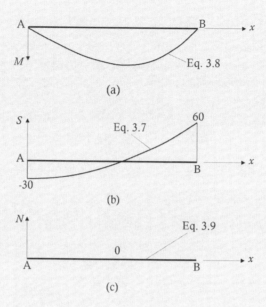

Figure 3.22 Internal actions for Worked Example 3.5. (a) BMD (kNm). (b) SFD (kN). (c) AFD (kN).

From Worked Examples 3.2 to 3.5, the observations below can be made with respect to beams and frames with straight segments.

SHEAR FORCE DIAGRAMS

(1) Where a concentrated load is applied normal (or transverse) to the axis of a member, a step occurs in the shear force diagram equal in magnitude to the concentrated load.

(2) Between points of load application, the shear force is constant.

(3) In regions of a beam or frame subjected to uniformly distributed transverse load, the shear force diagram is linear.

(4) In regions of a beam or frame subjected to a linearly varying transverse load, the shear force diagram is parabolic.

(5) The points on the shear force diagram where the shear force is zero correspond to the points where the bending moment is either a maximum or a minimum.

BENDING MOMENT DIAGRAMS

(1) Between points of transverse load application, the bending moment diagram is linear.

(2) In regions of a beam subjected to uniformly distributed load, the bending moment diagram is parabolic.

(3) In regions of a beam subjected to a linearly varying load, the bending moment diagram is cubic.

(4) At the points where concentrated transverse loads are applied to a beam, the bending moment diagram changes direction (kinks).

(5) A step occurs in the bending moment diagram at points where a couple is applied and the step is equal in magnitude to the applied couple.

(6) The bending moment diagram reaches a maximum or minimum at points where the shear force is zero.

PROBLEMS

3.1 Consider the beams shown below and specify whether they are statically determinate, indeterminate, or unstable.

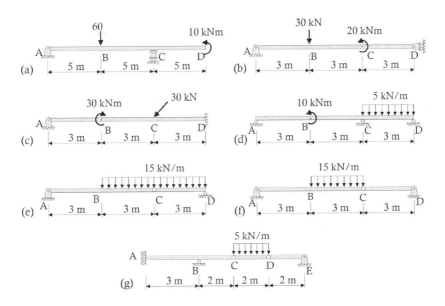

3.2 For the frames illustrated below, specify whether they are statically determinate, indeterminate or unstable.

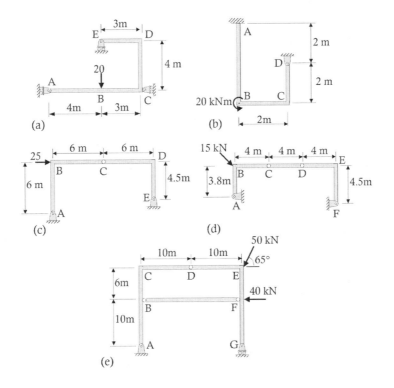

3.3   For the simply-supported beam shown below, determine the expressions for the internal actions (*N*, *S* and *M*) and plot the axial force, shear force and bending moment diagrams.

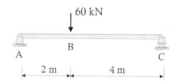

3.4   Evaluate the expressions for the internal actions (*N*, *S* and *M*) for the simply-supported beam illustrated, and draw the axial force, shear force and bending moment diagrams.

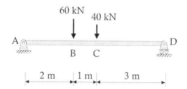

3.5   Plot the axial force, shear force and bending moment diagrams for the simply-supported beam shown below.

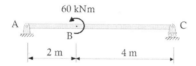

3.6   Consider the beam shown and draw the axial force, shear force and bending moment diagrams.

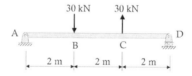

3.7   Determine the expressions for the internal actions (*N*, *S* and *M*) and draw the axial force, shear force and bending moment diagrams for the structure shown below.

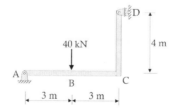

3.8   Consider the structure illustrated and plot the axial force, shear force and bending moment diagrams.

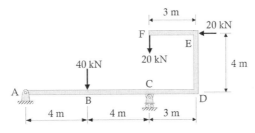

3.9   For the frame shown, calculate N, S and M in each region of the beam and draw the axial force, shear force and bending moment diagrams.

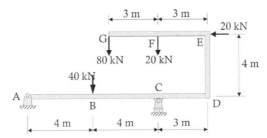

3.10 Determine the expressions for the internal actions N, S and M in each region of the beam shown below and plot the axial force, shear force and bending moment diagrams.

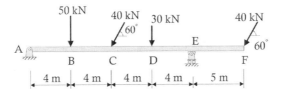

3.11 For the beam ABCDEF shown below, calculate the expressions for the internal actions N, S and M, and plot their diagrams along the member length.

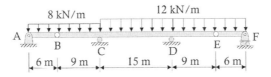

3.12 Consider the two beams shown and plot the axial force, shear force and bending moment diagrams.

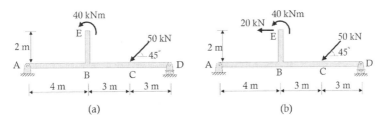

3.13 For the beam shown below, determine expressions for the internal actions (N, S and M) in each segment of the beam and plot the axial force, shear force and bending moment diagrams.

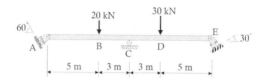

3.14 Determine the expressions for the internal actions (N, S and M) for the beam illustrated and plot the axial force, shear force and bending moment diagrams.

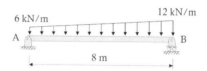

3.15 Evaluate the expressions for the internal actions (N, S and M) for the inclined beam and sketch the axial force, shear force and bending moment diagrams. In particular, calculate the axial force, shear force and bending moment at mid-span of the beam.

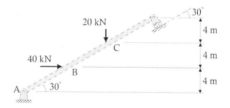

3.16 For the beam shown, write expressions for the internal actions (N, S and M) and plot the axial force, shear force and bending moment diagrams.

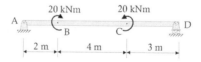

3.17 Draw the axial force, shear force and bending moment diagrams for the frame shown.

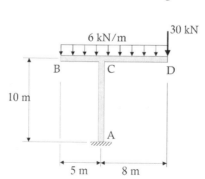

3.18 For the bent beam shown below, plot the axial force, shear force and bending moment diagrams, and calculate N, S and M at the mid-points of segments AB, BC, CD, ED and FE.

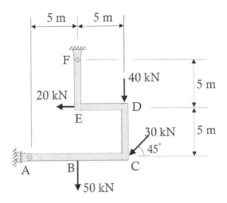

3.19 Consider the beam shown below and draw the axial force, shear force and bending moment diagrams.

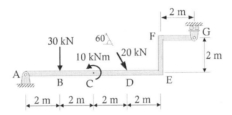

3.20 For the structure shown, determine the expressions for the internal actions (N, S and M) and plot the axial force, shear force and bending moment diagrams.

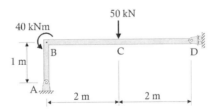

3.21 Consider the beam shown below and plot the axial force, shear force and bending moment diagrams.

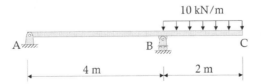

3.22 For the structure shown, draw the axial force, shear force and bending moment diagrams.

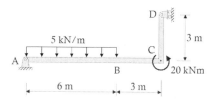

3.23 For the portal frame ABCDE shown, determine the expressions for the internal actions (N, S and M) and draw the axial force, shear force and bending moment diagrams.

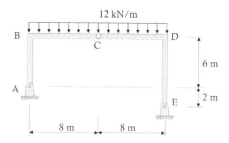

3.24 For the frame illustrated below, plot the axial force, shear force and bending moment diagrams.

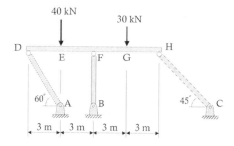

3.25 Consider the portal frame shown and determine the expressions for the internal actions (N, S and M). Plot the axial force, shear force and bending moment diagrams.

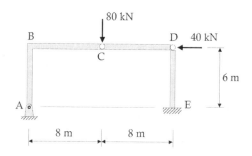

3.26 Evaluate the expressions for the internal actions (N, S and M) for the frame illustrated and plot the axial force, shear force and bending moment diagrams.

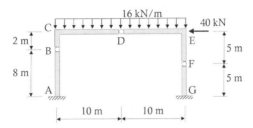

3.27 The frame ABCDE shown is pinned at A and E, and has a roller support at D. For the loading shown, find the five reaction components given that the bending moments in the frame at B and at D are $M_B = -80$ kNm and $M_D = -60$ kNm (i.e. tension is on the outside of the frame at B and D).

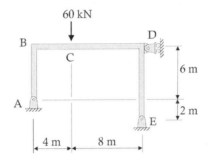

3.28 The frame shown has a pinned support at A, a roller support at G and an internal hinge at D. It is prevented from collapsing by the tie BF connected to the frame by means of end hinges as shown. Determine the tension force in the tie BF and draw its axial force diagram.

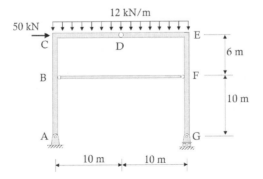

3.29 Consider the cantilever beams illustrated below and plot qualitative diagrams outlining the variations of the axial force, shear force and bending moment along the beam length. In these diagrams, specify the order of the polynomial describing the internal actions, for example, stating whether the curves are linear, parabolic or cubic.

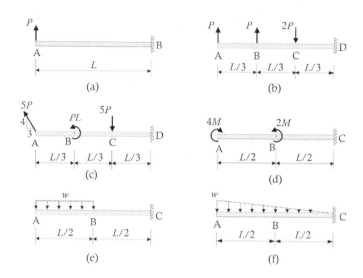

3.30 For the structures shown, draw qualitative diagrams of the axial force, shear force and bending moment. In these diagrams, specify the order of the polynomial describing the internal actions, for example, stating whether the curves are linear, parabolic or cubic.

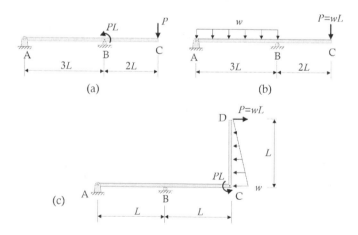

Chapter 4

Statically determinate trusses

4.1 INTRODUCTION

A *truss* is a structure consisting of a number of rigid bars fastened together at their ends and arranged in such a way that the bars form a series of stable triangular units (see Figure 4.1). For the purpose of calculating the forces in the bars (called members of the truss), the connections at the ends of the members (called the nodes of the truss) are considered to be *pinned*. At a *pinned* connection, the end of each member is free to rotate independently of, and unrestrained by, the other members framing into the connection.

In reality, very few trusses are built with perfectly pinned connections. The connections in steel trusses, for example, are generally made by either welding or bolting, as shown in Figure 4.2. Some members may even be continuous through the connection. In practical trusses, therefore, the individual members framing into a connection are not generally free to rotate. However, the members of most trusses are relatively slender and any fixity within a connection has a minor effect upon the internal force system, provided the members of the truss are appropriately arranged. For trusses loaded at the nodes, the predominant internal action in each member is an *axial force* and there is little tendency for the ends of the members to rotate relative to each other. Assuming the connections are pinned not only simplifies the analysis but also results in a reasonably accurate assessment of the member forces.

From a qualitative viewpoint, trusses can be classified as *simple, compound* or *complex*. The simplest stable truss consists of the triangular arrangement depicted in Figure 4.1a. A truss is regarded as *simple* when it is built up by extending the basic triangle by adding two members and one node at a time (such as the trusses shown in Figure 4.1b through e). Two or more simple trusses can be appropriately combined to form a *compound truss*. This can occur, for example, by connecting two simple trusses by means of one member and one node (as shown in Figure 4.3a) or by inserting three non-parallel

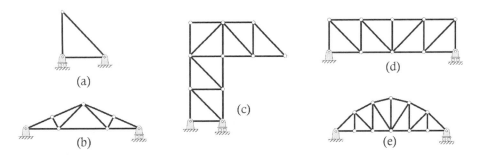

Figure 4.1 Typical truss layouts.

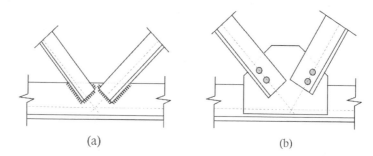

Figure 4.2 Typical member connections in light steel trusses. (a) Welded connection. (b) Bolted connection.

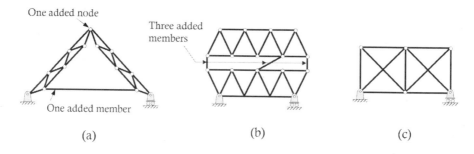

Figure 4.3 Examples of compound and complex trusses. (a–b) Compound truss. (c) Complex truss.

members between two simple trusses (as shown in Figure 4.3b). Trusses that do not fall into the categories of simple and compound trusses are defined as *complex* trusses, such as the one shown in Figure 4.3c.

In this chapter, we are primarily concerned with the determination of the axial forces in simple, statically determinate, *pin-jointed trusses* with loads applied only at the nodes. Two methods of analysis are presented, called the *methods of joints* and the *method of sections*, and each is illustrated by a number of worked examples. Procedures for the evaluation of the stability of particular truss layouts and for the determination of the degree of statical indeterminacy of a truss are also outlined. The methods of analysis are first presented for two-dimensional or plane trusses and are then extended to cover three-dimensional or space trusses.

4.2 ASSUMPTIONS FOR TRUSS ANALYSIS

The analysis of *simple* trusses relies on the assumptions listed below.

1. The nodes in a truss are pinned, i.e. the connections are assumed to possess no rotational rigidity and cannot transmit moments.
2. All external actions and support reactions consist of forces applied at the truss nodes.
3. All truss members are straight and the line connecting their end nodes coincides with the centroidal axis of the member.

As a consequence of these first three assumptions, the members of the truss are subjected to axial loads only, either compressive or tensile, and therefore there are no shear forces and no moments in any member.

4. All displacements are small in comparison with the lengths of the members of the truss.

If the deformation of each member of the truss is small, the undeformed geometry of the truss can be used to write the equilibrium equations without significant errors. For a truss with slender and flexible elements, this assumption may be unreasonable and the equilibrium equations may need to be rewritten in the displaced configuration.

A truss complying with the above assumptions is considered to be *ideal*. Of course, no real truss has perfectly pinned connections or is made up of perfectly straight members. In addition, the self-weight of truss members acts as a distributed load along the member and will produce both moments and shear forces. Often, uniformly distributed floor, roof or ceiling loads act on truss members. These distributed loads are generally included in the analysis by means of statically equivalent point loads applied at the node at each end of the member in an *ideal* truss. To avoid confusion when dealing with internal actions, member axial forces calculated under the *ideal* truss assumptions 1–4 are referred to here as *primary forces*. Other internal actions (i.e. axial forces, shear forces and moments) generated by other effects not complying with the above assumptions are referred to as *secondary forces*.

The analysis procedures presented in this chapter are useful for the calculation of the primary forces. In Section 4.9, we will consider trusses for which assumptions 2 and 3 are not satisfied and discuss simple ways of determining the resulting secondary forces. In situations where the secondary forces are likely to be significant and to lead to pronounced deformations, the general methods of analysis outlined in subsequent chapters are more appropriate.

4.3 SIGN CONVENTION AND NOTATION

Tensile axial forces are taken to be positive and compressive forces are negative (see Section 3.3). When considering the members and nodes of a truss, individual members are numbered and nodes are assigned an alphabetic character (upper case). A member is referred to using either its member number or the letters associated with its end nodes. For example, in Figure 4.4, the member connecting nodes B and E is referred to as member 5 or simply BE.

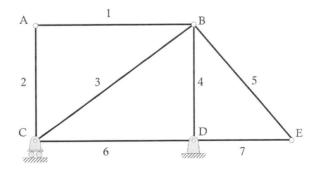

Figure 4.4 Element numbering and node labelling.

4.4 AN INTRODUCTION TO THE METHOD OF JOINTS

The analysis of a truss under a particular set of nodal loads generally involves the calculation of the reactions at each support using the equations of equilibrium, followed by the
determination of the axial force in each member.

Consider the statically determinate truss shown in Figure 4.5a carrying a single horizontal applied load P at node C. The corresponding free-body diagram is shown in Figure
4.5b and includes the two unknown reaction components at the pinned support at A (i.e.
H_A and V_A) and the vertical reaction at the roller support at B (i.e. V_B) in addition to the
applied load P.

The three unknown reactions are determined by enforcing the three equilibrium equations:

Horizontal equilibrium:

$$\xrightarrow{+} \quad \Sigma F_H = 0: \quad P + H_A = 0 \quad \therefore H_A = -P \tag{4.1a}$$

Rotational equilibrium about the support at A:

$$\curvearrowleft + \quad \Sigma M_A = 0: \quad -hP + LV_B = 0 \quad \therefore V_B = \frac{h}{L}P \tag{4.1b}$$

Vertical equilibrium:

$$\uparrow + \quad \Sigma F_V = 0: \quad V_A + V_B = 0 \quad \therefore V_A = -V_B = -\frac{h}{L}P \tag{4.1c}$$

Clearly, both H_A and V_A act in directions opposite to those shown in Figure 4.5b (assuming P to be positive).

For convenience, rotational equilibrium has been calculated with respect to the support
at A to limit the number of unknowns in the equation to one. In this simple example,
rotational equilibrium about the support at B would have provided a similar advantage, as
the lines of action of both H_A and V_B pass through node B. Other points could have been

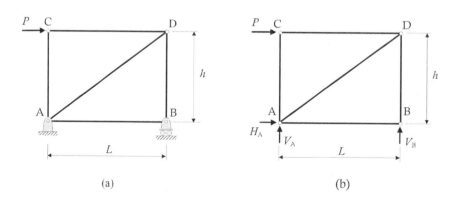

Figure 4.5 Statically determinate trusses. (a) Truss layout and loading. (b) Free-body diagram.

selected to enforce rotational equilibrium, but would have produced an equation with more than one unknown. For example, calculating rotational equilibrium with respect to point D produces an equation containing two unknowns H_A and V_A:

$$\sum M_D = 0: \quad hH_A - LV_A = 0$$

Having obtained the reactions from Equations 4.1, we may now determine the axial force in each member. A suitable method for this task is the *method of joints*.

The free-body diagram of each member and each node of the truss is shown in Figure 4.6. At each node or member end, tensile member forces act away from the node or member end, and initially all member forces are shown as tensile. For statically determinate simple trusses, it is possible to determine all the unknown member forces by satisfying the requirements of equilibrium at each node.

For the truss of Figure 4.6, node C is a convenient starting point as there are only two unknown forces acting on the node, i.e. N_{AC} and N_{CD}. By imposing equilibrium in the direction of the x-axis at node C (i.e. horizontal equilibrium), the axial force in member CD is readily determined:

$$\overset{+}{\to} \quad \sum F_x = 0 \text{ at node C:} \quad P + N_{CD} = 0 \quad \therefore N_{CD} = -P \tag{4.2}$$

The force in CD (N_{CD}) is therefore compressive (assuming P to be positive).

Similarly, N_{AC} can be calculated from equilibrium along the y-axis (i.e. vertical equilibrium) at node C:

$$\uparrow + \quad \sum F_y = 0 \text{ at node C:} \quad \therefore N_{AC} = 0 \tag{4.3}$$

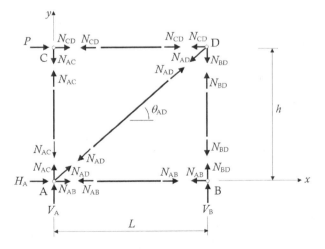

Figure 4.6 Free-body diagrams of each node and each member of the truss.

REFLECTION ACTIVITY 4.1

Is it possible to determine from visual inspection of the truss in Figure 4.7a that the axial forces in members AC and BE, i.e. N_{AC} and N_{BE}, are both zero?

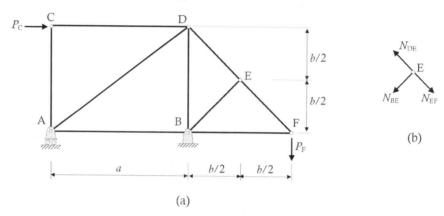

Figure 4.7 Truss for Reflection Activity 4.1. (a) Truss layout and loading. (b) Free-body diagram of node E.

Yes. Since N_{AC} represents the only force acting in the y direction at node C and being the only term contained in the vertical equilibrium equation at C, it must be equal to zero. Similarly, the member BE in Figure 4.7 is unloaded. Writing the equilibrium equation at node E in the direction perpendicular to member DEF leads to $N_{BE} = 0$. Of course, the axial force in BE will not be zero if a vertical (or horizontal) nodal load was applied to the truss at E.

We will now continue the analysis of the truss of Figure 4.6 and consider equilibrium in both the x and y directions at node D. The unknown member forces N_{AD} and N_{BD} can be obtained:

$$\xrightarrow{\pm} \ \Sigma F_x = 0 \text{ at node D:} \quad -N_{CD} - N_{AD}\cos\theta_{AD} = 0 \quad \therefore N_{AD} = -\frac{N_{CD}}{\cos\theta_{AD}} = \frac{P}{\cos\theta_{AD}} \tag{4.4}$$

$$\uparrow+ \ \Sigma F_y = 0 \text{ at node D:} \quad -N_{BD} - N_{AD}\sin\theta_{AD} = 0$$

$$\therefore N_{BD} = -N_{AD}\sin\theta_{AD} = -P\frac{\sin\theta_{AD}}{\cos\theta_{AD}} = -P\tan\theta_{AD} = -P\frac{b}{L} \tag{4.5}$$

and the last unknown N_{AB} is calculated from equilibrium along the x-axis at node B:

$$\xrightarrow{\pm} \ \Sigma F_x = 0 \text{ at node B:} \quad \therefore N_{AB} = 0 \tag{4.6}$$

REFLECTION ACTIVITY 4.2

In the determination of the unknown axial forces in each of the five truss members of Figures 4.5 and 4.6 (i.e. N_{AB}, N_{BD}, N_{CD}, N_{AC} and N_{AD}), we have used only five equilibrium equations out of the eight available (noting that two equilibrium equations are available at each of the four nodes). The three unused equations could be adopted to define vertical equilibrium at B and both vertical and horizontal equilibrium at A. Why do we have these extra equations?

In reality, we don't have three extra equilibrium equations. In fact, the total number of unknowns in this problem is eight, which includes the axial forces in each of the five members and the three reactions. With N_{BD} obtained from Equation 4.5, considering vertical equilibrium at node B allows the vertical reaction at B (V_B) to be determined. In this example, the value of V_B had previously been calculated by considering equilibrium of the entire structure. Similarly, with all the member forces known, the vertical equilibrium equation at node A may be used to determine V_A and the horizontal equilibrium equation at A may be use to find H_A.

For ease of solution, the order of nodes followed in the solution process using the method of joints should, where possible, be such that only two unknown forces act on the node being considered. Other node orders could be adopted, where three or more unknown forces may exist at a particular node, and although the solution process will be more tedious, it will lead to the same result.

For the *statically determinate truss* of Figure 4.8, there are more than two unknowns at every node. However, if we first calculate the reactions at A and C by applying the equilibrium equations to the entire truss, the number of unknowns at both node A and node C reduces to two and either node is a convenient starting point.

SUMMARY OF STEPS 4.1: Method of joints — Steps in solution procedure

1. Calculate the unknown reactions by enforcing equilibrium of the entire truss.

2. Draw a free-body diagram for each node of the truss, initially assuming that all member forces are tensile, i.e. acting away from the node.

3. By considering the number of unknown forces at each node, determine an appropriate nodal sequence.

4. Apply the equilibrium equations at each node in two non-coincident directions (often the directions of the coordinate axes) and solve the resulting equations to determine the unknown member axial forces.

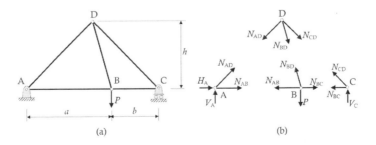

Figure 4.8 Truss and free-body diagrams of its nodes.

WORKED EXAMPLE 4.1

Determine the unknown reactions and member axial forces for the truss of Figure 4.9a using the method of joints based on the solution procedure detailed in Summary of Steps 4.1. The free-body diagram of each node is shown in Figure 4.9b.

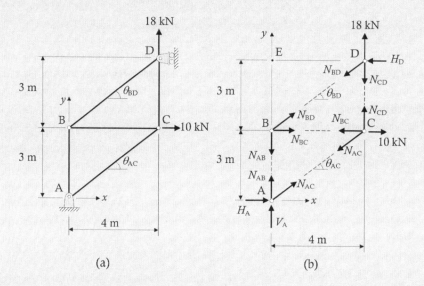

(a)   (b)

Figure 4.9 Truss and free-body diagrams for Worked Example 4.1.

The three unknown reactions, i.e. H_A, V_A and H_D, can be readily calculated by applying the three equations of equilibrium to the entire structure. The horizontal reaction at D (H_D) is obtained by considering rotational equilibrium about the support A $\left(\Sigma M_A = 0\right)$. By setting the sum of all the horizontal forces to zero $\left(\Sigma F_x = 0\right)$, the reaction H_A is obtained and setting the sum of the vertical forces to zero $\left(\Sigma F_y = 0\right)$ will yield the remaining reaction V_A. Alternatively, rotational equilibrium with respect to point E (Figure 4.9b), being the intersection of the lines of actions of V_A and H_D, could have been used as a starting point to determine H_A.

Remember that when placing the reactions on the free-body diagram, their positive directions are arbitrary and their actual directions are defined by the calculated signs for each reaction. The three equilibrium equations are written as:

$$\curvearrowright \quad \Sigma M_A = 0: \quad 18 \times 4 - 10 \times 3 + H_D \times 6 = 0 \quad \therefore H_D = -7 \text{ kN}$$

$$\xrightarrow{\pm} \quad \Sigma F_x = 0: \quad H_A + 10 - H_D = 0 \quad \therefore H_A = -17 \text{ kN}$$

$$\uparrow + \quad \Sigma F_y = 0: \quad V_A + 18 = 0 \quad \therefore V_A = -18 \text{ kN}$$

By inspection, there are two unknown member forces at nodes A and D, while nodes B and C each have three unknowns. A suitable nodal sequence is:

(1) Node A (to obtain N_{AB} and N_{AC})

(2) Node D (to obtain N_{BD} and N_{CD})

(3) Node B (to get N_{BC})

(4) Node C (where we can use the two available equilibrium equations to check that the calculated member forces are correct)

From the geometry of the truss: $\cos \theta_{AC} = \cos \theta_{BD} = 0.8$ and $\sin \theta_{AC} = \sin \theta_{BD} = 0.6$.

(1) Node A:

$$\overset{+}{\rightarrow} \quad \Sigma F_x = 0: \quad -17 + N_{AC}\frac{4}{5} = 0 \quad \therefore N_{AC} = 21.25 \text{ kN}$$

$$\uparrow + \quad \Sigma F_y = 0: \quad -18 + N_{AC}\frac{3}{5} + N_{AB} = 0 \quad \therefore N_{AB} = 5.25 \text{ kN}$$

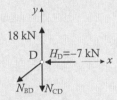

(2) Node D:

$$\overset{+}{\rightarrow} \quad \Sigma F_x = 0: \quad -N_{BD}\frac{4}{5} + 7 = 0 \quad \therefore N_{BD} = 8.75 \text{ kN}$$

$$\uparrow + \quad \Sigma F_y = 0: \quad 18 - N_{BD}\frac{3}{5} - N_{CD} = 0 \quad \therefore N_{CD} = 12.75 \text{ kN}$$

(3) Node B:

$$\overset{+}{\rightarrow} \quad \Sigma F_x = 0: \quad N_{BD}\frac{4}{5} + N_{BC} = 0 \quad \therefore N_{BC} = -7 \text{ kN}$$

(4) Node C:

With all bar forces now determined, we can check that the results are correct using the unused equilibrium equations at Node C:

$$\uparrow+ \quad \Sigma F_y = 0: \quad N_{CD} - N_{AC}\frac{3}{5} = 0 \quad \therefore 12.75 - 21.25 \times 0.6 = 0 \quad \therefore OK$$

$$\xrightarrow{+} \quad \Sigma F_x = 0: \quad 10 - N_{AC}\frac{4}{5} - N_{BC} = 0 \quad \therefore 10 - 21.25 \times 0.8 + 7 = 0 \quad \therefore OK$$

The method of joints is very useful for calculating the forces of a simple truss by hand. In addition, because of the well-structured set of steps, the method can be expressed in matrix form for easy implementation in computer software.

4.5 METHOD OF JOINTS IN MATRIX FORM

When using the method of joints, we have seen that at each node of a truss there are two equilibrium equations relating to the sum of forces in two non-coincident directions. In the following, we will refer to each direction at each node as a *degree of freedom (dof)* or simply a *freedom*. Let us now consider the truss of Figure 4.10a, with each *freedom* labeled with a different integer. For example, the directions along the positive x- and y-axes at node B are referred to as *dof* 3 and *dof* 4, respectively.

The equilibrium equations associated with each degree of freedom are written, including the components of all unknown member forces, applied loads and reactions acting in that direction at the node under consideration. This can be facilitated using the appropriate free-body

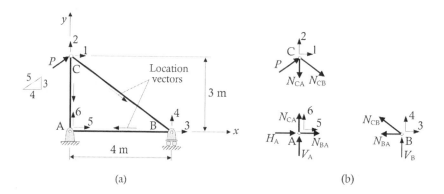

Figure 4.10 Truss layout, degrees of freedom, location vectors and free-body diagrams. (a) Truss layout with freedoms and location vectors. (b) Free-body diagram of each node.

diagram of the relevant node (as shown in Figure 4.10b). For example, the equilibrium equation along *dof* 1 (i.e. horizontal equilibrium along the *x*-axis at C) can be written as:

$$dof\ 1:\quad \Sigma F_x = 0:\quad N_{CBx} + N_{CAx} + P_x = 0 \qquad (4.7)$$

where the additional subscript "*x*" identifies the force component parallel to the *x*-axis.

To simplify the calculation of the force projections, we introduce a positive direction along each member, which defines a starting node (also referred to as node 1 in the following) and an ending node (node 2). This direction is depicted by a *location vector* assigned to each truss member pointing from node 1 to node 2 as shown in Figure 4.10a. The choice of the positive direction of the location vector is arbitrary but, once assigned, must be maintained throughout the analysis. For example, on the basis of the location vector specified in Figure 4.10a, nodes 1 and 2 for member CB are C and B, respectively.

For practical purposes, it is useful to express the projections of the member forces (i.e. the components in the *x* and *y* directions) in terms of their direction cosines calculated with respect to the coordinate axes or relevant freedoms. Direction cosines were introduced earlier in Section 2.3. Considering the force vector N in Figure 4.11, the projections (or components) parallel to the *x*- and *y*-axes are:

$$N_x = N \cos \theta_x = lN \qquad \text{where} \qquad l = \cos \theta_x \qquad (4.8a)$$

$$N_y = N \sin \theta_x = N \cos \theta_y = mN \qquad \text{where} \qquad m = \cos \theta_y \qquad (4.8b)$$

and θ_x and θ_y represent the angles formed between the force vector and the *x*- and *y*-axes, respectively, while *l* and *m* are their direction cosines. On the basis of the positive direction specified by the location vector, the direction cosines of a member in a truss are:

$$l = \frac{x_{node2} - x_{node1}}{L} \quad \text{and} \quad m = \frac{y_{node2} - y_{node1}}{L} \qquad (4.9a,b)$$

in which the coordinates of nodes 1 and 2 are depicted by (x_{node1}, y_{node1}) and (x_{node2}, y_{node2}), and *L* is the length of the member.

It is now possible to express the projections of the member forces included in the equilibrium equations by means of the direction cosines and, consequently, as a function of the node coordinates (using Equations 4.9). Remember that the axial forces in a member of a truss applied to the free-body diagrams at each of its two end nodes have opposite

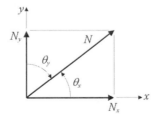

Figure 4.11 Projections of a force vector.

directions. For example, the force in member CB (N_{CB}) acting at node C in Figure 4.10b is pointing in the opposite direction to the same force at node B.

Using the sign convention adopted earlier, a member force applied at its first node has the identical positive direction as the member location vector and, because of this, its direction cosines can be used to define the force projections. For example, the direction cosines of N_{CB} applied at node C in Figure 4.10b are:

$$l_{CB,1} = l_{CB} \quad \text{and} \quad m_{CB,1} = m_{CB} \tag{4.10a,b}$$

where l_{CB} and m_{CB} are the direction cosines of the location vector adopted for CB (and determined using Equations 4.9), while $l_{CB,1}$ and $m_{CB,1}$ define the direction cosines for the axial force N_{CB} applied at its first node C (where the additional subscript "1" denotes the first node).

Considering the second node of the member, the positive direction of N_{CB} applied at B is opposite to the positive direction of the location vector. In this case, the direction cosines of N_{CB} applied at the second node of CB are:

$$l_{CB,2} = -l_{CB} \quad \text{and} \quad m_{CB,2} = -m_{CB} \tag{4.11a,b}$$

where the additional subscript "2" depicts the second node specified by the location vector.

Equilibrium at node C in the direction of freedom 1, previously written as Equation 4.7, can now be re-written in terms of the direction cosines related to the axial and external forces applied at C as follows:

$$l_{CB,1} N_{CB} + l_{CA,1} N_{CA} + l_p P = 0 \tag{4.12}$$

where l_p represents the direction cosine l related to the applied load P at node C and, from Figure 4.10a, $l_p = 4/5 = 0.8$, and the direction cosines of the members at node C are:

$$l_{CB,1} = \frac{x_B - x_C}{L_{CB}} = \frac{4-0}{5} = 0.8 \tag{4.13a}$$

$$l_{CA,1} = \frac{x_A - x_C}{L_{CA}} = \frac{0-0}{3} = 0 \tag{4.13b}$$

and the equilibrium equation at node C in the direction of freedom 1 (Equation 4.12) becomes:

$$0.8 N_{BC} + 0.8P = 0 \tag{4.14}$$

REFLECTION ACTIVITY 4.3

Write the equilibrium equations along freedoms 3 and 4 for the truss shown in Figure 4.10 using the direction cosines introduced in Equations 4.8 through 4.11 and comment on the possible advantages of their use.

All dimensions in the solutions are in metres.

Freedoms 3 and 4 correspond to the horizontal and vertical directions at node B.

For member BA: (Node 1: B (x_B, y_B) = (4, 0); Node 2: A (x_A, y_A) = (0, 0); and L_{BA} = 4)

$$l_{BA} = \frac{x_A - x_B}{L_{BA}} = \frac{0-4}{4} = -1; \quad m_{BA} = \frac{y_A - y_B}{L_{BA}} = \frac{0-0}{4} = 0$$

For member CB: (Node 1: C (x_C, y_C) = (0, 3); Node 2: B (x_B, y_B) = (4, 0); and L_{CB} = 5)

$$l_{CB} = \frac{x_B - x_C}{L_{CB}} = \frac{4-0}{5} = 0.8; \quad m_{CB} = \frac{y_B - y_C}{L_{BC}} = \frac{0-3}{5} = -0.6$$

The direction cosines for V_B are l_{V_B} = 0 and m_{V_B} = 1.
Equilibrium in the direction of freedom 3 at node B can be expressed as:

$$l_{BA,1}N_{BA} + l_{CB,2}N_{CB} + l_{V_B}V_B = -N_{BA} - 0.8N_{CB} = 0 \qquad (4.15a)$$

where from Equations 4.10 and 4.11: $l_{BA,1} = l_{BA} = -1$ and $l_{BC,2} = -l_{BC} = -0.8$.
Similarly, enforcing equilibrium along freedom 4 leads to:

$$m_{BA,1}N_{BA} + m_{CB,2}N_{CB} + m_{V_B}V_B = 0.6N_{CB} + V_B = 0 \qquad (4.15b)$$

where $m_{BA,1} = m_{BA} = 0$ and $m_{CB,2} = -m_{CB} = 0.6$.
While the use of direction cosines is tedious for hand calculations, it is preferred here for the matrix representation of the method of joints, which lends itself to computer programming.

It is usually convenient to tabulate the information related to each member, i.e. its location vectors and its direction cosines, before writing the equilibrium equations. The relevant information related to the members of the truss shown in Figure 4.10 is given in Table 4.1.

The equilibrium equations associated with each freedom of the truss of Figure 4.10 are as follows:

$$dof\ 1: \quad l_{CB,1}N_{CB} + l_{CA,1}N_{CA} + l_P P = 0 \qquad (4.16a)$$

$$dof\ 2: \quad m_{CB,1}N_{CB} + m_{CA,1}N_{CA} + m_P P = 0 \qquad (4.16b)$$

$$dof\ 3: \quad l_{BA,1}N_{BA} + l_{CB,2}N_{CB} + l_{V_B}V_B = 0 \qquad (4.16c)$$

$$dof\ 4: \quad m_{BA,1}N_{BA} + m_{CB,2}N_{CB} + m_{V_B}V_B = 0 \qquad (4.16d)$$

$$dof\ 5: \quad l_{BA,2}N_{BA} + l_{CA,2}N_{CA} + l_{H_A}H_A + l_{V_A}V_A = 0 \qquad (4.16e)$$

$$dof\ 6: \quad m_{BA,2}N_{BA} + m_{CA,2}N_{CA} + m_{H_A}H_A + m_{V_A}V_A = 0 \qquad (4.16f)$$

Table 4.1 Example of location vectors and direction cosines for a truss

Member	Node 1	Node 2	L (m)	l	m
BA	B	A	4	−1	0
CB	C	B	5	0.8	−0.6
CA	C	A	3	0	−1

Equations 4.16 can be re-arranged in matrix form as:

$$
\begin{bmatrix}
0 & l_{CB,1} & l_{CA,1} & 0 & 0 & 0 \\
0 & m_{CB,1} & m_{CA,1} & 0 & 0 & 0 \\
l_{BA,1} & l_{CB,2} & 0 & 0 & 0 & l_{V_B} \\
m_{BA,1} & m_{CB,2} & 0 & 0 & 0 & m_{V_B} \\
l_{BA,2} & 0 & l_{CA,2} & l_{H_A} & l_{V_A} & 0 \\
m_{BA,2} & 0 & m_{CA,2} & m_{H_A} & m_{V_A} & 0
\end{bmatrix}
\begin{bmatrix}
N_{BA} \\ N_{CB} \\ N_{CA} \\ H_A \\ V_A \\ V_B
\end{bmatrix}
+
\begin{bmatrix}
l_p P \\ m_p P \\ 0 \\ 0 \\ 0 \\ 0
\end{bmatrix}
=
\begin{bmatrix}
0 \\ 0 \\ 0 \\ 0 \\ 0 \\ 0
\end{bmatrix}
\tag{4.17}
$$

or, in more compact form, as:

$$\mathbf{A}\mathbf{p} + \mathbf{f} = 0 \tag{4.18}$$

where the direction cosines used to determine axial force projections are collected in $\mathbf{A}$, the vector $\mathbf{p}$ includes the six unknowns (three member forces and three reactions) and the external actions are grouped in vector $\mathbf{f}$.

The unknown reactions and axial forces can be obtained solving the system of equations specified in Equation 4.18:

$$\mathbf{p} = -\mathbf{A}^{-1}\mathbf{f} \tag{4.19}$$

where $\mathbf{A}^{-1}$ depicts the inverse of matrix $\mathbf{A}$. The solution of Equation 4.19 can be carried out using one of the solution methods outlined in Appendix C.

Matrix $\mathbf{A}$ is reproduced in Table 4.2 to highlight how its rows refer to the freedoms of the truss and how its columns refer to the unknown bar forces and reactions.

When the freedoms and unknowns are ordered in the rows and columns, respectively, the relevant direction cosines can be specified by inspection of the truss and the $\mathbf{A}$ matrix can be readily determined. For any freedom number (i.e. any row in the matrix), a direction cosine is assigned only to the columns associated with the unknown member forces and reactions acting at the node considered.

The overall procedure required by the method of joints when applied in matrix form is summarised below followed by a worked example.

Table 4.2 Example of the terms included in matrix $\mathbf{A}$

Freedoms	Member					
	N_{BA}	N_{CB}	N_{CA}	H_A	V_A	V_B
1		$l_{CB,1}$	$l_{CA,1}$			
2		$m_{CB,1}$	$m_{CA,1}$			
3	$l_{BA,1}$	$l_{CB,2}$				l_{V_B}
4	$m_{BA,1}$	$m_{CB,2}$				m_{V_B}
5	$l_{BA,2}$		$l_{CA,2}$	l_{H_A}	l_{V_A}	
6	$m_{BA,2}$		$m_{CA,2}$	m_{H_A}	m_{V_A}	

SUMMARY OF STEPS 4.2: Method of joints in matrix form — Solution procedure

The following are the main steps required when applying the method of joints in matrix form:

1. Specify a reference system to define the coordinates of each node.

2. Introduce two freedoms at each node, preferably pointing in the same positive directions as the coordinate axes, and number the freedoms sequentially moving from node to node.

3. Specify an arbitrary location vector for each member and tabulate node 1 and node 2, as well as the member length and direction cosines, of each member (as an example, see Table 4.1).

4. Define the equilibrium matrix **A** (following the layout of Table 4.2) with rows corresponding to the freedom numbers and columns corresponding to the vector **p** of member axial forces and reactions.

5. Determine the vector **f** of external loads.

6. Solve the equilibrium equations (as an example, see Equations 4.18 and 4.19) to determine the unknown member forces and reactions.

WORKED EXAMPLE 4.2

Consider the truss in Figure 4.12 and calculate the unknown member axial forces and reactions using the matrix form of the method of joints following the procedure detailed in Summary of Steps 4.2.

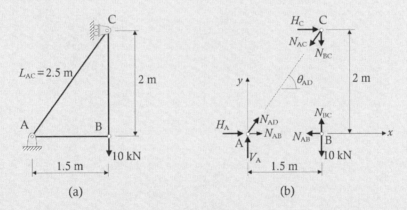

Figure 4.12 Truss and free-body diagrams for Worked Example 4.2. (a) Truss layout. (b) Free-body diagrams.

(1–3) The reference system adopted in the solution is shown in Figure 4.13, together with the assigned freedom numbers and location vectors. The origin of the reference system is at support A. The x- and y-axes are directed in the horizontal and vertical directions, respectively, and the coordinates of each node are:

$$(x_A, y_A) = (0,0); \quad (x_B, y_B) = (1.5,0); \quad (x_C, y_C) = (1.5,2)$$

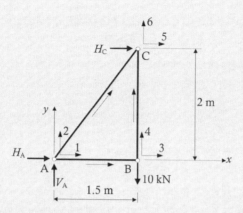

Figure 4.13 Numbering of the freedoms.

(4–5) The direction cosines related to each member are determined using Equations 4.9:

$$l_{AB} = \frac{x_B - x_A}{L_{AB}} = \frac{1.5 - 0}{1.5} = 1; \quad m_{AB} = \frac{y_B - y_A}{L_{AB}} = \frac{0 - 0}{1.5} = 0; \quad l_{AC} = \frac{x_C - x_A}{L_{AC}} = \frac{1.5 - 0}{2.5} = 0.6$$

$$m_{AC} = \frac{y_C - y_A}{L_{AC}} = \frac{2 - 0}{2.5} = 0.8; \quad l_{BC} = \frac{x_C - x_B}{L_{BC}} = \frac{1.5 - 1.5}{2} = 0 \quad \text{and} \quad m_{BC} = \frac{y_C - y_B}{L_{BC}} = \frac{2 - 0}{2} = 1$$

These results can be summarised as follows:

Member	Node 1	Node 2	L (m)	l	m
AB	A	B	1.5	1	0
AC	A	C	2.5	0.6	0.8
BC	B	C	2	0	1

From geometric considerations, the direction cosines related to the reactions are as follows:

For H_A: $l_{H_A} = 1$ and $m_{H_A} = 0$ For V_A: $l_{V_A} = 0$ and $m_{V_A} = 1$

For H_C: $l_{H_C} = 1$ and $m_{H_C} = 0$

Matrix **A** defining the orientation of the unknown axial forces and reactions can be obtained using the layout introduced in Table 4.2. Initially, it is useful to create an empty table with rows referring to the freedoms of the truss and columns to the unknown forces and reactions. This can then be populated considering one freedom after another. For example, the forces that might contribute to the equilibrium along freedom 1 are those acting at node A. These include the member forces N_{AB} and N_{AC} as well as the reactions H_A and V_A. Depending on their inclinations, these might, or might not, influence equilibrium in the direction of freedom 1. Considering that A is node 1 for both members AB and AC, the direction cosines to be used for N_{AB} and N_{AC}

in this case are $l_{AB,1}$ and $l_{AC,1}$, respectively, while the appropriate projections of the reactions are obtained from l_{H_A} and l_{V_A}. The first row of the table depicting matrix $\mathbf{A}$ can be filled as follows:

Freedoms	N_{AB}	N_{AC}	N_{BC}	H_A	V_A	H_C
1	$l_{AB,1}$	$l_{AC,1}$		l_{H_A}	l_{V_A}	

Equilibrium along freedom 2 can be easily obtained from the terms identified for freedom 1 as its equation considers different projections of the same forces. Based on this, the second row of the table becomes:

Freedoms	N_{AB}	N_{AC}	N_{BC}	H_A	V_A	H_C
1	$l_{AB,1}$	$l_{AC,1}$		l_{H_A}	l_{V_A}	
2	$m_{AB,1}$	$m_{AC,1}$		m_{H_A}	m_{V_A}	

Similarly, the relevant coefficients defining equilibrium along the remaining freedoms can be obtained as follows:

Freedoms	N_{AB}	N_{AC}	N_{BC}	H_A	V_A	H_C
1	$l_{AB,1}$	$l_{AC,1}$		l_{H_A}	l_{V_A}	
2	$m_{AB,1}$	$m_{AC,1}$		m_{H_A}	m_{V_A}	
3	$l_{AB,2}$		$l_{BC,1}$			
4	$m_{AB,2}$		$m_{BC,1}$			
5		$l_{AC,2}$	$l_{BC,2}$			l_{H_C}
6		$m_{AC,2}$	$m_{BC,2}$			m_{H_C}

Substituting the appropriate numerical values on the basis of Equations 4.9 through 4.11, with $l_1 = l$, $m_1 = m$, $l_2 = -l$ and $m_2 = -m$, the above table can be rewritten as:

Freedoms	N_{AB}	N_{AC}	N_{BC}	H_A	V_A	H_C
1	1	0.6		1	0	
2	0	0.8		0	1	
3	−1		0			
4	0		1			
5		−0.6	0			1
6		−0.8	−1			0

The matrix $\mathbf{A}$ and the vectors $\mathbf{p}$ and $\mathbf{f}$, which are populated based on the order adopted in the table above, are:

$$\mathbf{A} = \begin{bmatrix} 1 & 0.6 & 0 & 1 & 0 & 0 \\ 0 & 0.8 & 0 & 0 & 1 & 0 \\ -1 & 0 & 0 & 0 & 0 & 0 \\ 0 & 0 & 1 & 0 & 0 & 0 \\ 0 & -0.6 & 0 & 0 & 0 & 1 \\ 0 & -0.8 & -1 & 0 & 0 & 0 \end{bmatrix}; \quad \mathbf{p} = \begin{bmatrix} N_{AB} \\ N_{AC} \\ N_{BC} \\ H_A \\ V_A \\ H_C \end{bmatrix}; \quad \text{and} \quad \mathbf{f} = \begin{bmatrix} 0 \\ 0 \\ 0 \\ -10 \\ 0 \\ 0 \end{bmatrix}.$$

(6) The corresponding system of equation can be rewritten as (Equation 4.18):

$$N_{AB} + 0.6N_{AC} + H_A = 0$$

$$0.8N_{AC} + V_A = 0$$

$$-N_{AB} = 0$$

$$N_{BC} + (-10) = 0$$

$$-0.6N_{AC} + H_C = 0$$

$$-0.8N_{AC} - N_{BC} = 0$$

which can be solved producing the following results (in kN):

$$N_{AB} = 0; \quad N_{AC} = -12.5; \quad N_{BC} = 10; \quad H_A = 7.5; \quad V_A = 10; \quad H_C = -7.5$$

These can be collected in vector **p** as:

$$\mathbf{p} = \begin{bmatrix} N_{AB} \\ N_{AC} \\ N_{BC} \\ H_A \\ V_A \\ H_C \end{bmatrix} = \begin{bmatrix} 0 \\ -12.5 \\ 10 \\ 7.5 \\ 10 \\ -7.5 \end{bmatrix} \text{ in kN}$$

4.6 METHOD OF SECTIONS

In some cases, it is only necessary to calculate the axial forces of a limited number of truss members and the *method of sections* can be very useful in these instances. The idea on the basis of this approach is very simple: the truss is subdivided into two free-body diagrams making sure when producing these subdivisions to cut no more than three members with unknown axial forces.

For example, let us assume that we only need to calculate the axial forces in members BC, BG and FG of the cantilevered truss shown in Figure 4.14a. In order to use the method of sections, we need to perform a cut through these members that generates two free-body diagrams, one to the left of the cut and one to the right, as shown in Figure 4.14b. The number of unknown forces acting on the left free-body diagram of Figure 4.14b is six, including three unknown reactions H_A, H_E and V_E and the three unknown member forces. For the right free-body diagram, the three unknown axial forces are the only unknowns.

Applying the three equilibrium equations to the right-hand free-body diagram, the unknown forces N_{BC}, N_{BG} and N_{FG} can be calculated as follows:

Rotational equilibrium at G: (+ $\sum M_G = 0$: $3 \times N_{BC} - 3 \times 5 = 0$ $\therefore N_{BC} = 5 \, kN$

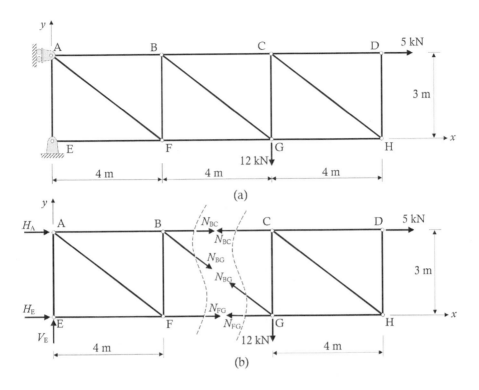

Figure 4.14 Truss and free-body diagrams—method of sections.

Vertical equilibrium: $\uparrow+$ $\Sigma F_y = 0$: $0.6N_{BG} - 12 = 0$ $\therefore N_{BG} = 20\,\text{kN}$

Horizontal equilibrium: $\xrightarrow{+}$ $\Sigma F_x = 0$: $5 - N_{BC} - 0.8N_{BG} - N_{FG} = 0$

$$5 - 5 - 0.8 \times 20 - N_{FG} = 0 \quad \therefore N_{FG} = -16\,\text{kN}$$

For convenience, rotational equilibrium has been determined with respect to G to reduce the number of unknowns in the equation to one, since G is the intersection of the lines of action of the unknowns N_{BG} and N_{FG}. A similar advantage would have been given by taking moment about the point B as, at this point, the lines of action of N_{BC} and N_{BG} meet. Other points could have been selected but would have resulted in a larger number of unknowns in the equation.

The unknown forces N_{BC}, N_{BG} and N_{FG} could also be calculated using the left free-body diagram shown in Figure 4.14b. In this case, the solution would be slightly longer, since the unknown reactions would first need to be calculated using the free-body of the whole truss.

The main steps required in the use of the method of sections are outlined below and then illustrated by another worked example.

SUMMARY OF STEPS 4.3: Method of sections — Solution procedure

The use of the method of sections is based on the steps outlined below.

1. Consider possible cuts passing through the truss elements for which member forces are sought. Select a cut that produces no more than three unknown axial forces.

2. Draw the two free-body diagrams of the truss after the cut is performed and count the number of unknowns on each (the difference is due to the presence of the unknown reactions). Choose the free-body diagram with the lowest number of unknowns and, before proceeding to step 3, calculate the unknown reactions acting on the selected free-body diagram (by consideration of an appropriate free-body diagram of the structure).

3. Apply the equilibrium equations to the selected free-body diagram to determine the unknown member forces.

WORKED EXAMPLE 4.3

For the truss of Figure 4.15, calculate the axial forces of members CD and BD using the method of sections.

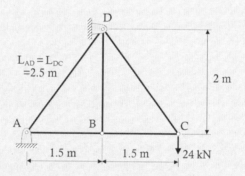

Figure 4.15 Truss for Worked Example 4.3.

The possible cuts that can be performed through members CD and BD are outlined in Figure 4.16 and referred to as cuts 1 and 2, respectively. Considering cut 1, the number of unknowns present in the top and bottom free-body diagrams are four and five, respectively. In the case of cut 2, the number of unknowns becomes six for the top left free-body diagram and three for the bottom right one.

For its lower number of unknowns, the bottom right free-body diagram of cut 2 (Figure 4.17) is used to determine N_{CD} and N_{BD}. By enforcing moment equilibrium at node B, N_{CD} is determined and N_{BD} is then calculated based on vertical equilibrium:

Rotational equilibrium at B: $(+\quad \Sigma M_B = 0: \quad 0.8N_{CD} \times 1.5 - 24 \times 1.5 = 0 \quad \therefore N_{CD} = 30\ \text{kN}$

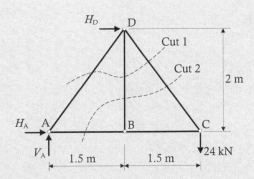

Figure 4.16 Cuts 1 and 2 for Worked Example 4.3.

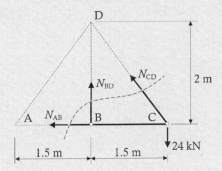

Figure 4.17 Free-body diagram of cut 2 for Worked Example 4.3.

Vertical equilibrium: ↑+ $\Sigma F_y = 0$: $N_{BD} + 0.8N_{CD} - 24 = 0$ ∴ $N_{BD} = 0\,kN$

Recalling Reflection Activity 4.1, N_{BD} is clearly zero (from vertical equilibrium at node B).

REFLECTION ACTIVITY 4.4

Re-consider the free-body diagram of Figure 4.17 and revisit the calculation of the contribution of force N_{CD} to the rotational equilibrium about node B.

There are two approaches to determine the moment about B caused by the member force N_{CD} in the free-body diagram of Figure 4.17, referred to as Options 1 and 2 below.

Option 1: The moment is calculated as the product of N_{CD} and the perpendicular distance from its line of action to node B, as shown in Figure 4.18a. This distance is referred to as L_{BE} and $L_{BE} = L_{BC} \sin \theta = 1.2$ m, since $\sin \theta = L_{BD}/L_{CD} = 0.8$. In this case, the moment is equal to $L_{BE}N_{CD} = 1.2N_{CD}$.

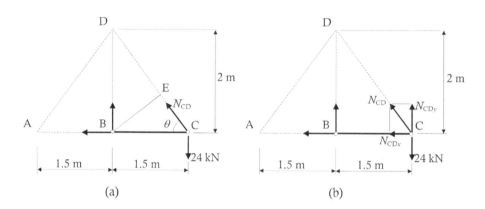

Figure 4.18 Free-body diagrams for Reflection Activity 4.4. (a) Option 1. (b) Option 2.

Option 2: It is convenient at times to calculate the moment due to N_{CD} as the sum of the moments produced by its horizontal and vertical components (or components in other directions), referred to as N_{CDx} and N_{CDy}, respectively, in Figure 4.18b. From the geometry of the truss, the force components become:

$$N_{CDx} = N_{CD} \cos \theta = N_{CD} \times L_{BC}/L_{CD} = 0.6 \, N_{CD} \quad \text{and} \quad N_{CDy} = N_{CD} \sin \theta = 0.8 \, N_{CD}$$

As the line of action of N_{CDx} passes through the node B, this component does not contribute to the moment about B. Therefore, the moment is calculated as $N_{CDy} L_{BC} = 0.8N_{CD} \times 1.5 = 1.2N_{CD}$ (which is identical to the solution obtained for Option 1, as expected).

REFLECTION ACTIVITY 4.5

Consider the truss shown in Figure 4.19a and comment whether the method of sections can be used for the determination of the axial forces resisted by members BF and BG.

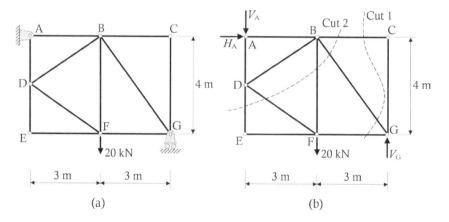

Figure 4.19 Truss and possible cuts for Reflection Activity 4.5. (a) Truss layout. (b) Some cuts through BF and BG.

By inspection, there is no cut that can be applied to the truss that cuts members BF and BG, and that intersects just three members with unknown forces. In cases like this, the method of sections needs to be applied in more than one stage.

First, the unknown forces N_{BG} and N_{BC} can be determined by considering the right free-body diagram produced by cut I. This requires the calculation of the reaction V_G, which can be carried out by enforcing moment equilibrium of the entire truss about node A (being the intersection of the lines of actions of H_A and V_A).

N_{BF} can then be obtained by considering the free-body diagram on the bottom right side of cut 2.

In some instances, it is more efficient to combine the use of both methods of joints and sections. In fact, applying Reflection Activity 4.1 to the horizontal and vertical directions at nodes C and E (applying the method of joints), it can be observed that members BC, CG, DE and EF are all unloaded. With this observation, the unknown forces N_{BG} and N_{BF} can be calculated by applying the method of sections to the bottom right side of cut 2.

4.7 STATICAL INDETERMINACY AND STABILITY OF TRUSSES

Statical indeterminacy provides a useful means of classifying all types of structures, including trusses. For a plane truss, with j nodes, b members and r reaction components, there are $2j$ equilibrium equations available and a total of $r + b$ unknown forces. Accordingly, plane trusses may be classified as follows:

Statically determinate when: $b + r = 2j$ (4.20a)

Statically indeterminate when: $b + r > 2j$ (4.20b)

Unstable when: $b + r < 2j$ (4.20c)

Statically indeterminate trusses are said to have a certain *degree of indeterminacy D* given by:

$$D = b + r - 2j \qquad (4.21)$$

Trusses satisfying the conditions of Equations 4.20a and b can still be *unstable* when the reactions are parallel or concurrent as, for example, shown in Figure 4.20b, or when an inadequate arrangement of the members is specified as, for example, shown in Figure 4.20c (see also Section 2.9). Recalling the categories of trusses introduced in Section 4.1, only compound and complex trusses can contain mechanisms. Simple trusses are always stable.

Comparing the trusses shown in Figure 4.20, the total number of unknowns and reactions are identical in all three trusses despite the fact that only the first truss is stable. These mechanisms need to be identified either by inspection (even if this is not always simple, especially when dealing with complex truss systems) or by calculating the rank of the matrix **A** (previously introduced for the writing of the equilibrium equations using the method of

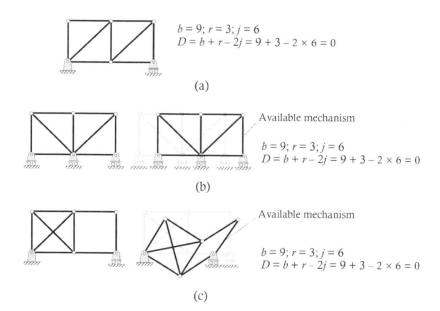

$b = 9; r = 3; j = 6$
$D = b + r - 2j = 9 + 3 - 2 \times 6 = 0$

(a)

Available mechanism

$b = 9; r = 3; j = 6$
$D = b + r - 2j = 9 + 3 - 2 \times 6 = 0$

(b)

Available mechanism

$b = 9; r = 3; j = 6$
$D = b + r - 2j = 9 + 3 - 2 \times 6 = 0$

(c)

Figure 4.20 Statically determinate and unstable trusses. (a) Statically determinite truss. (b) Unstable truss (concurrent reactions). (c) Unstable truss.

joints). The procedure for the calculation of the rank of a matrix is outlined in Appendix C. If the rank of **A** is smaller than the number of freedoms (n_{DOF}), then the truss contains a mechanism.

SUMMARY OF STEPS 4.4: Statical determinacy and instability

The classification of a truss on the basis of its degree of statical indeterminacy D and its possible instability is carried out as detailed below.

1. Count the number of members b, the number of reaction components r and the number of joints j in the truss.

2. Calculate the degree of statical indeterminacy D where $D = b + r - 2j$.

3. If $D = 0$ the truss is statically determinate.
   If $D > 0$ the truss is statically indeterminate.
   If $D < 0$ the truss is unstable.

Even if $D \geq 0$, the truss might still be unstable. This is the case if the rank of **A** is smaller than the total number of freedoms n_{DOF}, i.e. rank(**A**) $< n_{DOF}$. See Appendix C for the procedure required for the calculation of the rank of a matrix.

WORKED EXAMPLE 4.4

Consider the trusses shown in Figure 4.21. Determine for each truss whether it is statically determinate, statically indeterminate or unstable following the procedure detailed in Summary of Steps 4.4.

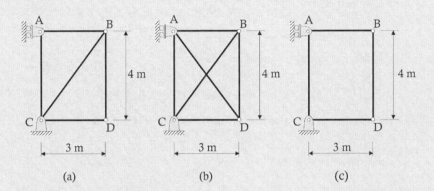

(a)   (b)   (c)

Figure 4.21 Trusses for Worked Example 4.4.

The free-body diagram of each truss is presented in Figure 4.22, highlighting the coordinate system, freedom numbering and location vectors adopted for each member.

<u>Truss A (Figure 4.22a):</u>

$b = 5$; $r = 3$; $j = 4$ from which $D = b + r - 2j = 5 + 3 - 2 \times 4 = 0$

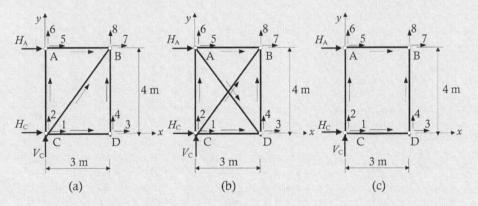

(a)   (b)   (c)

Figure 4.22 Free-body diagrams showing coordinate system, freedom numbering and location vectors. (a) Truss A. (b) Truss B. (c) Truss C.

The direction cosines related to the adopted location vectors are tabulated as:

Member	Node 1	Node 2	L (m)	l	m
CD	C	D	3	1	0
BD	D	B	4	0	1
AB	A	B	3	1	0
AC	C	A	4	0	1
BC	C	B	5	0.6	0.8

From geometric considerations, the direction cosines related to the reactions are:

$l_{H_A} = 1$ and $m_{H_A} = 0$ for H_A; $l_{H_C} = 1$ and $m_{H_C} = 0$ for H_C;   and $l_{V_C} = 0$ and $m_{V_C} = 1$ for V_C.

Matrix **A** is then evaluated based on the layout previously proposed in Table 4.2:

Freedoms	N_{CD}	N_{BD}	N_{AB}	N_{AC}	N_{BC}	H_A	H_C	V_C
1	$l_{CD,1}$			$l_{AC,1}$	$l_{BC,1}$		l_{H_C}	l_{V_C}
2	$m_{CD,1}$			$m_{AC,1}$	$m_{BC,1}$		m_{H_C}	m_{V_C}
3	$l_{CD,2}$	$l_{BD,1}$						
4	$m_{CD,2}$	$m_{BD,1}$						
5			$l_{AB,1}$	$l_{AC,2}$		l_{H_A}		
6			$m_{AB,1}$	$m_{AC,2}$		m_{H_A}		
7		$l_{BD,2}$	$l_{AB,2}$		$l_{BC,2}$			
8		$m_{BD,2}$	$m_{AB,2}$		$m_{BC,2}$			

That is:

Freedoms	N_{CD}	N_{BD}	N_{AB}	N_{AC}	N_{BC}	H_A	H_C	V_C
1	1			0	0.6		1	0
2	0			1	0.8		0	1
3	-1	0						
4	0	1						
5			1	0		1		
6			0	-1		0		
7		0	-1		-0.6			
8		-1	0		-0.8			

and therefore:

$$
\mathbf{A} =
\begin{bmatrix}
1 & 0 & 0 & 0 & 0.6 & 0 & 1 & 0 \\
0 & 0 & 0 & 1 & 0.8 & 0 & 0 & 1 \\
-1 & 0 & 0 & 0 & 0 & 0 & 0 & 0 \\
0 & 1 & 0 & 0 & 0 & 0 & 0 & 0 \\
0 & 0 & 1 & 0 & 0 & 1 & 0 & 0 \\
0 & 0 & 0 & -1 & 0 & 0 & 0 & 0 \\
0 & 0 & -1 & 0 & -0.6 & 0 & 0 & 0 \\
0 & -1 & 0 & 0 & -0.8 & 0 & 0 & 0
\end{bmatrix}
$$

The rank of **A** is 8 (see Appendix C). As the rank is identical to the total number of freedoms ($n_{DOF} = 8$), there are no mechanisms in Truss A and it is classified as statically determinate because $D = 0$.

<u>Truss B (Figure 4.22b):</u>

$b = 6$; $r = 3$; $j = 4$ $\therefore D = b + r - 2j = 6 + 3 - 2 \times 4 = 1$

Following the same procedure as for Truss A:

Member	Node 1	Node 2	L (m)	l	m
CD	C	D	3	1	0
BD	D	B	4	0	1
AB	A	B	3	1	0
AC	C	A	4	0	1
BC	C	B	5	0.6	0.8
AD	A	D	5	0.6	−0.8

Freedoms	N_{CD}	N_{BD}	N_{AB}	N_{AC}	N_{BC}	N_{AD}	H_A	H_C	V_C
1	1			0	0.6			1	0
2	0			1	0.8			0	1
3	−1	0				−0.6			
4	0	1				0.8			
5			1	0		0.6	1		
6			0	−1		−0.8	0		
7		0	−1		−0.6				
8		−1	0		−0.8				

The rank of **A** is 8 (= n_{DOF}) (see Appendix C), which ensures that there are no mechanisms. It can then be concluded that Truss B is statically indeterminate with $D = 1$.

<u>Truss C (Figure 4.22c):</u>

$b = 4$; $r = 3$; $j = 4$ $\therefore D = b + r - 2j = 4 + 3 - 2 \times 4 = -1$

As $D < 0$, Truss C is classified as unstable.

REFLECTION ACTIVITY 4.6

Reconsider the truss layout depicted in Figure 4.20c (reproduced for ease of reference in Figure 4.23a). Evaluate whether the procedure detailed in Summary of Steps 4.4 is capable of depicting the mechanism shown in Figure 4.20c.

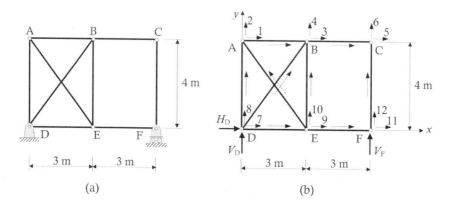

Figure 4.23 Truss and free-body diagram for Reflection Activity 4.6. (a) Truss layout. (b) Free-body diagram.

For the truss of Figure 4.23a: $b = 9$; $r = 3$; $j = 6$ and $D = b + r - 2j = 9 + 3 - 2 \times 6 = 0$ and, on this basis only, the truss appears to be statically determinate.

The direction cosines related to the adopted location vectors are tabulated as:

Member	Node 1	Node 2	L (m)	l	m
AB	A	B	3	1	0
BC	B	C	3	1	0
DE	D	E	3	1	0
EF	E	F	3	1	0
AD	D	A	4	0	1
BE	E	B	4	0	1
CF	F	C	4	0	1
AE	E	A	5	-0.6	0.8
BD	D	B	5	0.6	0.8

Matrix **A** is then evaluated based on the layout previously proposed in Table 4.2:

Freedoms	N_{AB}	N_{BC}	N_{DE}	N_{EF}	N_{AD}	N_{BE}	N_{CF}	N_{AE}	N_{BD}	H_D	V_D	V_F
1	1				0			0.6				
2	0				−1			−0.8				
3	−1	1				0			−0.6			
4	0	0				−1			−0.8			
5		−1					0					
6		0					−1					
7			1		0				0.6	1	0	
8				0	1				0.8	0	1	
9			−1	1		0		−0.6				
10			0	0		1		0.8				
11				−1			0					0
12				0			1					1

The rank of the **A** matrix is 11. This is smaller than $n_{DOF} = 12$ and indicates the presence of the mechanism. Despite having $D = 0$, this truss is not stable.

4.8 DEFORMATION OF TRUSSES

Figure 4.24a shows a portion of a straight truss member of length L deforming under an external axial tensile load P. The line of action of the axial force passes through the centroid of the cross-section illustrated in Figure 4.24b (see Appendix A). Figure 4.24c shows a thin slice through the member bounded by cross-sections A and B at a distance dx apart, with the resultant internal axial force N (= P) acting along the x-axis of the member at the centroid of each cross-section. Cross-sections A and B are parallel before loading and remain parallel after the axial load is applied, but every fibre of the cross-section elongates by an amount de and the cross-section at B moves relative to cross-section A as shown in Figure 4.24c. Strain at any point on the cross-section ε is defined as the deformation per unit length. For the axially loaded bar, the strain at every point on cross-section B is the same and is equal to the deformation of the thin slice de (in Figure 4.24c) divided by the original length dx. The

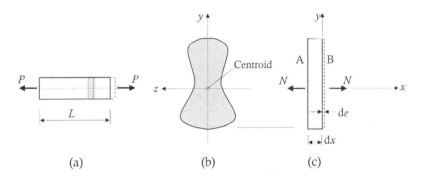

(a) (b) (c)

Figure 4.24 Portion of a straight truss member subjected to tension. (a) Member in axial tension. (b) Cross-section. (c) Slice elevation.

normal stress σ on the cross-section is also uniform and is equal to the axial force N divided by the area of the cross-section A:

$$\varepsilon = \frac{de}{dx} \quad \text{and} \quad \sigma = \frac{N}{A} \tag{4.22a,b}$$

Of course, the stress may vary along the member if the cross-sectional area changes. For a *linear–elastic* material, for which stresses and strains are related by the elastic modulus E ($\sigma = E\varepsilon$), the elongation of the thin slice in Figure 4.24c is:

$$de = \varepsilon\, dx = \frac{\sigma\, dx}{E} = \frac{N\, dx}{EA} \tag{4.23}$$

and the elongation of the member is obtained by integration:

$$e = \int_0^L \varepsilon\, dx = \int_0^L \frac{N\, dx}{EA} \tag{4.24}$$

Noting that the axial force N is constant along the bar and equal to P, if the cross-sectional area A of the straight bar is also constant (i.e. the bar is prismatic), the elongation becomes:

$$e = \frac{PL}{EA} \tag{4.25}$$

Each bar of a truss is subject to either an axial tensile or an axial compressive force and, therefore, each loaded bar either extends or contracts when the truss is loaded. The axial deformation of the bars in a truss may be determined using Equation 4.25 provided the material behaviour is linear–elastic.

The relationship between the applied axial load P and the resulting axial deformation e can be alternatively expressed as either:

$$e = fP \quad \text{where} \quad f = \frac{L}{EA} \tag{4.26}$$

or

$$P = ke \quad \text{where} \quad k = \frac{EA}{L} \tag{4.27}$$

where f is the extension caused by a unit force and is called the *axial flexibility coefficient*, and k is the force required to produce a unit extension and is called the *axial stiffness coefficient*. Clearly, $f = 1/k$.

The deformation of the bars of the truss will cause movement of the truss nodes and therefore a change in the geometry of the truss. For most trusses, certainly for trusses that are serviceable, the displacements of the nodes are very small compared to the dimensions of the structure and it is reasonable to assume that the direction of each bar before deformation is the same after deformation. This greatly simplifies the calculation of the node displacements and usually results in negligible error.

The *geometry of small displacements* is best illustrated by a simple example. Let us consider the truss shown in Figure 4.15 and analysed in Worked Example 4.3 (reproduced here as Figure 4.25 with the axial force in each bar shown in the figure).

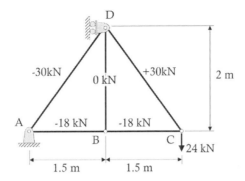

Figure 4.25 Internal axial forces calculated in Worked Example 4.3.

Suppose that the truss is fabricated from timber with E = 12,000 MPa and that the cross-sectional area of the bars in compression is 15,000 mm² (i.e. bars AB, BC and AD) and the cross-sectional area of the remaining bars is 5000 mm². The flexibility coefficients for each bar are calculated as $f_{AB} = f_{BC}$ = 8.33 × 10⁻⁶ mm/N, f_{AD} = 13.89 × 10⁻⁶ mm/N, f_{BD} = 33.33 × 10⁻⁶ mm/N and f_{CD} = 41.66 × 10⁻⁶ mm/N. The elongations of the bars are then obtained as the product of the flexibility coefficients and axial forces, and can be collected in vector form as:

$$\mathbf{e} = \begin{bmatrix} e_{AB} \\ e_{BC} \\ e_{AD} \\ e_{BD} \\ e_{CD} \end{bmatrix} = \begin{bmatrix} f_{AB}N_{AB} \\ f_{BC}N_{BC} \\ f_{AD}N_{AD} \\ f_{BD}N_{BD} \\ f_{CD}N_{CD} \end{bmatrix} = \begin{bmatrix} 8.33\times10^{-6}\times(-18000) \\ 8.33\times10^{-6}\times(-18000) \\ 13.89\times10^{-6}\times(-30000) \\ 33.33\times10^{-6}\times0 \\ 41.66\times10^{-6}\times30000 \end{bmatrix} = \begin{bmatrix} -0.15 \\ -0.15 \\ -0.417 \\ 0 \\ +1.25 \end{bmatrix} \text{mm}$$

For simple trusses containing a small number of members, like the one of Figure 4.25, the displacement of the nodes can be readily determined by simple geometry. For larger trusses, other available methods are more convenient and will be dealt with in subsequent chapters. The displacements of a node of the truss in the directions parallel to the x- and y-axes are here given the symbols u and v, respectively.

In the truss of Figure 4.25, node A is pinned and does not move (i.e. $u_A = v_A$ = 0). Node D is constrained by the roller support to only move vertically, and therefore u_D = 0. Since AD shortens by e_{AD} = −0.417 mm, by the geometry of small displacements, node D moves vertically (from D to D′) by an amount v_D, as shown in Figure 4.26a, where:

$$v_D = \frac{e_{AD}}{\cos\theta} = \frac{-0.417}{0.8} = -0.521 \text{ mm (i.e. downward)}$$

The vertical displacement of node B is the sum of v_D and e_{BD}. Since member BD is unloaded and undeformed in this example (i.e. e_{BD} = 0), we have:

$$v_B = v_D + e_{BD} = v_D = -0.521 \text{ mm (i.e. downward)}$$

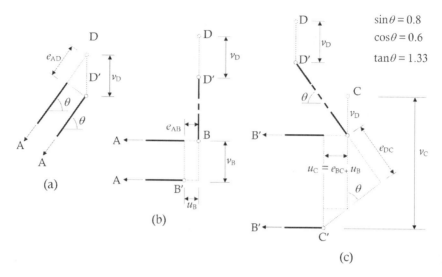

Figure 4.26 Nodal displacements. (a) Node D. (b) Node B. (c) Node C.

Member AB shortens by e_{AB} = −0.15 mm, so node B moves to the left (from B to B′) by 0.15 mm as shown in Figure 4.26b:

$$u_B = e_{AB} = -0.15 \text{ mm (i.e. to the left)}$$

With regard to node C (see Figure 4.26c), member BC shortens by e_{BC} = −0.15 mm and so node C moves closer to node B by 0.15 mm. The horizontal displacement of node C is therefore:

$$u_C = u_B + e_{BC} = -0.15 - 0.15 = -0.3 \text{ mm (i.e. to the left)}$$

The member CD extends by e_{CD} = +1.25 mm and the vertical displacement of node C is shown in Figure 4.26c and is equal to the sum of the magnitudes of v_D, ($e_{CD}/\sin\theta$) and ($e_{BC} + u_B$)/$\tan\theta$:

$$v_C = -\left[\left|-0.521\right| + \left|\frac{1.25}{0.8}\right| + \left|\frac{-0.15 - 0.15}{1.33}\right|\right] = -2.308 \text{ mm (i.e. downward)}$$

Because of the applied load, node C moves vertically downward by an amount of 2.308 mm and horizontally to the left by 0.3 mm. The vector of nodal displacements **d** is given by:

$$\mathbf{d} = \begin{bmatrix} u_A \\ v_A \\ u_B \\ v_B \\ u_C \\ v_C \\ u_D \\ v_D \end{bmatrix} = \begin{bmatrix} 0 \\ 0 \\ -0.15 \\ -0.521 \\ -0.3 \\ -2.308 \\ 0 \\ -0.521 \end{bmatrix} \text{ (in mm)}$$

4.9 TRUSSES WITH LOADED MEMBERS

If an external load is not applied directly to the nodes of a truss, but to a member AB of the truss, *secondary forces* in the form of moments and shear and axial forces occur in the loaded member in addition to the primary axial forces. The loaded member performs a dual function. First, it acts as a beam between the points A and B, and serves to transmit the load to these adjacent nodes. In this capacity, the member will be subjected to bending moments, shear forces and, in some cases, axial forces. Second, it acts as a member of the truss, carrying an axial force that is determined by the methods described earlier in this chapter.

If the loads acting on a particular member of a truss are replaced by a statically equivalent set of forces at the end nodes, the forces in all other members of the truss will be unaffected. Consider a member EF taken from a truss and loaded by a force P acting at some point along its length, as shown in Figure 4.27. The free-body diagram of the member is in equilibrium under the action of the load P and the bar forces N_1, N_2, N_3, N_4 and N_5 exerted by other members of the truss connected to nodes E and F. If the force P is replaced by forces R_1 and R_2 acting at nodes E and F, respectively, whose resultant is P, then the forces N_1, N_2, N_3, N_4 and N_5 remain unchanged. With R_1 and R_2 acting at nodes E and F, the truss is now loaded at the nodes only and may be analysed by either the method of joints or the method of sections. The axial force thus calculated in member EF, along with the bending moment, shear force and axial force determined by considering EF as a pin-ended beam carrying load P, is combined to determine the internal actions in the member EF (primary plus secondary forces). This procedure relies on the use of the principle of superposition where the sum of the effects of different loading conditions is equivalent to the effects of all loads applied at once. For the principle of superposition to be valid, the following conditions must be satisfied: (i) material behaviour is linear–elastic; (ii) displacements of the structure remain small enough, so that the equilibrium equations can be based on the geometry of the undisplaced structure; and (iii) there is no interaction between axial and the flexural actions caused by member loads. With regard to condition (iii), the member displacements must be small so that the axial force will not induce significant additional moments and deflections. Such additional actions and deformations could even lead to possible instability problems if not adequately accounted for (see Chapter 14).

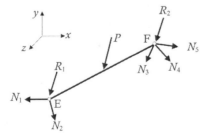

Figure 4.27 Free-body diagram of element EF.

WORKED EXAMPLE 4.5

For the truss shown in Figure 4.28, determine the maximum axial force, shear force and bending moment in the top chord member HI with the 80 kN load at its mid-point.

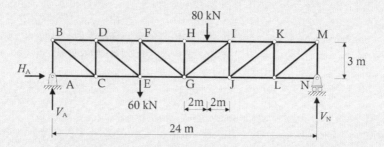

Figure 4.28 Truss for Worked Example 4.5.

(1) Find the reactions:

$$\Sigma M_A = 0: \quad V_N \times 24 - 80 \times 14 - 60 \times 8 = 0 \qquad \therefore V_N = 66.67 \text{ kN}$$

$$\Sigma M_N = 0: \quad -V_A \times 24 + 80 \times 10 + 60 \times 16 = 0 \qquad \therefore V_A = 73.33 \text{ kN}$$

Check: $\Sigma F_y = 0: \quad V_A + V_N - 80 - 60 = 0 \quad \therefore \text{OK}$

$\Sigma F_x = 0: \quad H_A = 0$

(2) Replace the 80 kN force with the statically equivalent forces at H and I (in this case, a vertical 40 kN downward force at each node) as shown below:

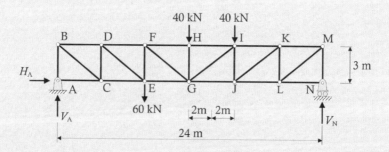

(3) Find the member force in HI:

Using the method of sections, Figure 4.29a shows a free-body diagram to the right of a cut made through members HI, GI and GJ.

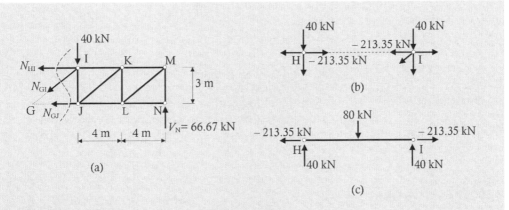

Figure 4.29 Free-body diagrams for Worked Example 4.5.

Taking moments about G:

$$(\curvearrowleft + \quad N_{HI} \times 3 + 66.67 \times 12 - 40 \times 4 = 0 \quad \therefore N_{HI} = -213.35 \text{ kN}$$

(4) Find the forces acting at H and I:
These forces are shown in Figure 4.29b. Although two of the forces acting at H (and three of the forces acting at I) have not been calculated, we know that they must equilibrate the two known forces at each node and hence must be equivalent to the two components shown at each end of the free-body diagram of member HI, shown in Figure 4.29c.

(5) Find the maximum internal actions N_{max}, S_{max} and M_{max} along member HI:
By inspection of the free-body in Figure 4.29c:

$$N_{max} = -213.35 \text{ kN}, S_{max} = 40 \text{ kN}, \text{ and } M_{max} = 40 \times 2 = 80 \text{ kNm}$$

This approach is also applicable for all types of member loads, including the self-weight of the truss members. The effects produced by these member loads are superimposed following the procedure outlined in Figure 4.27 for the point load. Obviously, this approach is valid as long as the member loads do not excessively deform the truss element, in which case the axial force applied at the nodes would be resisted by both the axial and flexural rigidities of the truss members, therefore affecting the analysis of the entire truss. The effects produced by axial forces in out-of-straight members are discussed in Chapter 14.

Occasionally, it may be necessary to analyse a truss having a member that is either curved or kinked, such as AB in Figure 4.30. If the out-of-straightness is sufficiently small, its effects could be accounted for by applying equilibrium to the free-body diagram of the member

Figure 4.30 Member with small out-of-straightness (deformations exaggerated in the figure for clarity).

when subjected to the end actions calculated from the truss analysis. In this manner, the moment and shear in the member are evaluated from statics and added to the axial force obtained from the truss analysis. This procedure is applicable when the out-of-straightness of the member is small, in which case it is acceptable to use the axial rigidity of the ideal straight member in the truss analysis. For cases with significant out-of-straightness, it might be necessary in the analysis of the truss to account for the effects of the internal axial force on the deflected shape and on the internal moment in the member. Significant inaccuracies may result by simply using superposition in the post-processing of the results and ignoring these effects. The coupling of the axial force, bending moment and deformed shape is discussed in more detail in Chapter 14 when dealing with instability problems.

4.10 SPACE TRUSSES

The analysis of three-dimensional space trusses can be carried out following the same procedures as for plane trusses, but introducing additional considerations for the additional third dimension. The simplest space truss with a stable structural solution is formed by six members located on the edges of a tetrahedron, as shown in Figure 4.31. From this basic module, it is possible to extend the truss by adding three bars and one node at a time. Trusses built following this procedure are referred to as *simple space trusses*.

Different types of support conditions can be provided, ranging from pinned supports, in which a node is held in position in three orthogonal directions, to conditions where the node is free to move in one or more directions. In Figure 4.32 (and in subsequent figures), the presence of a support is depicted by the restraining reactions shown as thick arrows.

Figure 4.31 Simplest stable space truss.

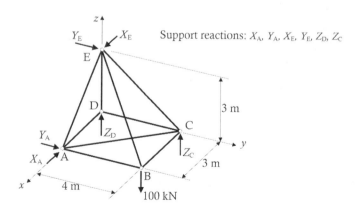

Figure 4.32 Free-body diagram of a space truss.

For example, at node E in Figure 4.32, the displacements along the x and y directions are prevented and restraining forces (reactions) X_E and Y_E, respectively, will develop. This way of visualising support conditions is commonly used in analysis software as it quickly provides a clear understanding of the restrained directions.

The use of the method of joints for the analysis of space trusses follows the same steps detailed in Summaries of Steps 4.1 and 4.2 with the only difference that now there are three equilibrium equations available at each node (instead of the two for plane trusses). The sum of the force components in the directions of each of the three coordinate axes is zero.

Similar considerations apply to the method of sections, which can rely on the steps already detailed in Summary of Steps 4.3 while making use of the six equilibrium equations available in three-dimensional structures (i.e. three force equilibrium equations in the direction of the three coordinate axes and three moment equilibrium equations with respect to rotation about each of the three axes). Because of this, each cut in the method of sections should not include more than six unknown axial forces.

WORKED EXAMPLE 4.6

Determine the reactions and member forces for the truss shown in Figure 4.32 using the method of joints as detailed in Summary of Steps 4.1.

There are six unknown reactions in the free-body diagram of Figure 4.32 that need to be determined before the evaluation of the nine unknown member forces. This is carried out by enforcing the six equilibrium equations.

The order of the equilibrium equations used for the calculation of the reactions is chosen to minimise the number of unknowns in each equation. For example, only one unknown reaction (X_E) features in the moment equilibrium equation about the y-axis through node D:

$$\sum M_{y(D)} = 0: \quad 3X_E + 3 \times 100 = 0 \qquad \therefore X_E = -100\,\text{kN}$$

Continuing to the remaining equilibrium equations:

$$\sum M_{z(D)} = 0: \quad 3Y_A = 0 \qquad \therefore Y_A = 0\,\text{kN}$$

$$\sum F_y = 0: \quad Y_A + Y_E = 0 \qquad \therefore Y_E = 0\,\text{kN}$$

$$\sum F_x = 0: \quad -X_A + X_E = 0 \qquad \therefore X_A = -100\,\text{kN}$$

$$\sum M_{x(D)} = 0: \quad -3Y_E + 4Z_C - 4 \times 100 = 0 \quad \therefore Z_C = 100\,\text{kN}$$

$$\sum F_z = 0: \quad Z_D + Z_C - 100 = 0 \qquad \therefore Z_D = 0\,\text{kN}$$

By inspection, there are four unknown member forces at nodes A, C and E, while nodes B and D each have three unknowns. Recalling that we have three equilibrium equations available at each node, a suitable nodal sequence is:

(1) Node D (to obtain N_{AD}, N_{CD} and N_{DE})
(2) Node A (N_{AE}, N_{AC} and N_{AB})
(3) Node C (N_{CE} and N_{BC})
(4) Node B (N_{BE})

The unused equations at nodes C and B may be used to verify that the calculated truss forces are correct.

From geometry: $L_{AB} = 4$ m, $L_{AC} = 5$ m, $L_{AD} = 3$ m, $L_{AE} = 4.243$ m, $L_{BC} = 3$ m, $L_{BE} = 5.831$ m, $L_{CD} = 4$ m, $L_{CE} = 5$ m and $L_{DE} = 3$ m.

Node D:

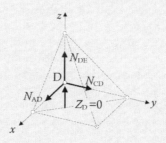

$\sum F_x = 0:\quad N_{AD} = 0$

$\sum F_y = 0:\quad N_{CD} = 0$

$\sum F_z = 0:\quad Z_D + N_{DE} = 0 \quad \therefore N_{DE} = 0\,\text{kN}$

Node A:

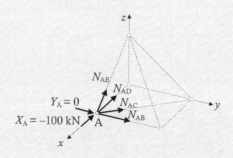

$\sum F_z = 0:\quad N_{AE} \times (3/L_{AE}) = 0 \quad \therefore N_{AE} = 0\,\text{kN}$

$\sum F_x = 0:\quad -X_A - N_{AD} - N_{AE} \times (3/L_{AE}) - N_{AC} \times (3/L_{AC}) = 0 \quad \therefore N_{AC} = 166.7\,\text{kN}$

$\sum F_y = 0:\quad N_{AB} + N_{AC} \times (4/L_{AC}) + Y_A = 0 \quad \therefore N_{AB} = -133.3\,\text{kN}$

Node C:

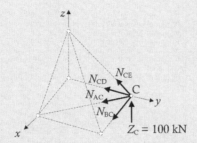

$$\sum F_y = 0: \quad -N_{CD} - N_{AC} \times (4/L_{AC}) - N_{CE} \times (4/L_{CE}) = 0 \quad \therefore N_{CE} = -166.7 \text{ kN}$$

$$\sum F_x = 0: \quad N_{BC} + N_{AC} \times (3/L_{AC}) = 0 \quad \therefore N_{BC} = -100 \text{ kN}$$

Node B:

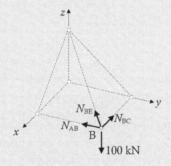

$$\sum F_z = 0: \quad -100 + N_{BE} \times (3/L_{BE}) = 0 \quad \therefore N_{BE} = 194.4 \text{ kN}$$

The remaining unused equilibrium equations may be used for checking purposes:
Node C:

$$\sum F_z = 0: \quad Z_C + N_{CE} \times (3/L_{CE}) = 0 \quad \therefore 100 + (-166.67) \times (3/5) = 0 \quad \therefore \text{OK}$$

Node B:

$$\sum F_x = 0: \quad -N_{BE} \times (3/L_{BE}) - N_{BC} = 0 \quad \therefore -194.4 \times (3/5.831) - (-100) = 0 \quad \therefore \text{OK}$$

$$\sum F_y = 0: \quad -N_{BE} \times (4/L_{BE}) - N_{AB} = 0 \quad \therefore -194.4 \frac{4}{5.831} - (-133.3) = 0 \quad \therefore \text{OK}$$

Node E:

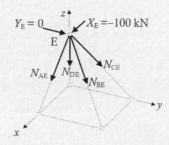

$$\sum F_x = 0: \ X_E + N_{AE} \times (3/L_{AE}) + N_{BE} \times (3/L_{BE}) = 0 \ \ \therefore -100 + 0 + 194.4 \times (3/5.831) = 0 \ \ \therefore OK$$

$$\sum F_y = 0: \ Y_E + N_{CE} \times (4/L_{CE}) + N_{BE} \times (4/L_{BE}) = 0 \ \ \therefore 0 - 166.7 \times (4/5) + 194.4 \times (4/5.831) = 0 \ \ \therefore OK$$

$$\sum F_z = 0: \ -N_{DE} - N_{AE} \times (3/L_{AE}) - N_{CE} \times (3/L_{CE}) - N_{BE} \times (3/L_{BE}) = 0$$

$$\therefore 0 - 0 - (-166.7)(3/5) - 194.4 \times (3/5.831) = 0 \ \ \therefore OK$$

REFLECTION ACTIVITY 4.7

The method of joints in matrix form, previously presented in Section 4.5 for the analysis of plane trusses (Summary of Steps 4.2), is to be extended for space trusses. The space truss shown in Figure 4.32 is to be used as an example, and the unknown reactions and member forces are to be recalculated.

The steps specified in Summary of Steps 4.2 for plane trusses are also valid for the analysis of space trusses except that equilibrium is enforced along the directions of the three coordinate axes (instead of the two directions necessary for plane trusses).

The projections of the forces acting at each node along the different axes are determined by recalling the definition of the direction cosines. For example, the components of force N shown in Figure 4.33 along the x-, y- and z-axes are calculated as follows:

$$N_x = N \cos \theta_x = lN, \quad \text{where} \quad l = \cos \theta_x \tag{4.28a}$$

$$N_y = N \cos \theta_y = mN, \quad \text{where} \quad m = \cos \theta_y \tag{4.28b}$$

$$N_z = N \cos \theta_z = nN, \quad \text{where} \quad n = \cos \theta_z \tag{4.28c}$$

where θ_x, θ_y and θ_z represent the angles formed between the force vector and the x-, y- and z-axes, respectively, and l, m and n are the direction cosines.

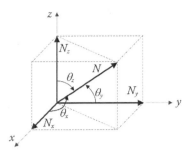

Figure 4.33 Projections of a force vector.

On the basis of the positive direction specified by the location vector, direction cosines of a member are calculated as:

$$l = \frac{x_{node2} - x_{node1}}{L}; \quad m = \frac{y_{node2} - y_{node1}}{L}; \quad n = \frac{z_{node2} - z_{node1}}{L} \qquad (4.29a–c)$$

where $(x_{node1}, y_{node1}, z_{node1})$ and $(x_{node2}, y_{node2}, z_{node2})$ are the coordinates of nodes 1 and 2, respectively, and L is the length of the member.

As for the case of a plane truss, the direction cosines to be used to evaluate the projections of the member force at node 1 of a member correspond to the actual values for l, m and n determined from the adopted location vector, as defined in Equations 4.29. When the projection is sought at node 2, the negative values of the direction cosines defined for the location vector need to be used. For example, reconsidering the truss of Figure 4.32 and specifying the arbitrary location vector shown in Figure 4.34a, the direction cosines to be applied for N_{BE} at node B (node 1 of BE) are:

$$l_{BE,1} = l_{BE}; \quad m_{BE,1} = m_{BE}; \quad n_{BE,1} = n_{BE} \qquad (4.30a–c)$$

where l_{BE}, m_{BE} and n_{BE} define the direction cosines of the location vector adopted for BE, and $l_{BE,1}$, $m_{BE,1}$ and $n_{BE,1}$ represent the direction cosines for the axial force N_{BE} applied at its first node B. When considering the same member at node E (node 2 of BE), the required direction cosines to be used in the equilibrium equations at node E are:

$$l_{BE,2} = -l_{BE}; \quad m_{BE,2} = -m_{BE}; \quad n_{BE,2} = -n_{BE} \qquad (4.31a–c)$$

For example, equilibrium at node B along the x-axis can be expressed in terms of the direction cosines related to the member axial forces and the external forces as follows:

$$l_{BC,1}N_{BC} + l_{AB,2}N_{AB} + l_{BE,1}N_{BE} + l_p P = 0 \qquad (4.32)$$

where l_p represents the direction cosine of the applied load P at node B with respect to the x-axis (in this case, $P = 100$ kN and $l_p = 0$), and the direction cosines of the members at node B are:

$$l_{BC,1} = \frac{x_C - x_B}{L_{BC}} = \frac{0-3}{3} = -1; \quad l_{AB,2} = -\frac{x_B - x_A}{L_{AB}} = -\frac{3-3}{4} = 0; \quad l_{BE,1} = \frac{x_E - x_B}{L_{BE}} = \frac{0-3}{5.831} = -0.514$$

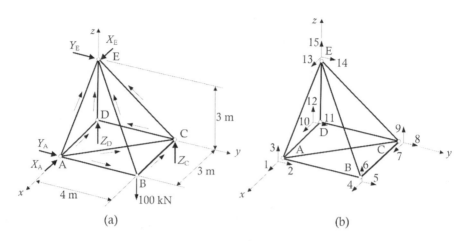

Figure 4.34 Free-body diagram and freedoms for Reflection Activity 4.7. (a) Free-body diagram with location on vectors. (b) Numbering of the freedoms.

and the equilibrium equation at node B in the x direction can be written as:

$$-1N_{BC} - 0.514N_{BE} = 0 \tag{4.33}$$

which is identical to the equilibrium equation already obtained in Worked Example 4.6 when checking the correctness of the solution.

The method of joints in matrix form can then be applied by expressing the equilibrium equations along the three coordinate axes at each node. Before deriving the coefficients included in the matrix **A** and vectors **p** and **f**, the direction cosines related to each truss member and to the reactions are calculated as:

Member	Node 1	Node 2	L (m)	l	m	n
AB	A	B	4	0	1	0
BC	B	C	3	−1	0	0
CD	C	D	4	0	−1	0
DE	D	E	3	0	0	1
AC	A	C	5	−0.6	0.8	0
AD	A	D	3	−1	0	0
AE	A	E	4.243	−0.707	0	0.707
BE	B	E	5.831	−0.514	−0.686	0.514
CE	C	E	5	0	−0.8	0.6

Reaction	l	m	n
X_A	1	0	0
Y_A	0	1	0
Z_C	0	0	1
Z_D	0	0	1
X_E	1	0	0
Y_E	0	1	0

Matrix **A** specifying the orientation of the unknown axial forces and reactions can then be determined on the basis of the layout already introduced in Table 4.2 and previously used for the analysis of plane trusses. For clarity, the coefficients related to the member forces and reactions are listed separately.

Freedoms	N_{AB}	N_{BC}	N_{CD}	N_{DE}	N_{AC}	N_{AD}	N_{AE}	N_{BE}	N_{CE}
1	$l_{AB,1}$				$l_{AC,1}$	$l_{AD,1}$	$l_{AE,1}$		
2	$m_{AB,1}$				$m_{AC,1}$	$m_{AD,1}$	$m_{AE,1}$		
3	$n_{AB,1}$				$n_{AC,1}$	$n_{AD,1}$	$n_{AE,1}$		
4	$l_{AB,2}$	$l_{BC,1}$						$l_{BE,1}$	
5	$m_{AB,2}$	$m_{BC,1}$						$m_{BE,1}$	
6	$n_{AB,2}$	$n_{BC,1}$						$n_{BE,1}$	
7		$l_{BC,2}$	$l_{CD,1}$		$l_{AC,2}$				$l_{CE,1}$
8		$m_{BC,2}$	$m_{CD,1}$		$m_{AC,2}$				$m_{CE,1}$
9		$n_{BC,2}$	$n_{CD,1}$		$n_{AC,2}$				$n_{CE,1}$
10			$l_{CD,2}$	$l_{DE,1}$		$l_{AD,2}$			
11			$m_{CD,2}$	$m_{DE,1}$		$m_{AD,2}$			
12			$n_{CD,2}$	$n_{DE,1}$		$n_{AD,2}$			
13				$l_{DE,2}$			$l_{AE,2}$	$l_{BE,2}$	$l_{CE,2}$
14				$m_{DE,2}$			$m_{AE,2}$	$m_{BE,2}$	$m_{CE,2}$
15				$n_{DE,2}$			$n_{AE,2}$	$n_{BE,2}$	$n_{CE,2}$

Freedoms	X_A	Y_A	Z_C	Z_D	X_E	Y_E
1	l_{X_A}	l_{Y_A}				
2	m_{X_A}	m_{Y_A}				
3	n_{X_A}	n_{Y_A}				
4						
5						
6						
7			l_{Z_C}			
8			m_{Z_C}			
9			n_{Z_C}			
10				l_{Z_D}		
11				m_{Z_D}		
12				n_{Z_D}		
13					l_{X_E}	l_{Y_E}
14					m_{X_E}	m_{Y_E}
15					n_{X_E}	n_{Y_E}

The corresponding vector of unknown axial forces and reactions is:

$$\mathbf{p} = [N_{AB}\ N_{BC}\ N_{CD}\ N_{DE}\ N_{AC}\ N_{AD}\ N_{AE}\ N_{BE}\ N_{CE}\ X_A\ Y_A\ Z_C\ Z_D\ X_E\ Y_E]^T$$

and the external load vector **f** is

$$\mathbf{f} = [0\ 0\ 0\ 0\ 0\ -100\ 0\ 0\ 0\ 0\ 0\ 0\ 0\ 0\ 0]^T$$

The unknowns of the problem are then calculated by solving the following system of equations:

$$-0.6N_{AC} - N_{AD} - 0.707N_{AE} - X_A = 0 \tag{1}$$

$$N_{AB} + 0.8N_{AC} + Y_A = 0 \tag{2}$$

$$0.707N_{AE} = 0 \tag{3}$$

$$-N_{BC} - 0.514N_{BE} = 0 \tag{4}$$

$$-N_{AB} - 0.686N_{BE} = 0 \tag{5}$$

$$0.514N_{BE} - 100 = 0 \tag{6}$$

$$N_{BC} + 0.6N_{AC} = 0 \tag{7}$$

$$-N_{CD} - 0.8N_{AC} - 0.8N_{CE} = 0 \tag{8}$$

$$0.6N_{CE} + Z_C = 0 \tag{9}$$

$$N_{AD} = 0 \tag{10}$$

$$N_{CD} = 0 \tag{11}$$

$$N_{DE} + Z_D = 0 \tag{12}$$

$$0.707N_{AE} + 0.5145N_{BE} + X_E = 0 \tag{13}$$

$$0.686N_{BE} + 0.8N_{CE} + Y_E = 0 \tag{14}$$

$$-N_{DE} - 0.707N_{AE} - 0.5145N_{BE} - 0.6N_{CE} = 0 \tag{15}$$

and this gives the following results (with all forces expressed in kilonewtons):

$$\mathbf{p} = [-133.3 \ -100 \ 0 \ 0 \ 166.7 \ 0 \ 0 \ 194.4 \ -166.7 \ -100 \ 0 \ 100 \ 0 \ -100 \ 0]^T$$

REFLECTION ACTIVITY 4.8

Reconsider the procedure outlined in Summary of Steps 4.4 and modify it for the classification of the *statical indeterminacy* of three-dimensional trusses.

The steps described in Summary of Steps 4.4 are also applicable for the classification of space trusses with the only difference that three equilibrium equations are now available at each node, i.e. $3j$ equations instead of the $2j$ equations used for plane trusses.

Based on this, the degree of statical indeterminacy D is calculated from $D = b + r - 3j$.

If $D = 0$ the truss is statically determinate.
If $D > 0$ the truss is statically indeterminate.
If $D < 0$ the truss is unstable.

Even if $D \geq 0$, the truss might still be unstable and this possibility can be verified by calculating the rank of matrix **A** (just as for plane trusses). If the rank of **A** is smaller than the total number of freedoms n_{DOF}, the truss is classified as unstable.

PROBLEMS

4.1   Calculate the reactions and member axial forces for the truss shown below using the method of joints. State if the members are in tension or compression.

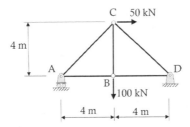

4.2   For the truss shown, determine the reactions and member axial forces using the method of joints. Clarify whether the members are in tension or compression.

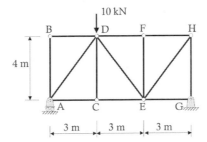

4.3   Evaluate the reactions and member axial forces for the truss shown using the method of joints.

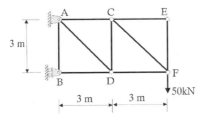

4.4   Calculate the reactions and member axial forces for the truss shown below using the method of joints. State if the members are in tension or compression.

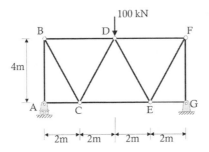

4.5   Evaluate reactions and member axial forces for the truss shown below using the method of joints. State if the members are in tension or compression.

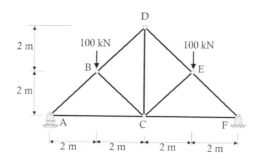

4.6   Determine reactions and member axial forces for the truss shown below using the method of joints. State if the members are in tension or compression.

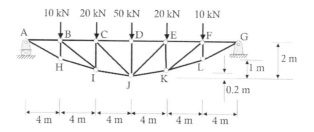

4.7   Find the axial forces in the members of the truss shown below if the three applied loads are at right angles to the top chord ABDG. Use the method of joints.

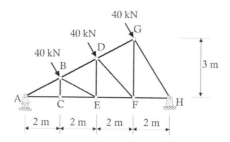

4.8   Find the axial forces in the members of the truss shown using the method of joints. All applied loads are vertical.

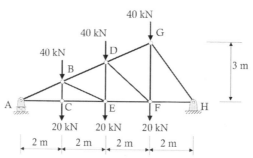

4.9   Find the member axial forces in the truss shown using the method of joints.

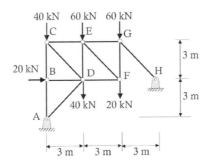

4.10 For the truss shown below:
a.   calculate the reactions and member forces using the method of joints; and
b.   recalculate the reactions and member forces using the method of joints in matrix form.

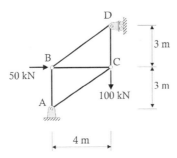

4.11 Reconsider the truss of Problem 4.1, and recalculate the reactions and member forces using the method of joints in matrix form.

4.12 Reconsider the truss of Problem 4.5, and recalculate the reactions and member forces using the method of joints in matrix form.

4.13 Determine the axial forces in members EG, DG and DF of the truss shown using the method of sections. Also determine the axial forces in members CE, CD and BD. State if the members are in tension or compression.

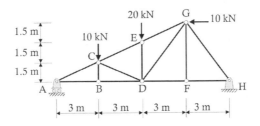

4.14 By first considering equilibrium at Node C, determine the member forces in BC and CD in the truss shown. Next, calculate the forces in BD and BE using the method of sections.

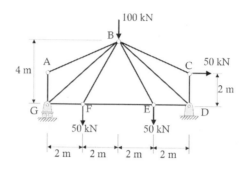

4.15 Reconsider the truss of Problem 4.6 and determine the axial forces in members BI, CJ and DJ using the method of sections. State if the members are in tension or compression.

4.16 Find the unknown forces F_1, F_2 and F_3 if the free-body diagram shown below is in equilibrium.

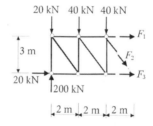

4.17 Find the unknown forces F_1, F_2 and F_3 if the free-body diagram shown below is in equilibrium.

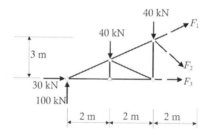

4.18 Find the unknown forces F_1, F_2, F_3 and F_4 if the free-body diagram shown is in equilibrium.

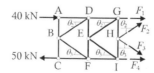

4.19 For the truss shown below, find:
  a. the reactions at A and J as well as the resultant force passing through the pin at F;
  b. the axial forces in members BD, BE and CE using the method of sections; and
  c. the axial forces in members FH, FI and IG using the method of sections.

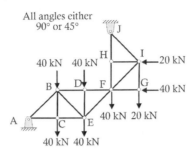

4.20 Reconsider the truss of Problem 4.7 and determine the axial forces in members DG, DF and EF using the method of sections. State if the members are in tension or compression.

4.21 For the truss shown below, the top chord members AB = BC = CD = DE = EF = FG = GH = HI = 7 m and the bottom chord members IJ = JK = KL = LE = EM = MN = NO = OA = 6.25 m. For the loading shown, determine:
  a. the reactions at A and I, as well as the resultant force passing through the pin at E; and
  b. the axial force in each member of the truss.

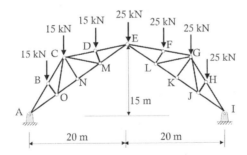

4.22 For the truss shown, the node B is mid-way between the nodes A and D and the node F is mid-way between the nodes D and G. If the top chords ABD and DFG are inclined at 16° to the horizontal, determine the forces in each of the truss members.

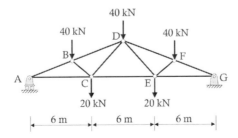

4.23 For the parallel chord truss shown below, the nodes on the top chord are 3 m apart and the nodes on the bottom chord are 1.5 m apart. All inclined members are at 45° to the horizontal. If P_1 = 20 kN and P_2 = 10 kN, determine the maximum force:

   i. in any top chord member;
   ii. in any bottom chord member;
   iii. in any inclined member; and
   iv. in any vertical member.

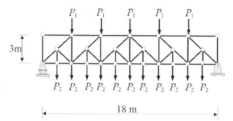

4.24 Find the axial forces in the bars of the truss shown.

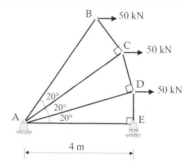

4.25 Determine whether the trusses shown are statically determinate, indeterminate, or unstable.

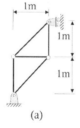
(a)

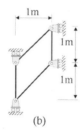
(b)

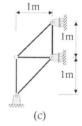
(c)

4.26 Determine the six reactions R_1 to R_6 and the member axial forces for the truss shown using the method of joints.

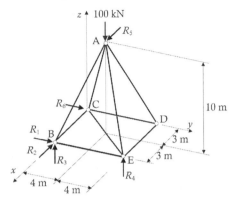

4.27 Determine the six reactions R_1 to R_6 and the member axial forces for the truss shown below using the method of joints. State if the members are in tension or compression.

The 100 kN load lies in the x–y plane and acts in the direction shown. The nodes C, D, E and F also lie in the x–y plane.

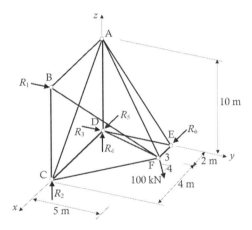

4.28 Reconsider the truss of Problem 4.26, and recalculate reactions and member forces using the method of joints in matrix form.

4.29 For the truss shown below, a 20 kN vertical load is applied at the midpoint of bar DE. Find the maximum axial force, shear force and bending moment in bar DE.

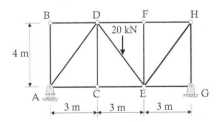

4.30 For the truss shown, each top chord member is loaded at the nodes and mid-way between the nodes. Find the primary axial forces in each member of the truss and the additional secondary forces in each top chord member.

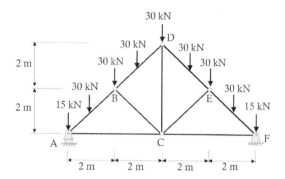

Chapter 5

Euler–Bernoulli beam model

5.1 INTRODUCTION

This chapter presents the derivation of the system of differential equations governing the behaviour of beams. The mathematical model is known as the *Euler–Bernoulli beam model* and is suitable for the analysis of statically determinate and indeterminate beams. The derivation of the governing differential equations is performed combining three sets of equations, namely, the *equilibrium, constitutive* and *kinematic equations*. The *equilibrium equations* have already been discussed in previous chapters and state the relationship between internal and external actions. The *constitutive equations* depend on the properties of the materials from which the beam is constructed and describe the relationship between uniaxial stress and strain. The *kinematic equations* provide a representation of how the deformations undergone by parts of the structure relate to the displacements of the whole structure. These are derived under the assumptions of what is known as the Euler–Bernoulli beam theory (see Chapter 13 for more details), in which plane sections perpendicular to the member axis remain plane and perpendicular to the member axis after deformation. Individual cross-section are therefore assumed not to deform in their plane. We will also assume that the displacements of the structure are small in comparison with the dimensions of the structure.

5.2 EQUILIBRIUM OF A SMALL LENGTH OF BEAM

For a straight beam loaded normal to its longitudinal x-axis, simple relationships exist between the axial force N, the shear force S, the bending moment M and the applied loads. These relationships can be developed by considering equilibrium of a small length of the beam. Let us consider a segment of a beam carrying longitudinal and transverse distributed loads that vary from point to point along the beam, as shown in Figure 5.1. The *intensity* of the distributed loads, denoted by $w(x)$ for the loading perpendicular to the member length and $n(x)$ for the longitudinal load (both defined as a load per unit length), is expressed as a function of the distance x along the beam. The distributed loads are also referred to as w and n for ease of notation. For the sign convention, the x and y directions shown in Figure 5.1 are taken as positive for all quantities, including the applied loads.

Figure 5.2 shows a beam element of infinitesimal length dx (also shown in Figure 5.1) isolated by two cuts dx apart. The total external loads applied on the small length dx are wdx (perpendicular to the beam length) and ndx (longitudinal to the member axis). On the left-hand cut surface, the internal actions are N, S and M, as shown, and on the right-hand cut surface, the internal actions have changed by small amounts dN, dS and dM, respectively, as shown.

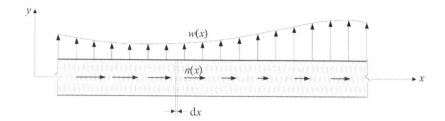

Figure 5.1 Beam segment carrying longitudinal and transverse distributed loads.

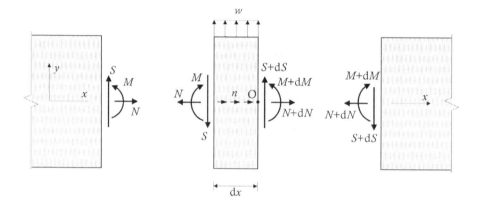

Figure 5.2 Free-body diagram of a small segment cut from the beam length.

Enforcing vertical equilibrium of the element of width dx enables us to write an expression that relates the variation of the shear force S to the transverse (vertical) distributed load w:

$$+\uparrow \quad (S + \mathrm{d}S) + w\mathrm{d}x - S = 0 \tag{5.1}$$

Rearranging gives:

$$\frac{\mathrm{d}S}{\mathrm{d}x} + w = 0 \quad \text{or} \quad S' + w = 0 \tag{5.2a,b}$$

where the prime denotes differentiation with respect to x.

Taking moments with respect to point O (i.e. about the centre of the right face of element dx), we obtain an equation establishing the dependency between the moment M, the shear force S and the transverse (vertical) distributed load w:

$$\left(\!\!\!\!\begin{array}{c}+\\ \end{array}\right. \quad (M + \mathrm{d}M) + S\,\mathrm{d}x - w\mathrm{d}x\frac{\mathrm{d}x}{2} - M = 0 \tag{5.3}$$

Neglecting the product of the two infinitesimally small quantities dx, Equation 5.3 becomes:

$$\frac{\mathrm{d}M}{\mathrm{d}x} + S = 0 \quad \text{or} \quad M' + S = 0 \tag{5.4a,b}$$

A relationship between the axial force N and the tangential distributed force n is defined by enforcing equilibrium in the longitudinal (horizontal) direction:

$$\overset{+}{\rightarrow} \quad N + dN + ndx - N = 0 \tag{5.5}$$

or

$$\frac{dN}{dx} + n = 0 \quad \text{or} \quad N' + n = 0 \tag{5.6a,b}$$

The above equations assume that the distributed external loads w and n (which depend on x and vary along the member length) are constant over the infinitesimal segment dx, because we neglect products of two infinitesimally small quantities (i.e. products between infinitesimal length dx and infinitesimal variations of the applied loads).

It is possible to integrate Equations 5.2, 5.4 and 5.6 to obtain the expressions for S, M and N as:

$$S = -\int w dx + C_S \qquad M = -\int S dx + C_M \qquad N = -\int n dx + C_N \tag{5.7a,b,c}$$

where C_S, C_M and C_N are constants of integration, determined from the boundary conditions of the problem considered. The relevant steps will be illustrated subsequently in Worked Example 5.1.

5.3 KINEMATIC (OR STRAIN–DISPLACEMENT) EQUATIONS

The relationship between the deformations undergone by a member at a particular cross-section and its corresponding displacements is defined by the *kinematic equations*, sometimes referred to as *strain–displacement equations*.

For beams and frames, there are two types of deformations that usually govern their response, the axial deformations and the bending deformations. For clarity, these are introduced separately below, and their effects are combined in Section 5.3.3. Other deformations, such as the shear deformations, are dealt with later in the book (see Chapter 13). The Euler–Bernoulli model may therefore not be able to accurately predict the response of beams that exhibit significant shearing deformations, such as beams that are very short in length, and other beam models or approaches should be used in these circumstances.

5.3.1 Axial deformations and displacements

Figure 5.3a shows a portion of a straight structural member of length L deforming under an external axial tensile load P, with its line of action passing through the centroid of the cross-section, as illustrated in Figure 5.3b. It is assumed that the longitudinal x-axis also passes through the centroid of the cross-section.

A function $u(x)$, also referred to as u, is introduced to describe the axial displacements undergone by points lying on the x-axis. In this manner, the axial displacement of a point B (located at x_B) can be easily calculated substituting $x = x_B$ into the expression for u (Figure 5.3c). Obviously, the expression for u depends on the beam layout, including loading and support conditions. It is not known a priori and needs to be evaluated from the analysis.

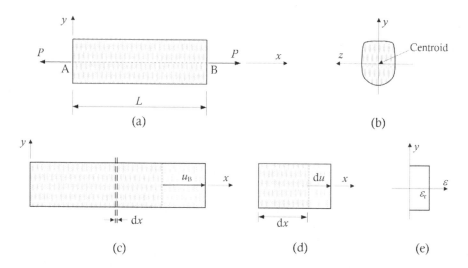

Figure 5.3 Deformations of a member owing to axial loading. (a) Layout of axial loading. (b) Cross-sectional layout. (c) Deformed shape. (d) Infinitesimal length of the deformed element. (e) Strain diagram.

Figure 5.3d shows an infinitesimal slice through the member (of length dx). In a straight beam, the two end cross-section of segment dx are parallel before loading and remain parallel after the axial load is applied, with every fibre of the cross-section elongating by an infinitesimal amount du. All points on the cross-section are deformed by the same amount with the strain at each point equal to the deformation du divided by the original length dx. This is illustrated in the strain diagram in Figure 5.3e, where the strain at the level of the reference axis is denoted as ε_r. Based on this, the strain at an arbitrary point on the cross-section ε is calculated as:

$$\varepsilon = \varepsilon_r \quad \text{or} \quad \varepsilon = \frac{du}{dx} = u' \tag{5.8a,b}$$

and the variations for ε_r (and u') along the member depends on the geometry, loading, and support conditions of the problem being analysed. The subscript r included in ε_r refers to the reference axis. There is no need to add a subscript to u' because the function u has been introduced to describe displacements only at the level of the reference axis.

From Equation 5.8a, it is possible to determine the expression for u:

$$u = \int \varepsilon_r \, dx + C_u \tag{5.9}$$

in which the constant of integration C_u is evaluated from the end or support conditions of the member being analysed.

The elongation of the member e_{AB} between two points A and B (with coordinates x_A and x_B) is obtained from:

$$e_{AB} = u(x_B) - u(x_A) \tag{5.10}$$

or e_{AB} could also be obtained integrating the strain at the level of the reference axis:

$$e_{AB} = \int_{x_A}^{x_B} \varepsilon_r \, dx \tag{5.11}$$

5.3.2 Bending (flexural) deformations and displacements

We will consider an initially straight beam that is subjected to a constant moment as illustrated in Figure 5.4a. It is assumed that the cross-section is symmetric about the vertical y-axis and that the positive bending moment M acts about the horizontal z-axis passing through the centroid of the cross-section (Figure 5.4b). The displaced shape of the beam after bending is depicted in Figure 5.4c. Let us now consider the infinitesimal segment of beam of length dx shown in Figures 5.4c and d. The sides of the segment dx, initially parallel before bending, rotate with respect to each other by an angle dθ after bending. This rotation produces strains that vary linearly with y over the depth of the cross-section. The corresponding strain diagram is shown in Figure 5.4e, and the slope of the strain diagram is the curvature κ.

The deflection of a beam at a point along its length depends on the support conditions and the *curvature* at each point along the beam. Referring to the slice of beam deformed in bending in Figure 5.4d, the curvature κ is the angular change of the beam axis per unit length and is given by:

$$\kappa = \frac{\mathrm{d}\theta}{\mathrm{d}x} = \theta' = \frac{1}{r} \tag{5.12}$$

where r is the radius of curvature illustrated in Figure 5.4d and dx is the undeformed width of the infinitesimal slice at the level of the centroid of the cross-section, which is related to the rotation by means of the radius of curvature as d$x = r$ dθ.

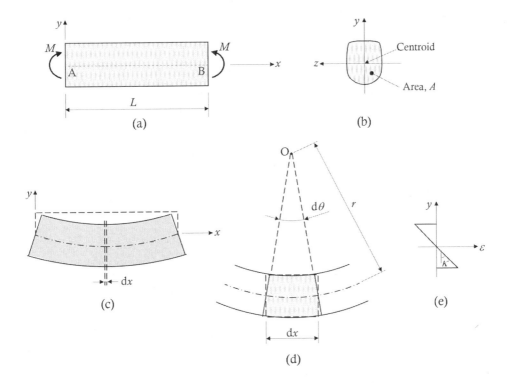

Figure 5.4 Deformations of a member produced by an applied moment. (a) Applied moment. (b) Cross-sectional layout. (c) Deformed shape. (d) Infinitesimal length of the deformed element. (e) Strain diagram.

The rotation undergone along the beam length can be calculated as the first derivative of the deflection v:

$$\theta = \frac{dv}{dx} = v' \tag{5.13}$$

which enforces plane sections to remain perpendicular to the member axis after deformation (see Chapter 13).

Substituting Equation 5.13 into Equation 5.12 gives the relationship between the deflection and the curvature:

$$\kappa = \theta' = v'' \tag{5.14}$$

It is usually convenient to derive the expression for the rotation θ and deflection v along the member by integrating the function describing the variation of the curvature κ:

$$\theta = \int \kappa \, dx + C_{v1} \tag{5.15}$$

$$v = \int\int \kappa \, dx \, dx + C_{v1}x + C_{v2} = \int \theta \, dx + C_{v2} \tag{5.16}$$

The constants of integration C_{v1} and C_{v2} can be evaluated from the end and support conditions of the member.

Consider a loaded beam divided into many small slices. Each slice subjected to bending will deform similarly to the deformed slice shown earlier in Figure 5.4d. If all the deformed slices are put back together, they will form a bent beam and an initially straight beam will deflect. The final deflection depends not only on the deformation of each small slice but on the support conditions as well. Figure 5.5 illustrates two beams with the same bending moment diagram (i.e. constant over the length) but with different support conditions. The deflected shape of each is quite different even though the deformation of any small slice taken from either beam is identical.

From geometry, the strain at any point on a particular cross-section at a distance y from the member axis can be calculated as (refer to Figure 5.4d and e):

$$\varepsilon = -y\frac{d\theta}{dx} = -y\kappa \tag{5.17}$$

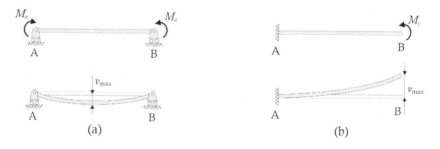

Figure 5.5 Examples of flexural actions and deformations.

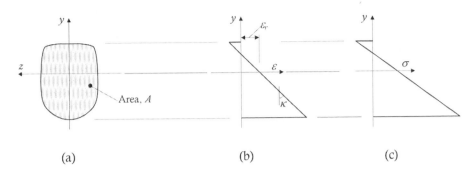

Figure 5.6 Typical stress and strain diagrams. (a) Cross-section. (b) Strain diagram. (c) Stress diagram.

in which the minus sign accounts for the fact that, in sagging moment regions (i.e. regions of positive moment), the curvature κ is assumed to be positive and, because of this, the top fibres of the cross-section with positive y are compressed and the strain must be negative (i.e. compressive based on our sign convention). In a similar manner, the bottom fibres of the section are stretched and the strain must be positive (i.e. tensile).

5.3.3 Combining axial and flexural deformations

In the previous sections, we derived the relationships between the displacements and strains for a beam segment subjected to either axial or flexural deformations. In reality, beams and frames are often subjected to both types of deformations. Combining Equations 5.8 and 5.17 gives:

$$\varepsilon = \varepsilon_r - y\kappa = u' - yv'' \tag{5.18}$$

The strain distribution on the cross-section shown in Figure 5.6a caused by an axial force and a bending moment is shown in Figure 5.6b. The stress diagram is shown in Figure 5.6c, where the stress and strain at any point are assumed to be related linearly. The stress–strain relationship is discussed in the following section. The level on the cross-section where the stresses are zero is usually referred to as the *neutral axis*.

5.4 CONSTITUTIVE EQUATIONS

The relationship between the deformations and stresses induced in a structure at a particular point is described by the *constitutive equations*, also referred to as *stress–strain relationships*. In most of this book, we will deal with linear–elastic materials that follow Hooke's law. That is:

$$\sigma = E\varepsilon \tag{5.19}$$

where σ and ε are the stress and strain at a particular point, respectively, and E is the elastic modulus of the material.

When dealing with beams and frames, it is sometimes more convenient to express the constitutive equations of a material (i.e. stress–strain relationship) in terms of the cross-sectional resultants (axial force N and moment M) and cross-sectional deformations (axial strain at the level of the reference axis ε_r and curvature κ). These relationships are applicable for members satisfying the assumptions of the Euler–Bernoulli beam theory, which have been used in the previous sections to derive the relevant strain–displacement equations.

Substituting Equation 5.18 into the Equation 5.19 leads to an expression relating internal stresses to the two variables describing the cross-sectional strain distribution, i.e. the curvature κ and the strain measured at the level of the reference axis ε_r (Figure 5.6b):

$$\sigma = E(\varepsilon_r - y\kappa) \tag{5.20}$$

The geometric properties of the cross-section of Figure 5.6a are denoted by A, B and I, which depict the area and the first and second moments of area of the cross-section about the z-axis, respectively (see Appendix A).

The internal axial force and moment about the reference axis resisted by the cross-section are denoted N and M, respectively. The axial force N is obtained as the integral of the stresses resisted by the cross-section and is given by:

$$N = \int_A \sigma \, dA = \int_A E(\varepsilon_r - y\kappa) \, dA = EA\varepsilon_r - EB\kappa \tag{5.21}$$

in which the product of the area A and the elastic modulus E represents the *axial rigidity* (*EA*) of the cross-section. The product BE is the rigidity related to the first moment of area of the cross-section about the z-axis.

Similarly, the equation for the internal moment M may be expressed as:

$$M = -\int_A y\sigma \, dA = -\int_A Ey(\varepsilon_r - y\kappa) \, dA = -EB\varepsilon_r + EI\kappa \tag{5.22}$$

where the product EI is known as the *flexural rigidity* of the cross-section and provides a measure of the flexibility (or stiffness) of the member in bending, while the negative sign is required to ensure that a stress distribution, with stresses varying from compressive (negative) at the top of the section to tensile (positive) at the bottom, produces a positive moment M in accordance with the adopted sign convention.

Considering a centroidal reference system (i.e. the origin of the adopted coordinate system coincides with the centroid of the cross-section), Equations 5.21 and 5.22 can be simplified to:

$$N = EA\varepsilon_r \quad \text{or} \quad \varepsilon_r = \frac{N}{EA} \tag{5.23a,b}$$

and

$$M = EI\kappa \quad \text{or} \quad \kappa = \frac{M}{EI} \tag{5.24a,b}$$

in which the terms including the first moment of area B have disappeared because B is zero when calculated about the centroidal reference axis (see Appendix A).

The stress distribution produced by the presence of N and M can be calculated after the internal actions are determined by substituting Equations 5.23b and 5.24b into Equation 5.20 as follows:

$$\sigma = E(\varepsilon_r - y\kappa) = \frac{N}{A} - \frac{My}{I} \qquad (5.25)$$

When only an axial force is resisted by the cross-section (i.e. $M = 0$), Equation 5.25 simplifies to:

$$\sigma = \frac{N}{A} \qquad (5.26a)$$

When the member is subjected to bending moment only (i.e. $N = 0$), Equation 5.25 becomes:

$$\sigma = -\frac{My}{I} \qquad (5.26b)$$

The use of the equilibrium, constitutive and kinematic equations is illustrated in Worked Example 5.1 and the calculation of the stress distribution over a cross-section is demonstrated in Worked Example 5.2.

WORKED EXAMPLE 5.1

Consider the two beams in Figure 5.7 that carry a linearly varying distributed load, ranging from 0 kN/m at A to −18 kN/m at B (negative because it is pointing in the negative direction of y). Beam 1 is a cantilever beam, fixed at end B and free at end A. Beam 2 is a propped cantilever, fixed at B and with a roller support at A. For both cases, using the equilibrium, constitutive and kinematic equations, determine the expressions for:

(i) the internal actions N and M

(ii) strain at the level of the reference axis ε_r and curvature κ

(iii) rotation, deflection, and axial displacement along the member length

Plot the variation for the internal actions (N and M) and displacements (u, v and θ). The cross-sections of the beams are uniform throughout their length with area A and second moment of area I. Assume the material is linear–elastic with elastic modulus E.

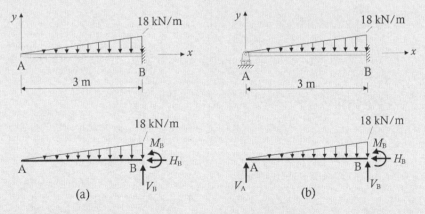

Figure 5.7 Beams and free-body diagrams for Worked Example 5.1. (a) Beam 1. (b) Beam 2.

Beam I

(I) Equilibrium Equations

Equations 5.7 are applied for the calculation of the internal actions along the cantilever beam of Figure 5.7a. On the basis of the support and loadings conditions, the shear force, axial force and bending moment are all zero at node A, as can be seen by enforcing vertical, horizontal and rotational equilibrium at node A (i.e. at $x = 0$ m in Figure 5.8a). It is not possible to enforce boundary conditions related to the internal actions at node B because they are all related to the as yet undetermined reactions, as depicted in Figure 5.7a.

(a)   (b)

Figure 5.8 Free-body diagram at point A (i.e. at $x = 0$ m). (a) Beam 1. (b) Beam 2.

The expression for the linearly varying distributed load in terms of x is established from the load intensity at A and B, i.e. $w_A = 0$ and $w_B = -18$ kN/m, respectively:

$$w = w_A + \frac{w_B - w_A}{L_{AB}} x = 0 + \frac{-18 - 0}{3} x = -6x \qquad (5.27)$$

where the negative sign for load indicates that the load is applied downward (i.e. in the negative direction of y).

The variation of the shear force is obtained from Equation 5.7a:

$$S = -\int w \, dx + C_S = -\int -6x \, dx + C_S = 3x^2 + C_S$$

Since $S = 0$ kN at $x = 0$, the constant of integration C_S is zero; therefore:

$$S = 3x^2 \qquad (5.28)$$

The moment is next obtained from Equation 5.7b:

$$M = -\int S \, dx + C_M = -\int 3x^2 \, dx + C_M = -x^3 + C_M$$

Since $M = 0$ kN at $x = 0$, the constant of integration C_M is zero and:

$$M = -x^3 \qquad (5.29)$$

We know that at $x = 0$ the axial force $N = 0$, and therefore C_N in Equation 5.7c is zero. There are no distributed axial loads on this beam (i.e. $n = 0$ throughout) and therefore Equation 5.7c reduces to:

$$N = 0 \qquad (5.30)$$

(2) Constitutive Equations

The strain at the level of the reference (centroidal) axis ε_r and curvature κ is determined by substituting Equations 5.29 and 5.30 into Equations 5.23b and 5.24b:

$$\kappa = \frac{M}{EI} = \frac{-x^3}{EI}; \quad \varepsilon_r = \frac{N}{EA} = 0 \qquad (5.31a,b)$$

(3) Kinematic (Strain–Displacement) Equations

The expressions for the deflection and rotation can be determined from Equations 5.15 and 5.16:

$$\theta = \int \kappa \, dx + C_{v1} = \int \frac{-x^3}{EI} \, dx + C_{v1} = \frac{-x^4}{4EI} + C_{v1}$$

$$v = \int \left(\frac{-x^4}{4EI} + C_{v1} \right) dx + C_{v2} = \frac{-x^5}{20EI} + C_{v1}x + C_{v2}$$

We know that at the fixed end of the cantilever (end B), both the slope and the deflection are zero, i.e. at $x = 3$ m we have $\theta_B = 0$ and $v_B = 0$. On the basis of these boundary conditions, the constants of integration are:

$$\theta(x = 3) = \frac{-81}{4EI} + C_{v1} = 0 \qquad \therefore C_{v1} = \frac{81}{4EI}$$

$$v(x = 3) = \frac{-243}{20EI} + C_{v1} \times 3 + C_{v2} = 0 \quad \therefore C_{v2} = -\frac{243}{5EI}$$

and the expressions for the rotation and deflection can be rewritten as:

$$\theta = \frac{-x^4}{4EI} + \frac{81}{4EI} \qquad v = \frac{-x^5}{20EI} + \frac{81}{4EI}x - \frac{243}{5EI}$$

The variation of the axial displacement u is determined by substituting Equation 5.31b into Equations 5.9 as:

$$u = \int \varepsilon_r \, dx + C_u = \int 0 \, dx + C_u = C_u$$

As $u = 0$ at $x = 3$ m, we see that $C_u = 0$ and therefore $u = 0$ at all points along the beam. As expected, there is no shortening or elongation at the level of the centroidal axis (as the axial force is everywhere zero).

All results are summarised in Figure 5.9.

Beam 2

The solution for beam 2 follows the same steps adopted for beam 1. The only difference between the two beams is that beam 2 has a roller support at A (not present for beam 1).

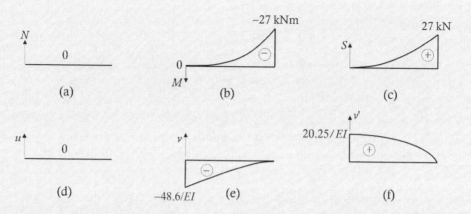

Figure 5.9 Response of beam 1 — internal actions and deformations. (a) N (kN). (b) M (kNm) (drawn on tension side). (c) S (kN). (d) u. (e) v. (f) v' (or θ).

(I) Equilibrium Equations

As for beam 1:

$$w = -6x$$

In the evaluation of the boundary conditions to be used to determine the constants of integration related to the internal actions, it is convenient to use the condition of zero axial force and zero moment at node A, as depicted by the free-body diagram of node A in Figure 5.8b. For the boundary condition to be used for the shear force at node A, we need to account for the presence of the unknown reaction V_A. This is a consequence of the fact that the structure being analysed is statically indeterminate (with four unknown reactions), whereas beam 1 was statically determinate (with only three unknown reactions).

The evaluation of the expressions for the internal actions is then carried out as for beam 1. The expression for shear force is obtained from Equation 5.7a:

$$S = -\int w\,dx + C_S = -\int -6x\,dx + C_S = 3x^2 + C_S$$

with $C_S = -V_A$, since $S = -V_A$ at $x = 0$. Therefore:

$$S = 3x^2 - V_A \tag{5.32}$$

From Equation 5.7b:

$$M = -\int S\,dx + C_M = -\int \left(3x^2 - V_A\right)dx + C_M = -x^3 + V_A x + C_M$$

and $C_M = 0$ since $M = 0$ at $x = 0$. Therefore:

$$M = -x^3 + V_A x \tag{5.33}$$

As for beam 1, the axial force is everywhere zero:

$$N = 0 \tag{5.34}$$

(2) Constitutive Equations

From Equations 5.23b and 5.24b:

$$\kappa = \frac{M}{EI} = \frac{-x^3 + V_A x}{EI}; \quad \varepsilon_r = \frac{N}{EA} = 0 \tag{5.35a,b}$$

(3) Kinematic (Strain–Displacement) Equations

From Equations 5.15 and 5.16, the rotation and deflection are:

$$\theta = \int \kappa \, dx + C_{v1} = \int \frac{-x^3 + V_A x}{EI} \, dx + C_{v1} = -\frac{x^4}{4EI} + \frac{V_A x^2}{2EI} + C_{v1}$$

$$v = \int \left(-\frac{x^4}{4EI} + \frac{V_A x^2}{2EI} + C_{v1} \right) dx + C_{v2} = -\frac{x^5}{20EI} + \frac{V_A x^3}{6EI} + C_{v1} x + C_{v2}$$

Applying the boundary conditions at the roller support at A ($v_A = 0$) and at the fixed support at B ($v_B = 0$ and $\theta_B = 0$), we get:

At $x = 0$: $v = 0$ $\qquad\qquad\qquad\qquad\qquad \therefore C_{v2} = 0$

At $x = 3$: $\theta = -\dfrac{81}{4EI} + \dfrac{9V_A}{2EI} + C_{v1} = 0$

At $x = 3$: $v = -\dfrac{243}{20EI} + \dfrac{9V_A}{2EI} + 3C_{v1} + C_{v2} = 0$ $\quad \therefore C_{v1} = -\dfrac{81}{20EI}, \quad V_A = 5.4 \, \text{kN}$

which leads to the following expressions for the rotation and deflection:

$$\theta = -\frac{x^4}{4EI} + \frac{27x^2}{10EI} - \frac{81}{20EI} \qquad v = -\frac{x^5}{20EI} + \frac{27x^3}{30EI} - \frac{81}{20EI}x$$

This illustrates that the calculation of the reactions for a statically indeterminate beam cannot be performed on the basis of equilibrium considerations alone, but also requires knowledge of the material properties (constitutive equations) and of the relationship between the displacements and the strains (kinematic equations). In fact, the reaction V_A was evaluated on the basis of a boundary condition related to the vertical displacement of the support at A.

The expression for axial displacement u is identical to that calculated for beam 1, i.e. $u = 0$.

The internal actions can now be rewritten by substituting the calculated value for V_A into Equations 5.32, 5.33 and 5.34:

$$S = 3x^2 - 5.4; \quad M = -x^3 + 5.4x; \quad N = 0$$

The calculated results are plotted in Figure 5.10.

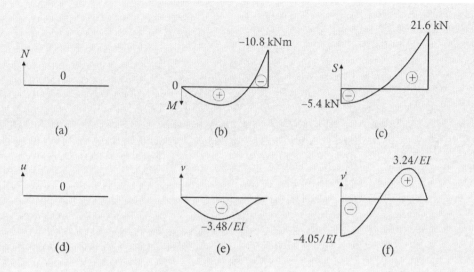

Figure 5.10 Response of beam 2 — internal actions and deformations. (a) N (kN). (b) M (kNm) (drawn on tension side). (c) S (kN). (d) u. (e) v. (f) v' (or θ).

WORKED EXAMPLE 5.2

If the beam analysed in Worked Example 5.1 and illustrated in Figure 5.7a is fabricated from steel with the I-shaped cross-section shown in Figure 5.11, calculate the maximum compressive and maximum tensile stresses in the steel at the cross-section at B. The cross-section is symmetrical about the vertical y-axis, with the second moment of area $I = 32.51 \times 10^6$ mm⁴ (calculated with respect to the centroidal axis located at 84.5 mm from the bottom of the section).

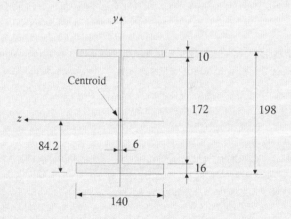

Figure 5.11 Cross-section for Worked Example 5.2.

In Worked Example 5.1, the bending moment at B was evaluated from Equation 5.29 as equal to −27 kNm (see Figure 5.9b). Because the bending moment is negative, the top flange of the beam is in tension and the bottom flange is in compression.

The maximum tensile stress occurs at the top fibre of the cross-section at the support B, where $y = 198 − 84.2 = 113.8$ mm. Using Equation 5.26b:

$$\sigma_{top} = -\frac{(-27 \times 10^6) \times 113.8}{32.51 \times 10^6} = 94.5 \text{ MPa}$$

The maximum compressive stress occurs at the bottom fibre, where $y = −84.2$ mm. Using Equation 5.26b:

$$\sigma_{bottom} = -\frac{(-27 \times 10^6) \times (-84.2)}{32.51 \times 10^6} = -69.9 \text{ MPa}$$

5.5 METHOD OF DOUBLE INTEGRATION

On the basis of the kinematic equations derived in the previous section, the deflection along a member can be calculated by integrating the curvature twice with respect to x, because $\kappa = v''$ (Equations 5.14 through 5.16). Combining these equations with the constitutive equation related to the flexural response (Equations 5.24), the relationship between the moment and the deflection is established. This forms the basis for the procedure usually referred to as the *method of double integration*. Under the assumption of linear–elastic material properties, we can write:

$$\kappa = \frac{M}{EI} \tag{5.36}$$

$$\theta = \int \kappa \, dx + C_{v1} = \int \frac{M}{EI} \, dx + C_{v1} \tag{5.37}$$

$$v = \int \int \kappa \, dx \, dx + C_{v1} x + C_{v2} = \int \int \frac{M}{EI} \, dx \, dx + C_{v1} x + C_{v2} \tag{5.38}$$

These equations are very useful, in particular, when dealing with statically determinate beams, where the expression of the internal moment can be determined from equilibrium considerations. In these cases, the expression for the curvature is simply the expression for the bending moment divided by the flexural rigidity EI. It is possible to apply the method of double integration to statically indeterminate structures, but unknown terms are included in the expressions for the moment. For these cases, other methods such as the governing differential equations presented in the next section might be more useful and efficient in achieving the solution.

The steps involved in the method of double integration are summarised below and then illustrated in Worked Example 5.3.

SUMMARY OF STEPS 5.1: Method of double integration — Steps in solution procedure

1. Determine the expression for the internal moment over the member length.

2. Insert the expression for the moment in Equation 5.36 and integrate it once to get the expression for the rotation (Equation 5.37) and twice for the deflection (Equation 5.38).

3. Determine the constants of integration by applying the boundary conditions relevant to the support conditions of the particular problem under consideration.

WORKED EXAMPLE 5.3

For the beam of Figure 5.12, calculate the expressions for the rotation and deflection along its length. Assume that the cross-sectional area A and second moment of area I remain constant throughout the member and that the material behaviour is linear–elastic with elastic modulus E.

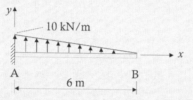

Figure 5.12 Beam for Worked Example 5.3.

---

(1) Determine the expression for bending moment

The expression for the linearly varying upward distributed load is:

$$w = w_A + \frac{w_B - w_A}{L_{AB}} x = 10 + \frac{0 - 10}{6} x = 10 - \frac{5}{3} x$$

A cut is performed between A and B, subdividing the structure into two free-body diagrams, referred to as 1 and 2 in Figure 5.13b and c, respectively.

Before the expression for M can be evaluated using free-body diagram 1, it is necessary to calculate the three unknown reactions (or at least V_A and M_A). With the use of free-body diagram 2, it is possible to determine M without the calculation of the reactions. For completeness, the evaluation of M is carried out using both free-body diagrams.

Free-body diagram 1:

The unknown reactions are calculated from statics (Figure 5.13a):

$$V_A + \frac{10 + 0}{2} 6 = 0 \qquad \therefore V_A = -30 \text{ kN}; \quad H_A = 0 \text{ kN}$$

$$M_A - \frac{10 + 0}{2} 6 \frac{1}{3} 6 = 0 \quad \therefore M_A = 60 \text{ kNm}$$

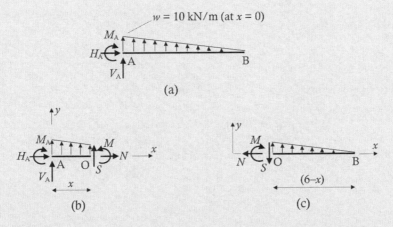

Figure 5.13 Free-body diagrams for Worked Example 5.3. (a) Free-body diagram of the entire structure. (b) Free-body diagram 1. (c) Free-body diagram 2.

The expression for *M* is then evaluated from rotational equilibrium of free-body diagram 1 (in Figure 5.13b):

$$M - M_A - V_A x - \frac{10 - \left(10 - \frac{5}{3}x\right)}{2} \times \frac{2}{3}x - \left(10 - \frac{5}{3}x\right) \times \frac{x}{2} = 0$$

$$\therefore M = 60 - 30x + 5x^2 - \frac{5}{18}x^3$$

Free-body diagram 2:
Applying rotational equilibrium to free-body diagram 2 of Figure 5.13c gives:

$$M - \frac{\left(10 - \frac{5}{3}x\right)}{2}(6-x)\frac{(6-x)}{3} = 0 \quad \therefore M = 60 - 30x + 5x^2 - \frac{5}{18}x^3$$

which, as expected, yields the same expression obtained with free-body diagram 1.

(2) Integrate twice the expression for curvature
Substituting the expression for *M* into Equation 5.36, we get:

$$\kappa = \frac{M}{EI} = \frac{60 - 30x + 5x^2 - \frac{5}{18}x^3}{EI}$$

and this can be integrated to produce the rotation and deflection:

$$\theta = \int \kappa \, dx + C_{vI} = \frac{1}{EI} \int \left(60 - 30x + 5x^2 - \frac{5}{18}x^3\right) dx + C_{vI}$$

$$= \frac{1}{EI}\left(60x - 15x^2 + \frac{5}{3}x^3 - \frac{5}{72}x^4\right) + C_{vI}$$

$$v = \int \theta\, dx + C_{v2} = \int \left[\frac{1}{EI} \left(60x - 15x^2 + \frac{5}{3}x^3 - \frac{5}{72}x^4 \right) + C_{v1} \right] dx + C_{v2}$$

$$= \frac{1}{EI} \left(30x^2 - 5x^3 + \frac{5}{12}x^4 - \frac{1}{72}x^5 \right) + C_{v1}x + C_{v2}$$

(3) Calculate the constants of integration

The two constants of integration C_{v1} and C_{v2} can be calculated from the conditions of zero deflection and zero rotation at the fixed support at A ($x = 0$), i.e. $v_A = 0$ and $\theta_A = 0$:

At $x = 0$: $\theta = 0$ $\qquad\qquad$ $\therefore C_{v1} = 0$

At $x = 0$: $v = 0$ $\qquad\qquad$ $\therefore C_{v2} = 0$

The expressions for θ and v can be rewritten as:

$$\theta = \frac{1}{EI} \left(60x - 15x^2 + \frac{5}{3}x^3 - \frac{5}{72}x^4 \right) \qquad v = \frac{1}{EI} \left(30x^2 - 5x^3 + \frac{5}{12}x^4 - \frac{1}{72}x^5 \right)$$

5.6 GOVERNING DIFFERENTIAL EQUATIONS (AS A FUNCTION OF DISPLACEMENTS)

The three sets of equations presented in the previous sections, i.e. equilibrium, kinematic and constitutive equations, form the basis for the analysis of any type of structure. On the basis of the assumptions already introduced, these can be combined into a system of two governing differential equations expressed in terms of the vertical and axial displacements (u and v).

Substituting the constitutive equations (Equations 5.23 and 5.24) into the kinematic equations (Equations 5.8 and 5.14) gives:

$$N = EAu' \tag{5.39a}$$

$$M = EIv'' \tag{5.39b}$$

which relate the internal actions N and M to the displacements u and v, respectively. Equation 5.39b was utilised in the previous section for the derivation of the method of double integration.

Differentiating Equation 5.39b with respect to x ($M' = EIv'''$) and substituting it into Equation 5.4b ($M' + S = 0$) leads to an expression relating shear force and vertical deflection v:

$$S = -M' = -EIv''' \tag{5.39c}$$

Differentiating Equation 5.4b with respect to x ($M'' + S' = 0$) and substituting it into Equation 5.2b ($S' + w = 0$) produces the following relationship between the moment and the applied transverse load w:

$$M'' - w = 0 \tag{5.40}$$

After differentiating Equation 5.39b twice with respect to x ($M'' = EIv^{IV}$), it can be substituted into Equation 5.40 to establish a relationship between the displacement v and the load w as:

$$EIv^{IV} - w = 0 \tag{5.41a}$$

We can differentiate Equation 5.39a once with respect to x ($N' = EAu''$) and substitute it into Equation 5.6b. This leads to:

$$EAu'' + n = 0 \tag{5.41b}$$

Equations 5.41 form the governing differential equations defined in terms of displacements describing the behaviour of a beam or frame member.

Expressions for u and v can be obtained by integrating Equations 5.41 with respect to x. In particular, v is evaluated by integrating Equation 5.41a four times:

$$v^{IV} = \frac{w}{EI} \tag{5.42a}$$

$$v''' = \int \frac{w}{EI} dx + C_{v1} \tag{5.42b}$$

$$v'' = \int \int \frac{w}{EI} dx\, dx + C_{v1}x + C_{v2} \tag{5.42c}$$

$$v' = \int \int \int \frac{w}{EI} dx\, dx\, dx + C_{v1}\frac{x^2}{2} + C_{v2}x + C_{v3} \tag{5.42d}$$

$$v = \int \int \int \int \frac{w}{EI} dx\, dx\, dx\, dx + C_{v1}\frac{x^3}{6} + C_{v2}\frac{x^2}{2} + C_{v3}x + C_{v4} \tag{5.42e}$$

and u can be determined by integrating Equation 5.41b twice:

$$u'' = -\frac{n}{EA} \tag{5.43a}$$

$$u' = -\int \frac{n}{EA} dx + C_{u1} \tag{5.43b}$$

$$u = -\int \int \frac{n}{EA} dx\, dx + C_{u1}x + C_{u2} \tag{5.43c}$$

The constants of integration C_{u1}, C_{u2} for the axial displacement and C_{v1}, C_{v2}, C_{v3}, C_{v4} for the deflection can be determined from six boundary conditions for the problem. These are outlined in the following sub-sections, with the cases of axial displacement and deflection considered separately.

5.6.1 Boundary conditions for the axial displacement

When dealing with the axial displacements, two possible boundary conditions can be encountered at supports. The first is when the axial displacement is restrained (i.e. $u = 0$). The second is when the axial displacement is unrestrained and the structure is free to move axially (and the boundary condition is governed by Equation 5.39a: $N = EAu'$).

The boundary conditions related to the axial displacement that are applied for the different support conditions are outlined in the following:
- fixed or pinned support (Figure 5.14a): $u = 0$
- roller support or free edge without external axial load (Figure 5.14b): $u' = 0$
- roller support or free edge with external axial load applied P (Figure 5.14c): $u' = \pm(P/EA)$, where the sign depends on whether P is consistent with the sign convention adopted for the internal axial force N (because this boundary condition is based on $N = EAu'$)

The reaction corresponding to the restrained axial displacement may be calculated, once u is determined, by applying axial equilibrium to the free-body diagram at the support. This is illustrated in Worked Examples 5.4 through 5.6.

5.6.2 Boundary conditions for the vertical displacement

The boundary conditions at supports related to the vertical displacements involve both deflection (v) and rotation (v' or θ), and can be applied as follows:

- fixed support (Figure 5.15a): $v = 0$ and $v' = 0$
- pinned and roller supports without external moment (Figure 5.15b): $v = 0$ and $v'' = 0$
- pinned support and roller supports with external moment M_e applied (Figure 5.15c): $v = 0$ and $v'' = \pm M_e/EI$ (where the sign depends on whether M_e is positive in the same direction of the internal bending moment M, because this boundary condition is based on $M = EIv''$)
- free edge without external loads (Figure 5.15d): $v'' = 0$ and $v''' = 0$ (derived from the condition of zero internal bending moment $M = EIv'' = 0$ and zero internal shear force at the free edge $S = -M' = -EIv''' = 0$)
- free edge with external moment M_e and vertical load applied P (Figure 5.15e): $v'' = \pm M_e/EI$ and $v''' = \pm P/EI$ (obtained from $M = EIv''$ and $S = -EIv'''$)

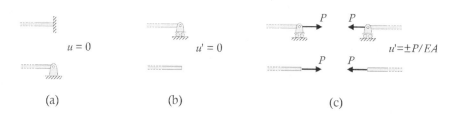

Figure 5.14 Boundary conditions for the axial displacement. (a) Fixed or pinned support. (b) Roller support or free edge. (c) Roller support or free edge with external axial force.

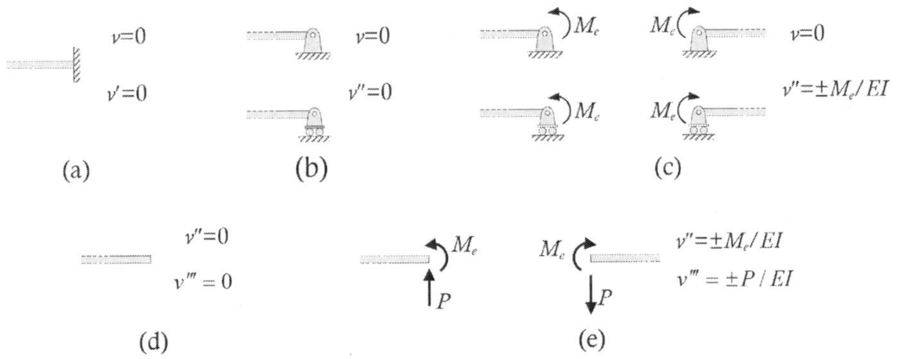

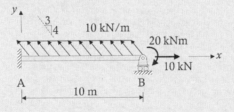

Figure 5.15 Boundary conditions for the vertical displacement. (a) Fixed support. (b) Pinned or roller support. (c) Pinned or roller support with an external applied moment. (d) Free edge. (e) Free edge with external applied moment and force.

When the expression for v has been determined, the calculation of the reactions at each support may be carried out on the basis of equilibrium consideration applied to the free-body diagram of the support. This is illustrated in Worked Examples 5.4 through 5.6.

The use of the differential equations (Equations 5.41) and corresponding boundary conditions are now outlined with worked examples.

WORKED EXAMPLE 5.4

The propped cantilever beam shown in Figure 5.16 is loaded with an inclined uniformly distributed load of 10 kN/m (in the direction shown), as well as with a couple and a horizontal load applied at support B. Use the differential equations expressed in terms of displacements (Equations 5.41) to determine the expressions for u and v, and calculate the reactions at the supports. Assume constant cross-sectional properties, with area A and second moment of area I, and linear–elastic material behaviour, with elastic modulus E.

Figure 5.16 Beam for Worked Example 5.4.

(1) Identify boundary conditions

The boundary conditions to be used for the evaluation of the expressions of the axial displacement are as follows:

− $u = 0$ at $x = 0$, i.e. the axial displacement is restrained by the fixed support.

− $u' = 10/EA$ at $x = 10$ m, i.e. point B is unrestrained axially and subjected to 10 kN. The boundary condition is applied with a positive value because the internal axial force at B is $N = 10$ kN (see the free-body diagram of point B in Figure 5.17b).

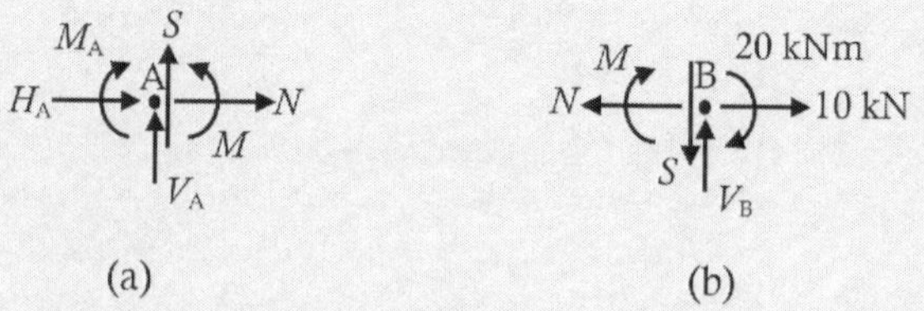

(a)   (b)

Figure 5.17 Free-body diagrams of beam support nodes. (a) Free-body diagram of support A. (b) Free-body diagram of support B.

The boundary conditions to be used for the determination of the deflection and rotation are as follows:

$- v = 0$ at $x = 0$, i.e. the vertical displacement is restrained at the fixed support.

$- v' = 0$ at $x = 0$, i.e. the rotation is restrained at the fixed support.

$- v = 0$ at $x = 10$ m, i.e. the vertical displacement is restrained at the roller support.

$- v'' = -20/EI$ at $x = 10$ m, i.e. rotation is unrestrained at the roller support and node B is subjected to an external moment. A negative value is introduced for the moment as the internal moment at B is $M = -20$ kNm (see Figure 5.17b).

(2) Determine the constants of integration and the expressions for deflection, rotation and axial displacement

The uniformly distributed load can be replaced by its vertical and horizontal components (on the basis of the inclination specified in Figure 5.16) with intensity:

$$w = 8 \text{ kN/m} \quad \text{and} \quad n = -6 \text{ kN/m}$$

The expressions related to both deflection and axial displacement present in the two differential equations (Equations 5.41a and b) are integrated considering the applied loads (which are constant over the member length) and can be written as (Equations 5.42a through e and 5.43a through c):

$$v^{IV} = \frac{8}{EI}; \quad v''' = \frac{8}{EI}x + C_{v1}; \quad v'' = \frac{4}{EI}x^2 + C_{v1}x + C_{v2}$$

$$v' = \frac{4}{3EI}x^3 + C_{v1}\frac{x^2}{2} + C_{v2}x + C_{v3}; \quad v = \frac{1}{3EI}x^4 + C_{v1}\frac{x^3}{6} + C_{v2}\frac{x^2}{2} + C_{v3}x + C_{v4}$$

$$u'' = \frac{6}{EA}; \quad u' = \frac{6}{EA}x + C_{u1}; \quad u = \frac{3}{EA}x^2 + C_{u1}x + C_{u2}$$

We first apply the boundary conditions for the vertical displacement v:

$$v(x = 0) = \frac{1}{3EI}0^4 + C_{v1}\frac{0^3}{6} + C_{v2}\frac{0^2}{2} + C_{v3} \times 0 + C_{v4} = 0 \quad \therefore C_{v4} = 0$$

$$v'(x = 0) = \frac{4}{3EI}0^3 + C_{v1}\frac{0^2}{2} + C_{v2} \times 0 + C_{v3} = 0 \quad \therefore C_{v3} = 0$$

$$v(x=10) = \frac{1}{3EI}10^4 + C_{v1}\frac{10^3}{6} + C_{v2}\frac{10^2}{2} + C_{v3}\times 10 + C_{v4} = 0$$

$$v''(x=10) = \frac{4}{EI}10^2 + C_{v1}\times 10 + C_{v2} = -\frac{20}{EI}$$

and with $C_{v3} = C_{v4} = 0$, solving these last two equations simultaneously gives:

$$C_{v1} = -\frac{53}{EI} \quad \text{and} \quad C_{v2} = \frac{110}{EI}$$

The deflection v can now be written as: $v = \frac{1}{3EI}x^4 - \frac{53}{6EI}x^3 + \frac{55}{EI}x^2$.

The constants of integration related to u can be obtained as follows:

$$u(x=0) = \frac{3}{EA}0^2 + C_{u1}\times 0 + C_{u2} = 0 \quad \therefore C_{u2} = 0$$

$$u'(x=10) = \frac{6}{EA}10 + C_{u1} = \frac{10}{EA} \quad \therefore C_{u1} = -\frac{50}{EA}$$

The expression for u becomes: $u = \frac{3}{EA}x^2 - \frac{50}{EA}x$.

(3) Calculate reactions

Free-body diagrams of the support points A and B are shown in Figure 5.17. The reactions are calculated by ensuring these free-body diagrams are in equilibrium.

By considering equilibrium at support A (Figures 5.14, 5.15 and 5.17a), it is possible to determine H_A, V_A and M_A:

$$H_A = -N(x=0) = -EAu'(x=0) = -EA\left(-\frac{50}{EA}\right) = 50 \text{ kN}$$

$$V_A = -S(x=0) = EIv'''(x=0) = EI\left(-\frac{53}{EI}\right) = -53 \text{ kN}$$

$$M_A = M(x=0) = EIv''(x=0) = EI\frac{110}{EI} = 110 \text{ kNm}$$

Likewise, the reaction at B (i.e. V_B) is obtained from equilibrium considerations applied at node B (in Figures 5.15e and 5.17b):

$$V_B = S(x=10) = -EIv'''(x=10) = -EI\left(\frac{8}{EI}10 - \frac{53}{EI}\right) = -27 \text{ kN}$$

WORKED EXAMPLE 5.5

Consider the fixed-ended beam shown in Figure 5.18 subjected to a uniformly distributed vertical load w. Determine the expressions for the deflection and axial displacement. Assume that the cross-sectional area A and second moment of area I remain constant throughout the member and the material is linear–elastic with elastic modulus E.

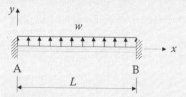

Figure 5.18 Beam for Worked Example 5.5.

---

(1) Identify boundary conditions
As both ends of the beam are fixed, all displacements at each support are restrained:

$$u(x = 0) = u(x = L) = 0; \quad v(x = 0) = v(x = L) = 0; \quad v'(x = 0) = v'(x = L) = 0$$

(2) Determine constants of integration and expressions for deflection, rotation, and axial displacement
Equations 5.41 are integrated on the basis of the applied load of the problem:

$$v^{IV} = \frac{w}{EI}; \quad v''' = \frac{w}{EI}x + C_{v1}; \quad v'' = \frac{w}{2EI}x^2 + C_{v1}x + C_{v2};$$

$$v' = \frac{w}{6EI}x^3 + C_{v1}\frac{x^2}{2} + C_{v2}x + C_{v3}; \quad v = \frac{w}{24EI}x^4 + C_{v1}\frac{x^3}{6} + C_{v2}\frac{x^2}{2} + C_{v3}x + C_{v4};$$

$$u'' = 0; \quad u' = C_{u1}; \quad u = C_{u1}x + C_{u2}$$

Applying the boundary conditions identified at point I:

$$v(x = 0) = 0 \qquad \therefore C_{v4} = 0$$

$$v'(x = 0) = 0 \qquad \therefore C_{v3} = 0$$

$$v(x = L) = \frac{w}{24EI}L^4 + C_{v1}\frac{L^3}{6} + C_{v2}\frac{L^2}{2} = 0$$

$$v'(x = L) = \frac{w}{6EI}L^3 + C_{v1}\frac{L^2}{2} + C_{v2}L = 0$$

and solving these two simultaneous equations gives $C_{v1} = -\dfrac{wL}{2EI}$ and $C_{v2} = \dfrac{wL^2}{12EI}$.

$u(x = 0) = 0$ $\qquad\qquad$ $\therefore C_{u2} = 0$

$u(x = L) = C_{u1}L + C_{u2} = 0$ $\qquad$ $\therefore C_{u1} = 0$

The expressions for u and v are therefore:

$$u = 0 \quad \text{and} \quad v = \frac{w}{24EI}x^4 - \frac{wL}{2EI}\frac{x^3}{6} + \frac{wL^2}{12EI}\frac{x^2}{2}$$

(3) Calculate reactions

The free-body diagrams of the two end nodes (supports) of the beam are shown in Figure 5.19. We recall that the positive direction of each reaction, unlike the case for the internal actions, is arbitrary as the actual sign of the calculated value determines the direction.

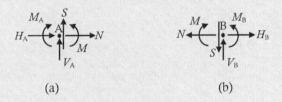

(a) $\qquad\qquad\qquad\qquad\qquad$ (b)

Figure 5.19 Free-body diagrams of beam support nodes. (a) Support A. (b) Support B.

Invoking the equilibrium equations for the free-body diagrams of Figure 5.19, we get:

$$H_A = -N(x = 0) = -EAu'(x = 0) = -EA \times 0 = 0$$

$$V_A = -S(x = 0) = EIv'''(x = 0) = EI\left(-\frac{wL}{2EI}\right) = -\frac{wL}{2}$$

$$M_A = M(x = 0) = EIv''(x = 0) = EI\frac{wL^2}{12IE} = \frac{wL^2}{12}$$

$$H_B = N(x = L) = EAu'(x = L) = EA \times 0 = 0$$

$$V_B = S(x = L) = -EIv'''(x = L) = -EI\frac{wL}{2EI} = -\frac{wL}{2}$$

$$R_B = M(x = L) = EIv''(x = 0) = EI\frac{wL^2}{12EI} = \frac{wL^2}{12}$$

The calculated reactions are plotted in Figure 5.20.

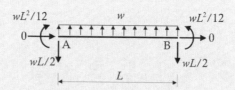

Figure 5.20 Free-body diagram for Worked Example 5.5.

WORKED EXAMPLE 5.6

Consider the unloaded fixed-ended beam shown in Figure 5.21. Determine the expressions for deflection and axial displacement for the following three cases:
(i) unit displacement of A in the positive direction of x
(ii) unit displacement of A in the positive direction of y
(iii) unit anti-clockwise rotation at A

For each case, calculate the support reactions induced by the enforced displacement. Assume constant cross-sectional and material properties with area A, second moment of area I, and elastic modulus E.

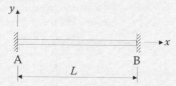

Figure 5.21 Unloaded fixed-ended beam for Worked Example 5.6.

The three induced displacements are considered separately in the following.

(i) Unit displacement at A in the positive direction of x
(1) Boundary conditions:

$$u(x = 0) = 1 \quad u(x = L) = 0$$

$$v(x = 0) = v(x = L) = 0$$

$$v'(x = 0) = v'(x = L) = 0$$

(2) Determine constants of integration and expressions for displacements:

$$v^{IV} = 0; \quad v''' = C_{v1}; \quad v'' = C_{v1}x + C_{v2}$$

$$v' = C_{v1}\frac{x^2}{2} + C_{v2}x + C_{v3}; \quad v = C_{v1}\frac{x^3}{6} + C_{v2}\frac{x^2}{2} + C_{v3}x + C_{v4}$$

$$u'' = 0; \quad u' = C_{u1}; \quad u = C_{u1}x + C_{u2}$$

Applying relevant boundary conditions:

$$v(x = 0) = 0 \qquad\qquad \therefore C_{v4} = 0$$

$$v'(x = 0) = 0 \qquad\qquad \therefore C_{v3} = 0$$

$$v(x = L) = C_{v1}\frac{L^3}{6} + C_{v2}\frac{L^2}{2} = 0$$

$$v'(x = L) = C_{v1}\frac{L^2}{2} + C_{v2}L = 0 \quad \therefore C_{v1} = 0, \quad C_{v2} = 0$$

$$u(x = 0) = 1 \qquad\qquad \therefore C_{u2} = 1$$

$$u(x = L) = C_{u1}L + C_{u2} = 0 \qquad \therefore C_{u1} = -\frac{1}{L}$$

The expressions for u and v become: $u = -\dfrac{1}{L}x + 1$ and $v = 0$.

(3) The reactions are calculated from equilibrium at the two end nodes (Figure 5.22a and b) and are summarised in Figure 5.22c (remember that the direction of the reaction vector is arbitrary).

$$H_A = -N(x = 0) = -EAu'(x = 0) = \frac{EA}{L}; \quad V_A = -S(x = 0) = EIv'''(x = 0) = 0$$

$$M_A = -M(x = 0) = -EIv''(x = 0) = 0; \quad H_B = N(x = L) = EAu'(x = L) = -\frac{EA}{L}$$

$$V_B = S(x = L) = -EIv'''(x = L) = 0; \quad M_B = M(x = L) = EIv''(x = 0) = 0$$

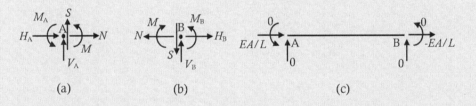

(a)   (b)   (c)

Figure 5.22 Free-body diagrams for case i.

(ii) Unit displacement at A in the positive direction of y
(1) Boundary conditions:

$$u(x = 0) = 0 \qquad u(x = L) = 0$$

$$v(x = 0) = 1 \qquad v(x = L) = 0$$

$$v'(x = 0) = 0 \qquad v'(x = L) = 0$$

(2) Determine constants of integration and expressions for displacements:

$$v(x = 0) = 1 \qquad\qquad \therefore C_{v4} = 1$$

$$v'(x = 0) = 0 \qquad\qquad \therefore C_{v3} = 0$$

$$v(x = L) = C_{v1}\frac{L^3}{6} + C_{v2}\frac{L^2}{2} = 0$$

$$v'(x = L) = C_{v1}\frac{L^2}{2} + C_{v2}L = 0 \qquad \therefore C_{v1} = \frac{12}{L^3}, \quad C_{v2} = -\frac{6}{L^2}$$

$$u(x = 0) = 0 \qquad\qquad \therefore C_{u2} = 0$$

$$u(x = L) = C_{u1}L + C_{u2} = 0 \qquad \therefore C_{u1} = 0$$

The expressions for u and v become: $\quad u = 0$ and $v = \dfrac{2}{L^3}x^3 - \dfrac{3}{L^2}x^2 + 1$.

(3) The reactions are calculated from equilibrium considerations (see Figure 5.22a and b) and are summarised in Figure 5.23a.

$$H_A = -N(x = 0) = -EAu'(x = 0) = 0; \quad V_A = -S(x = 0) = EIv'''(x = 0) = \frac{12EI}{L^3}$$

$$M_A = -M(x = 0) = -EIv''(x = 0) = \frac{6EI}{L^2}; \quad H_B = N(x = L) = EAu'(x = L) = 0$$

$$V_B = S(x = L) = -EIv'''(x = L) = -\frac{12EI}{L^3}; \quad M_B = M(x = L) = EIv''(x = 0) = \frac{6EI}{L^2}$$

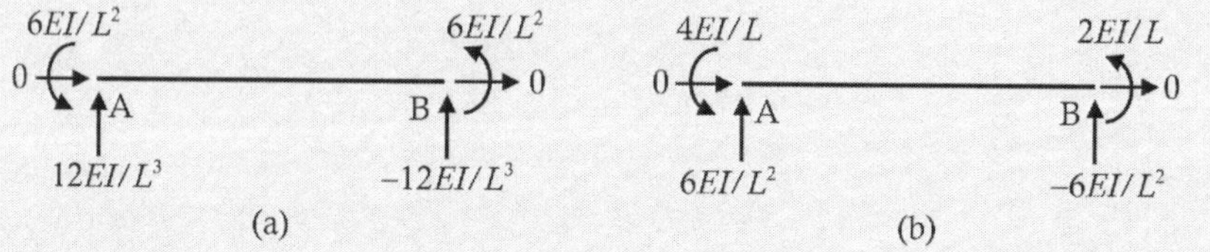

Figure 5.23 Free-body diagrams. (a) Case ii. (b) Case iii.

(iii) Unit anti-clockwise rotation at A

(1) Boundary conditions:

$$u(x = 0) = 0; \qquad u(x = L) = 0$$

$$v(x = 0) = 0; \qquad v(x = L) = 0$$

$$v'(x = 0) = 1; \qquad v'(x = L) = 0$$

(2) Determine constants of integration and expressions for displacements:

$$v(x = 0) \quad \therefore C_{v4} = 0; \qquad v'(x = 0) = 1 \quad \therefore C_{v3} = 1$$

$$v(x = L) = C_{v1}\frac{L^3}{6} + C_{v2}\frac{L^2}{2} + L = 0$$

$$v'(x = L) = C_{v1}\frac{L^2}{2} + C_{v2}L + 1 = 0 \quad \therefore C_{v1} = \frac{6}{L^2}, \quad C_{v2} = -\frac{4}{L}$$

$$u(x = 0) = 0 \quad \therefore C_{u2} = 0 \qquad u(x = L) = C_{u1}L + C_{u2} = 0 \quad \therefore C_{u1} = 0$$

The expressions for u and v become: $u = 0$ and $v = \frac{1}{L^2}x^3 - \frac{2}{L}x^2 + x$.

(3) The reactions are calculated from equilibrium considerations (see Figure 5.22a and b) and are summarised in Figure 5.23b.

$$H_A = -N(x = 0) = -EAu'(x = 0) = 0 \qquad V_A = -S(x = 0) = EIv'''(x = 0) = \frac{6EI}{L^2}$$

$$M_A = -M(x = 0) = -EIv''(x = 0) = \frac{4EI}{L} \qquad H_B = N(x = L) = EAu'(x = L) = 0$$

$$V_B = S(x = L) = -EIv'''(x = L) = -\frac{6EI}{L^2} \qquad M_B = M(x = L) = EIv''(x = 0) = \frac{2EI}{L}$$

5.7 RELATIONSHIP BETWEEN BENDING MOMENT, SHEAR FORCE AND MEMBER LOADING

In Chapter 3, we saw that shear force and bending moment diagrams could be readily determined for any segment of a beam or frame by cutting the segment and considering equilibrium of the free-body diagram on either side of the cut. At the end of Chapter 3, we made a number of observations regarding the shape of the shear force and bending moment diagrams and the relationship between them for different load configurations. We will now revisit the dependency between the member load w, the shear force S and the moment M relying on a number of equations derived in this chapter and reproduced here for ease of reference:

$$S' = -w \tag{5.2b}$$

$$M'' = w \tag{5.40}$$

$$S = -M' \tag{5.39c}$$

The main aspects relating w, S and M will be highlighted in the following by means of simple examples to keep the complexity of the analysis to a minimum. For this purpose, the shear force and bending moment diagrams associated with a number of loading conditions applied to a cantilever beam are shown in Figures 5.24 and 5.25.

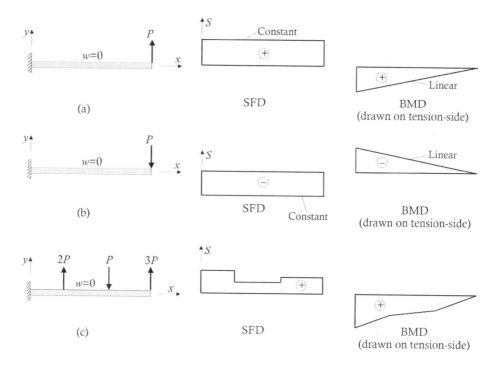

Figure 5.24 Shear force and bending moment diagrams due to point loads.

In the case of a cantilever with a transverse load P applied at its tip, the shear force diagram is constant and the bending moment diagram is linear, as illustrated in Figure 5.24a and b. These results reflect the fact that, when the beam is not subjected to a member load w (i.e. $w = 0$), the shear force needs to be constant because its slope is zero (according to Equation 5.2b where $S' = -w = 0$). In a similar manner, the distribution of the moment is linear as its first derivative equals the constant $-S$ (from Equation 5.39c). This case is shown for both an upward load (Figure 5.24a) and a downward load (Figure 5.24b) to highlight how the load direction affects the signs of the shear force and moment for the support and loading conditions considered. The bending moment diagrams are plotted on the tension side of the member.

In Figure 5.24c, a similar beam subjected to three transverse loads applied along the member length is considered. This example is useful to show that, where a concentrated load is applied transverse to the axis of the beam, a step occurs in the shear force diagram equal in magnitude to the concentrated load. Between points of load application, the shear force is constant and the bending moment diagram is linear, with its slope varying according to the intensity of the shear force. In fact, at the points where concentrated transverse loads are applied to a beam, the bending moment diagram will change direction and the diagram will kink at these points.

The case of a uniformly distributed load is considered in Figures 5.25a and b, with the member load being positive in Figure 5.25a and negative in Figure 5.25b. For a uniformly distributed loading (i.e. $w = $ constant), the shear force diagram is linear (Equation 5.2b) and the bending moment diagram is parabolic (Equations 5.39c and 5.40).

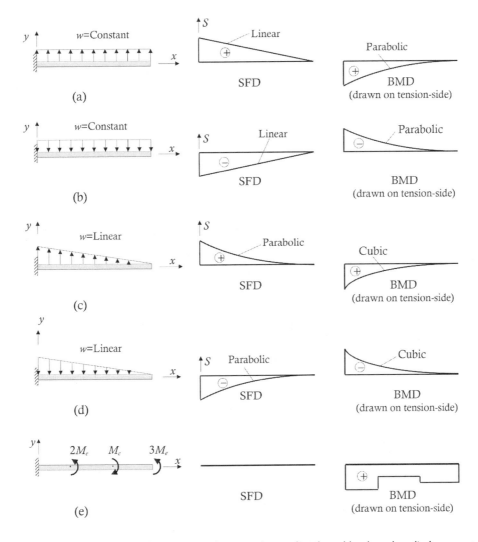

Figure 5.25 Shear force and bending moment diagrams due to distributed loads and applied moments.

A linearly varying transverse load leads to a parabolic distribution of the shear force and to a cubic variation for the bending moment diagram, as shown in Figures 5.25c and d. It is noted that the change in shear from one point to another along the beam is numerically equal to the total load between the two points (as also pointed out in Equation 5.7a).

When the external couples are applied to the beam, a step occurs in the bending moment diagram at their point of application, with the step being equal in magnitude to the applied couple (as shown in Figure 5.25e).

Recalling that the maximum or the minimum of a function occurs when its first derivative is equal to zero, the points on the shear force diagram where the shear force is zero correspond to the points where the bending moment is either a maximum or a minimum (based on Equation 5.39c). At these locations, the slope of the bending moment diagram is zero. This can be easily observed in Figure 5.26 for the simply-supported beam subjected

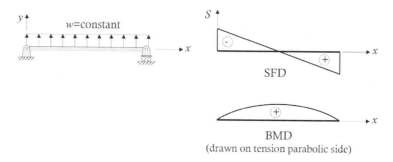

SFD

BMD
(drawn on tension parabolic side)

Figure 5.26 Shear force and bending moment diagrams for a simply-supported beam subjected to a uniformly distributed load.

to a uniformly distributed load, with the maximum moment at mid-span where the shear force is zero.

The relationships between load, shear force, bending moment, rotation and deflection can be written sequentially in terms of the operations of integration or differentiation. These have been summarised for ease of reference in Table 5.1.

Table 5.1 Relationship between member load w, shear force S, bending moment M, rotation θ and deflection v

Variable	By integration (constants of integration omitted for clarity)	By differentiation
Applied load w	w	$w = \dfrac{d^2}{dx^2}\left(EI\dfrac{d^2v}{dx^2}\right)$ or $w = EI\dfrac{d^4v}{dx^4}$ (when EI is constant)
Internal shear force S	$S = -\displaystyle\int w\,dx$	$S = -\dfrac{d}{dx}\left(EI\dfrac{d^2v}{dx^2}\right)$ or $S = -EI\dfrac{d^3v}{dx^3}$ (when EI is constant)
Internal moment M	$M = -\displaystyle\int S\,dx$ or $M = \displaystyle\int\int w\,dx\,dx$	$M = EI\dfrac{d^2v}{dx^2}$
Rotation θ	$\theta = \displaystyle\int \dfrac{M}{EI}\,dx$ or $\theta = \displaystyle\int\int\int \dfrac{w}{EI}\,dx\,dx\,dx$	$\theta = \dfrac{dv}{dx}$
Deflection v	$v = \displaystyle\int \theta\,dx$ or $v = \displaystyle\int\int\int\int \dfrac{w}{EI}\,dx\,dx\,dx\,dx$	v

WORKED EXAMPLE 5.7

Reconsider the beams of Worked Example 5.1 and verify that the expressions specified in Equations 5.2b, 5.39c and 5.40 relating the member load w, shear force S and bending moment M are satisfied. Both Beams 1 and 2 in Worked Example 5.1 are subjected to a linearly varying distributed load of $w = -6x$. The expressions for S and M previously obtained for the two beams are as follows:

Beam 1: $S = 3x^2$ and $M = -x^3$ Beam 2: $S = 3x^2 - 5.4$ and $M = -x^3 + 5.4x$

Beam 1

The first derivative of the shear force is $S' = 6x$. Considering that the applied load is $w = -6x$, Equation 5.2b ($S' = -w$) is satisfied.
The first derivative of the moment is $M' = -3x^2$, which is equal and opposite to the shear S as specified by Equation 5.39c ($S = -M'$).
Differentiating the equation for M twice with respect to x leads to $M'' = -6x$, which is identical to the load w as required by Equation 5.40.

Beam 2

Similar conclusions are obtained for the results of Beam 2:
$S' = 6x$ which satisfies Equation 5.2b ($S' = -w$).
$M' = -3x^2 + 5.4$ is equal and opposite to S ($= 3x^2 - 5.4$) as required by Equation 5.39c.
$M'' = -6x$ equals w as specified in Equation 5.40.

REFLECTION ACTIVITY 5.1

For the beams in Figure 5.27, determine the locations where the shear force, bending moment, and axial force have a maximum or a minimum.

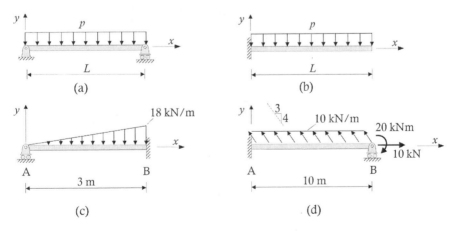

Figure 5.27 Beams for Reflection Activity 5.1. (a) Simply-supported beam. (b) Cantilever beam. (c) Beam 2 from Worked Example 5.1. (d) Beam from Worked Example 5.4.

(a) Simply-supported beam: For the simply-supported beam of Figure 5.27a, the expressions for S, M and N are obtained from statics (see Chapter 3):

$$S = px - \frac{pL}{2}; \quad M = -\frac{p}{2}x^2 + \frac{pL}{2}x; \quad N = 0$$

The uniformly distributed load is defined by $w = -p$, as w is assumed to be positive in the positive direction of y.

The locations of maximum or minimum values for a function are found at the positions where the first derivative of the function is zero. Based on this, the location of the minimum or maximum shear force is found where $S' = 0$. In this case: $S' = p$, which means that the slope of the shear force diagram is constant throughout the length of the beam. This should have been expected as the function for S is linear and, because of this, it does not possess a minimum or maximum. In these situations, the maximum and minimum are at the boundary of our domain of interest, i.e. at $x = 0$ and $x = L$. In particular, $S(x = 0) = -pL/2$ and $S(x = L) = pL/2$ represent the minimum and maximum for the shear force in our problem (assuming p to be positive).

The maximum or minimum for M is found by solving $M' = -px + pL/2 = 0$, which is satisfied when $x = L/2$, i.e. at mid-span. Substituting this back into the expression for M, we obtain: $M(x = L/2) = pL^2/8$. Because the value for $M(x = L/2)$ is positive, the identified point represents a maximum point. At the two supports: $M(x = 0) = M(x = L) = 0$. Based on these results, M has a minimum at the supports, i.e. at $x = 0$ and $x = L$, and a maximum at $x = L/2$.

The axial force N is zero everywhere and, therefore, has no minimum or maximum values.

(b) Cantilever beam: The functions for S, M and N of the beam shown in Figure 5.27b are as follows: $S = px - pL; M = -\frac{p}{2}x^2 + pLx - \frac{pL^2}{2}; N = 0.$

As in the previous case, the diagram of S is represented by a straight line and the maximum and minimum occur at the boundaries of our domain: $S(x = 0) = -pL$ and $S(x = L) = 0$.

The first derivative of the moment is zero ($M' = -px + pL = 0$) for $x = L$. At this location, the value of the moment is zero, i.e. $M(x = L) = 0$. We know that the moment at the fixed support is $-pL^2/2$, which is certainly the moment with the largest magnitude along the beam length and the value that an engineer would use to size the member. This location was not detected by equating the first derivative of M to zero because M' considers all values for x (from −infinity to +infinity). In this case, the expression for M represents a parabola, as shown in Figure 5.28, which has zero slope at $x = L$. The

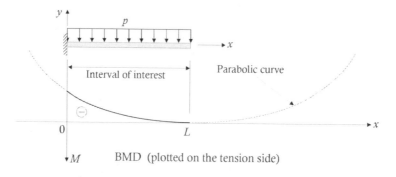

Figure 5.28 Bending moment diagram for a uniformly loaded cantilever beam.

use of the derivative is suitable to detect the presence of a maximum or minimum within a specified interval (say between $x = 0$ and $x = L$, i.e. $0 < x < L$), but it does not provide information related to its boundaries. The possibility of evaluating whether a minimum or maximum is located at the boundaries must be checked separately by simply calculating the values of M at these two locations and comparing these values with the values at the locations where the first derivative becomes zero (if any). For the particular case under consideration, no maximum or minimum were found within the internal span (i.e. in the interval $0 < x < L$, M' did not equal zero). The actual points of maximum or minimum for the moment M occurred at the two member ends (assuming p positive):
– minimum value of $-pL^2/2$ at $x = 0$
– maximum value of 0 at $x = L$
The axial force N is zero everywhere and, as such, has no maximum or minimum.

(c) Beam 2 from Worked Example 5.1: The expressions for S, M and N for the beam depicted in Figure 5.27c are $S = 3x^2 - 5.4$; $M = -x^3 + 5.4x$; and $N = 0$.

The location at which $S'(= 6x) = 0$ is $x = 0$. Considering that the first derivative did not identify a minimum or maximum along the beam (i.e. within the interval between $x = 0$ and $x = L$), we now calculate S at these locations: $S(x = 0) = -5.4$ kN and $S(x = 3) = 21.6$ kN. Based on this, S has a minimum at $x = 0$ and a maximum at $x = 3$.

The first derivative for M (i.e. $M' = -3x^2 + 5.4$) is zero at $x = \pm1.34$ m. The solution $x = -1.34$ m is disregarded because it is outside the interval of interest. The value for M at $x = 1.34$ m is 4.83 kNm. We still need to check values for M at $x = 0$ and $x = L$, which are $M(x = 0) = 0$ and $M(x = 3) = -10.8$ kNm. Based on this, M has a minimum at $x = 3$ m (equal to -10.8 kNm) and a maximum point at $x = 1.34$ m (equal to 4.83 kNm).

The axial force N is zero everywhere and, as such, has no minimum or maximum values.

(d) Beam from Worked Example 5.4: The functions for S, M and N are obtained for the beam shown in Figure 5.27d from the expressions derived in Worked Example 5.4 for the axial displacement and deflection. Substituting these into Equation 5.39a ($N = AEu'$), Equation 5.39b ($M = EIv''$) and Equation 5.39c ($S = -IEv'''$) produces:

$$S = -EI\left(\frac{8}{EI}x - \frac{53}{EI}\right) = -8x + 53$$

$$M = EI\left(\frac{4}{EI}x^2 - \frac{53}{EI}x + \frac{110}{EI}\right) = 4x^2 - 53x + 110$$

$$N = EA\left(\frac{6}{EA}x - \frac{50}{EA}\right) = 6x - 50$$

There is no minimum or maximum for S, as S is linear. Because of this, its maximum and minimum occur at the supports: $S(x = 0) = 53$ kN and $S(x = 10) = -27$ kN.

With regard to the moment, $M' = -S = 8x - 53 = 0$ at $x = 6.625$ m. The values for M calculated at $x = 0$, 6.625 and 10 m are $M(x = 0) = +110$ kNm, $M(x = 6.625) = -65.56$ kNm and $M(x = 10) = -20$ kNm. We can conclude that M has a minimum at $x = 6.625$ m (equal to -65.56 kNm) and a maximum at $x = 0$ m (equal to 110 kNm).

As the expression for N is linear, the maximum and minimum values occur at the supports: $N(x = 0) = -50$ kN and $N(x = 10) = 10$ kN.

REFLECTION ACTIVITY 5.2

Consider a simply-supported beam with a cross-section of height h and symmetric with respect to the y-axis, as shown in Figure 5.29a. This cross-section is subjected to a change in temperature described by the linearly varying distribution of Figure 5.29b. This is defined in terms of temperature changes taking place at the top and bottom fibres of the cross-section, referred to as ΔT_1 and ΔT_2, respectively. The thermal expansion induced by a change in temperature ΔT is represented by a thermal strain ε_T calculated as $\varepsilon_T = \alpha \Delta T$, where α is the coefficient of thermal expansion of the material. On the basis of this relationship, the linearly varying temperature distribution of Figure 5.29b produces the linearly varying thermal strain distribution shown in Figure 5.29c. This distribution is defined by the two variables $\varepsilon_{T,r}$ and κ_T, where $\varepsilon_{T,r}$ is the thermal strain at the level of the reference axis and κ_T is the slope of the thermal strain diagram, respectively.

The linear temperature variation causes deformation of every cross-section of the beam and, if the beam is unrestrained (i.e. free to elongate and free to deflect and rotate at its supports), the deformation of the beam will change but there will be no change in the internal actions. If this is the case, the total strain at any point on a cross-section ε will equal the stress-dependent strain (caused by the internal actions) plus the thermal strain. If the material behaviour is linear–elastic, with elastic modulus E, the total strain is therefore given by $\varepsilon = \sigma/E + \varepsilon_T$ and the stress σ is $\sigma = E(\varepsilon - \varepsilon_T)$.

If the thermal change is constant along the length of the beam, explain how you would account for these thermal effects using (i) the method of double integration and (ii) the governing system of equations expressed in terms of the displacements.

A simply-supported beam is free to elongate and free to rotate at its supports, so the deformation caused by the uniform temperature change will not be restrained and the internal actions on each cross-section will not change. In this case, the thermal effects influence only the constitutive equations (Equations 5.21 through 5.24), while the equilibrium (Equations 5.2, 5.4 and 5.6) and kinematic equations (Equations 5.8 and 5.14) remain unchanged. The case of varying temperature distributions over the beam length could also be dealt with using the following procedure with minor modifications.

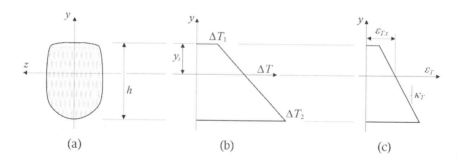

(a)   (b)   (c)

Figure 5.29 Temperature distribution and strain through a cross-section. (a) Cross-section. (b) Temperature variation. (c) Thermal strain ε_T.

The thermal strain (i.e. strain that would occur in the material if free to deform) is given by:

$$\varepsilon_T = \alpha \Delta T \tag{5.44}$$

where ΔT depicts the temperature change at a particular level of the cross-section. For ease of reference, Equation 5.44 can be expressed in terms of $\varepsilon_{T,r}$ and κ_T, as shown in Figure 5.29c:

$$\varepsilon_T = \varepsilon_{T,r} - y\kappa_T \tag{5.45}$$

in which:

$$\varepsilon_{T,r} = \left[\Delta T_1 + (\Delta T_2 - \Delta T_1)\frac{y_t}{h} \right] \qquad \kappa_T = \alpha\frac{\Delta T_2 - \Delta T_1}{h} \tag{5.46a,b}$$

Based on this, the stress–strain relationship of the linear–elastic material accounting for thermal effects can be written as:

$$\sigma = E\varepsilon - E\varepsilon_T = E(\varepsilon_r - y\kappa) - E(\varepsilon_{T,r} - y\kappa_T) \tag{5.47}$$

Substituting Equation 5.47 into the integrals defining the internal axial force N (Equation 5.21) and moment M (Equation 5.22) leads to:

$$N = \int_A \sigma\, dA = \int_A \left[E(\varepsilon_r - y\kappa) - E(\varepsilon_{T,r} - y\kappa_T) \right] dA =$$
$$= EA\varepsilon_r - EB\kappa - EA\varepsilon_{T,r} + EB\kappa_T \tag{5.48a}$$

$$M = -\int_A y\sigma\, dA = -\int_A y\left[E(\varepsilon_r - y\kappa) - E(\varepsilon_{T,r} - y\kappa_T) \right] dA =$$
$$= -EB\varepsilon_r + EI\kappa + EB\varepsilon_{T,r} - EI\kappa_T \tag{5.48b}$$

where A is the area of the cross-section, and B and I are the first and second moments of the cross-sectional area about the reference axis. That is:

$$B = \int_A y\, dA \quad \text{and} \quad I = \int_A y^2\, dA$$

If we adopt a centroidal reference system (i.e. with origin in the centroid of the cross-section), the term B is zero, and the revised relationships between the cross-sectional resultants (N and M) and cross-sectional deformations (ε_r and κ) are:

$$N = EA\varepsilon_r - EA\varepsilon_{T,r} \quad \text{or} \quad \varepsilon_r = \frac{N}{EA} + \varepsilon_{T,r} \tag{5.49a,b}$$

$$M = EI\kappa - EI\kappa_T \quad \text{or} \quad \kappa = \frac{M}{EI} + \kappa_T \tag{5.50a,b}$$

The inclusion of these thermal effects in the method of double integration and the system of differential equations in terms of displacements is now considered.

(i) Method of double integration

The expression for the curvature is obtained by substituting Equation 5.50b ($\kappa = M/EI + \kappa_T$) into Equation 5.14 ($\kappa = v''$):

$$v''\kappa = \frac{M}{EI} + \kappa_T \tag{5.51a}$$

which can be integrated once and twice to obtain the functions describing the rotation and deflection, respectively, as:

$$\theta = \int \kappa\, dx + C_{v1} = \int \left(\frac{M}{EI} + \kappa_T\right) dx + C_{v1} \tag{5.51b}$$

$$v = \int\int \kappa\, dx\, dx + C_{v1}x + C_{v2} = \int\int \left(\frac{M}{EI} + \kappa_T\right) dx\, dx + C_{v1}x + C_{v2} \tag{5.51c}$$

(ii) System of differential equations written in terms of displacements

The system of differential equations remains unchanged from that outlined in Section 5.6 and specified in Equations 5.41. The only difference is related to the boundary conditions associated with the axial displacement and deflection to be used in the solution process and these are illustrated in Figures 5.30 and 5.31, respectively. The relevant boundary conditions shown in each figure are obtained from the following relationships between displacements and cross-sectional resultants:

$$N = EAu' - EA\varepsilon_{T,r} \quad \text{or} \quad u' = \frac{N}{EA} + \varepsilon_{T,r} \tag{5.52a,b}$$

$$M = EIv'' - EI\kappa_T \quad \text{or} \quad v'' = \frac{M}{EI} + \kappa_T \tag{5.53a,b}$$

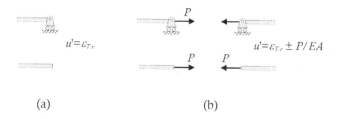

(a)   (b)

Figure 5.30 Boundary conditions for the axial displacement. (a) Roller support or free edge. (b) Roller support or free edge with external axial force.

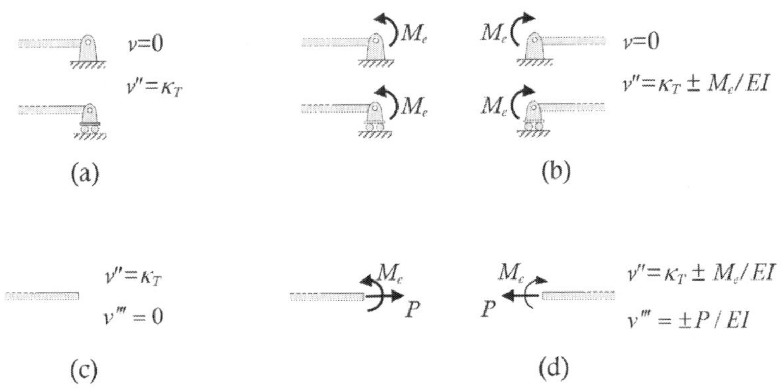

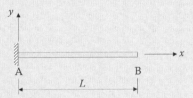

Figure 5.31 Boundary conditions for the vertical displacement. (a) Pinned or roller support. (b) Pinned or roller support with an external applied moment. (c) Free edge. (d) Free edge with external applied moment and force.

WORKED EXAMPLE 5.8

If the beam shown in Figure 5.32 is subjected to the temperature change described in Figures 5.29b and c and in Equations 5.45 and 5.46 over its entire length, calculate the deflected shape using:
(i) the method of double integration
(ii) the system of differential equations in terms of displacements

Use the expressions derived in Reflection Activity 5.2 in your solution. Assume the beam is uniform throughout, with area A, second moment of area I, elastic modulus E and coefficient of thermal expansion α.

Figure 5.32 Beam for Worked Example 5.8.

(i) Method of double integration

The expression for the internal moment is $M = 0$ as no external loads are applied to the cantilever beam. Based on Equation 5.50b, the curvature along the length of the beam can be written as:

$$\kappa = \frac{0}{EI} + \kappa_T = \kappa_T \tag{5.54a}$$

from which the expressions for the rotation and deflection become

$$\theta = \int \kappa_T \, dx + C_{v1} = \kappa_T x + C_{v1} \tag{5.54b}$$

$$v = \int\int \kappa \, dx \, dx + C_{v1}x + C_{v2} = \frac{\kappa_T}{2}x^2 + C_{v1}x + C_{v2} \tag{5.54c}$$

Constants of integration C_{v1} and C_{v2} are obtained by applying the following boundary conditions: $v(x = 0) = 0$ and $v'(x = 0) = 0$, which lead to $C_{v1} = 0$ and $C_{v2} = 0$. Based on these, the expression for deflection becomes:

$$v = \frac{\kappa_T}{2}x^2 \tag{5.55}$$

(ii) System of differential equations expressed in terms of displacements

The six constants of integration included in the system of differential equation (Equations 5.41) are determined applying the following boundary conditions at the ends of the beam (using Equations 5.42 and 5.43) and recalling that the member is unloaded, i.e. $w = n = 0$ (see Figures 5.30 and 5.31):

$u(x = 0) = C_{u2} = 0;$ $u'(x = L) = C_{u1} = \varepsilon_{T,r}$

$v(x = 0) = C_{v4} = 0;$ $v'(x = 0) = C_{v3} = 0$

$v'''(x = L) = C_{v1}L = 0;$ $v''(x = L) = C_{v1}L + C_{v2} = \kappa_T$

from which $C_{v1} = C_{v3} = C_{v4} = C_{u2} = 0$, $C_{u1} = \varepsilon_{T,r}$ and $C_{v2} = \kappa_T$ and the expressions for u and v become:

$$u = \varepsilon_{T,r}x \qquad v = \frac{\kappa_T}{2}x^2 \tag{5.56a,b}$$

As expected, the results obtained for v from the two methods are identical, as observed by comparing Equations 5.55 and 5.56b.

WORKED EXAMPLE 5.9

Consider the fixed-ended beam illustrated in Figure 5.21 of Worked Example 5.6 and calculate the support reactions induced by temperature changes such as those described in Figure 5.29b and in Equations 5.45 and 5.46. Assume the beam is of uniform cross-section with area A, second moment of area I, elastic modulus E and coefficient of thermal expansion α.

The reactions are calculated by solving the system of differential equations (Equations 5.41) and enforcing equilibrium of the free-body diagrams of the two member end nodes.

(1) Boundary conditions: $u(x = 0) = u(x = L) = 0$

$$v(x = 0) = v(x = L) = 0 \quad v'(x = 0) = v'(x = L) = 0$$

(2) Determine constants of integration and expressions for displacements:

$$v^{IV} = 0; \quad v''' = C_{v1}; \quad v'' = C_{v1}x + C_{v2}$$

$$v' = C_{v1}\frac{x^2}{2} + C_{v2}x + C_{v3}; \quad v = C_{v1}\frac{x^3}{6} + C_{v2}\frac{x^2}{2} + C_{v3}x + C_{v4}$$

$$u'' = 0; \quad u' = C_{u1}; \quad u = C_{u1}x + C_{u2}$$

After applying the relevant boundary conditions:

$$v(x = 0) = C_{v4} = 0 \qquad \therefore C_{v4} = 0 \qquad v'(x = 0) = C_{v3} = 0 \qquad \therefore C_{v3} = 0$$

$$v(x = L) = C_{v1}\frac{L^3}{6} + C_{v2}\frac{L^2}{2} = 0 \ \therefore C_{v1} = 0 \qquad v'(x = L) = C_{v1}\frac{L^2}{2} + C_{v2}L = 0 \ \therefore C_{v2} = 0$$

$$u(x = 0) = C_{u2} = 0 \qquad \therefore C_{u2} = 0 \qquad u(x = L) = C_{u1}L + C_{u2} = 0 \quad \therefore C_{u1} = 0$$

we get the following expressions for u and v: $u = 0$ and $v = 0$.
Based on these, the variations of N and M become (Equations 5.52a and 5.53a)

$$N = -EA\varepsilon_{T,r}; \quad M = -EI\kappa_T \tag{5.57a,b}$$

(3) The reactions are calculated from equilibrium at the two end nodes (Figures 5.33a and b) and are summarised in Figure 5.33c.

$$H_A = -N(x = 0) = -(-EA\varepsilon_{T,r}) = EA\varepsilon_{T,r} \qquad V_A = -S(x = 0) = EIv'''(x = 0) = 0$$

$$M_A = -M(x = 0) = -(-EI\kappa_T) = EI\kappa_T \qquad H_B = N(x = L) = -EA\varepsilon_{T,r}$$

$$V_B = S(x = L) = -EIv'''(x = L) = 0 \qquad M_B = M(x = L) = -EI\kappa_T$$

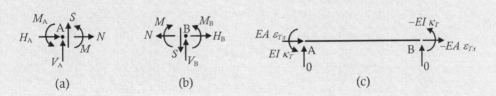

(a)   (b)   (c)

Figure 5.33 Free-body diagrams for Worked Example 5.8.

PROBLEMS

5.1   For the beam illustrated below, determine using the equilibrium, constitutive and kinematic equations, the expressions for (i) the internal actions, (ii) the strain variables, including the strain at the level of the reference axis and the curvature, and (iii) rotation, deflection and axial displacement. Plot all expressions along the member axis. The cross-section remains constant throughout the beam and is defined by area A and second moment of area I. Assume linear–elastic material properties with elastic modulus E.

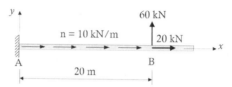

5.2   Derive the expressions for the internal actions and for the displacements (axial displacement and deflection) for the beam shown by means of the equilibrium, constitutive and kinematic equations. Plot all expressions along the member axis. The cross-section remains constant over the entire beam (with area A and second moment of area I). Assume linear–elastic material properties with elastic modulus E.

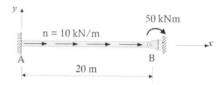

5.3   Consider the beam shown and determine the expressions for the internal actions, for the deformations (strain at the level of the reference axis and curvature) and for the displacements (axial displacement and deflection). Plot all expressions along the member axis. In your solution, use the equilibrium, constitutive and kinematic equations. Assume A, I and E are constant throughout.

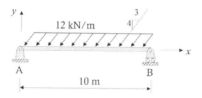

5.4   Use the equilibrium, constitutive and kinematic equations to obtain and plot the expressions for the internal actions, for the deformations (strain at the level of the reference axis and curvature) and for the displacements (axial displacement and deflection) for the beam illustrated below. Assume A, I and E are constant throughout.

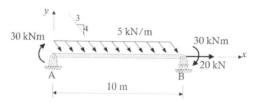

5.5 Calculate the stress distribution induced by a bending moment and axial force of 120 kNm and 90 kN, respectively. Assume the section to be I-shaped and doubly-symmetric with $A = 10,500$ mm^2, $I = 372 \times 10^6$ mm^4 and $E = 200,000$ MPa. Adopt a centroidal coordinate system with origin at mid-height of the section (of depth 460 mm).

5.6 For a cross-section with $A = 2820$ mm^2, $I = 15.3 \times 10^6$ mm^4 and $E = 200,000$ MPa, determine the magnitude of the moment to produce a stress of +180 MPa in the top fibre of the section. Consider a doubly-symmetric I-shaped section with depth of 180 mm. Adopt a centroidal coordinate system.

5.7 Reconsider the section of Problem 5.6 and recalculate the moment assuming the presence of an axial force equal to 160 kN.

5.8 For the beam illustrated below, calculate the expressions for the rotation and deflection along its length using the method of double integration. Assume its area A and second moment of area I to remain constant throughout the member. Assume the material to have elastic modulus E.

5.9 Determine the expressions for the rotation and deflection for the beam shown using the method of double integration. Assume A, I and E are constant throughout.

5.10 Consider the beam shown and evaluate the expressions for the rotation and deflection along the member length using the method of double integration. Adopt linear–elastic material properties for the member with elastic modulus E. The beam is prismatic with area A and second moment of area I.

5.11 Evaluate the expressions for the rotation and deflection for the beam shown, assuming linear–elastic material properties (elastic modulus E) and constant cross-sectional properties (area A and second moment of area I). Use the method of double integration in your solution.

5.12 Derive the expressions describing the variations for the rotation and deflection for the beam shown using the method of double integration. The cross-sectional properties of the beam are defined by area A and second moment of area I, which are taken as constant throughout its length. The material is linear–elastic with elastic modulus E.

5.13 For the beam shown, calculate the expressions for the rotation and deflection along its length using the method of double integration. Assume its area A and second moment of area I to remain constant throughout the member. Assume the material to have elastic modulus E.

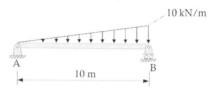

5.14 Determine the expressions for the rotation and deflection for the beam shown using the method of double integration. Assume constant A and I over the entire beam length and linear–elastic material properties with elastic modulus E.

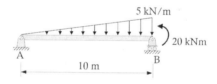

5.15 Consider the beam shown and calculate the expressions for the axial displacement and deflection solving the governing differential equations (in terms of displacements). Determine also the reactions at the supports. Assume A, I and E are constant throughout.

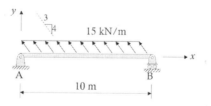

5.16 Evaluate the expressions for the axial displacement and deflection solving the governing differential equations (in terms of displacements) for the beam shown. Calculate the support reactions and assume linear–elastic material properties (elastic modulus E). Adopt constant cross-sectional properties for the beam with area A and second moment of area I.

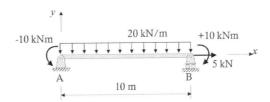

5.17 Determine the expressions describing the variations for the axial displacement and deflection for the beam shown using the differential equations expressed in terms of displacements. Calculate the support reactions. The cross-sectional properties of the beam are defined by area A and second moment of area I, which are taken as constant throughout its length. The material is linear–elastic with elastic modulus E.

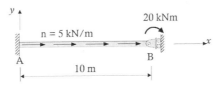

5.18 Determine the expressions for the deflection and axial displacement for the following three cases: (i) unit displacement of B in the positive direction of x, (ii) unit displacement of B in the positive direction of y and (iii) unit anti-clockwise rotation at B. For each case, calculate the support reactions induced by the enforced displacement. Assume A, I and E are constant throughout.

5.19 Use the differential equations expressed in terms of displacements to determine the expressions for u and v, and calculate the reactions at the supports for the beam depicted below. Assume A, I and E are constant throughout.

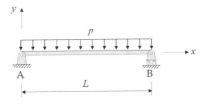

5.20 Calculate the expressions for the axial displacement and deflection solving the governing differential equations (in terms of displacements) for the beam shown below. Calculate the support reactions and assume linear–elastic material properties (elastic modulus E). Adopt constant cross-sectional properties for the beam with area A and second moment of area I.

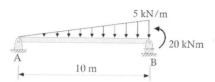

5.21 Consider the beam illustrated below and evaluate the expressions for the axial displacement and deflection solving the governing differential equations (in terms of displacements). Determine the support reactions. Adopt linear–elastic material properties (elastic modulus E) and constant cross-sectional properties for the beam with area A and second moment of area I.

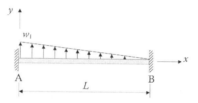

5.22 Solve the governing differential equations (in terms of displacements) for the beam shown and subjected to a parabolic distributed load (with values of 0 kN/m at the supports and 10 kN/m at mid-span). Determine the expressions for the axial displacement and deflection and the support reactions. Assume A, I and E are constant throughout.

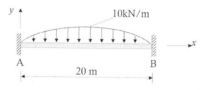

5.23 Determine the expressions for the deflection and axial displacement for the following three cases: (i) unit displacement of B in the positive direction of x, (ii) unit displacement of B in the positive direction of y and (iii) unit anti-clockwise rotation at B. For each case, calculate the support reactions induced by the enforced displacement. Assume A, I and E are constant throughout.

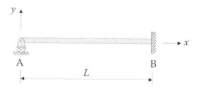

5.24 Reconsider Problem 5.3 and evaluate maximum or minimum values for the distributions of the shear force, moment and axial force, and locations along the member length where they occur.

5.25 Determine maximum or minimum values of the shear force, moment and axial force for the beam of Problem 5.16 as well as the locations where these values occur.

5.26 Calculate maximum or minimum values for the distributions of the shear force, moment and axial force. Determine locations of these values along the member length for the beam of Problem 5.17.

5.27 For the beam of Problem 5.21, evaluate maximum or minimum values for the distributions of the shear force, moment and axial force as well as the locations where these values occur.

5.28 Determine maximum or minimum values for the distributions of the shear force, moment and axial force for the beam of Problem 5.22 as well as the locations where these values occur.

5.29 By considering a small segment of a beam, derive the equilibrium equations assuming that the shear force is positive when it points downward on the right-hand side of the segment and upwards on the opposite left side (i.e. the opposite sign convention to what has been assumed elsewhere in the book). Comment on how this affects the relationships between member load, shear force and moment.

5.30 Derive the equilibrium, kinematic and constitutive equations if the y-axis is assumed to be positive downwards. Comment on how this influences the variables related to the internal actions, strain diagram and displacements.

5.31 Reconsider the beam of Problem 5.23 and determine the displacements and support reactions when subjected to a temperature distribution varying over the cross-section and constant along the member length. Assume the cross-section to have constant properties described by area A, second moment of area I, elastic modulus E and thermal expansion coefficient α.

Chapter 6

Slope-deflection methods

6.1 INTRODUCTION

In Chapter 5, we saw that when the material behaviour is linear–elastic, the stress at any point on a cross-section subjected to bending about the z-axis given by:

$$\sigma = E\varepsilon = -yE\frac{d\theta}{dx} = -\frac{My}{I} \tag{6.1}$$

and the curvature is calculated as:

$$\kappa(x) = \frac{d\theta}{dx} = \frac{d^2v(x)}{dx^2} = \frac{M(x)}{EI} \tag{6.2}$$

In addition, we saw that when the curvature can be expressed as a continuous function of x (i.e. the distance along the beam), the deflection of the beam can be obtained by twice integrating Equation 6.2:

$$v(x) = \int\int \kappa(x)\,dx\,dx = \int\int \frac{M(x)}{EI}\,dx\,dx \tag{6.3}$$

where, for each integration, a constant of integration must be introduced and evaluated from the support conditions or boundary conditions of the problem. For nonlinear material behaviour, the deflection can still be obtained by double integration of the curvature provided the variation of curvature can be expressed as a function of x. This will be illustrated in Chapter 15.

For all but the simplest of problems, the calculation of deflection by double integration of the curvature is time consuming and tedious. A number of methods, based on the same considerations of geometry, have been developed that simplify the mathematics and are useful for hand calculation. Several of these methods are presented in this chapter. Although these methods are not implemented in modern structural analysis software, they are still useful because they provide valuable insight into the behaviour of structures and form convenient tools for a quick determination of the slope and deflection of a beam or frame.

6.2 METHOD OF DOUBLE INTEGRATION WITH STEP FUNCTIONS

When the loading on a beam is not continuous or when the cross-section dimensions change abruptly, the curvature diagram is discontinuous and the expressions for curvature are different in different regions of the beam. By introducing so-called Macaulay terms or *step functions*, a single expression for the curvature at every point along a beam can be developed. A step function $<x - x_1>$ that is set to zero when the value of the expression is negative (or nil) and set to the expression itself when its value is positive:

$$<x - x_1> = \begin{cases} 0 & \text{for } x < x_1 \\ x - x_1 & \text{for } x \geq x_1 \end{cases} \tag{6.4}$$

By cutting the beam shown in Figure 6.1 at any point x from the support at A and considering the left-hand free-body diagram, the expression for the bending moment at any point along the beam can be written as

$$M = 24x - 24 <x-2> - \frac{8}{2} <x-4>^2 + 48 <x-8> \tag{6.5}$$

In Segment AB, x varies from 0 to 2 m, with $<x - 2> = <x - 4> = <x - 8> = 0$. In Segment BC, x varies from 2 to 4 m, with $<x - 2> = x - 2$ and $<x - 4> = <x - 8> = 0$. In Segment CD, x varies from 4 to 8 m, with $<x - 2> = x - 2$, $<x - 4> = x - 4$ and $<x - 8> = 0$. In Segment DE, x varies from 8 to 10 m, with $<x - 2> = x - 2$, $<x - 4> = x - 4$ and $<x - 8> = x - 8$.

To determine the deflection at any point in segment AB using the method of double integration (already introduced in Section 5.5 for beams with continuous expressions for the curvature and moment), we consider Equation 6.2 applied to segment AB and perform the double integration:

$$M_{AB} = EI \frac{d^2v}{dx^2} = 24x \tag{6.6a}$$

$$EI \frac{dv}{dx} = 12x^2 + C_1 \tag{6.6b}$$

$$EIv = 4x^3 + C_1x + C_2 \tag{6.6c}$$

We know that $v = 0$ at $x = 0$ and, therefore, from Equation 6.6c: $C_2 = 0$. The value of C_1 will be determined later.

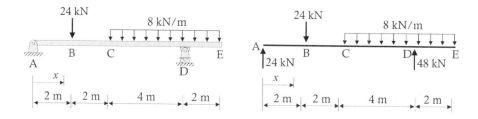

Figure 6.1 Beam layout and free-body diagram.

We next consider segment BC:

$$M_{BC} = EI \frac{d^2v}{dx^2} = 24x - 24 <x-2> \tag{6.7a}$$

$$EI \frac{dv}{dx} = 12x^2 - 12<x-2>^2 + C_3 \tag{6.7b}$$

$$EIv = 4x^3 - 4<x-2>^3 + C_3x + C_4 \tag{6.7c}$$

At point B, where $x = 2$ m, the slope from the expression for segment AB and the slope from the expression for segment BC are the same, i.e. $((dv/dx)_B)_{AB} = ((dv/dx)_B)_{BC}$, and therefore: $C_1 = C_3$. Similarly, the deflection at B (at $x = 2$ m) obtained from each of the above expressions is the same, i.e. $(v_B)_{AB} = (v_B)_{BC}$, and therefore: $C_4 = C_2 = 0$. It follows that the integration constants for each segment are identical. Therefore, when we get to segment CD, knowing that the deflection at the support D is zero, we can determine the unknown integration constant C_1. This procedure is illustrated in Worked Example 6.1.

WORKED EXAMPLE 6.1

For the beam shown in Figure 6.1, determine the deflection at points C and E using the method of double integration with step functions. Assume the beam has a uniform cross-section with $I = 50 \times 10^6$ mm⁴ and take $E = 2 \times 10^5$ MPa.

Using step functions, the expression for the moment at any point along the beam in Figure 6.1 is given by Equation 6.5 (reproduced here for convenience):

$$M = EI \frac{d^2v}{dx^2} = 24x - 24 <x-2> - \frac{8}{2} <x-4>^2 + 48 <x-8>$$

Integrating twice gives:

$$EI \frac{dv}{dx} = 12x^2 - 12 <x-2>^2 - \frac{4}{3} <x-4>^3 + 24<x-8>^2 + C_1$$

$$EIv = 4x^3 - 4 <x-2>^3 - \frac{1}{3} <x-4>^4 + 8 <x-8>^3 + C_1x + C_2 \tag{1}$$

Enforcing the boundary condition that $v = 0$ at $x = 0$ and with $<x-2> = <x-4> = <x-8> = 0$, C_2 is calculated using Equation 1 as $C_2 = 0$.
At $x = 8$, we have $v = 0$, $<x-2> = x-2$, $<x-4> = x-4$ and $<x-8> = 0$, and from Equation 1: $C_1 = -137.3$.
At Point C, where $x = 4$ m, $<x-2> = x-2$, $<x-4> = 0$ and $<x-8> = 0$, and with $EI = 10 \times 10^3$ kNm², the deflection is obtained from Equation 1:

$$v_C = \frac{1}{10 \times 10^3} (4 \times 4^3 - 4 \times (4-2)^3 - 137.3 \times 4) = -0.0325 \text{ m} = -32.5 \text{ mm}$$

which represents a downward deflection (on the basis of our sign convention).

At Point E, where $x = 10$ m, $\langle x - 2 \rangle = x - 2$, $\langle x - 4 \rangle = x - 4$, $\langle x - 8 \rangle = x - 8$ and with $EI = 10 \times 10^3$ kNm^2, the deflection is calculated from Equation 1 as:

$$v_E = \frac{1}{10 \times 10^3} \left(4 \times 10^3 - 4 \times (10 - 2)^3 - \frac{1}{3} \times (10 - 4)^4 + 8 \times (10 - 8)^3 - 137.3 \times 10 \right)$$

$$= +0.0211 \text{ m} = +21.1 \text{ mm (i.e. upward)}$$

6.3 MOMENT-AREA METHOD

The moment-area method is based on two theorems that relate the deflected shape of the beam to its curvature. They rely on the fact that the first integral of a function is the area under the graph of the function and the second integral of the function is the moment of that area about a particular reference line. In particular, the function to be integrated is the curvature whose first integral gives us information on the slope of the deflected shape. The second integral of the curvature gives us the deflection. We will consider the case where the flexural rigidity is constant throughout, i.e. where the cross-section is uniform and the material behaviour is linear–elastic. For this case, the curvature diagram has the same shape as the bending moment diagram, i.e. $\kappa = M/EI$. The method is also applicable to cases where there is a variation in the flexural rigidity of the member.

Moment-Area Theorem 1
The change in slope between any two points on a beam is equal to the area under the curvature diagram between those two points.

If the slope at any point on a beam is known, such as at a fixed support, then the slope at any other point on the beam can be readily determined using *Moment-Area Theorem 1* provided the curvature diagram is known. For example, consider the cantilever beam of Figure 6.2a, which is subjected to an anti-clockwise moment M_e at its tip. The corresponding curvature is positive and constant throughout its length, as illustrated in Figure 6.2b, and the deflected shape is shown in Figure 6.2c. In this particular case, the use of Moment-Area Theorem 1 would be useful for the evaluation of the rotation at B, which is calculated as:

$$\theta_B = \theta_{AB} = \int_A^B \kappa \, dx = \int_A^B \frac{M_e}{IE} \, dx = \frac{M_e}{IE} L \tag{6.8}$$

where θ_{AB} is the change in slope from A to B ($= \theta_B - \theta_A$). For the cantilever of Figure 6.2a, $\theta_A = 0$ and, therefore, $\theta_{AB} = \theta_B$. On the basis of the adopted sign convention, a positive rotation corresponds to an anti-clockwise rotation. It is not always necessary to calculate the integral of Equation 6.8 to evaluate the area under the curvature diagram, which can be determined based on considerations of geometry (see Figure A.4 for the analytical expressions for the areas of common shapes). Reconsidering the cantilever of Figure 6.2 and performing the integral from B to A, we would get a negative angle, which is consistent with our sign convention. In fact, the angle formed moving from the tangent to the deflected shape at B to the tangent at A is clockwise and is, therefore, negative.

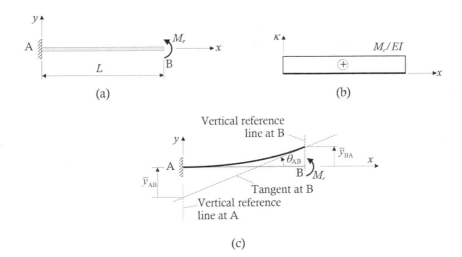

Figure 6.2 Beam layout and free-body diagram. (a) Cantilever beam. (b) Curvature diagram. (c) Deflected shape.

We will now consider the more complex problem of the beam shown in Figure 6.3. The curvature diagram and deflected shape for the beam are also shown in the figure. Moment-Area Theorem 1 implies that the slope of the tangent to the deflection curve at point F relative to the tangent to the curve at point B (i.e. $\theta_{BF} = \theta_F - \theta_B$) is equal to the shaded area under the curvature diagram between points B and F. In this case, the change in slope from B to F, i.e. $\theta_{BF} = \theta_F - \theta_B$, is shown as positive (because an anti-clockwise rotation is required to move from the slope at B to the slope at F).

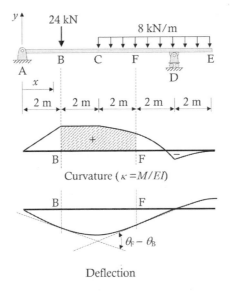

Figure 6.3 Curvature diagram and deflected shape for a beam subjected to transverse loads.

Moment-Area Theorem 2

The intercept $\bar{y}$ made on a vertical reference line by the tangents to the beam axis at two points B and F is equal to the first moment of the area under the curvature diagram between B and F about that reference line.

Referring to the curvature diagram and deflected shape shown in Figure 6.4 (reproduced from Figure 6.3), this theorem states that the intercept $\bar{y}_{BF}$ on the vertical y-axis made by the tangents to the deflected shape at B and F is equal to the first moment of the shaded area under the curvature diagram between B and F about the y-axis. Consider the small area dA under the curvature diagram in Figure 6.4. The first moment of the area dA about the y-axis is calculated as $x\,dA$. It follows that:

$$\bar{y}_{BF} = \int_{B}^{F} x\kappa\,dx \tag{6.9}$$

Clearly, $\bar{y}_{BF}$ is not necessarily the deflection at any point along the beam. However, *Theorem 2* can be used to determine deflection at a point, if a judicious choice of the position of the reference line is made. In order to highlight this, let us consider the case of the cantilever beam of Figure 6.5 subjected to a point load P at its tip. From statics:

$$M = Px \qquad \kappa = \frac{M}{EI} = \frac{Px}{EI}$$

We will now use *Moment-Area Theorem 2* to calculate the deflection v_A at point A. The vertical reference line (i.e. the y-axis) is taken to pass through A. The intercept $\bar{y}_{AB}$ is obtained as:

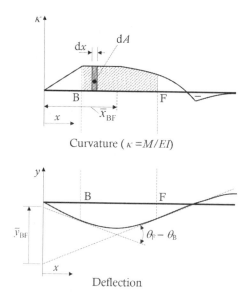

Figure 6.4 Curvature diagram and deflected shape.

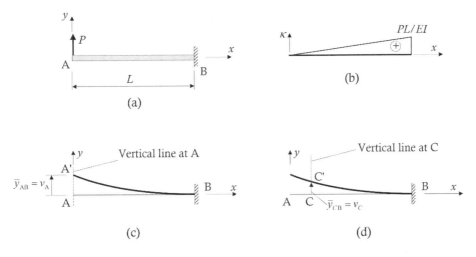

Figure 6.5 Deformation of a cantilever beam subjected to an end point load. (a) Cantilever beam. (b) Curvature diagram. (c) Deflected shape. (d) Deflected shape.

$$\bar{y}_{AB} = \int_A^B x\kappa\, dx = \int_A^B x\, \frac{xP}{EI}\, dx = \left[\frac{x^3}{3}\frac{P}{EI}\right]_A^B = \frac{x_B^3}{3}\frac{P}{EI} - \frac{x_A^3}{3}\frac{P}{EI} = \frac{PL^3}{3EI}$$

which equals v_A because the slope at B is zero. A positive value for $\bar{y}_{AB}$ describes the fact that we need to move upwards (i.e. in the positive direction of y) to go from A to A', as shown in Figure 6.5c. A negative value for $\bar{y}_{AB}$ would represent a downward movement.

Considering the complexity of the curvature diagrams in common structures, it is usually more practical to perform the integral of Equation 6.9 on the basis of geometric considerations:

$$\bar{y}_{BF} = \int_B^F x\kappa\, dx = \bar{x}_{BF} A_{BF} \tag{6.10}$$

where A_{BF} is the area under the curvature curve between B and F (the shaded area in Figure 6.4) and $\bar{x}_{BF}$ is the distance between the centroid of this area and the vertical reference line (in this case, the y-axis). It is noted that A_{BF} is in fact dimensionless.

It is common to encounter cases where we want to calculate the intercept $\bar{y}$ with respect to any vertical axis, not necessarily the y-axis, for example, if we want to calculate the deflection at point C along the cantilever of Figure 6.5d using Moment-Area Theorem 2. In this instance, the intercept $\bar{y}_{CB}$ is determined from:

$$\bar{y}_{CB} = \int_C^B (x - x_C)\kappa\, dx = \bar{x}_{CB} A_{CB} \tag{6.11}$$

where x_C is the x coordinate of C and defines the location of the vertical reference line, A_{CB} is the area under the curvature diagram between C and B, and $\bar{x}_{CB}$ is the distance from the centroid of A_{CB} to the vertical axis through C. A positive value for $\bar{y}_{CB}$ implies an upward movement from C to C' (see Figure 6.5d).

The main features of this approach are outlined in Summary of Steps 6.1 and its application is illustrated in Worked Examples 6.2 through 6.4.

SUMMARY OF STEPS 6.1: DETERMINATION OF SLOPE AND DEFLECTION OF A BEAM USING THE MOMENT-AREA METHOD

The following steps outline the application of the moment-area method.

1. Determine the bending moment diagram and, hence, the curvature diagram. If EI is constant, the curvature diagram has the same shape as the bending moment diagram (M/EI). In many cases, the curvature diagram will be a straight line (in unloaded regions of the beam), a parabola (under uniformly distributed loads), a cubic equation under linearly varying loads and so on. In these cases, the determination of the areas under the curvature diagram and the position of their centroids is simple (see Appendix A for geometric properties of common shapes).

2. Recalling that the deflection at each support is zero and that the slope at any fixed support is also zero, sketch the deflected shape of the beam. Sketch to an exaggerated scale, remembering that in regions of positive moment, the curvature is positive (sagging) and, in regions of negative moment, the curvature is negative (hogging). The points where either the slope or deflection of the beam is to be calculated should be identified.

3. Select the location of the vertical reference line, such that the moment-area theorems can be readily applied to determine the required information. The choice of the two points and the position of the reference axis are important when applying this approach. Since we already may know information about the slope or deflection at the supports, one of the two points is usually the point where the slope or deflection is required and the other is one of the supports.

4. Use Moment-Area Theorem 1 to determine the difference in slope between two points on the curve and Moment-Area Theorem 2 to find the vertical distance at a particular location between a point on the deflected shape to the extended tangent projected from a different point.

WORKED EXAMPLE 6.2

For the uniformly loaded, simply-supported beam shown in Figure 6.6a, use the moment-area theorems to determine the deflection at mid-span v_C and the slope of the beam θ_A at support A. Assume the flexural rigidity EI is constant throughout.

The bending moment at any distance x from the support A is:

$$M(x) = 0.5w_1 Lx - 0.5w_1 x^2$$

and the corresponding curvature is:

$$\kappa(x) = \frac{M(x)}{EI} = \frac{1}{2EI}(w_1 Lx - w_1 x^2)$$

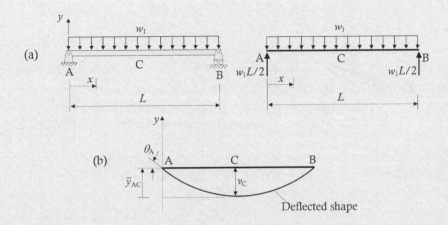

Figure 6.6 Beam for Worked Example 6.2.

Realising that the slope of the tangent to the deflection curve at mid-span is zero (owing to the symmetry of the support and loading conditions) and using Moment-Area Theorem 1, the slope at A is calculated as the area under the curvature diagram between A and C:

$$\theta_{AC} = \frac{1}{2EI} \int\limits_{0}^{L/2} (w_1 Lx - w_1 x^2)\,dx = \frac{1}{2EI}\left[\frac{w_1 Lx^2}{2} - \frac{w_1 x^3}{3}\right]_{0}^{L/2} = \frac{w_1 L^3}{24EI}$$

For the loading shown (where w_1 is positive downward), the angle θ_{AC} is positive, i.e. there is a positive (anti-clockwise) rotation of the beam between A and C. As the slope at C is zero, it follows that $\theta_A = -\theta_{AC}$.

Because the slope at C is zero, the projection of the tangents at A and C onto a vertical reference line through support A is equal in magnitude to the vertical deflection at C, as shown in Figure 6.6b. Using Moment-Area Theorem 2:

$$\bar{y}_{AC} = \int\limits_{0}^{L/2} \kappa(x)x\,dx = \frac{1}{2EI}\int\limits_{0}^{L/2}(w_1 Lx^2 - w_1 x^3)\,dx = \frac{1}{2EI}\left[\frac{w_1 Lx^3}{3} - \frac{w_1 x^4}{4}\right]_{0}^{L/2} = \frac{5}{384}\frac{w_1 L^4}{EI}$$

This implies that a positive upward movement $\bar{y}_{AC}$ is required to move from C to A and hence the deflection at C relative to A is downwards (because deflecting in the negative direction of the y-axis):

$$v_C = -\bar{y}_{AC} = -\frac{5}{384}\frac{w_1 L^4}{EI}$$

WORKED EXAMPLE 6.3

For the beam analysed in Worked Example 6.1 and shown in Figure 6.1, use the moment-area theorems to determine the deflection at point C and at point E.

The curvature diagram for the beam (M/EI) is shown in Figure 6.7, where the end A is taken as the origin. To facilitate the calculations, the area under the curvature diagrams has been divided into five regions (A_1 to A_5). In particular, the geometric properties of the areas under the curve are as follows:

- the triangular area between A and B: $A_1 = +4.8 \times 10^{-3}$ with centroid at $\bar{x}_1 = 1.333$ m
- rectangular area between B and C: $A_2 = +9.6 \times 10^{-3}$ with centroid at $\bar{x}_2 = 3.0$ m
- area between C and D is subdivided into A_3 and A_4, with:
  - the parabolic area $A_3 = +17.06 \times 10^{-3}$ with centroid at $\bar{x}_3 = 5.5$ m
  - the rectangular area $A_4 = -6.4 \times 10^{-3}$ with centroid at $\bar{x}_4 = 6.0$ m
- the parabolic area between D and E: $A_5 = -1.066 \times 10^{-3}$ with centroid at $\bar{x}_5 = 8.5$ m

These geometric properties have been obtained on the basis of the information provided in Appendix A.

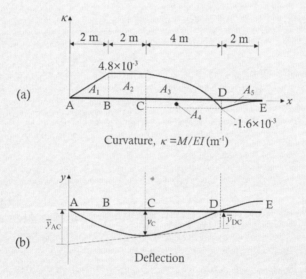

Figure 6.7 Curvature and deflected shape of beam for v_C in Worked Example 6.3.

Using Moment-Area Theorem 2, the projection of the tangent to the deflection curve at point C on a vertical reference line through A is shown in Figure 6.7 and is given by:

$$\bar{y}_{AC} = \int_A^C \kappa(x) x \, dx = A_1 \times \bar{x}_1 + A_2 \times \bar{x}_2 = 4.8 \times 10^{-3} \times 1.333 + 9.6 \times 10^{-3} \times 3.0 = 0.0352 \text{ m}$$

The projection of the tangent to the deflection curve at point C on a vertical reference line through D is shown in Figure 6.7 and is calculated as:

$$\bar{y}_{DC} = A_3 \times (8 - \bar{x}_3) + A_4 \times (8 - \bar{x}_4)$$
$$= 17.06 \times 10^{-3} \times (8 - 5.5) - 6.4 \times 10^{-3} \times (8 - 6.0) = 0.0298 \text{ m}$$

From the sketch of the deflected shape in Figure 6.7b, the deflection at point C, mid-way between A and D, is:

$$v_C = -\frac{\bar{y}_{AC} + \bar{y}_{DC}}{2} = -\frac{0.0352 + 0.0298}{2} = -0.0325 \text{ m} = -32.5 \text{ mm}$$

and, as shown in Figure 6.7, the deflection is downwards.

To determine the deflection at point E, we first determine the dimension $\bar{y}_{E2}$ shown in Figure 6.8b, where $\bar{y}_{E2}$ is the displacement to the point where the tangent to the deflection curve at C intersects the vertical reference line through E. From simple geometry:

$$\frac{\bar{y}_{AC} - \bar{y}_{DC}}{L_{AD}} = \frac{\bar{y}_{AC} - \bar{y}_{E2}}{L_{AE}} \quad \text{i.e.} \quad \frac{0.0352 - 0.0298}{8} = \frac{0.0352 - \bar{y}_{E2}}{10} \quad \therefore \bar{y}_{E2} = 0.0285 \text{ m}$$

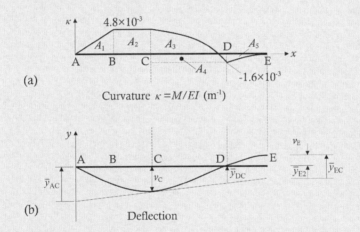

Figure 6.8 Curvature and deflected shape of beam for v_E in Worked Example 6.3.

The dimension $\bar{y}_{EC}$ in Figure 6.8b is obtained as the distance between the points where the tangents to the deflection curve at C and E intersect a vertical line through E and this can be obtained using Moment-Area Theorem 2. With $\bar{y}_{EC}$ so determined, the deflection at E can be obtained from geometry: $v_E = \bar{y}_{EC} - \bar{y}_{E2}$. Applying Moment-Area Theorem 2 to the area under the curvature diagram between C and E:

$$\bar{y}_{EC} = v_E + \bar{y}_{E2} = A_3 \times (10 - \bar{x}_3) + A_4 \times (10 - \bar{x}_4) + A_5 \times (10 - \bar{x}_5)$$
$$= 17.06 \times 10^{-3} \times (10 - 5.5) - 6.4 \times 10^{-3} \times (10 - 6.0) - 1.066 \times 10^{-3} \times (10 - 8.5)$$
$$= 0.0496 \text{ m}$$

From geometry (Figure 6.8b):

$$v_E = \bar{y}_{EC} - \bar{y}_{E2} = 0.0496 - 0.0285 = 0.0211\,\text{m} = 21.1\,\text{mm}$$

Confirming the deflected shape of Figure 6.8, the deflection at E is clearly upwards.

WORKED EXAMPLE 6.4

If the flexural rigidity EI is constant along the fixed-ended beam of Figure 6.9a, use the moment-area theorems to determine the reactions and the deflection at mid-span.

With horizontal displacement at support B permitted and no horizontal components of the applied loads, the horizontal reactions at A and B are both zero. The four unknown reactions for the statically indeterminate beam are shown in the free-body diagram of Figure 6.9d. This free-body diagram is statically equivalent to the free-body diagrams of a simply-supported beam carrying a uniformly distributed load (Figure 6.9b) superimposed on the free-body diagram of a simple beam subjected to end couples M_A and M_B (Figure 6.9c).

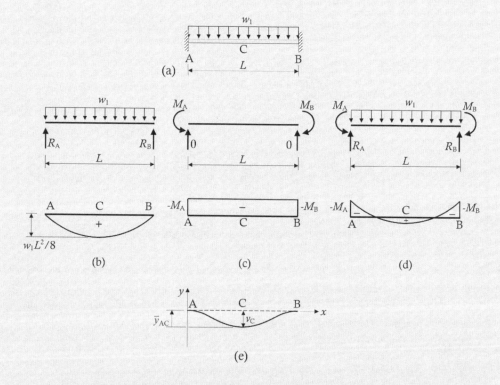

Figure 6.9 Beam for Worked Example 6.4. (a) Fixed-ended beam under UDL. (b) Free-body diagram of a simply-supported beam subjected to a uniformly distributed load. (c) Free-body diagram of a simply-supported beam subjected to end moments. (d) Free-body diagram of the fixed-ended beam. (e) Deflected shape.

For the simple span in Figure 6.9b, moment equilibrium about support A gives $R_B = w_lL/2$, and then summing the vertical forces gives $R_A = w_lL/2$. The bending moment diagram of the simple beam in Figure 6.9b is sometimes called the *free-span bending moment diagram* and the maximum moment at mid-span is called the *total static moment* M_0. For this uniformly loaded beam: $M_0 = w_lL^2/8$.

Symmetry dictates that the moments M_A and M_B are equal and moment equilibrium of the free-body diagram in Figure 6.9c tells us that the vertical reactions caused by M_A and M_B are both zero. It follows that the vertical reactions of the statically indeterminate beam are the same as those for the simple span of Figure 6.9b. In this case, M_A and M_B are reactions and, as such, their positive direction is arbitrary.

Noting that the slope of the beam at the fixed end A is zero and the slope of the beam at the point of maximum deflection (at the mid-span C) is also zero, Moment-Area Theorem 1 tells us that the area under the curvature diagram between A and C must be zero. Therefore, the sum of the area under the positive parabolic curvature diagram between A and C for the simple span of Figure 6.9b and the corresponding area under the negative curvature diagram caused by the end couples M_A and M_B must be zero. On the basis of the geometric properties provided in Appendix A, we have:

$$\frac{1}{EI}\left[\frac{2}{3}\frac{w_lL^2}{8}\frac{L}{2} - M_A\frac{L}{2}\right] = 0 \quad \therefore M_A = \frac{w_lL^2}{12}$$

and from symmetry: $M_B = M_A$. Both moment-area theorems are required to determine the reactions of a fixed-ended beam when the applied loading is not symmetrical.

To determine the deflection at the mid-span C, we use Moment-Area Theorem 2. Taking first moments of the area under the curvature diagram between A and C about the support at A gives the vertical distance $\bar{y}_{AC}$, which is equal in magnitude to the vertical deflection of the beam at C. Moving upwards from C to A by $\bar{y}_{AC}$ (as shown in Figure 6.9e) corresponds to a downward deflection of C relative to A, i.e., $v_C = -\bar{y}_{AC}$.

Using the area and dimensions given for a parabolic shape in Appendix A, we get:

$$\bar{y}_{AC} = \frac{1}{EI}\left[\frac{2}{3}\frac{w_lL^2}{8}\frac{L}{2}\frac{5L}{16} - \frac{w_lL^2}{12}\frac{L}{2}\frac{L}{4}\right] = \frac{1}{384}\frac{w_lL^4}{EI} \quad \therefore v_C = -\bar{y}_{AC} = -\frac{1}{384}\frac{w_lL^4}{EI}$$

For the loading shown with a positive value of w_l, v_C is negative (downwards), as shown in Figure 6.9e.

6.4 CONJUGATE BEAM METHOD

The conjugate beam method converts the problem of geometry into one of simple statics. In Chapter 5, we saw that the mathematical relationships between load, shear force and bending moment are of the same form as the relationships between curvature, slope and deflection (as summarised in Table 5.1). It follows then that the slope and deflection of a beam should be able to be calculated from the curvature in exactly the same way as the shear force and bending moments are calculated from the applied load.

If the curvature diagram is treated as a distributed load applied to a *conjugate beam*, the slope and deflection of the real beam at a particular point are the same as the shear force and bending moment, respectively, at that point in the conjugate beam. The relationships are summarised in Table 6.1 (see also Section 5.7).

Table 6.1 Relationship between real and conjugate beams

Real beam	Conjugate beam
$\kappa = M/EI$	w^c
$\theta = \int \kappa \, dx$	$S^c = -\int w^c \, dx$
$v = \int \theta \, dx$	$M^c = -\int S^c \, dx = \int\int w^c \, dx \, dx$

Note: a superscript 'c' is used for the variables related to the conjugate beam.

The supports of the conjugate beam must be chosen so that slope and deflection conditions of the real beam are accurately represented. For example, a fixed support in a real beam is such that the slope and deflection are both zero: $\theta = 0$ and $v = 0$. In the conjugate beam, both S^c and M^c must therefore be zero at this point, which can only occur at a free, unsupported end. A superscript 'c' is used for the variables related to the conjugate beam to distinguish these from the variables describing the behaviour of the real beam. Similarly, at a free end of a real beam, both θ and v are non-zero, and so in the conjugate beam, this end must be fixed so that both S^c and M^c are also non-zero.

At an exterior pinned or roller support in a real beam, $v = 0$ but θ is non-zero. In the conjugate beam, we require $M^c = 0$ (because v is conjugated to M^c, as shown in Table 6.1) but S^c must be non-zero (because θ is conjugated to S^c). These are the conditions provided by a pinned or roller support in the conjugate beam.

At an internal support (pin or roller) in a real beam, $v = 0$ but θ is non-zero. It follows that in the conjugate beam at this point, we must have $M^c = 0$ but S^c must be non-zero. These are the conditions for an internal hinge. Likewise at an internal hinge in a real beam, v and θ are both non-zero and therefore M^c and S^c must both be non-zero. These are the conditions of an internal support (pin or roller).

The support conditions to be used for the conjugate beam are summarised in Figure 6.10 where these are related to the support conditions of the real beam.

Let us use the conjugate beam method for the analysis of a simple beam, such as the cantilever shown in Figure 6.11a. The external moment M_e applied to its tip produces a constant positive moment throughout the member length, which corresponds to a constant positive curvature (Figure 6.11b). The uplift movement generated by the positive (anti-clockwise) M_e is described by a positive rotation (Figure 6.11c) and positive deflection (Figure 6.11d). In the case of the conjugate beam, the support conditions are outlined in Figure 6.11e, together with the uniform curvature applied as a uniformly distributed load. The load is upwards when the curvature is positive (and downwards if the curvature is negative). The consequent internal shear force S^c and bending moment M^c in the conjugate beam are obtained from statics and are plotted in Figures 6.11f and g. On the basis of the sign convention adopted for the internal actions, the rotation in the real beam and the shear force in the conjugate beam are equal in absolute value but opposite in sign because of the minus sign included in the integral for S^c, i.e. $S^c = -\int w^c \, dx$, but not present in the integral for θ, i.e. $\theta = \int \kappa \, dx$ (see Table 6.1). The deflection of the real beam is equal (also in sign) to the bending moment determined from the conjugate beam.

Real Beam	Conjugate Beam	Real Beam	Conjugate Beam
$\theta = 0$ $v = 0$	$S^c = 0$ $M^c = 0$	θ v	S^c M^c
Fixed support	Free end	Free end	Fixed support
θ $v = 0$	S^c $M^c = 0$	θ $v = 0$	S^c $M^c = 0$
Pinned support	Pinned support	Roller support	Roller support
θ $v = 0$	S^c $M^c = 0$	θ $v = 0$	S^c $M^c = 0$
Internal pinned support	Internal hinge	Internal roller support	Internal hinge
θ v	S^c M^c		
Internal hinge	Internal support		

Figure 6.10 Relationship between support conditions of real and conjugate beams.

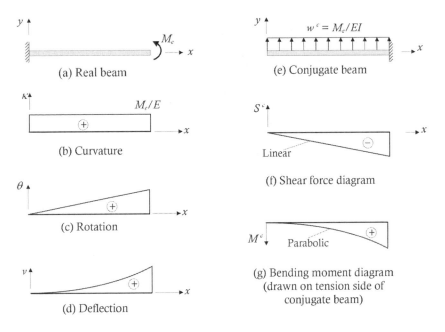

(a) Real beam

(b) Curvature

(c) Rotation

(d) Deflection

(e) Conjugate beam

(f) Shear force diagram

(g) Bending moment diagram (drawn on tension side of conjugate beam)

Figure 6.11 Cantilever beam subjected to an end moment: real and conjugate beams. (a) Real beam. (b) Curvature. (c) Rotation. (d) Deflection. (e) Conjugate beam. (f) Shear force diagram. (g) Bending moment diagram (drawn on tension side).

SUMMARY OF STEPS 6.2: Determination of slope and deflection of a beam using the conjugate beam method

The following steps outline the application of the conjugate beam method.

1. Determine the bending moment diagram and, hence, the curvature diagram. If EI is constant, the curvature diagram has the same shape as the bending moment diagram (M/EI).

2. Sketch the conjugate beam being careful to specify all appropriate support conditions as outlined in Figure 6.10.

3. Load the conjugate beam with the curvature diagram from the real beam, treating the curvature as a distributed load. Where the curvature is positive, the *load* on the conjugate beam is upwards and, where the curvature is negative, the *load* on the conjugate beam is downwards.

4. Determine the reactions at the supports of the conjugate beam using the equations of equilibrium.

5. Cut the conjugate beam at the point where the slope or deflection is required and, considering either the free-body diagram to the left or to the right of the cut, calculate the internal actions. The calculated shear force in the conjugate beam is equal in magnitude and opposite in sign to the slope of the real beam at that point, while the calculated moment is equal to the deflection of the real beam at that point.

WORKED EXAMPLE 6.5

For the uniformly loaded, simply-supported beam shown in Figure 6.12a (previously analysed in Worked Example 6.2), use the conjugate beam method to determine the deflection at mid-span and the slope at the supports. The flexural rigidity EI is constant throughout.

As calculated in Worked Example 6.2, the bending moment at any distance x from support A in Figure 6.12a is $M(x) = 0.5w_1Lx - 0.5w_1x^2$ and the corresponding curvature diagram is shown in Figure 6.12b. For this simply-supported beam, the support conditions for the conjugate beam are the same as for the real beam. The conjugate beam, loaded with the curvature diagram, is shown in Figure 6.12c. The reactions at each support of the symmetrically loaded conjugate beam are both equal to half the total load, which is calculated as:

$$R_A^c = R_B^c = \frac{1}{2}\frac{2}{3}\frac{w_1L^2}{8EI}L = \frac{w_1L^3}{24EI}$$

The shear force at support A of the conjugate beam is equal to R_A^c, i.e. $S_A^c = R_A^c$, which is equal to the opposite of the slope of the real beam at A. Based on this, the rotation at A can then be determined as:

$$\theta_A = -S_A^c = -\frac{w_1L^3}{24EI}$$

where the negative sign for θ_A implies that for a positive value for w_1 (causing positive bending and curvature), the rotation at A is clockwise (i.e. negative).

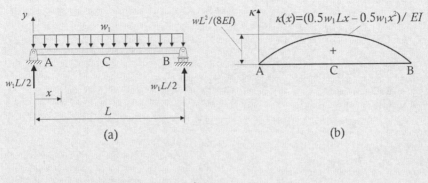

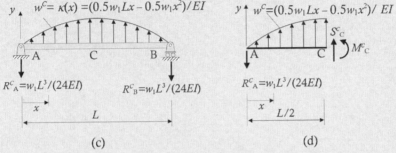

Figure 6.12 Beam for Worked Example 6.5. (a) Real beam. (b) Curvature. (c) Conjugate beam. (d) Free-body of conjugate beam.

Cutting the conjugate beam at mid-span (Figure 6.12d), the calculated moment at mid-span is:

$$M_C^c = -\frac{w_1 L^3}{24EI}\frac{L}{2} + \left(\frac{2}{3}\frac{w_1 L^2}{8EI}\frac{L}{2}\right)\left(\frac{3}{8}\frac{L}{2}\right) = -\frac{5}{384}\frac{w_1 L^4}{EI}$$

which corresponds to the mid-span deflection at C:

$$v_C = -\frac{5}{384}\frac{w_1 L^4}{EI}$$

The negative sign implies that for a positive value for w_1, the deflection is in the negative y direction, i.e. a download displacement.

WORKED EXAMPLE 6.6

For the beam analysed in Worked Example 6.1, use the conjugate beam method to determine the deflections at points C and E.

The real beam and the conjugate beam are shown in Figure 6.13. The free end at E of the real beam is replaced by a fixed support at E in the conjugate beam. The internal support at D is replaced by an internal hinge in the conjugate beam, while the pinned support at A in the real

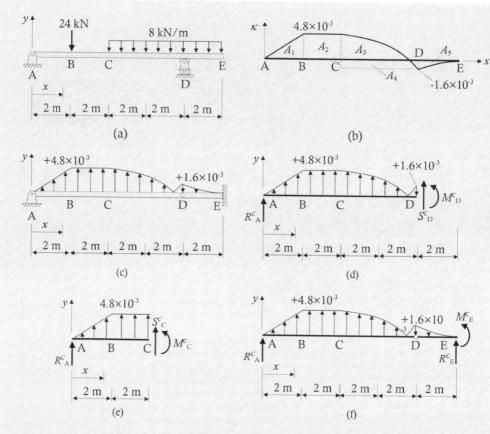

Figure 6.13 Beam for Worked Example 6.6. (a) Real beam. (b) Curvature $\kappa = M/EI$ (m⁻¹). (c) Conjugate beam. (d) Free-body diagram AD from conjugate beam. (e) Free-body diagram AC from conjugate beam. (f) Free-body diagram of conjugate beam.

beam remains a pinned support in the conjugate beam. The curvature distribution calculated in Worked Example 6.1 is:

$$\kappa = \frac{1}{EI}\left(24x - 24\langle x - 2\rangle - \frac{8}{2}\langle x - 4\rangle^2 + 48\langle x - 8\rangle\right)$$

and is shown in Figure 6.13b.

We start by calculating the reactions of the conjugate beam (Figure 6.13c) assuming it to be loaded with a distributed load defined by the curvature (in units of m⁻¹).

Consider a free-body diagram of the conjugate beam from A to D shown in Figure 6.13d. Taking moments about the internal hinge at D, we get the vertical reaction at A of the conjugate beam:

$$M_D^c = +R_A^c \times 8 + A_1 \times 6.667 + A_2 \times 5 + A_3 \times 2.5 + A_4 \times 2$$

$$= R_A^c \times 8 + 4.8 \times 10^{-3} \times 6.667 + 9.6 \times 10^{-3} \times 5 + 17.06 \times 2.5 - 6.4 \times 10^{-3} \times 2 = 0$$

$$\therefore R_A^c = -13.73 \times 10^{-3} \text{ m}^{-1} \text{ (i.e. downward)}$$

where all areas were already calculated in Worked Example 6.3.

We are now in a position to calculate the deflection at C and E of the real beam, because we can evaluate the moments at these locations in the conjugate beam. Cutting the conjugate beam at C and taking moment about C on the free-body diagram AC (Figure 6.13e) gives:

$$M_C^c = -13.73 \times 10^{-3} \times 4 + 4.8 \times 10^{-3} \times 2.667 + 9.6 \times 10^{-3} \times 1 = -0.0325 \text{ m}$$

which equals the deflection at C:

$$v_C = M_C^c = -0.0325 \text{ m} = -32.5 \text{ mm (i.e. downward)}$$

Considering the entire conjugate beam (Figure 6.13f), taking moments about E will give the moment at the fixed support and this of course is equal to the deflection at E (v_E) in the real beam:

$$M_E^c = -13.73 \times 10^{-3} \times 10 + 4.8 \times 10^{-3} \times 8.667 + 9.6 \times 10^{-3} \times 7 + 17.06 \times 10^{-3} \times 4.5$$
$$- 6.4 \times 10^{-3} \times 4 - 1.06 \times 10^{-3} \times 1.5$$
$$= +0.0211 \text{ m}$$
$$\therefore v_E = +0.0211 \text{ m} = +21.1 \text{ mm (i.e. upward)}$$

WORKED EXAMPLE 6.7

If the flexural rigidity EI is constant throughout the fixed-ended beam of Figure 6.14a, use the conjugate beam method to determine the reactions and the deflection at mid-span.

The real beam and the conjugate beam are shown in Figure 6.14. Based on Figure 6.10, the fixed supports at A and B in the real beam are replaced with free ends in the conjugate beam. The slope and deflection of the real beam at each support are both zero and so the shear force and bending moment in the conjugate beam at each end must both be zero. This means the conjugate beam is without supports, and so the loads on the conjugate beam must be self-equilibrating. That is, the resultant upward load on the conjugate beam (the area under the positive curvature diagram) must be equal in magnitude to the downward load on the conjugate beam (the area under the negative curvature diagram). The moment about any point on the conjugate beam must also be zero.

From symmetry in the real beam, $M_A = M_B$ (and these support reactions are shown in Figure 6.14b). Considering the conjugate beam shown in Figure 6.14d, the area under the parabolic positive curvature diagram over the length L of the conjugate beam is:

$$2/3 \times (w_1 L^2/8EI) \times L = w_1 L^3/12EI$$

and the area under the rectangular negative curvature diagram is $-M_A L/EI$.

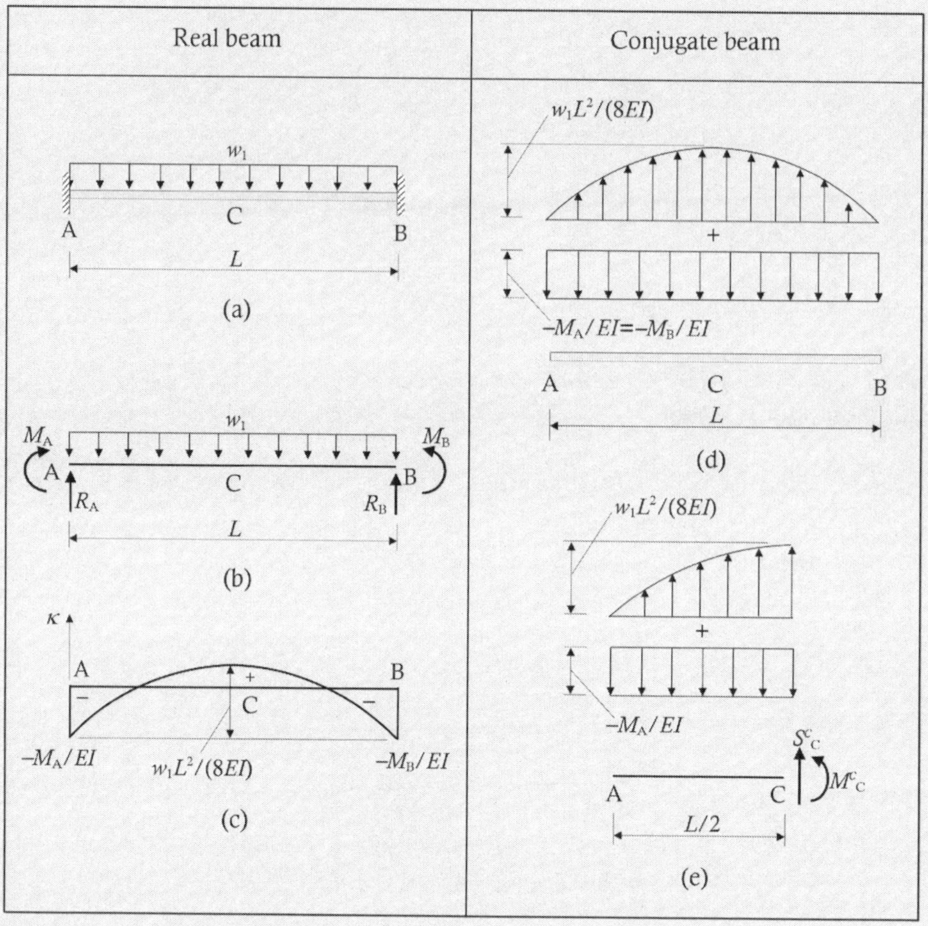

Real beam	Conjugate beam

Figure 6.14 Real and conjugate beams for Worked Example 6.7. (a) Elevation. (b) Free-body diagram. (c) Curvature diagram. (d) Elevation. (e) Free-body AC.

Enforcing the condition that the sum of the loads applied to the conjugate beam must be zero (i.e. vertical equilibrium of loads):

$$\frac{w_1 L^3}{12EI} - \frac{M_A L}{EI} = 0 \quad \text{from which:} \quad M_A = \frac{w_1 L^2}{12}$$

Cutting the conjugate beam at mid-span (Figure 6.14e), the calculated internal moment at mid-span is equal to the deflection of the real beam at mid-span. Therefore:

$$v_C = M_C^c = \frac{2}{3}\frac{w_1 L^2}{8EI}\frac{L}{2}\frac{3}{8}\frac{L}{2} - \frac{w_1 L^2}{12EI}\frac{L}{2}\frac{L}{4} = -\frac{1}{384}\frac{w_1 L^4}{EI}$$

WORKED EXAMPLE 6.8

For the beam shown in Figure 6.15a, determine the position and magnitude of the maximum deflection using the conjugate beam method. The flexural rigidity EI is constant throughout, with $E = 200,000$ MPa and $I = 40 \times 10^6$ mm^4.

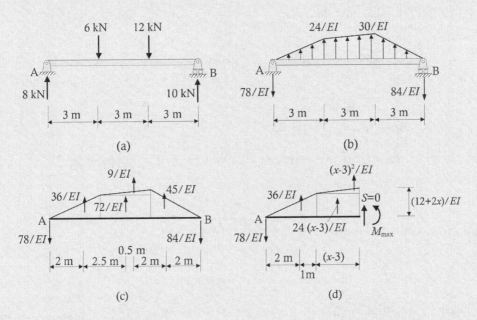

(a)

(b)

(c)

(d)

Figure 6.15 Real and conjugate beams for Worked Example 6.8. (a) Real beam. (b) Conjugate beam. (c) Free-body diagram of conjugate beam with resultant loads and reactions. (d) Free-body diagram of conjugate beam to left of point of maximum deflection.

The conjugate beam is shown in Figure 6.15b, where the distributed load corresponds to the curvature diagram of the real beam. The reactions at each end of the conjugate beam have been determined from statics and the resultant distributed loads on the conjugate beam (i.e. the areas under the curvature diagram) are shown in Figure 6.15c. The position of maximum deflection in the real beam corresponds to the point of maximum moment in the conjugate beam, i.e. the point of zero shear. From the free-body diagram of Figure 6.15c, this clearly occurs in the middle third of the span. The free-body diagram of the conjugate beam to the left of the point of maximum deflection is shown in Figure 6.15d and the location of the maximum moment is obtained by determining the value of x at which the shear force is zero.

By equating the expression for shear force S^c in Figure 6.15d to zero, we find the value of x corresponding to the point of maximum deflection in the real beam:

$$-\frac{36}{EI} - \frac{24(x-3)}{EI} - \frac{(x-3)^2}{EI} + \frac{78}{EI} = 0$$

from which $x^2 + 18x - 105 = 0$.

Solving this quadratic equation gives $x = 4.638$ m.

With $EI = 8 \times 10^{12}$ Nmm^2 = 8×10^3 kNm^2, the maximum deflection is found by considering the moment equilibrium of the free-body diagram of Figure 6.12d at $x = 4.638$ m:

$$M_{max}^c = \frac{36}{EI}(x-2) + \frac{24(x-3)}{EI}\frac{(x-3)}{2} + \frac{(x-3)^2}{EI}\frac{(x-3)}{3} - \frac{78}{EI}x$$

$$= \frac{1}{8 \times 10^3}[36 \times (4.638 - 2) + 12 \times (4.638 - 3)^2 + (4.638 - 3)^3/3 - 78 \times 4.638]$$

$$= -0.0291\,\text{m}$$

$$\therefore v_{max} = -0.0291\,\text{m} = -29.1\,\text{mm (i.e. downward)}$$

6.5 THE SLOPE-DEFLECTION EQUATIONS

6.5.1 Sign convention for support moments and rotations

The slope-deflection equations relate the deformation of a structural member, expressed in terms of its end slopes and displacements, to the loads and actions applied to the member. Consider the internal span of a continuous beam subjected to an arbitrary load distribution shown in Figure 6.16a. The free-body diagram of the span is shown in Figure 6.16b and the deflected shape of the span is shown in Figure 6.16c. The rotation of the beam at the supports A and B, θ_A and θ_B, respectively, are caused by the loads applied along the span of the beam and the moments applied at each end of the span (shown on the free-body diagram of Figure 6.16b as M_A and M_B). The slope-deflection equations are developed relating the internal moments (M_A and M_B) and the applied loads to the rotations at each support (θ_A and θ_B).

We will assume that external load components are positive when they occur in the positive x and y directions (where the x-axis is the longitudinal member axis). We will also assume that the external moments and rotations are positive when they act in an anticlockwise sense.

For the internal span shown in Figure 6.17a, the right-hand support moves by a positive (upwards) amount Δ. Additional moments (and shears) are induced in the beam (as shown on the free-body of Figure 6.17b) and additional rotation occurs at each end of the span (Figure 6.17c). The relationship between Δ and the end moments of the span can be determined

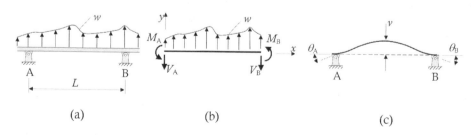

(a)   (b)   (c)

Figure 6.16 Displacements of an interior span caused by transverse load. (a) Interior span. (b) Free-body diagram. (c) Deflected shape.

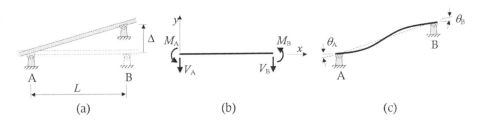

Figure 6.17 Displacements and actions caused by relative support settlement. (a) Interior span. (b) Free-body diagram. (c) Deflected shape.

conveniently using the conjugate beam method. For the deformation shown in Figure 6.17a, as the line AB rotates in an anti-clockwise direction, the rotation Δ/L is positive.

We will now derive expressions that relate the end moments M_A and M_B to the beam deformations (θ_A, θ_B, and Δ) and the applied member loads using the conjugate beam method. These will then be used for the analysis of statically indeterminate beams.

6.5.2 Rotation at support A, θ_A

The beam shown in Figure 6.18a is pinned at support A and fixed at support B. It is subjected to an applied moment M_A at support A sufficient to cause a rotation at that support of θ_A, as shown. According to our sign convention, M_A and θ_A are both positive in Figure 6.18a (i.e. both are anti-clockwise). A free-body diagram of the beam showing the reactions (M_B, V_B and V_A) caused by the applied moment M_A is shown in Figure 6.18b and the resulting curvature diagram is shown in Figure 6.18c. The relationship between M_A and θ_A can be found by analysing the conjugate beam shown in Figure 6.18d.

The vertical reaction at end A of the conjugate beam is equal to the rotation θ_A of the real beam at A, i.e. $V_A^c = \theta_A(=-S_A^c)$. The deflection of the beam at both supports is zero and

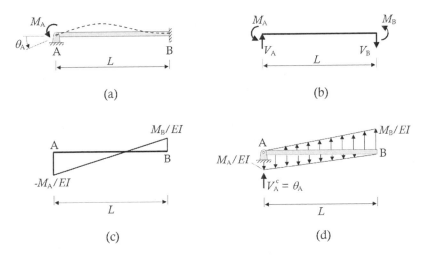

Figure 6.18 Real and conjugate beams for a propped cantilever with an applied moment at the pinned end. (a) Beam. (b) Free-body diagram. (c) Curvature diagram. (d) Conjugate beam.

therefore the moments about A and B in the conjugate beam must also be zero. Assuming anti-clockwise moments are positive, we get:

$$\Sigma M_\text{A}^\text{c} = 0: \quad -\left(0.5 \times \frac{M_\text{A}}{EI} \times L \times L/3\right) + \left(0.5 \times \frac{M_\text{B}}{EI} \times L \times 2L/3\right) = 0 \tag{6.12}$$

which simplifies to:

$$M_\text{A} = 2M_\text{B} \tag{1}$$

$$\Sigma M_\text{B}^\text{c} = 0: \quad +\left(0.5 \times \frac{M_\text{A}}{EI} \times L \times 2L/3\right) - \left(0.5 \times \frac{M_\text{B}}{EI} \times L \times L/3\right) - \theta_\text{A}L = 0 \tag{2}$$

Substituting Equation 1 into Equation 2 gives the relationships between the end moments M_A, M_B and θ_A:

$$M_\text{A} = \frac{4EI}{L}\theta_\text{A} \qquad M_\text{B} = \frac{2EI}{L}\theta_\text{A} \tag{6.13a,b}$$

6.5.3 Rotation at support B, θ_B

Similarly, if end B of the beam rotates by θ_B under the application of an applied moment M_B with end A fixed, the relationship between M_B and θ_B and the relationship between the reaction M_A and θ_B can be written as:

$$M_\text{B} = \frac{4EI}{L}\theta_\text{B} \qquad M_\text{A} = \frac{2EI}{L}\theta_\text{B} \tag{6.14a,b}$$

6.5.4 Fixed-end moments caused by applied loads

Most often, the rotation at supports is not caused by couples applied at the supports but by loads applied along the span. If we can find the reaction moments at the supports caused by any loads applied on the span, we can use Equations 6.13 and 6.14 to find the rotations at the supports. Let us consider the fixed-ended beam shown in Figure 6.19a subjected to a point

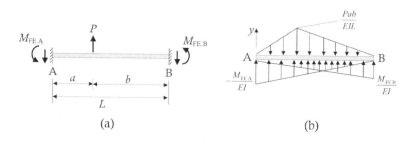

(a)   (b)

Figure 6.19 Real and conjugate beams for a fixed-ended beam subjected to applied loads. (a) Beam. (b) Conjugate beam.

load P. The load will produce *fixed-end moments* at A and B, referred to as $M_{FE.A}$ and $M_{FE.B}$ in Figure 6.19a (and both drawn as positive, i.e. anti-clockwise). To determine the fixed-end moments, we will analyse the conjugate beam shown in Figure 6.19b. Note that if the reaction moment $M_{FE.A}$ is in the direction shown, the internal bending moment in the beam at end A is $-M_{FE.A}$ and the curvature is $-M_{FE.A}/EI$.

As the real beam has zero slope and zero deflection at each end, the moment and shear at each end of the conjugate beam are also zero. Equating the sum of the vertical loads on the conjugate beam to zero, we get:

$$\sum F_y^c = 0: \quad \left[\frac{1}{2} \times \left(-\frac{M_{FE.A}}{EI}\right) \times L\right] + \left(\frac{1}{2} \times \frac{M_{FE.B}}{EI} \times L\right) - \left(\frac{1}{2} \times \frac{Pab}{EIL} L\right) = 0$$

$$-M_{FE.A}L + M_{FE.B}L - Pab = 0 \tag{1}$$

Summing the moments about end A of the conjugate beam gives:

$$\sum M_A^c = 0: \quad \left[\frac{1}{2} \times \left(-\frac{M_{FE.A}}{EI}\right) \times L \times \frac{L}{3}\right] + \left(\frac{1}{2} \times \frac{M_{FE.B}}{EI} \times L \times \frac{2L}{3}\right)$$

$$-\left(\frac{1}{2} \times \frac{Pab}{EIL} L \times \frac{b+2a}{3}\right) = 0$$

$$-M_{FE.A}L + 2M_{FE.B}L - \frac{Pab}{L}(b+2a) = 0 \tag{2}$$

Solving Equations 1 and 2 gives:

$$M_{FE.A} = -\frac{Pab^2}{L^2} \quad \text{and} \quad M_{FE.B} = +\frac{Pa^2b}{L^2} \tag{6.15a,b}$$

The fixed-end moments for other loadings are given in Appendix B for a span fixed at both ends and for a span fixed at the left support at A and pinned at the right support at B.

6.5.5 Support settlement Δ

Consider the fixed-ended beam shown in Figure 6.20a, where the right-hand support is displaced by an amount Δ with respect to the left support as shown. Following the sign convention adopted here, since the line AB rotates in an anti-clockwise sense, Δ is taken to be positive. A moment and a shear reaction develop at each support. From the deflected shape of the beam, it is clear that the internal bending moment at A is positive (sagging) and that at B is negative (hogging). The resulting curvature diagram is shown in Figure 6.20b and the conjugate beam is shown in Figure 6.20c. The moment at end B in the conjugate beam must equal the upward (positive) displacement of the real beam, i.e. $+\Delta$ as shown. Summing the forces on the conjugate beam in the vertical direction gives:

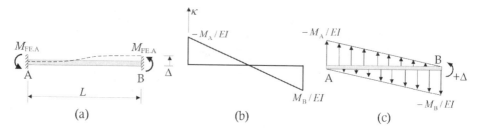

Figure 6.20 Real and conjugate beams for a fixed-end beam subjected to settlement of one support. (a) Beam. (b) Curvature diagram. (c) Conjugate beam.

$$\Sigma F_y^c = 0: \quad +\left(0.5 \times \frac{-M_{FE.A}}{EI} \times L\right) - \left(0.5 \times \frac{-M_{FE.B}}{EI} \times L\right) = 0 \quad \therefore M_{FE.A} = M_{FE.B} \tag{1}$$

and summing the moments about B, we get:

$$\Sigma M_B^c = 0: \quad -\left(0.5 \times \frac{-M_{FE.A}}{EI} \times L \times \frac{2L}{3}\right) + \left(0.5 \times \frac{-M_{FE.B}}{EI} \times L \times \frac{L}{3}\right) + \Delta = 0 \tag{2}$$

Substituting Equation 1 into Equation 2 gives:

$$M_{FE.A} = -\frac{6EI}{L^2}\Delta \quad \text{and} \quad M_{FE.B} = -\frac{6EI}{L^2}\Delta \tag{6.16a,b}$$

6.5.6 Slope-deflection equations

Span continuous at both ends: If we sum the moments at each end of a fixed-ended beam caused by each of the imposed displacements θ_A, θ_B and Δ (Equations 6.13, 6.14 and 6.16) and the fixed-end moments caused by the imposed loads, we get the slope-deflection equations for an interior span:

$$M_A = \frac{EI}{L}\left(4\theta_A + 2\theta_B - \frac{6\Delta}{L}\right) + M_{FE.A} \qquad M_B = \frac{EI}{L}\left(2\theta_A + 4\theta_B - \frac{6\Delta}{L}\right) + M_{FE.B} \tag{6.17a,b}$$

Span pinned at end B and continuous at end A: If we apply Equations 6.17 to an end span of a continuous beam where end A is continuous and end B is pinned, the moment at end B is zero and Equations 6.17 can be rearranged as:

$$M_A = \frac{EI}{L}\left(4\theta_A + 2\theta_B - \frac{6\Delta}{L}\right) + M_{FE.A} - \frac{1}{2}M_{FE.B} \qquad M_B = 0 \tag{6.18a,b}$$

where $M_{FE.A}$ and $M_{FE.B}$ are still the end moments of a fixed-ended beam previously introduced in Section 6.5.4.

It is more convenient when using the slope-deflection methods in spans pinned at one end and fixed at the other to rewrite Equation 6.18a in terms of the end moment of a propped cantilever as:

$$M_A = \frac{EI}{L}\left(3\theta_A - \frac{3\Delta}{L}\right) + M_{FE^*.A} \tag{6.19}$$

in which:

$$M_{FE^*.A} = M_{FE.A} - \frac{1}{2}M_{FE.B} \tag{6.20}$$

where $M_{FE^*.A}$ is the support moment of a propped cantilever whose values are tabulated in Appendix B (see right column of table in Appendix B).

WORKED EXAMPLE 6.9

For the continuous beam shown in Figure 6.21, the flexural rigidity EI is constant throughout. Calculate the reactions at each support using the slope-deflection equations and plot the shear force and bending moment diagrams.

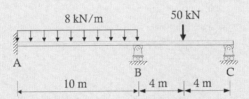

Figure 6.21 Beam for Worked Example 6.9.

Span AB: From Appendix B, the fixed-end moments at A and B produced by a uniformly distributed load are:

$$M_{FE.A} = -\frac{wL^2}{12} = -\frac{-8 \times 10^2}{12} = 66.67 \text{ kNm} \quad \text{and} \quad M_{FE.B} = \frac{wL^2}{12} = \frac{-8 \times 10^2}{12} = -66.67 \text{ kNm}$$

Since end A is fixed, θ_A is zero, and as there is no support settlement, Δ is zero. The slope-deflection equations (Equations 6.17) can then be written as:

$$(M_A)_{AB} = \frac{EI}{L}\left(4\theta_A + 2\theta_B - \frac{6\Delta}{L}\right) + M_{FE.A} = \frac{EI}{10}(2\theta_B) + 66.67 = 0.2EI\theta_B + 66.67 \tag{1}$$

$$(M_B)_{AB} = \frac{EI}{L}\left(2\theta_A + 4\theta_B - \frac{6\Delta}{L}\right) + M_{FE.B} = \frac{EI}{10}(4\theta_B) - 66.67 = 0.4EI\theta_B - 66.67 \qquad (2)$$

Span BC: From Appendix B, the fixed-end moments at B and C of this end span are:

$$M_{FE^*.B} = -\frac{3PL}{16} = -\frac{3\times(-50)\times 8}{16} = 75 \text{ kNm} \quad \text{and} \quad M_{FE^*.C} = 0$$

With end B the continuous end, Equation 6.19 becomes:

$$(M_B)_{BC} = \frac{EI}{L}\left(3\theta_B - \frac{3\Delta}{L}\right) + M_{FE^*.B} = \frac{EI}{8}(3\theta_B) + 75 = 0.375EI\theta_B + 75 \qquad (3)$$

Equilibrium at support B: For moment equilibrium at support B:

$$(M_B)_{AB} + (M_B)_{BC} = 0$$

and therefore:

$$0.4EI\theta_B - 66.67 + 0.375EI\theta_B + 75 = 0$$

$$\theta_B = -\frac{10.75}{EI}$$

which represents a negative (clockwise) rotation at B.
Substituting into Equations 1, 2 and 3 gives:

$$(M_A)_{AB} = +64.5 \text{ kNm}; \quad (M_B)_{AB} = -71.0 \text{ kNm}; \quad (M_B)_{BC} = +71.0 \text{ kNm}$$

Equilibrium: The unknown vertical reactions, required for the definition of the shear force diagram, are determined from equilibrium considerations. The moment reaction at A is a positive (anti-clockwise) couple $(M_A)_{AB}$ of magnitude 64.5 kNm, as shown on the free-body diagram in Figure 6.22a. For the free-body of the span BC in Figure 6.22b, the moment at end B is a positive (anti-clockwise) couple $(M_B)_{BC}$ of magnitude 71.0 kNm.

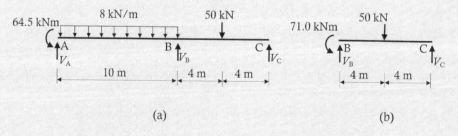

(a)   (b)

Figure 6.22 Free-body diagrams for Worked Example 6.9. (a) Free-body diagram of whole beam. (b) Free-body of span BC.

Summing the moments about B in Figure 6.22b:

$$-(V_C \times 8) + (50 \times 4) - 71.0 = 0 \qquad \qquad \therefore V_C = 16.1 \text{ kN}$$

Summing the moments about A in Figure 6.22a:

$$-(V_B \times 10) - (16.1 \times 18) + (50 \times 14) + (8 \times 10 \times 5) - 64.5 = 0 \qquad \therefore V_B = 74.5 \text{ kN}$$

Summing the vertical forces:

$$V_A + 74.5 + 16.1 - 50 - (8 \times 10) = 0 \qquad \qquad \therefore V_A = 39.4 \text{ kN}$$

These reactions can be readily checked by taking moments about any other point (e.g. support C in Figure 6.22a). The bending moment and shear force diagrams are readily determined from statics and are shown in Figure 6.23.

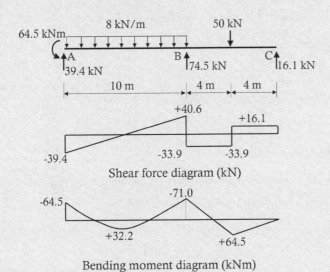

Figure 6.23 Reactions, shear force and bending moment diagrams for Worked Example 6.9.

WORKED EXAMPLE 6.10

For the beam of Worked Example 6.9 (in Figure 6.21), calculate the support reactions caused by a (downwards) support settlement of 60 mm at support B and draw the bending moment and shear force diagrams. Assume $EI = 12 \times 10^3 \text{ kNm}^2$ throughout.

Since both spans are unloaded except for the support settlement, the fixed-end moments are zero. The support settlement at B is 0.06 m downward and, as in Worked Example 6.9, θ_A is zero. The line through AB rotates in a clockwise direction, so Δ/L is negative. The slope-deflection equations for span AB are (Equations 6.17):

$$(M_A)_{AB} = \frac{EI}{L}\left(2\theta_B - \frac{6\Delta}{L}\right) = \frac{EI}{10}\left[2\theta_B - \frac{6\times(-0.06)}{10}\right] = EI(0.2\theta_B + 0.0036) \tag{1}$$

$$(M_B)_{AB} = \frac{EI}{L}\left(4\theta_B - \frac{6\Delta}{L}\right) = \frac{EI}{10}\left[4\theta_B - \frac{6\times(-0.06)}{10}\right] = EI(0.4\theta_B + 0.0036) \tag{2}$$

With CD rotating in an anti-clockwise sense (i.e. so Δ/L is positive), the slope-deflection equation for the end span BC is:

$$(M_B)_{BC}\frac{EI}{L}\left(3\theta_B - \frac{3\Delta}{L}\right) = \frac{EI}{8}\left(3\theta_B - \frac{3\times0.06}{8}\right) = EI(0.375\theta_B - 0.0028) \tag{3}$$

Equilibrium at support B: For moment equilibrium at support B:

$$(M_B)_{AB} + (M_B)_{BC} = 0$$

and therefore:

$$EI(0.4\theta_B + 0.0036) + EI(0.375\theta_B - 0.0028) = 0$$

$$\theta_B = -1.032 \times 10^{-3} \text{ rad}$$

Substituting into Equations 1, 2 and 3 gives:

$$(M_A)_{AB} = EI(0.2\theta_B + 0.0036) = 12 \times 10^3[0.2 \times (-0.001032) + 0.0036] = +40.72 \text{ kNm}$$

$$(M_B)_{AB} = EI(0.4\theta_B + 0.0036) = 12 \times 10^3[0.4 \times (-0.001032) + 0.0036] = +38.25 \text{ kNm}$$

$$(M_B)_{BC} = EI(0.375\theta_B - 0.0028) = 12 \times 10^3[0.375 \times (-0.001032) - 0.0028] = -38.25 \text{ kNm}$$

Equilibrium: The vertical reactions are calculated from equilibrium considerations. The moment reaction at A is an anti-clockwise couple $(M_A)_{AB}$ of magnitude 40.7 kNm, as shown on the free-body diagram in Figure 6.24a. For the free-body of the span BC in Figure 6.24b, the moment at end B is a clockwise couple $(M_B)_{BC}$ of magnitude 38.25 kNm.

Summing the moments about B in Figure 6.24b:

$$-(V_C \times 8) + 38.25 = 0 \quad \therefore V_C = 4.78 \text{ kN}$$

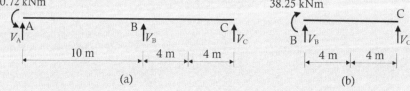

Figure 6.24 Free-body diagrams for Worked Example 6.10. (a) Free-body diagram of whole beam. (b) Free-body diagram of span BC.

Summing the moments about A in Figure 6.24a:

$-(V_B \times 10) - (4.78 \times 18) - 40.72 = 0 \quad \therefore V_B = -12.68$ kN

Summing the vertical forces:

$V_A - 12.68 + 4.78 = 0 \quad \therefore V_A = 7.90$ kN

The bending moment and shear force diagrams are readily determined from statics and are shown in Figure 6.25.

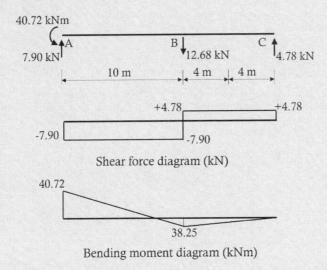

Shear force diagram (kN)

Bending moment diagram (kNm)

Figure 6.25 Reactions, shear force and bending moment diagrams for Worked Example 6.10.

6.5.7 Frames without sidesway

The approach included in the previous section for the analysis of continuous beams can be applied to frames that do not exhibit sidesway movements. A frame that is non-symmetrical or one that is loaded non-symmetrically will sway to the side, unless it is physically prevented from doing so, as is the case for the example in Figure 6.26a owing to the

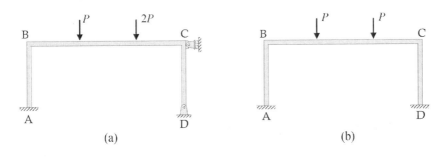

Figure 6.26 Frames without sidesway.

presence of the roller support at C. Any frame with loads and boundary conditions that are symmetrical, such as the frame in Figure 6.26b, will also not sway. This assumes that all members are inextensible; otherwise, the shortening or elongation of members could induce a small degree of sway. This assumption is usually acceptable as, in common structures, the flexural deformations are those governing the displaced shape, as the axial deformations are much smaller and have only a marginal effect on the calculated displacements.

WORKED EXAMPLE 6.11

For the frame shown in Figure 6.27, calculate the support reactions caused by the applied loads and draw the axial force, shear force and bending moment diagrams. Assume EI is constant throughout.

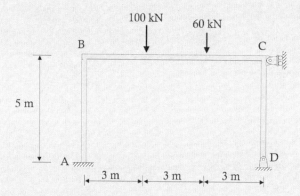

Figure 6.27 Frame for Worked Example 6.11.

The analysis of the frame is carried out assuming that no sidesway can occur because of the restraining action provided by the roller support.

Member AB: As the member is not subjected to external loads, the fixed-end moments are zero. With $\theta_A = 0$, the slope-deflection equations for member AB are (Equations 6.17):

$$(M_A)_{AB} = \frac{EI}{L}\left(4\theta_A + 2\theta_B - \frac{6\Delta}{L}\right) = \frac{EI}{5}(2\theta_B) = EI(0.4\theta_B) \tag{1}$$

$$(M_B)_{AB} = \frac{EI}{L}\left(2\theta_A + 4\theta_B - \frac{6\Delta}{L}\right) = \frac{EI}{5}(4\theta_B) = EI(0.8\theta_B) \tag{2}$$

Member BC: From Appendix B, the fixed-end moments are:

$$M_{FE.B} = -\frac{-100 \times 3 \times 6^2}{9^2} - \frac{-60 \times 6 \times 3^2}{9^2} = +173.3 \text{ kNm}$$

$$M_{FE.C} = \frac{-100 \times 3^2 \times 6}{9^2} + \frac{-60 \times 6^2 \times 3}{9^2} = -146.6 \text{ kNm}$$

and the slope-deflection equations are:

$$(M_B)_{BC} = \frac{EI}{L}\left(4\theta_B + 2\theta_C - \frac{6\Delta}{L}\right) + M_{FE.B} = \frac{EI}{9}(4\theta_B + 2\theta_C) + 173.3$$

$$= EI(0.444\theta_B + 0.222\theta_C) + 173.3 \tag{3}$$

$$(M_C)_{BC} = \frac{EI}{L}\left(2\theta_B + 4\theta_C - \frac{6\Delta}{L}\right) + M_{FE.C} = \frac{EI}{9}(2\theta_B + 4\theta_C) - 146.6$$

$$= EI(0.222\theta_B + 0.444\theta_C) - 146.6 \tag{4}$$

Member CD: The slope-deflection equation for this member pinned at D is:

$$(M_C)_{CD} = \frac{EI}{L}\left(3\theta_C - \frac{3\Delta}{L}\right) + M_{FE*.C} = \frac{EI}{5}(3\theta_C) = 0.6EI\theta_C \tag{5}$$

Equilibrium at corner C: For moment equilibrium at corner C:

$(M_C)_{BC} + (M_C)_{CD} = 0$ and therefore:

$$EI(0.222\theta_B + 0.444\theta_C) - 146.6 + EI(0.6\theta_C) = 0$$

$$EI(0.222\theta_B + 1.044\theta_C) - 146.6 = 0 \tag{6}$$

Equilibrium at corner B: For moment equilibrium at corner B:

$(M_B)_{AB} + (M_B)_{BC} = 0$ and therefore:

$$EI(0.8\theta_B) + EI(0.444\theta_B + 0.222\theta_C) + 173.3 = 0$$

$$EI(1.244\theta_B + 0.222\theta_C) + 173.3 = 0 \tag{7}$$

Solving Equations 6 and 7 simultaneously, we get:

$$\theta_B = -\frac{170.8}{EI} \quad \text{and} \quad \theta_C = \frac{176.8}{EI}$$

which substituted into Equations 1 through 5 gives:

$$(M_A)_{AB} = EI(0.4\theta_B) = -68.3 \text{ kNm}$$

$$(M_B)_{AB} = EI(0.8\theta_B) = -136.7 \text{ kNm}$$

$(M_B)_{BC} = EI(0.444\theta_B + 0.222\theta_C) + 173.3 = +136.7$ kNm

$(M_C)_{BC} = EI(0.222\theta_B + 0.444\theta_C) - 146.6 = -106.1$ kNm

$(M_C)_{CD} = 0.6EI\theta_C = +106.1$ kNm

Equilibrium: The moment reaction at A is a negative (clockwise) couple $(M_A)_{AB}$ of magnitude 68.3 kNm, as shown on the free-body diagram in Figure 6.28a, together with the other five force reactions, H_A, V_A, H_C, H_D and V_D. Free-body diagrams of AB, BCD and CD are shown in Figure 6.28b, c and d, respectively, and are used to calculate the unknown reactions.

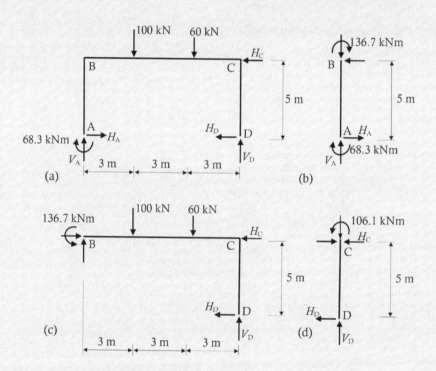

Figure 6.28 Free-body diagrams for Worked Example 6.11.

Summing the moments about C in Figure 6.28d:

$(H_D \times 5) - 106.1 = 0$ $\therefore H_D = +21.21$ kN (i.e. ←)

Summing the moments about B in Figure 6.28b:

$68.3 + 136.6 - (H_A \times 5) = 0$ $\therefore H_A = 41.00$ kN (i.e. →)

Summing the horizontal forces on Figure 6.28a:

$41.00 - 21.21 - H_C = 0$ $\therefore H_C = 19.79$ kN (i.e. ←)

Summing the moments about B in Figure 6.28c:

$$100 \times 3 + 60 \times 6 + (21.21 \times 5) - 136.7 - (V_D \times 9) = 0 \qquad \therefore V_D = +69.93 \text{ kN (i.e. } \uparrow)$$

Summing the vertical forces on Figure 6.28a:

$$V_A + 69.93 - 100 - 60 = 0 \qquad\qquad \therefore V_A = 90.07 \text{ kN (i.e. } \uparrow)$$

The above calculations can now be verified by checking moment equilibrium about the support at D in Figure 6.28a:

$$68.3 + (90.07 \times 9) - (100 \times 6) - (60 \times 3) - (19.79 \times 5)$$
$$= 68.3 + 810.6 - 600 - 180 - 98.9 = 0 \quad \therefore \text{OK}$$

The axial force, shear force and bending moment diagrams are readily determined from statics and are shown in Figure 6.29.

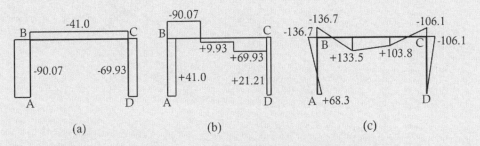

(a) (b) (c)

Figure 6.29 Distributions of internal actions for Worked Example 6.11. (a) Axial force diagram (kNm). (b) Shear force diagram (kNm). (c) Bending moment diagram (kNm).

6.5.8 Frames with sidesway

We saw in the previous section that an unbraced frame will sway to the side (see Figure 6.30) unless it is symmetrical, in terms of its geometry, support conditions and loading. Consider the displaced shape of the frame shown in Figure 6.30. In the analysis of such

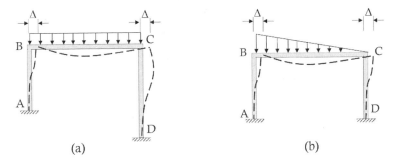

(a) (b)

Figure 6.30 Frames with sidesway. (a) Non-symmetrical frame with sidesway. (b) Sidesway caused by non-symmetrical loading.

frames, the slope-deflection equations for the columns must include the sidesway Δ, as demonstrated in the following worked example. For such frames, the corners of the frame at B and C are assumed to remain at right angles after deformation, i.e. $(\theta_B)_{AB} = (\theta_B)_{BC}$ and $(\theta_C)_{BC} = (\theta_C)_{CD}$.

WORKED EXAMPLE 6.12

For the frame shown in Figure 6.31, calculate the support reactions caused by the applied loads and draw the axial force, shear force and bending moment diagrams.

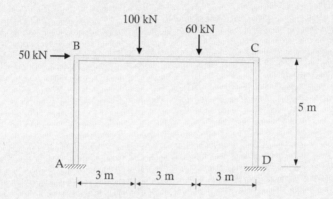

Figure 6.31 Frame for Worked Example 6.12.

The deflected shape of the frame is shown in Figure 6.32, with sidesway occurring because the applied loading is non-symmetric. Both columns AB and CD suffer the same transverse displacement Δ at the top, as shown, and with the axis of both columns rotating in a clockwise sense, Δ/L is negative for both columns.

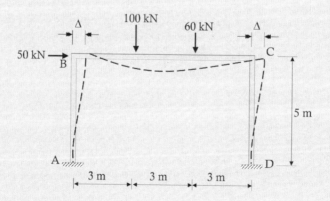

Figure 6.32 Sidesway of the frame for Worked Example 6.12.

Member AB: As this member is not subjected to external loads, the fixed-end moments are zero. With $\theta_A = 0$, the slope-deflection equations for member AB are (Equations 6.17):

$$(M_A)_{AB} = \frac{EI}{L}\left(2\theta_B - \frac{6\Delta}{L}\right) = \frac{EI}{5}\left(2\theta_B - \frac{6\Delta}{5}\right) = EI(0.4\theta_B - 0.24\Delta) \tag{1}$$

$$(M_B)_{AB} = \frac{EI}{L}\left(4\theta_B - \frac{6\Delta}{L}\right) = \frac{EI}{5}\left(4\theta_B - \frac{6\Delta}{5}\right) = EI(0.8\theta_B - 0.24\Delta) \tag{2}$$

Member BC: From Appendix B, the fixed-end moments are:

$$M_{FE.B} = -\frac{-100\times3\times6^2}{9^2} - \frac{-60\times6\times3^2}{9^2} = +173.3 \text{ kNm}$$

$$M_{FE.C} = +\frac{-100\times3^2\times6}{9^2} + \frac{-60\times6^2\times3}{9^2} = -146.6 \text{ kNm}$$

and, for member BC, $\Delta = 0$. The slope-deflection equations are:

$$(M_B)_{BC} = \frac{EI}{L}\left(4\theta_B + 2\theta_C - \frac{6\Delta}{L}\right) + M_{FE.B} = \frac{EI}{9}(4\theta_B + 2\theta_C) + 173.3$$
$$= EI(0.444\theta_B + 0.222\theta_C) + 173.3 \tag{3}$$

$$(M_C)_{BC} = \frac{EI}{L}\left(2\theta_B + 4\theta_C - \frac{6\Delta}{L}\right) + M_{FE.C} = \frac{EI}{9}(2\theta_B + 4\theta_C) - 146.6$$
$$= EI(0.222\theta_B + 0.444\theta_C) - 146.6 \tag{4}$$

Member CD: As the member is not subjected to external loads, the fixed-end moments are zero. With $\theta_D = 0$, the slope-deflection equations for member CD are:

$$(M_C)_{CD} = \frac{EI}{L}\left(4\theta_C - \frac{6\Delta}{L}\right) = \frac{EI}{5}\left(4\theta_C - \frac{6\Delta}{5}\right) = EI(0.8\theta_C - 0.24\Delta) \tag{5}$$

$$(M_D)_{CD} = \frac{EI}{L}\left(2\theta_C - \frac{6\Delta}{L}\right) = \frac{EI}{5}\left(2\theta_C - \frac{6\Delta}{5}\right) = EI(0.4\theta_C - 0.24\Delta) \tag{6}$$

Equations 1 through 6 contain nine unknowns. We get two more equations by considering moment equilibrium of corners B and C:

$$(M_B)_{AB} + (M_B)_{BC} = 0 \tag{7}$$

$$(M_C)_{BC} + (M_C)_{CD} = 0 \tag{8}$$

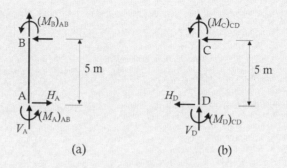

Figure 6.33 Free-body diagrams of segments AB and CD.

and the ninth equation can be obtained by considering horizontal equilibrium of the frame. Figure 6.33a shows free-body diagrams of column AB. By considering moment equilibrium about B, we can express the reaction H_A in terms of the end moments $(M_A)_{AB}$ and $(M_B)_{AB}$:

$$(H_A \times 5) + (M_A)_{AB} + (M_B)_{AB} = 0 \quad \therefore H_A = -\frac{(M_A)_{AB} + (M_B)_{AB}}{5}$$

Similarly, moment equilibrium about C on a free-body of CD gives (Figure 6.33b):

$$-(H_D \times 5) + (M_D)_{CD} + (M_C)_{CD} = 0 \quad \therefore H_D = \frac{(M_D)_{CD} + (M_C)_{CD}}{5}$$

The horizontal force equilibrium equation for the frame is:

$$50 + H_A - H_D = 0 \quad \therefore 50 - \frac{(M_A)_{AB} + (M_B)_{AB}}{5} - \frac{(M_D)_{DC} + (M_C)_{DC}}{5} = 0 \tag{9}$$

By substituting Equations 2 and 3 into Equation 7, we get:

$$1.244\theta_B + 0.222\theta_C - 0.24\Delta = -173.3/EI \tag{10}$$

and substituting Equations 4 and 5 into Equation 8, we get:

$$0.222\theta_B + 1.244\theta_C - 0.24\Delta = +146.6/EI \tag{11}$$

Finally, substituting Equations 1, 2, 5 and 6 into Equation 9 gives:

$$-0.24\theta_B - 0.24\theta_C + 0.192\Delta = -50/EI \tag{12}$$

Equations 10, 11 and 12 represent three simultaneous equations in three unknowns (θ_B, θ_C and Δ). Solving, we get:

$$\theta_B = -244.0/EI; \quad \theta_C = +69.0/EI; \quad \text{and} \quad \Delta = -479.2/EI$$

and with these values, Equations 1 through 6 give:

$$(M_A)_{AB} = +17.39 \text{ kNm}; \quad (M_B)_{AB} = -80.20 \text{ kNm}; \quad (M_B)_{BC} = +80.23 \text{ kNm};$$

$$(M_C)_{BC} = -170.2 \text{ kNm}; \quad (M_C)_{CD} = +170.2 \text{ kNm}; \quad (M_D)_{CD} = +142.6 \text{ kNm}$$

Equilibrium: The free-body diagram in Figure 6.34a shows the moment reactions at A and D ($(M_A)_{AB}$ and $(M_D)_{DC}$, respectively), together with the other four force reactions, H_A, V_A, H_D and V_D. Free-body diagrams of columns AB and CD are shown in Figures 6.34b and c, respectively.

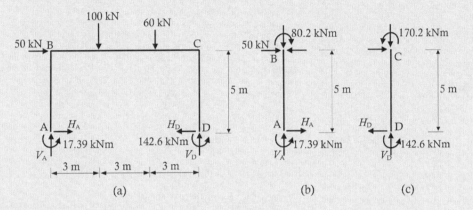

Figure 6.34 Free-body diagrams for Worked Example 6.12.

Summing the moments about C in Figure 6.34c:

$$(H_D \times 5) - 170.2 - 142.6 = 0 \qquad\qquad \therefore H_D = +62.56 \text{ kN (i.e. } \leftarrow)$$

Summing the moments about B in Figure 6.34b:

$$80.2 - 17.39 - (H_A \times 5) = 0 \qquad\qquad \therefore H_A = 12.56 \text{ kN (i.e. } \rightarrow)$$

Summing the moments about D in Figure 6.34a:

$$(V_A \times 9) - 17.39 - 142.6 + (50 \times 5) - (100 \times 6) - (60 \times 3) = 0 \quad \therefore V_A = 76.67 \text{ kN (i.e. } \uparrow)$$

Summing the moments about A in Figure 6.34a:

$$-(V_D \times 9) - 17.39 - 142.6 + (50 \times 5) + (100 \times 3) + (60 \times 6) = 0 \quad \therefore V_D = 83.33 \text{ kN (i.e. } \uparrow)$$

The axial force, shear force and bending moment diagrams are readily determined from statics and are shown in Figure 6.35.

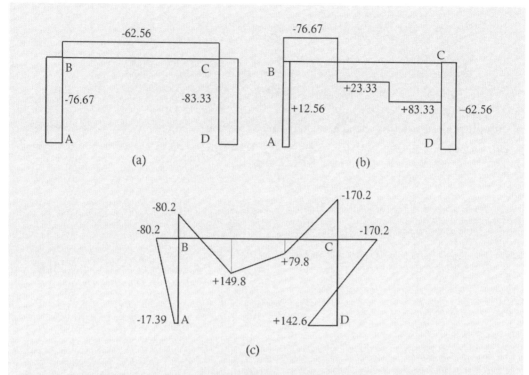

Figure 6.35 Distributions of internal actions for Worked Example 6.12. (a) Axial force diagram (kN). (b) Shear force diagram (kN). (c) Bending moment diagram (kNm).

PROBLEMS

6.1 If the flexural rigidity EI of each of the propped cantilevers shown is constant, use the moment-area method to determine (i) the reactions at A and B, (ii) the deflection at the mid-span and (iii) the rotation at support A of each beam in terms of the load (w or F), the span L and the flexural rigidity EI.

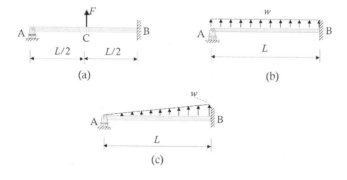

6.2 If the flexural rigidity *EI* of each of the fixed-ended beams shown is constant, use the moment-area method to determine (i) the reactions at A and B and (ii) the deflection at the mid-span of each beam in terms of the load (*w* or *F*), the span *L* and the flexural rigidity *EI*.

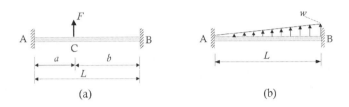

(a)   (b)

6.3 If *EI* is constant throughout the beam illustrated below, determine (i) the reactions at A and B induced by the couple M_A applied at the support A and (ii) the rotation θ_A at support A. Use the moment-area method.

6.4 For the beams of Problem 6.1, determine the deflection at mid-span using double integration. Assume *EI* is constant throughout.

6.5 For each of the beams shown in Problem 6.1, determine the reactions at supports A and B, the deflection at mid-span and the rotation at the roller support at A using the conjugate beam method. Assume *EI* is constant throughout.

6.6 For each of the fixed-ended beams shown in Problem 6.2, determine the reactions at supports A and B, and the deflection at mid-span using the conjugate beam method. Assume *EI* is constant throughout.

6.7 Solve Problem 6.3 using the conjugate beam method.

6.8 For the beams shown below, determine the support reactions and plot the shear forces and bending moment diagrams. Assume $EI = 25,000$ kNm² throughout and use the slope-deflection equations.

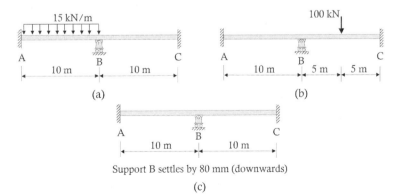

(a)   (b)

Support B settles by 80 mm (downwards)

(c)

6.9 The beam shown below is identical to that analysed in Problem 6.8, except that the loads on both spans are applied at the same time as the support at B settles by 80 mm. Determine the support reactions and plot the shear force and bending moment diagrams. As in the previous problem, assume $EI = 25,000$ kNm² throughout and use the slope-deflection equations. Verify that the results obtained here could have been obtained from Problem 6.8 by superposition.

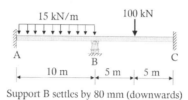

Support B settles by 80 mm (downwards)

6.10 For the beam shown, determine the reactions at A, B and C using the slope-deflection equations. Assume EI is constant.

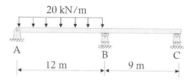

6.11 For the beam shown, determine the reactions at A, B and C using the conjugate beam method and draw the shear force and bending moment diagrams. Assume EI is constant.

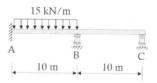

6.12 For the beam of Problem 6.11, determine the reactions at A, B and C using the slope-deflection equations. Assume EI is constant.

6.13 For the beam shown, calculate the reactions at supports A and B using the moment-area method. Also determine the vertical deflection at C. Assume $EI = 15,000$ kNm².

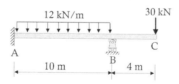

6.14 For the beam of Problem 6.13, recalculate the reactions at supports A and B using the slope-deflection equations.

6.15 For the beam of Problem 6.13, calculate the change in reactions at A and B, if the support at B settles by 80 mm (downwards). Assume $EI = 15,000$ kNm².

6.16 For the beam shown, calculate the reactions at supports A and B using the slope-deflection equations. Assume EI is constant throughout.

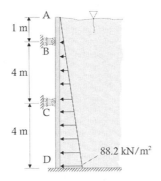

12 kN/m

A

B

C

D

10 m 12 m 12 m

6.17 For the beam of Problem 6.16, calculate the change in reactions at each support, if the support at C settles by 50 mm (downwards). Use the slope-deflection equations. Assume $EI = 30{,}000$ kNm² throughout.

6.18 The wall of a water tank is fixed at the base and supported as shown below. Determine the reactions at the supports per 1 m width of the wall under the linearly varying water pressure. Use the slope-deflection equations. Assume EI is constant.

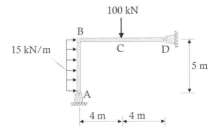

A

1 m

B

4 m

C

4 m

D

88.2 kN/m²

6.19 For the frame shown, determine the reactions at each support and the bending moment at B using the slope-deflection equations. Assume EI is constant.

100 kN

B

15 kN/m

C   D

5 m

A

4 m 4 m

6.20 If the supports at A and D of the frame of Problem 6.19 are fixed (instead of pinned), determine the reactions at each support using the slope-deflection equations. Assume EI is constant.

6.21 If *EI* is constant throughout the frame illustrated below, determine the reactions at A and D using the slope-deflection equations and draw the bending moment diagram.

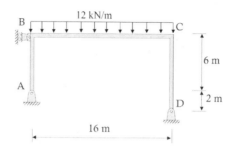

6.22 For the frame shown, the 60 kN force is applied at the mid-point of BC. Find the reactions and draw the bending moment diagram. Use the slope-deflection equations. Assume *EI* is constant.

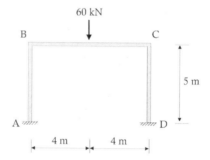

6.23 For the frame of Problem 6.22, if the 60 kN force on member BC is moved 2 m to the left of the position shown (i.e. 2 m from B), find the reactions and draw the bending moment diagram. Use the slope-deflection equations. Assume *EI* is constant.

6.24 For the frame shown, a 50 kN horizontal force is applied at B and a 60 kN vertical force is applied at the mid-point of BC. Find the reactions and draw the bending moment diagram. Use the slope-deflection equations. Assume *EI* is constant.

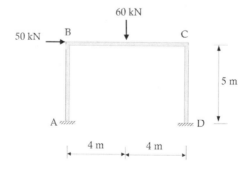

6.25 Reanalyse the frame of Problem 6.24, if the supports at A and D are pinned rather than fixed. Use the slope-deflection equations.

6.26 For the frame shown, find the reactions and draw the bending moment diagram. Assume EI is constant. Use the slope-deflection equations.

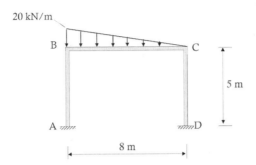

6.27 If the frame of Problem 6.26 has pinned supports (instead of fixed), find the reactions and draw the bending moment diagram. Assume EI is constant. Use the slope-deflection equations.

6.28 If EI is constant throughout the frame illustrated below, determine the reactions at A and D and draw the bending moment diagram. Use the slope-deflection equations.

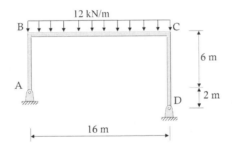

6.29 If the frame of Problem 6.28 has fixed supports (instead of pinned), find the reactions and draw the bending moment diagram. Assume EI is constant. Use the slope-deflection equations.

Chapter 7

Work–energy methods

7.1 STRAIN ENERGY

When a structural member is loaded, it deforms and, as it deforms, work is done. Figure 7.1a shows the load-deformation response of a bar of length L in axial tension. The shape of the curve depends on the size of the bar and the material from which it is made. The work done as an axial load P_B is gradually applied to the bar is the area under the curve from $P = 0$ to $P = P_B$ shown as the shaded area OABC in Figure 7.1a. If the bar is subsequently unloaded, some of the energy associated with this work may be recovered. The unloading curve is represented by the dashed line from B to D. For the materials commonly used in structures, the unloading path BD is a straight line approximately parallel to the tangent to the curve at the origin (i.e. parallel to the line OA in Figure 7.1a). The area of the triangular region CBD represents the energy recovered during unloading and is called *elastic strain energy* U_e. The energy that is not recoverable during unloading has been used to cause inelastic deformation of the material (*inelastic strain energy*).

For many common structural materials, as the bar is loaded and the tensile force P increases, the curve in Figure 7.1a is initially linear (from O to A). We have seen that in this linear range, the elongation of the bar e is equal to PL/EA, where L is the bar length, A is the cross-sectional area of the bar and E is the elastic modulus of the bar material. If the bar is loaded within the linear range and then unloaded, for most materials, the elongation will be recovered (i.e. the unloading path follows the loading path and after unloading $e = 0$). This behaviour is said to be *elastic*.

In structural analysis, it is common to assume that the response of a structural member is linear–elastic, i.e. the load on the member is proportional to the deformation caused by it. This assumption is valid if the maximum applied load is less than the load corresponding to the proportional limit (P_A in Figure 7.1a). In this proportional load range, all the work associated with loading from $P = 0$ to $P = P_1$ ($<P_A$) as the member deforms from $e = 0$ to $e = e_1$ is stored as elastic strain energy and is given by:

$$U_e = \frac{P_1 e_1}{2} \tag{7.1}$$

The elastic strain energy is shown as the shaded area in Figure 7.1b, which is the load-deformation response of a linear–elastic member. Further considerations on materials not behaving in a linear–elastic manner are provided in Chapter 15, but in this chapter, we will only be concerned with linear–elastic material behaviour.

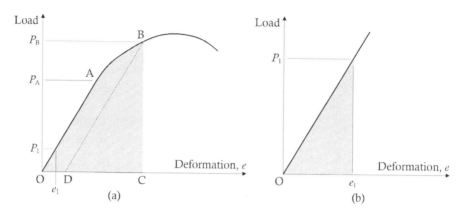

Figure 7.1 Load versus deformation.

7.1.1 Axially loaded members

In Equations 4.26 and 4.27, the flexibility coefficient, $f = L/EA$, and the stiffness coefficient, $k = EA/L$, relating axial force P and elongation e of an axially loaded bar of length L were introduced. If a tension member is loaded with an axial force P_1 and it suffers an extension of e_1 (as shown in Figure 7.1b): $e_1 = fP_1$ and $P_1 = ke_1$. The elastic strain energy (Equation 7.1) can therefore be expressed as:

$$U_e = \frac{P_1 e_1}{2} = \frac{f P_1^2}{2} = \frac{k e_1^2}{2} \tag{7.2}$$

The incremental elastic strain energy dU_e in an infinitesimal length of bar dx subjected to an axial force N is given by:

$$dU_e = \frac{N^2 dx}{2EA} \tag{7.3}$$

and, if the internal axial force N varies along the bar, the elastic strain energy stored in the bar is calculated as:

$$U_e = \int_0^L \frac{N^2 dx}{2EA} \tag{7.4}$$

7.1.2 Beams in bending

In Section 5.3.2, we examined the deformation of a segment of a straight beam in bending (see Figure 5.4) and this is revisited here. Consider the length of beam shown in Figure 7.2a. The cross-section shown in Figure 7.2b is symmetric about the vertical y-axis and is subjected to a positive bending moment M acting about the horizontal z-axis that passes through its centroid. The small length of beam shown in Figure 7.2c is bounded by cross-sections A and B at a distance dx apart. In a straight beam, cross-sections A and B are initially parallel

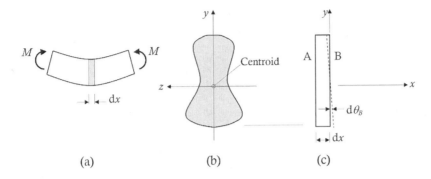

Figure 7.2 Typical deformations caused by bending. (a) Beam length in bending. (b) Cross-section. (c) Slice elevation.

before bending, but after bending, the cross-section at B rotates with respect to the cross-section at A by an angle $d\theta$, as shown.

A positive bending moment M produces stresses that are compressive above the z-axis and tensile below it. Rotation of cross-section B with respect to A by $d\theta$ produces strains that vary linearly with y (as indicated in Equation 5.17 and reproduced here as Equation 7.5).

$$\varepsilon = -y\frac{d\theta}{dx} = -y\kappa \tag{7.5}$$

In the linear–elastic range, the moment M is proportional to the deformation (rotation) $d\theta$ and the *elastic strain energy* done by the moment is the area under the moment versus rotation graph. In this case, the incremental elastic strain energy dU_e can be calculated as:

$$dU_e = \frac{Md\theta}{2} \tag{7.6}$$

In Equation 5.24a, we saw that for a linear–elastic member, the moment M and curvature $(d\theta/dx)$ are related by the flexural rigidity EI as follows:

$$M = EI\kappa = EI\frac{d\theta}{dx}$$

where I is the second moment of area about the centroidal axis of bending. The increment of elastic strain energy stored in the thin slice of beam of length dx can therefore be expressed as the area under the linear moment–curvature graph up to a moment M:

$$dU_e = \frac{M\kappa}{2}dx = \frac{M^2dx}{2EI} = \frac{EI}{2}\kappa^2dx \tag{7.7}$$

Recalling that the curvature may be expressed as the second derivative of the deflection v, i.e. $\kappa = \theta' = v''$, Equation 7.7 can be rewritten as (Equation 5.14):

$$dU_e = \frac{EI}{2}\left(\frac{d^2v}{dx^2}\right)^2dx$$

The elastic strain energy caused by bending in a member of length L can be calculated using:

$$U_e = \int_0^L \frac{M^2}{2EI}\,dx \quad \text{or} \quad U_e = \int_0^L \frac{EI}{2}\left(\frac{d^2v}{dx^2}\right)^2 dx \tag{7.8a,b}$$

The principle of conservation of energy requires that the energy produced by the internal actions is equal to the energy associated with the external actions as the member deforms. By equating the internal and external energy, we can write a relationship that can be used to calculate, for example, the unknown displacement at a location of interest. This procedure is outlined in Worked Example 7.1.

WORKED EXAMPLE 7.1

Calculate the elastic strain energy caused by bending in the beam shown in Figure 7.3 and derive the expression for mid-span deflection caused by the central point load. Ignore the self-weight of the beam and assume the flexural rigidity EI is constant throughout.

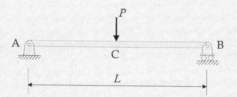

Figure 7.3 Beam for Worked Example 7.1.

The bending moment diagram is discontinuous at C, but symmetrical about C, so that the elastic strain energy in segment AC is the same as that in segment CB. The moment in AC at any distance x from A is $M = Px/2$ and, from Equation 7.8a, the elastic strain energy in the whole beam is:

$$(U_e)_{AB} = (U_e)_{AC} + (U_e)_{CB} = 2\int_0^{L/2} \frac{M^2}{2EI}\,dx = 2\int_0^{L/2} \frac{P^2x^2}{8EI}\,dx = \frac{P^2L^3}{96EI}$$

The external work done by the applied load P as the beam displaces by u_C in the direction of P is $1/2\,Pu_C$ and equating the external work done with the elastic strain energy gives:

$$u_C = \frac{PL^3}{48EI}$$

7.2 THE WORK THEOREM

From basic physics, we know that when a force P is applied to an object and the object moves by a distance δ, the work done is $P\delta$. However, in structural engineering, when a force causes a displacement in the direction of the force, the force is rarely constant during that displacement. In Figure 7.1, to cause the displacement e_1, the force P was increased from 0 to P_1 and, for the internal stresses to remain in equilibrium with the external load, the strains, and hence elongation of the bar, must also increase at the same rate as P. In this case, each infinitesimal increment of force is associated with a different level of elongation and the full force P_1 undergoes an average displacement $e_1/2$.

Consider a structure being acted on by a load P. If we define *work* as the product of a force and its displacement (sometimes called the *work product*), the external work (or the external work product) associated with the structure is the product of the load and its displacement: $W_{\text{ext}} = P\delta$. The internal work W_{int} is the work done in deforming the structure. In the case of a pin-jointed truss, the internal work is the sum of the product of axial force N and elongation e for each of the n members of the truss and can be written as:

$$W_{\text{int}} = \sum_{i=1}^{n} N_i e_i \tag{7.9}$$

In the case of a beam, the internal work caused by bending is the sum of the product of moment and curvature on each infinitesimal length $\mathrm{d}x$ of the beam. That is:

$$W_{\text{int}} = \int_0^L M(\mathrm{d}\theta/\mathrm{d}x)\,\mathrm{d}x = \int_0^L M\left(\frac{\mathrm{d}^2 v}{\mathrm{d}x^2}\right)\mathrm{d}x = \int_0^L M\kappa\,\mathrm{d}x \tag{7.10}$$

The work theorem states:
If a structure is in equilibrium (i.e. the external forces acting on the structure are in equilibrium with the internal actions), then for any geometrically consistent displacement field, the external work is equal to the internal work.

A geometrically consistent displacement field is one where the displacements of the nodes and the supports are consistent with the deformations and the geometry of the individual members in the structure. The work theorem applies even when the geometrically consistent displacement field is not produced by the forces applied to the structure.

Consider the simple truss shown in Figure 7.4a. The free-body diagram in Figure 7.4b shows the applied load, the reactions and the axial force in each bar and represents the equilibrium condition. Figures 7.4c through e show three different consistent displacement fields (none of which are likely to be the displacements caused by the applied load). For each displacement field, the extensions of each bar may be calculated from geometry and are given in the figure. Because the displacements are very small compared to the truss dimensions, the bar deformations may be calculated by assuming the geometry of small displacements, i.e. the direction of each bar is the same before and after the displacement.

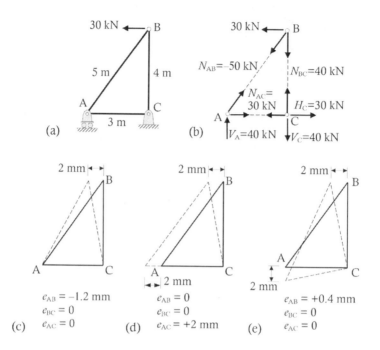

Figure 7.4 Examples of consistent displacement fields.

For displacement field 1 (Figure 7.4c), only the 30 kN applied force contributes to the external work, as the displacement of each reaction component is zero. The external work is therefore:

$$W_{\text{ext}} = 30 \times 2 = 60 \text{ kNmm}$$

and the internal work is obtained from Equation 7.9 as:

$$W_{\text{int}} = \sum_{i=1}^{3} N_i e_i = -50 \times (-1.2) + 40 \times 0 + 30 \times 0 = 60 \text{ kNmm}$$

For displacement field 2 (Figure 7.4d), once again only the 30 kN applied force contributes to the external work. The external work and the internal work are therefore:

$$W_{\text{ext}} = 30 \times 2 = 60 \text{ kNmm} \quad \text{and} \quad W_{\text{int}} = \sum_{i=1}^{3} N_i e_i = -50 \times 0 + 40 \times 0 + 30 \times 2 = 60 \text{ kNmm}$$

For displacement field 3 (Figure 7.4e), both the 30 kN applied force and the vertical reaction at A contribute to the external work. The external work and the internal work are therefore:

$$W_{\text{ext}} = 30 \times 2 - 40 \times 2 = -20 \text{ kNmm} \quad \text{and}$$

$$W_{int} = \sum_{i=1}^{3} N_i e_i = -50 \times 0.4 + 40 \times 0 + 30 \times 0 = -20 \text{ kNmm}$$

For each consistent displacement field the work theorem is satisfied.

WORKED EXAMPLE 7.2

The cantilever beam, analysed as beam 1 in Worked Example 5.1 and shown in Figure 5.7a, is re-examined here to demonstrate the validity of the work theorem. The beam carries a linearly varying distributed load, as shown in Figure 7.5, and has a uniform cross-section with $EI = 10,000$ kNm². Show that the work theorem is satisfied by considering the equilibrium force field and the displacement field produced by it.

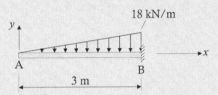

Figure 7.5 Beam for Worked Example 7.2.

As determined in Worked Example 5.1, at any cross-section x (in m) from A, the load intensity and the bending moment are given by (Equations 5.27 and 5.29, respectively)

$w(x) = -6x$ and $M(x) = -x^3$

The curvature and slope were determined in Worked Example 5.1 as:

$$\kappa(x) = M(x)/EI = -10^{-4}x^3 \quad \text{and} \quad \theta = \frac{-x^4}{4EI} + \frac{81}{4EI} = -0.25 \times 10^{-4} x^4 + 0.002025$$

and the deflection was determined as:

$$v = \frac{-x^5}{20EI} + \frac{81}{4EI}x - \frac{243}{5EI} = -0.05 \times 10^{-4} x^5 + 0.002025x - 0.00486$$

The external work is the product of the load and deflection given by:

$$W_{ext} = \int_0^3 w(x)v(x)dx = \int_0^3 (+0.3 \times 10^{-4} x^6 - 12.15 \times 10^{-3} x^2 + 29.16 \times 10^{-3} x)dx$$

$$= \left[0.0429 \times 10^{-4} x^7 - 4.05 \times 10^{-3} x^3 + 14.58 \times 10^{-3} x^2 \right]_0^3 = 0.0312 \text{ kNm}$$

The internal work is obtained from Equation 7.10:

$$W_{int} = \int_0^L M\kappa\, dx = \int_0^3 (-x^3) \times (-10^{-4} x^3)\ dx = \left[\frac{10^{-4}}{7} x^7 \right]_0^3$$

$$= 0.0312 \text{ kNm}$$

As expected, the internal work W_{int} equals the external work W_{ext} and the work theorem is satisfied.

7.3 VIRTUAL WORK

In the preceding section, we saw that the work theorem is valid even when the force field and the consistent displacement field are unrelated, i.e. even if one is not caused by the other. This fact is fundamental to the concept of *virtual work*, which is one of the most powerful tools available in structural analysis. *Virtual* work means *imaginary* work and the concept has two different forms:

   i. *Virtual forces* – where the real displacement field is combined with a virtual (or imaginary) force field
   ii. *Virtual displacements* – where the force field is real and the displacements are virtual

In the former, the virtual forces are introduced so that information can be gathered about the real displacements, while in the latter, virtual displacements are selected in order to obtain information about the real forces. In both approaches, the virtual external work ($\bar{W}_{ext}$) equals the virtual internal work ($\bar{W}_{int}$)

$$\bar{W}_{ext} = \bar{W}_{int} \tag{7.11}$$

The bar over the symbols in Equation 7.11 indicates that the external and internal work are virtual.

7.4 VIRTUAL WORK APPLIED TO TRUSSES

7.4.1 Principle of virtual forces

Virtual forces are commonly used to determine the joint displacements resulting from known member deformations. In Section 4.8, we calculated the displacements of the nodes of a simple truss relying only on geometry and the known member deformations. While this approach is suitable for simple trusses, its use for the analysis of large trusses with many members and joints is tedious. By introducing an appropriate virtual force field and applying the *principle of virtual forces*, the tedious problem of geometry is transformed into a much more tractable problem of statics. The process is illustrated in Worked Example 7.3, where the simple truss previously analysed in Section 4.8 is reconsidered.

WORKED EXAMPLE 7.3

Calculate the vertical and horizontal displacements at node C of the truss shown in Figure 7.6. The bar forces, previously shown in Figure 4.25, are also given in Figure 7.6, along with the resulting bar extensions (previously calculated in Section 4.8) and the reactions.

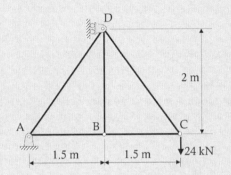

Bar forces N (kN) and bar extensions e (mm):

$N_{AD} = -30$ kN $e_{AD} = -0.417$ mm
$N_{AB} = -18$ kN $e_{AB} = -0.15$ mm
$N_{BD} = 0$ kN $e_{BD} = 0$ mm
$N_{BC} = -18$ kN $e_{BC} = -0.15$ mm
$N_{CD} = +30$ kN $e_{CD} = +1.25$ mm

Reactions:

$H_A = +36$ kN $\rightarrow$ $V_A = +24$ kN $\uparrow$
$H_D = +36$ kN $\leftarrow$

Figure 7.6 Truss for Worked Example 7.3.

In the method of virtual forces, only one joint displacement can be determined for each virtual force field.

(i) Determine the vertical displacement v_C at C

A vertical virtual force of unit magnitude is introduced at node C. The corresponding virtual bar forces $\bar{N}_i$ and reactions are shown in Figure 7.7a. The internal virtual work is the sum of the virtual work in each bar (i.e. the virtual force times the real extension in each bar):

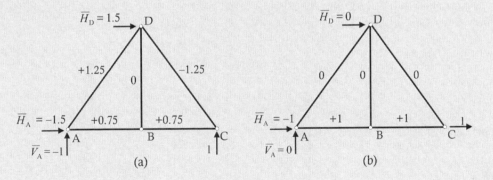

Figure 7.7 Virtual forces $\bar{N}$ (kN).

$$\overline{W}_{int} = \overline{N}_{AD}e_{AD} + \overline{N}_{AB}e_{AB} + \overline{N}_{BD}e_{BD} + \overline{N}_{BC}e_{BC} + \overline{N}_{CD}e_{CD}$$
$$= 1.25 \times (-0.417) + 0.75 \times (-0.15) + 0 \times 0 + 0.75 \times (-0.15) - 1.25 \times 1.25$$
$$= -2.31 \text{ kNmm}$$

and the external work is the product of the unit virtual force and the real vertical displacement:

$$\overline{W}_{ext} = 1 \times v_C$$

Equating the internal and external virtual work gives $v_C = -2.31$ mm (i.e. downwards).

(ii) Determine horizontal displacement u_C at C

A horizontal virtual force of unit magnitude is introduced at node C. The corresponding virtual bar forces $\overline{N}_i$ and reactions are shown in Figure 7.7b. The internal virtual work is

$$\overline{W}_{int} = \overline{N}_{AD}e_{AD} + \overline{N}_{AB}e_{AB} + \overline{N}_{BD}e_{BD} + \overline{N}_{BC}e_{BC} + \overline{N}_{CD}e_{CD}$$
$$= 0 \times (-0.417) + 1.0 \times (-0.15) + 0 \times 0 + 1.0 \times (-0.15) - 0 \times 1.25 = -0.3 \text{ kNmm}$$

and the external work is:

$$\overline{W}_{ext} = 1 \times u_C$$

Equating the internal and external virtual work gives $u_C = -0.3$ mm (i.e. to the left).
As expected, the joint displacements u_C and v_C calculated here are the same as those determined from geometry in Section 4.8.

WORKED EXAMPLE 7.4

The members AB, BC, CD and DE of the truss shown in Figure 7.8 suffer a temperature rise, while members AF and FE are cooled down. If the temperature-induced changes in length of each member are as shown, calculate the horizontal displacement at support A.

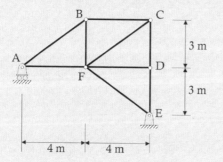

Member extensions e (mm) due to temperature change:

$e_{AB} = +1.5$ mm $\qquad e_{BC} = +1.2$ mm
$e_{CD} = +0.9$ mm $\qquad e_{DE} = +0.9$ mm
$e_{EF} = -1.0$ mm $\qquad e_{AF} = -0.8$ mm
$e_{DF} = e_{CF} = e_{BF} = 0.$

Figure 7.8 Truss for Worked Example 7.4.

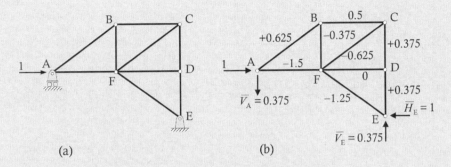

Figure 7.9 Virtual forces $\bar{N}$ (kN).

A horizontal virtual force of unit magnitude is introduced at node A, as shown in Figure 7.9a, and the corresponding virtual bar forces $\bar{N}_i$ and reactions are shown in Figure 7.9b. The internal virtual work is given by:

$$\bar{W}_{int} = \bar{N}_{AB}e_{AB} + \bar{N}_{BC}e_{BC} + \bar{N}_{CD}e_{CD} + \bar{N}_{DE}e_{DE} + \bar{N}_{EF}e_{EF} + \bar{N}_{AF}e_{AF} + \bar{N}_{DF}e_{DF} + \bar{N}_{CF}e_{CF} + \bar{N}_{BF}e_{BF}$$
$$= (0.625 \times 1.5) + (0.5 \times 1.2) + (0.375 \times 0.9) + (0.375 \times 0.9) + [-1.25 \times (-1.0)] + [-1.5 \times (-0.8)]$$
$$+ 0 \times 0 - 0.625 \times 0 - 0.375 \times 0$$
$$= +4.66 \text{ kNmm}$$

and the external work is the product of the unit virtual force and the real displacement at A:
$$\bar{W}_{ext} = \bar{1} \times u_A.$$
Equating the internal and external virtual work gives $u_A = +4.66$ mm (i.e. to the right).

WORKED EXAMPLE 7.5

The steel truss shown in Figure 7.10a is fabricated from members all with a cross-sectional area of $A = 4000$ mm². Assuming all members are stressed in the linear–elastic range, calculate the relative movement of C and E, caused by the applied loads at B, C, D, E and F. The elastic modulus for steel is $E = 200,000$ MPa. The reactions and member forces caused by the applied loads are shown in Figure 7.10b.

As each member is stressed in the linear–elastic range, the member extensions are determined from Equation 4.25 (i.e. $e = NL/EA$) and are given in Table 7.1.
To determine the relative horizontal displacement between joints C and E, introduce a pair of unit virtual forces, as shown in Figure 7.11. The virtual reactions and member forces caused by the pair of virtual loads are also shown in Figure 7.11. The internal virtual work is the sum of the virtual force in each bar multiplied by the extension of the bar caused by the applied loads:

$$\bar{W}_{int} = (1.333 \times 0.6) + (1.333 \times 0.8) + [-1.886 \times (-0.225)] + (0 \times -0.6) + [-1.667 \times (-1.25)]$$
$$+ (-1.667 \times 0.625) + (0 \times 0.3) + [-1.886 \times (-3.375)] + [1.333 \times (-0.4)] + [1.333 \times (-0.3)]$$
$$= +8.76 \text{ kNmm}$$

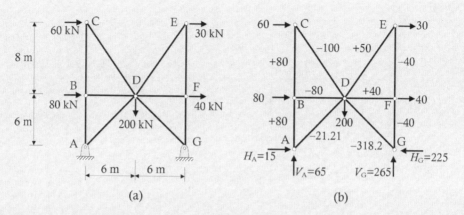

Figure 7.10 Truss for Worked Example 7.5. (a) Truss layout with applied loads. (b) Applied loads, reaction and member forces (kN).

Table 7.1 Bar extensions caused by applied loads

Member	AB	BC	AD	BD	CD	DE	DF	DG	EF	FG
Length	6000	8000	8485	6000	10,000	10,000	6000	8485	8000	6000
N (kN)	+80	+80	−21.21	−80	−100	+50	+40	−318.2	−40	−40
e (mm)	+0.6	+0.8	−0.225	−0.6	−1.25	+0.625	+0.3	−3.375	−0.4	−0.3

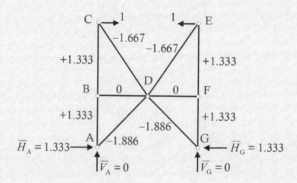

Figure 7.11 Virtual forces $\bar{N}$ (kN).

and the external work is the product of the unit virtual force times the relative horizontal displacement between C and E: $\bar{W}_{ext} = 1 \times u_{CE}$.

Equating the internal and external virtual work gives:

u_{CE} = +8.76 mm

(i.e. C and E move closer together, in the positive direction of the unit forces, by 8.76 mm).

7.4.2 Principle of virtual displacements

The virtual displacement method is illustrated in Worked Example 7.6. It can be used to solve problems in statics. The method is useful for solving more advanced problems in structural analysis, but for simple problems, other methods are usually more convenient.

WORKED EXAMPLE 7.6

For the truss shown in Figure 7.I2a (and previously analysed in Worked Examples 7.3), determine the horizontal reaction at the support D using the method of virtual displacements.

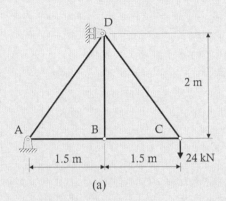
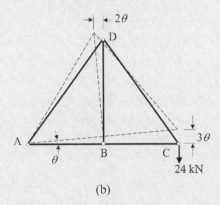

(a)   (b)

Figure 7.I2 Virtual forces (kN).

To determine the unknown reaction at D, we must introduce a virtual displacement for the truss that will cause only that reaction to do work and no other.

If we rotate the truss about the support at A through an angle θ, as shown in Figure 7.I2b, the virtual vertical displacement at C (in the direction of the applied load) is 3θ (upwards) and the virtual horizontal displacement at B (in the direction of the unknown reaction H_D) is 2θ (to the left). Since no truss members undergo any change in length, the internal work is zero. The external work, therefore, must also be zero and is given by:

$$\bar{W}_{ext} = -24 \times 3\theta + H_D \times 2\theta = 0$$

which can be solved to give H_D = +36 kN (i.e. acting to the left in the direction of the virtual displacement at D).

Of course, the same result is readily obtained by taking moments about the support at A.

7.4.3 Transfer coefficients

From the previous considerations on the principle of virtual work, there is clearly a close relationship between the geometry of the displacement field and the equilibrium of the corresponding force field. A convenient way to express this relationship is in terms of *transfer coefficients*. Consider the truss shown in Figure 7.13. Each joint is associated with two numbered displacement directions, also referred to as freedoms in Chapter 4.

If we apply a unit force to the truss at any joint in one of the numbered directions, that force will induce forces in each member of the truss. For example, let us consider the case where a unit force is applied at node F in direction 12. We will refer to the consequent axial

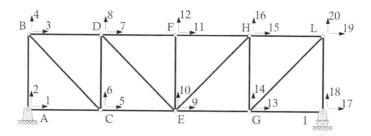

Figure 7.13 Numbering of truss freedoms.

force induced in member HG, for example, as $C_{HG,12}$. The axial force produced along bar HG caused by an applied load P_{12} at F in direction 12 is:

$$N_{HG} = C_{HG,12}P_{12} \tag{7.12}$$

where the coefficient $C_{HG,12}$ is called a *transfer coefficient* because it transfers the force applied at a node in a particular direction into a particular member of the truss.

There is a corresponding relationship between the deformation of that same member and the resulting displacement at the joint in question. Consider again the truss shown in Figure 7.13 loaded with a single vertical force at joint F of magnitude P_{12}. If member HG now suffers an elongation e_{HG} (perhaps owing to a temperature rise of that member) with all other members remaining undeformed, the elongation e_{HG} will result in a vertical displacement at joint F of u_{12}. The resulting external work is $W_{ext} = P_{12}u_{12}$ and the internal work is $W_{int} = N_{HG}e_{HG}$. By applying the principle of work and including Equation 7.12, we get:

$$P_{12}u_{12} = N_{HG}e_{HG} = C_{HG,12}P_{12}e_{HG}$$

from which:

$$u_{12} = C_{HG,12}e_{HG} \tag{7.13}$$

Evidently, in addition to transferring forces from node direction 12 to the member HG, the transfer coefficient $C_{HG,12}$ also transfers displacement from the member HG to the displacement of node F in direction 12.

In summary, a unit force at node F in direction 12 causes a force in member HG of $C_{HG,12}$ and a unit deformation of member HG causes a displacement at node F in direction 12 of $C_{HG,12}$.

7.5 VIRTUAL WORK APPLIED TO BEAMS AND FRAMES

For the purposes of calculating the internal work in the previous section, a truss was subdivided into its various members and the internal work in each member was summed to obtain the internal work in the truss. In the case of beams and frames, it is usual to determine the internal work in small elements of length dx and obtain the internal work in a member by integration. The internal work caused by bending in a beam or frame of length L is the integral of the product of moment and rotation as expressed in Equation 7.10.

7.5.1 Principle of virtual forces

Virtual forces are commonly used to determine the deformation at any point along a beam or frame resulting from a known distribution of curvatures. If the displacement of a beam or frame caused by bending is to be determined at a particular point A, it is convenient to apply a unit virtual force to the structure at the point A in the direction of the desired displacement. The external virtual work is the product of the unit virtual force and the actual displacement at A, while the internal work is the integral over the length of the beam of the virtual moment $\bar{M}(x)$ and the actual curvature $\kappa(x)$ and calculated as:

$$\bar{W}_{int} = \int_0^L \bar{M}(x)\kappa(x)\,dx = \int_0^L \bar{M}(d\theta/dx)\,dx = \int_0^L \bar{M}\left(\frac{d^2v}{dx^2}\right)dx \tag{7.14}$$

WORKED EXAMPLE 7.7

Using the method of virtual forces, determine the displacement u_C at the mid-span of the uniformly loaded, simply-supported beam shown in Figure 7.14. The flexural rigidity EI is constant throughout and the total load on the beam is w_lL. Determine also the rotation at support B (θ_B).

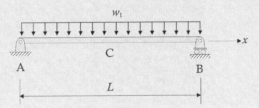

Figure 7.14 Beam for Worked Example 7.7.

The bending moment M at any point x from support A is

$M = 0.5w_lLx - 0.5w_lx^2$

and the curvature caused by the uniformly distributed load w_l (per unit length) is $\kappa = M/EI$.

(i) Calculate u_C

If a unit virtual force is applied at the mid-span C as shown in Figure 7.15a, the virtual bending moment diagram is shown in Figure 7.15b and the virtual bending moment at x from A (when $x \le L/2$) is $\bar{M} = 0.5x$. Because of the discontinuity in the virtual bending moment a mid-span, the internal virtual work is calculated in each half span separately.

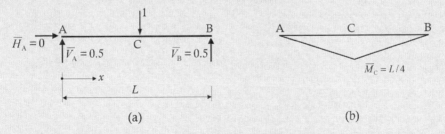

(a)   (b)

Figure 7.15 Unit virtual force at C — reactions and virtual bending moments. (a) Unit virtual force at C and reactions. (b) Virtual bending moment diagram.

In this case, both the virtual moment diagram and the actual curvature diagram are symmetrical about the mid-span and, therefore, the internal virtual work for the entire beam is twice the internal virtual work in the length of beam from A to C:

$$\bar{W}_{int} = 2\int_0^{L/2} \bar{M}\kappa\, dx = 2\int_0^{L/2} 0.5x \left[\frac{1}{EI}(0.5w_1Lx - 0.5w_1x^2)\right]dx$$

$$= \frac{1}{2EI}\int_0^{L/2}(w_1Lx^2 - w_1x^3)dx = \frac{1}{2EI}\left[\frac{w_1L^4}{24} - \frac{w_1L^4}{64}\right] = \frac{5}{384}\frac{w_1L^4}{EI}$$

The external virtual work is $\bar{W}_{ext} = 1 \times u_C$ and the work theorem gives:

$$u_C = \frac{5}{384}\frac{w_1L^4}{EI}$$

(ii) Calculate θ_B

To determine the rotation at support B, a unit virtual couple is applied at support B. The couple and the corresponding reactions are shown in Figure 7.16a and the virtual bending moment diagram is shown in Figure 7.16b. The virtual bending moment at any point along the span at x from A is $\bar{M}(x) = x/L$. The internal virtual work for the span AB is:

$$\bar{W}_{int} = \int_0^L \bar{M}\kappa\, dx = \frac{1}{LEI}\int_0^L \left(0.5w_1Lx^2 - 0.5w_1x^3\right)dx$$

$$= \frac{1}{LEI}\left[\frac{w_1L^4}{6} - \frac{w_1L^4}{8}\right] = \frac{w_1L^3}{24EI}$$

The external work is the product of the virtual couple at B and the rotation at B (θ_B). That is, $\bar{W}_{ext} = 1 \times \theta_B$, and the work theorem gives:

$$\theta_B = \frac{w_1L^3}{24EI}$$

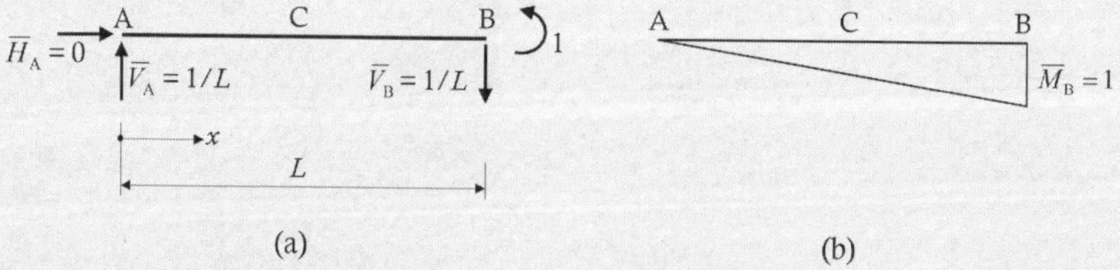

Figure 7.16 Unit virtual couple at B — reactions and virtual bending moment. (a) Unit virtual couple at B and reactions. (b) Virtual bending moment diagram.

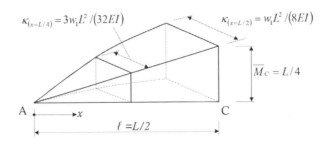

Figure 7.17 Volume integral for half-span AC in Worked Example 7.7.

The definite integral of Equation 7.14 involves the product of two functions of x (namely, the virtual moment $\overline{M}(x)$ and the actual curvature $\kappa(x)$). In many practical problems, these two functions are usually quite simple ($\overline{M}(x)$ is usually linearly varying, while $\kappa(x)$ is often constant, linearly varying or parabolic) and the evaluation of the internal virtual work $\overline{W}_{\text{int}}$ is straightforward.

The integral $\int \overline{M}(x)\kappa(x)\,\mathrm{d}x$ over a length ℓ of a beam or frame can be visualised as the volume of an object of length ℓ, with a rectangular varying cross-section. The plan view of the object is $\kappa(x)$, while the elevation is $\overline{M}(x)$. Figure 7.17 shows the object for the half-span AC of Worked Example 7.7, with the virtual moment varying linearly from A to C and the actual curvature varying parabolically from A to C. The volume of this object is the internal virtual work in the half-span (in this case, $\ell = L/2$). In fact, over any length ℓ of a beam or frame, whenever the curvature diagram is constant, linear or parabolic, and when the virtual moment varies linearly, the internal virtual work can be calculated as follows:

$$\overline{W}_{\text{int}} = \int_0^\ell \overline{M}(x)\kappa(x)\,\mathrm{d}x = \frac{\ell}{6}(\overline{M}_0\kappa_0 + 4\overline{M}_{\ell/2}\kappa_{\ell/2} + \overline{M}_\ell\kappa_\ell) \tag{7.15}$$

where the subscripts 0, $\ell/2$ and ℓ indicate the values of virtual moment and actual curvature at $x = 0$, $x = \ell/2$ and $x = \ell$, respectively.

WORKED EXAMPLE 7.8

The frame in Figure 7.18a was analysed in Worked Example 3.4 and the bending moment diagram is reproduced in Figure 7.18b. Assuming the flexural rigidity is uniform with $EI = 60 \times 10^{12}$ Nmm², the curvature diagram is shown in Figure 7.18c. Using the method of virtual forces, determine the horizontal displacement at point D caused by bending.

The reactions caused by a horizontal unit virtual force acting to the right at D and the corresponding virtual bending moment diagram are shown in Figures 7.19a and b, respectively.

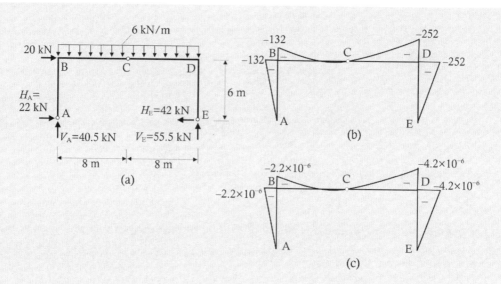

Figure 7.18 Frame for Worked Example 7.8. (a) Frame dimensions and loadings. (b) Bending moment diagram (kNm). (c) Curvature diagram, κ (mm^{-1}).

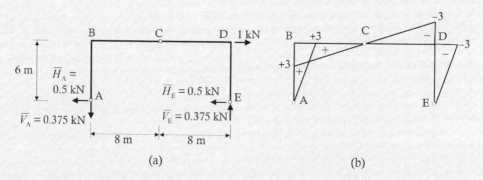

Figure 7.19 Virtual actions for Worked Example 7.8. (a) Virtual force and virtual reactions. (b) Virtual bending moment diagram (kNm).

The internal virtual work in each segment of the frame is calculated using Equation 7.15 as follows:

Segment AB: $L = 6000$ mm; $\bar{M}_0 = 0$; $\bar{M}_{L/2} = +1.5$ kNm; $\bar{M}_L = +3$ kNm; $\kappa_0 = 0$; $\kappa_{L/2} = -1.1 \times 10^{-6}$ mm^{-1}; and $\kappa_L = -2.2 \times 10^{-6}$ mm^{-1}. Therefore:

$$(\bar{W}_{int})_{AB} = \frac{6000}{6}\left[0 \times 0 + 4 \times 1.5 \times 10^6 \times (-1.1 \times 10^{-6}) + 3 \times 10^6 \times (-2.2 \times 10^{-6})\right]$$

$$= -13{,}200 \text{ Nmm}$$

Segment BD: L = 16,000 mm; $\bar{M}_0$ = +3 kNm; $\bar{M}_{L/2}$ = 0; $\bar{M}_L$ = −3 kNm; κ_0 = −2.2 × 10^{-6} mm^{-1}; $\kappa_{L/2}$ = 0; and κ_L = −4.2 × 10^{-6} mm^{-1}. Therefore:

$$(\bar{W}_{int})_{BD} = \frac{16,000}{6}\left[3\times10^6 \times(-2.2\times10^{-6})+4\times0\times0\times(-3\times10^6)\times(-4.2\times10^{-6})\right]$$

$$= +16,000 \text{ Nmm}$$

Segment DE: L = 6000 mm; $\bar{M}_0$ = −3 kNm; $\bar{M}_{L/2}$ = −1.5 kNm; $\bar{M}_L$ = 0; κ_0 = −4.2 × 10^{-6} mm^{-1}; $\kappa_{L/2}$ = −2.1 × 10^{-6} mm^{-1}; and κ_L = 0. Therefore:

$$(\bar{W}_{int})_{DE} = \frac{6000}{6}\left[(-3\times10^6)\times(-4.2\times10^{-6})+4\times(-1.5\times10^6)\times(-2.1\times10^{-6})+0\times0\right]$$

$$= +25,200 \text{ Nmm}$$

The total internal virtual work is therefore:

$$\bar{W}_{int} = (\bar{W}_{int})_{AB} + (\bar{W}_{int})_{BD} + (\bar{W}_{int})_{DE} = +28,000 \text{ Nmm}$$

The external virtual work is the product of the 1 kN (1000 N) virtual force and the lateral deflection u_D:

$$\bar{W}_{ext} = 1000 \times u_D$$

and, from the work theorem $\bar{W}_{ext} = \bar{W}_{int}$, we can calculate u_D to be:

$$\therefore u_D = +28.0 \text{ mm}$$

(in the same direction as the virtual load).

7.5.2 Principle of virtual displacements

When using the principle of virtual displacements, a virtual displacement field is combined with a real force field to determine information about the force field. For example, unknown internal actions may be determined if the external loads are known. Alternatively, if the internal actions are known, the loads on a structure can be determined. For simple structures, it is often far simpler to obtain information about the force field using the principles of statics, but for more advanced analysis, the method of virtual displacements becomes useful.

In Worked Example 7.10, we made use of our knowledge of the bending moment diagram to identify the loading pattern on the beam and the magnitudes of the loads. If we did not identify the loading pattern in this way, or if the bending moment diagram was of a more general shape, the method of virtual displacements could still be used to identify the loads on the beam. By dividing the beam into n small segments, and postulating n virtual displacement fields such that only one small element underwent a

WORKED EXAMPLE 7.9

Using the method of virtual displacements, determine the bending moment at the point D mid-way between the supports of the beam shown in Figure 7.20.

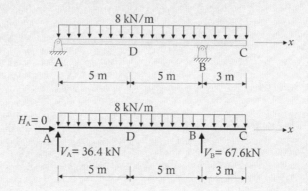

Figure 7.20 Beam for Worked Example 7.9.

The virtual displacement should be such as to cause internal deformation only at the point under consideration (i.e. at point D). Introduce the virtual displacement shown in Figure 7.21, where the member is kinked at point D and the length of beam DC is rotated through an angle $\bar{\theta}$ with respect to AD, as shown. As lengths DC and AD remain straight and undeformed, internal virtual work only occurs at point D and is equal to $\overline{W}_{int} = +M_D\bar{\theta}$.

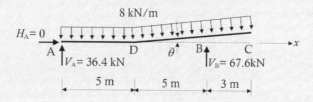

Figure 7.21 Virtual displacement for Worked Example 7.9.

The 8 kNm uniformly distributed load on 8 m length of DC undergoes external virtual work, with the resultant downward load on DC (acting 4 m from D) moving through an upward virtual displacement of $4\bar{\theta}$. The upward reaction at B also causes external virtual work moving through an upward virtual displacement of $5\bar{\theta}$. The total external work is therefore:

$$\overline{W}_{ext} = 67.6 \times 5\bar{\theta} - 8 \times 8 \times 4\bar{\theta} = 82\bar{\theta}$$

Equating internal and external work gives:

$$M_D = 82 \text{ kNm}$$

WORKED EXAMPLE 7.10

The bending moment diagram for a straight beam ABCD is shown in Figure 7.22. Using the method of virtual displacements, determine the loads on the beam, including the reactions.

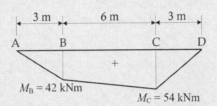

Figure 7.22 Bending moment diagram for the beam of Worked Example 7.10.

We saw in Section 3.5 that the bending moment diagram in a straight beam is linear between points of load application and the bending moment diagram changes direction (kinks) at points where concentrated transverse loads are applied to the beam. If we make use of these observations, it is clear that the beam is loaded with concentrated loads at A, B, C and D, as shown in Figure 7.23a. To determine the four unknown loads P_A, P_B, P_C and P_D, the four virtual displacement fields shown in Figures 7.23b through e are employed. For each displacement field, only one of the external loads produces external virtual work.

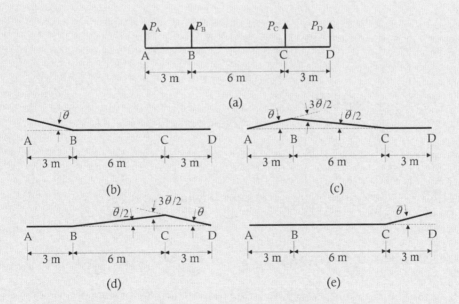

Figure 7.23 Virtual displacements for Worked Example 7.10. (a) Loads on beam ABCD. (b) Virtual displacement field 1. (c) Virtual displacement field 2. (d) Virtual displacement field 3. (e) Virtual displacement field 4.

(i) Displacement Field 1:

$$\bar{W}_{int} = +M_B\bar{\theta} = 42\bar{\theta} \qquad \bar{W}_{ext} = +P_A3\bar{\theta}$$

$$\therefore P_A = \frac{42\bar{\theta}}{3\bar{\theta}} = +14 \text{ kN (i.e. upwards)}$$

(ii) Displacement Field 2:

$$\bar{W}_{int} = +M_A\bar{\theta} + M_B\times(-3\bar{\theta}/2) + M_C\bar{\theta}/2$$
$$= 0\times\bar{\theta} + 42\times(-3\bar{\theta}/2) + 54\times\bar{\theta}/2 = -36\bar{\theta}$$

$$\bar{W}_{ext} = +P_B3\bar{\theta} \quad \therefore P_B = \frac{-36\bar{\theta}}{3\bar{\theta}} = -12 \text{ kN (i.e. downwards)}$$

(iii) Displacement Field 3:

$$\bar{W}_{int} = M_B(\bar{\theta}/2) + M_C(-3\bar{\theta}/2) + M_D\bar{\theta}$$
$$= 42\times(\bar{\theta}/2) + 54\times(-3\bar{\theta}/2) + 0\times\bar{\theta} = -60\bar{\theta}$$

$$\bar{W}_{ext} = +P_C3\bar{\theta} \quad \therefore P_C = \frac{-60\bar{\theta}}{3\bar{\theta}} = -20 \text{ kN (i.e. downwards)}$$

(iv) Displacement Field 4:

$$\bar{W}_{int} = +M_C\bar{\theta} = 54\bar{\theta} \qquad \bar{W}_{ext} = +P_D3\bar{\theta}$$

$$\therefore P_D = \frac{54\bar{\theta}}{3\bar{\theta}} = +18 \text{ kN (i.e. upwards)}$$

virtual displacement in each displacement field, the load on each small element could then be determined.

7.6 CASTIGLIANO'S THEOREM

If a structure behaves in a linear–elastic manner, the deflection or slope at any point resulting from applied loads may be determined using a technique known as *Castigliano's theorem* (also called the *method of least work*). To calculate a displacement, a force P is applied to the structure at the point in question and in the direction of the required displacement. The magnitude of the displacement u is equal to the first partial derivative of the strain energy in the structure with respect to P. That is:

$$u = \frac{\partial U_e}{\partial P} \tag{7.16}$$

Similarly, to calculate a slope, a moment M is applied to the structure at the point in question and in the direction of rotation. The magnitude of the slope is equal to the first partial derivative of the strain energy in the structure with respect to M.

7.6.1 Application to trusses

For a truss member of uniform cross-section, length L and carrying an axial force N, the elastic strain is obtained from Equation 7.4 as:

$$U_e = \frac{N^2 L}{2EA}$$

Substituting this into Equation 7.16 and summing over all members of the truss gives:

$$u = \frac{\partial}{\partial P} \Sigma \frac{N^2 L}{2EA}$$

For most problems, it is more convenient to differentiate the strain energy in each bar before the summation. When the elastic modulus E and the cross-sectional area A of a member are constant, the displacement may be determined from:

$$u = \Sigma \frac{\partial N}{\partial P} \frac{NL}{EA} = \Sigma \frac{\partial N}{\partial P} e \tag{7.17}$$

where u is the joint displacement in the truss in the direction of the applied load P, N is the axial load in a member caused by both the external loads on the truss and the load P, and e is the elongation of the member caused by the external loads, i.e. when P is infinitesimally small. This equation is similar to that developed in the method of virtual forces $\left(1u = \Sigma \bar{N}e = \Sigma \bar{N} \, NL/EA\right)$, with the partial derivative $\partial N/\partial P$ replacing the virtual force in each bar $\bar{N}$, and is simply a different way of looking at the same thing, with both terms representing the change in the internal member force caused by a unit load P.

WORKED EXAMPLE 7.11

Using Castigliano's theorem, calculate the vertical and horizontal displacements at node C of the truss shown in Figure 7.6 (previously analysed using virtual work in Worked Example 7.3). The reactions and bar forces caused by the applied load are also shown in Figure 7.6, together with the resulting bar extensions (determined earlier in Section 4.8). As before, the truss is fabricated from timber with $E = 12{,}000$ MPa and the cross-sectional areas of the bars are 15,000 mm² for bars AB, BC and AD and 5000 mm² for bars BD and CD.

(i) Determine the vertical displacement v_C at C

A vertical upward force P is applied to node C (in addition to the 24 kN applied load) as shown in Figure 7.24a. The resulting reactions are shown in Figure 7.24b.

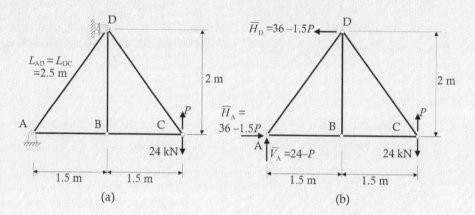

Figure 7.24 Truss and free-body diagram for Worked Example 7.11 — part i.

Using the method of joints, the forces in each member of the truss in Figure 7.24b are calculated in terms of the applied loads and P. The partial differential $\partial N/\partial P$ is determined, as are the elongations of the member caused by the applied loads (with $P = 0$), and the summation of Equation 7.17 is performed.

Member	L (mm)	A (mm²)	N (kN)	$\frac{\partial N}{\partial P}$	N (kN) (P = 0)	e = NL/EA (mm)	$\frac{\partial N}{\partial P} e$ (mm)
AB	1500	15,000	−18 + 0.75P	+0.75	−18	−0.15	−0.113
AD	2500	15,000	−30 + 1.25P	+1.25	−30	−0.417	−0.521
BC	1500	15,000	−18 + 0.75P	+0.75	−18	−0.15	−0.113
BD	2000	5000	0	0	0	0	0
CD	2500	5000	+30 − 1.25P	−1.25	+30	+1.25	−1.563
						Σ	−2.31

From Equation 7.17, the vertical displacement at C in the direction of P is:

$$v_C = -2.31 \text{ mm (downwards)}$$

which is the same as obtained using the virtual work in Worked Example 7.3.

(ii) Determine the horizontal displacement u_C at C

A horizontal force P (acting to the right) is applied to node C (in addition to the 24 kN applied load) as shown in Figure 7.25a. The resulting reactions are shown in Figure 7.25b.

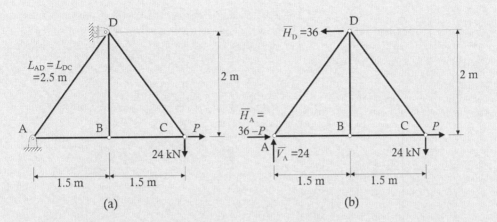

Figure 7.25 Truss and free-body diagram for Worked Example 7.11 — part ii.

Member	L (mm)	A (mm²)	N (kN)	$\dfrac{\partial N}{\partial P}$	N (kN) (P = 0)	e = NL/EA (mm)	$\dfrac{\partial N}{\partial P}$ e (mm)
AB	1500	15,000	−18 + P	+1	−18	−0.15	−0.15
AD	2500	15,000	−30	0	−30	−0.417	0
BC	1500	15,000	−18 + P	+1	−18	−0.15	−0.15
BD	2000	5000	0	0	0	0	0
CD	2500	5000	+30	0	+30	+1.25	0
						Σ	−0.3

From Equation 7.17, the horizontal displacement at C in the direction of P is:

$$u_C = -0.3 \text{ mm}$$

and this is the same as that obtained using the principle of virtual work in Worked Example 7.3.

WORKED EXAMPLE 7.12

For the steel truss shown in Figure 7.10 and previously analysed in Worked Example 7.5, calculate the relative movement of C and E, caused by the applied loads, using Castigliano's theorem. All data are as specified in Worked Example 7.5.

As calculated previously, the member extensions caused by the applied loads are as follows:

Member	AB	BC	AD	BD	CD	DE	DF	DG	EF	FG
e (mm)	+0.6	+08	−0.225	−0.6	−1.25	+0.625	+0.3	−3.375	−0.4	−0.3

To determine the relative horizontal displacement between joints C and E, a pair of forces of magnitude P is introduced at C and E, as shown in Figure 7.26b, together with the resulting member forces. The internal actions caused by the applied loads, previously calculated in Worked Example 7.5, are shown in Figure 7.26a.

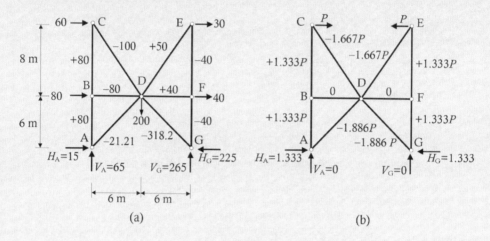

(a)   (b)

Figure 7.26 Truss for Worked Example 7.12. (a) Applied loads, reactions and member forces (kN). (b) Internal forces caused by loads P.

The terms included in the right-hand side of Equation 7.17 are evaluated in the table below.

Member	$e = NL/EA$ (mm)	N (kN)	$\dfrac{\partial N}{\partial P}$	$\dfrac{\partial N}{\partial P}e$ (mm)
AB	+0.6	$80 + 1.333P$	+1.333	+0.800
BC	+0.8	$80 + 1.333P$	+1.333	+1.066
AD	−0.225	$-21.21 - 1.886P$	−1.886	+0.424
BD	−0.6	−80	0	0
CD	−1.25	$-100 - 1.667P$	−1.667	+2.083
DE	+0.625	$+50 - 1.667P$	−1.667	−1.042
DF	+0.3	+40	0	0
DG	−3.375	$-318.2 - 1.886P$	−1.886	+6.365
EF	−0.4	$-40 + 1.333P$	+1.333	−0.533
FG	−0.3	$-40 + 1.333P$	+1.333	+0.400

Points C and E move toward each other, i.e. in the direction of the applied loads P, by an amount $u_{CE} = +8.76$ mm, which is the same as the displacement obtained using virtual work in Worked Example 7.5.

7.6.2 Application to beams and frames

For a beam or frame, the elastic strain caused by bending is given by Equation 7.8a, i.e. $U_e = \int_0^L (M^2/2EI)\,dx$. Substituting this into Equation 7.16 leads to:

$$u = \frac{\partial}{\partial P}\int_0^L [M^2/(2EI)\,dx$$

Differentiating before the integration and assuming EI remains constant throughout, we get:

$$u = \int_0^L \left(\frac{\partial M}{\partial P}\right)\frac{M}{EI}\,dx = \int_0^L \left(\frac{\partial M}{\partial P}\right)\kappa\,dx \tag{7.18a}$$

where u is the displacement caused by the real loads in the direction of and at the point of application of the external force P; M is the bending moment in the beam or frame caused by both the external loads and the force P, expressed as a function of the distance x along the axis of the member; and $\kappa = (M/EI)$ is the curvature caused by the external loads, i.e. when P is infinitesimally small.

This equation is similar to that developed in the method of virtual forces, namely, $\overline{1}u = \int_0^L \overline{M}(x)(M/EI)\,dx$, except that the partial derivative $\partial M/\partial P$ has replaced the virtual moment $\overline{M}(x)$.

If the slope θ is required at a particular point, a couple M^* is applied at the point of interest and Equation 7.18a becomes:

$$\theta = \int_0^L \left(\frac{\partial M}{\partial M^*}\right)\frac{M}{EI}\,dx = \int_0^L \left(\frac{\partial M}{\partial M^*}\right)\kappa\,dx \tag{7.18b}$$

It is noted that these equations account only for deformation caused by bending strain energy, which is usually responsible for the majority of deformations in most beams and frames. Deformations resulting from the strain energies caused by shear, axial force and torsion are not included here, but may be determined from expressions similar to Equations 7.18a and b.

WORKED EXAMPLE 7.13

Using Castigliano's theorem, determine the deflection v_C at the mid-span and the rotation at the support B (θ_B) of the uniformly loaded, simply-supported beam shown in Figure 7.14 and analysed previously in Worked Example 7.7.

(i) Calculate v_C

Applying a downward load P at mid-span, the bending moment M caused by the actual uniformly distributed load w_l and the load P at any point x from support A is:

$$M = 0.5w_l Lx + (Px/2) - 0.5w_l x^2 \qquad \text{for } x \le L/2$$

$$M = 0.5w_l Lx - P(x - L)/2 - 0.5w_l x^2 \qquad \text{for } x > L/2$$

The partial derivative $\partial M/\partial P$ is therefore:

$\partial M/\partial P = x/2$ for $x \le L/2$

$\partial M/\partial P = -(x - L)/2$ for $x > L/2$

The vertical deflection at mid-span caused by the applied load is obtained from Equation 7.18a:

$$u_C = \int_0^L \left(\frac{\partial M}{\partial P}\right)\kappa\, dx = \frac{1}{4EI}\int_0^{L/2}(w_1 L x^2 - w_1 x^3)\,dx + \frac{1}{4EI}\int_{L/2}^L(-w_1 L x^2 + w_1 L^2 x + w_1 x^3 - w_1 L x^2)\,dx$$

$$= \frac{5}{768}\frac{w_1 L^4}{EI} + \frac{5}{768}\frac{w_1 L^4}{EI} = \frac{5}{384}\frac{w_1 L^4}{EI}$$

(ii) Calculate θ_B

Applying a counterclockwise couple M^* at B, the bending moment caused by the uniformly distributed load w_1 and the couple M^* at any point x from A is:

$M = 0.5 w_1 L x + x M^*/L - 0.5 w_1 x^2$ for $0 \le x \le L$

The partial derivative $\partial M/\partial M^*$ is therefore x/L and the rotation at support B caused by the applied load can be obtained using Equation 7.18b as:

$$\theta_B = \int_0^L \left(\frac{\partial M}{\partial M^*}\right)\kappa\, dx = \frac{1}{2EI}\int_0^L\left(w_1 x^2 - \frac{w_1 x^3}{L}\right)dx = \frac{1}{2EI}\left(\frac{w_1 L^3}{3} - \frac{w_1 L^3}{4}\right) = \frac{w_1 L^3}{24EI}$$

The expressions for the deflection and the rotation determined here are identical to those derived using virtual work in Worked Example 7.7.

WORKED EXAMPLE 7.14

Using Castigliano's theorem, determine the horizontal displacement at point D caused by bending for the frame in Figure 7.18a (previously analysed in Worked Example 7.8). The flexural rigidity $EI = 60 \times 10^{12}$ Nmm^2 ($= 60 \times 10^3$ kNm^2) is constant throughout.

A horizontal load P is applied at D (acting to the right). The bending moment diagrams caused by the applied loads and by the force P are shown in Figure 7.27. These are combined and expressed algebraically, together with the partial derivative $\partial M/\partial P$, for each segment of the frame, as outlined below.

Segment AB: With x measured vertically from the support at A, the bending moment caused by both the applied loads and P and its derivative are:

$M_{AB} = -22x + 0.5Px$ and $\partial M_{AB}/\partial P = 0.5x$ (for $x = 0$ to 6 m)

Segment BD: With x measured horizontally from the frame corner at B, the bending moment caused by applied loads and P and its derivative are:

$M_{BD} = (40.5 - 0.375P)x - (132 - 3P) - 3x^2$ and $\partial M_{BD}/\partial P = 3 - 0.375x$ (for $x = 0$ to 16 m)

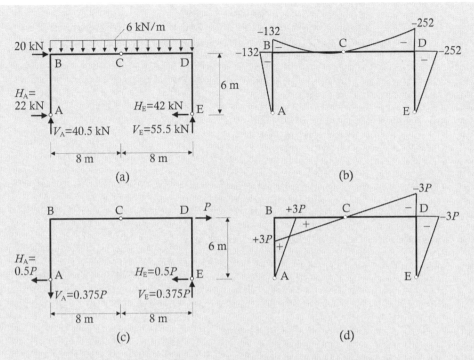

Figure 7.27 Frame for Worked Example 7.14. (a) Frame dimensions and loading. (b) Bending diagram due to applied loads (kNm). (c) Force P and resulting reactions. (d) Bending moment due to P.

Segment ED: With x measured vertically from the support at E, the bending moment caused by applied loads and P and its derivative are:

$M_{ED} = -42x - 0.5Px$ and $\partial M_{ED}/\partial P = -0.5x$ (for x = 0 to 6 m)

The horizontal deflection at D u_D: is calculated applying Equation 7.18a to each segment:

$$u_D = \int_0^6 \left(\frac{\partial M_{AB}}{\partial P}\right)\kappa_{AB}\,dx + \int_0^{16} \left(\frac{\partial M_{BD}}{\partial P}\right)\kappa_{BD}\,dx + \int_0^6 \left(\frac{\partial M_{ED}}{\partial P}\right)\kappa_{ED}\,dx$$

$$= \frac{1}{EI}\int_0^6 0.5x \times (-22x)\,dx + \frac{1}{EI}\int_0^{16}(3 - 0.375x) \times (40.5x - 132 - 3x^2)\,dx$$

$$+ \frac{1}{EI}\int_0^6 -0.5x \times (-42x)\,dx$$

$$= \frac{1}{EI}\left[-3.667x^3\right]_0^6 + \frac{1}{EI}\left[-396x + 85.5x^2 - 8.0625x^3 + 0.28125x^4\right]_0^{16}$$

$$+ \frac{1}{EI}\left[+7.00x^3\right]_0^6$$

$$= \frac{1}{60 \times 10^3}(-792 + 960 + 1512)$$

$$= 0.028 \text{ m} = 28 \text{ mm}$$

which is identical to the displacement calculated in Worked Example 7.8.

PROBLEMS

7.1 Determine expressions for the deflection and the slope at the free end of the cantilevers shown. Assume *EI* is constant. Use the principle of virtual work.

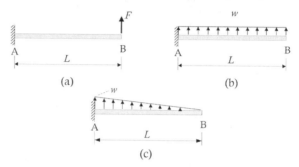

7.2 Solve Problem 7.1 using Castigliano's theorem.

7.3 If the flexural rigidity *EI* of each of the propped cantilevers shown is constant, determine the expression for the deflection at the mid-span of each beam in terms of the load, the span *L* and flexural rigidity using the principle of virtual work.

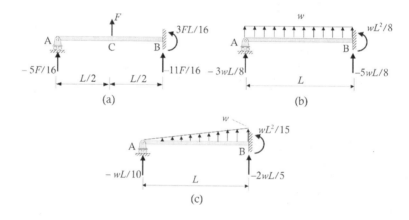

7.4 Solve Problem 7.3 using Castigliano's theorem.

7.5 For the truss shown, calculate the vertical displacement of the joint E using the principle of virtual work. Take $EA = 60 \times 10^6$ N for each member of the truss.

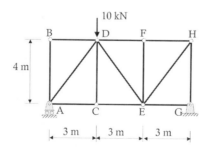

7.6 For the truss shown, determine the vertical displacement of joint D and the horizontal displacement of the roller support at A using the principle of virtual work. Take $EA = 100 \times 10^6$ N for each member of the truss.

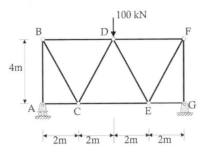

7.7 Solve Problem 7.6 using Castigliano's theorem.

7.8 Using the principle of virtual work, determine the vertical displacement at joint C of the truss shown. Take $EA = 50 \times 10^6$ N for each member of the truss.

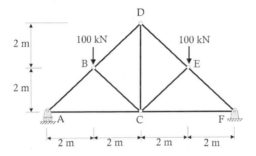

7.9 Solve Problem 7.8 using Castigliano's theorem.

7.10 For the beam shown, if $M_1 = 50$ kNm and $M_2 = 70$ kNm (in the directions shown), determine the vertical deflection at C using the principle of virtual work. Take $EI = 10 \times 10^3$ kNm² throughout.

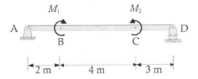

7.11 Solve Problem 7.10 using Castigliano's theorem.

7.12 Determine the vertical displacement of joint J of the truss shown using the principle of virtual work. Take $EA = 5 \times 10^8$ N for each member of the truss.

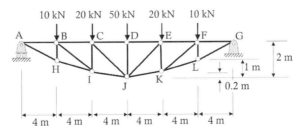

7.13 Solve Problem 7.12 using Castigliano's theorem.

7.14 Determine the vertical displacement of joint E and the horizontal displacement of the roller support at A for the truss shown using the principle of virtual work. Take $EA = 50 \times 10^6$ N for each member of the truss. The point loads are perpendicular to segment ABDG.

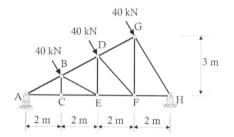

7.15 Solve Problem 7.14 using Castigliano's theorem.

7.16 Using virtual forces, calculate the horizontal and vertical displacements of joint F in the timber truss shown. Take $E = 12,000$ MPa and $A = 10,000$ mm² for each member.

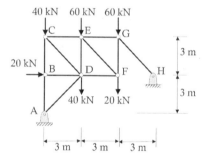

7.17 The beam shown below is supported by a pin at A and a roller at D. Using virtual work, determine the vertical deflection at B and the horizontal movement at E caused by the applied loads. Take $EI = 5 \times 10^3$ kNm² throughout.

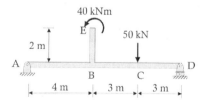

7.18 The truss members shown have cross-sectional area $A = 15{,}000$ mm^2 and elastic modulus $E = 12{,}000$ MPa. Calculate the horizontal and vertical displacements u_C and v_C of node C using the principle of virtual work.

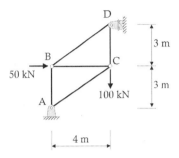

7.19 The beam ABCDEF shown is pinned at A, supported on rollers at C, D and F, and has internal hinges at B and E as shown. Determine the deflection at the internal hinge at E. Assume $EI = 3000 \times 10^3$ kNm^2 throughout. Use the principle of virtual work.

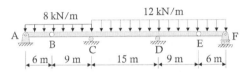

7.20 Calculate the vertical deflection at joint D of the truss shown below using the principle of virtual work. Take $E = 10{,}000$ MPa and $A = 4000$ mm^2 for each member.

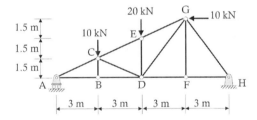

7.21 Solve Problem 7.20 using Castigliano's theorem.

7.22 If all the horizontal and vertical members of the truss shown have $L = 2$ m, determine the vertical displacement of joint F using the principle of virtual work. Take $E = 10{,}000$ MPa and $A = 6000$ mm^2 for each truss member.

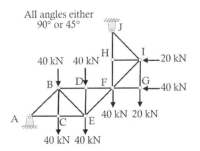

7.23 The cantilevered frame ABCD shown is part of the support structure for a grandstand. For the loading shown, determine the vertical and horizontal deflection at D using the principle of virtual work. Take $EI = 2000 \times 10^3$ kNm² throughout.

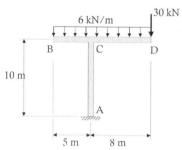

7.24 For the portal frame ABCDE, determine the vertical deflection at C using the principle of virtual work. Take $EI = 2000 \times 10^3$ kNm² throughout.

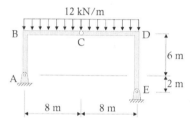

7.25 For the portal frame ABCDE, determine the vertical deflection at C using the principle of virtual work. Take $EI = 2000 \times 10^3$ kNm² throughout.

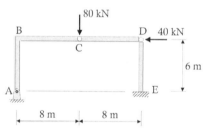

7.26 For the truss shown, $A = 6000$ mm² and $E = 20,000$ MPa for each truss member. Using the principle of virtual forces, determine:
  i. the displacement of joint A in the z direction; and
  ii. the displacement of joint D in the y direction.

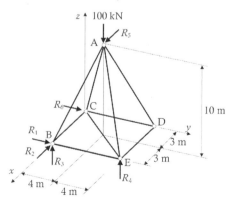

Chapter 8

The force method

8.1 INTRODUCTION

We have seen that for statically determinate structures, deformations of individual cross-sections can take place without restraints being introduced at the supports, and reactions and internal actions can be determined using only the principles of statics. For any set of loads on a statically determinate structure, there is one set of reactions and internal actions that satisfies equilibrium, i.e. there is a single load path.

In Sections 2.9, 3.4 and 4.7, we also saw that a statically indeterminate structure is one where the number of unknown reactions is greater than the number of equilibrium equations available for the analysis. For any set of applied loads, there are an infinite number of sets of reactions and internal actions that satisfy equilibrium, but only one set that also satisfies geometric compatibility and the stress–strain relationships for the constituent materials at each point in the structure.

We have seen in Chapter 6 that the reactions and internal actions in a statically indeterminate structure depend on the relative stiffness of the individual parts of the structure. The stiffness of any cross-section may change as the load level increases and may also change with time and, consequently, so too will the reactions and internal actions. Imposed deformations, such as may occur owing to a temperature change or a support settlement, will cause changes in the reactions and internal actions in an indeterminate structure. If the indeterminate structure is made of reinforced concrete, for example, cracking of the concrete in one region causes a sudden change in stiffness and results in a change in the reactions and internal actions.

In this chapter, we turn our attention to the analysis of statically indeterminate structures. There are two broad approaches:

    i. the *force method* (also called the *flexibility method*)
    ii. the displacement method (also called the stiffness method)

For one- or two-fold indeterminate structures, the force method is a convenient approach. It involves satisfying equations that enforce geometric compatibility of the structure, i.e. equations that ensure that the individual parts of the structure fit together after deformation without discontinuities, and then solving these equations to determine the redundant forces. With the redundant forces determined, the remaining unknown forces can be calculated from statics. The force method applied to statically indeterminate trusses, beams and frames is discussed in this chapter.

For indeterminate structures with more than one or two redundant reactions, the displacement method is usually more convenient. Various forms of the displacement method are discussed subsequently in Chapters 9 through 13. The force–displacement relationships are developed for the members of the structure and the unknown displacements are obtained by satisfying the requirements of equilibrium. With the displacements so determined, the forces are readily calculated from the appropriate stress–strain relationships.

Most structures are in fact statically indeterminate. Building frames, for instance, usually have many more reactions than available equilibrium equations. The columns and beams in reinforced concrete buildings are usually poured on-site as continuous members, and rotations and movements at the supports and connections are usually at least partially restrained. Reinforced concrete structures are rarely statically determinate.

Although indeterminate structures require more effort for the analysis, they have many advantages over statically determinate members. The magnitudes of the internal actions and deformations are often much smaller. For a beam made continuous over several spans, the maximum bending moments and mid-span deflections are significantly smaller than those that would occur in simply-supported beams over the same spans. For example, the magnitude of the maximum moment in a uniformly loaded fixed-ended beam is two-thirds of that in a similarly loaded simply-supported beam of the same span, while the magnitude of the maximum deflection is only one-fifth. The reduced demand for strength and the increased stiffness permit shallower member cross-sections, often providing both significant savings in material and construction costs and a more slender aesthetically pleasing structure. Continuity also provides increased resistance to transient loads and to progressive collapse that could result, for example, from earthquake, blast or impact loading.

In statically indeterminate structures, failure of one member or cross-section does not necessarily jeopardise the entire structure, as a redistribution of internal actions may occur provided the structure is sufficiently ductile and an alternative load path is available.

Notwithstanding the advantages, statically indeterminate structures may have certain disadvantages over determinate structures. For some types of construction, additional costs associated with providing continuity at the connections and supports may outweigh the material savings associated with the more slender members. Differential displacement of the supports of indeterminate structures will introduce internal actions into the structure and, if sufficiently large, these could compromise the strength or the serviceability of the structure.

8.2 THE FORCE METHOD APPLIED TO TRUSSES

8.2.1 Determination of member forces in an *n*-fold indeterminate truss

In Section 4.7, we saw that for a plane truss, with j nodes, b members and r reaction components, there are $2j$ equilibrium equations available and a total of $r + b$ unknown forces. Accordingly, plane trusses are classified as statically indeterminate if the number of unknown forces exceeds the number of equilibrium equations.

A truss is said to be *n*-fold indeterminate if $(r + b) - 2j = n$ (see Section 4.7). To analyse such a truss using the force method, n of the members of the truss (or redundant reactions) are selected and then removed or disconnected from the truss to create a statically determinate truss known as *the primary truss*. The selection of the redundant forces is arbitrary, except that the primary truss and its reactions must be stable. Not only must the primary truss carry the external loads, it must also carry the unknown force in each redundant truss member and the unknown redundant reactions. For each redundant truss member, a pair of unknown member forces is applied to the primary truss, one force applied in the direction of the member at each of the disconnected nodes.

Consider the statically indeterminate truss of Figure 8.1a. There are 6 nodes ($j = 6$), 10 members ($b = 10$) and 4 reactions ($r = 4$). Therefore, $(r + b) - 2j = 2$ and the truss is 2-fold

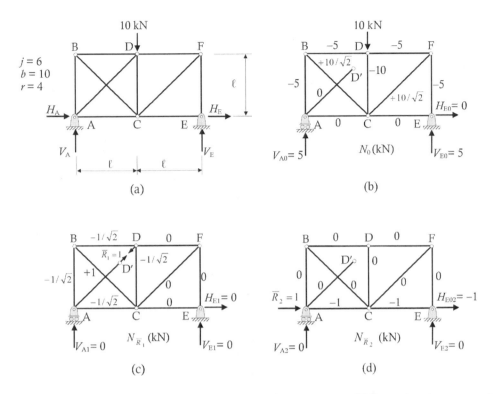

Figure 8.1 Twofold indeterminate truss. (a) Statistically in determine truss. (b) Forces in primary truss (N_0) carrying external loads. (c) Forces in primary truss $\left(N_{\bar{R}_1}\right)$ due to unit force in redundant member AD $\left(\bar{R}_1\right)$. (d) Forces in primary truss $\left(N_{\bar{R}_2}\right)$ due to a unit horizontal force at A $\left(\bar{R}_2\right)$.

indeterminate. For the purposes of this illustrative example, let us assume that the truss is fabricated from steel with $E = 200$ kN/mm², the dimension $\ell = 2000$ mm, the cross-sectional area of members AD, BC and CF is 600 mm² and, for all other truss members, $A = 2000$ mm². The axial flexibility for each bar ($f = L/EA$) is shown in column 3 of Table 8.1.

The member AD in the truss of Figure 8.1a is selected as a redundant member (redundant 1) and the horizontal reaction at A is selected as a redundant reaction (redundant 2). With member AD disconnected at D and reaction H_A removed, the stable primary truss is shown in Figure 8.1b through d. The reactions and member forces N_0 in the primary truss caused by the external applied loads are shown in Figure 8.1b. It is convenient to analyse the primary truss for unit values $\bar{R}_1$ and $\bar{R}_2$ of each of the unknown redundant forces $R_1 = N_{AD}$ and $R_2 = H_A$. The reactions and member forces in the primary truss caused by the unit force $\bar{R}_1$ in member AD are shown in Figure 8.1c, while the reactions and member forces in the primary truss caused by a unit horizontal force $\bar{R}_2$ at A are shown in Figure 8.1d. The bar forces N_0 caused by the external loads and the bar forces $N_{\bar{R}_1}$ and $N_{\bar{R}_2}$ caused by $\bar{R}_1$ and $\bar{R}_2$, respectively, are given in columns 4, 5 and 6 of Table 8.1.

We know that in the statically indeterminate truss, D and D′ are at the same point both before and after deformation, and therefore the gap between D and D′ is zero, i.e. $u_{DD'} = u_1 = 0$. The subscript "1" highlights the fact that the displacement u is in the direction of the redundant force R_1. Similarly, in the statically indeterminate truss, the displacement

Table 8.1 Determination of displacements for the truss of Figure 8.1

(1) Member	(2) Length (mm)	(3) $f = L/EA$ (mm/kN)	(4) N_0 (kN)	(5) $N_{\bar{R}_1}$ (kN)	(6) $N_{\bar{R}_2}$ (kN)	(7) $N_{\bar{R}_1} e_0$ (kNmm)	(8) $N_{\bar{R}_2} e_0$ (kNmm)	(9) $N_{\bar{R}_1} e_{\bar{R}_1}$ (kNmm)	(10) $N_{\bar{R}_2} e_{\bar{R}_1}$ (kNmm)	(11) $N_{\bar{R}_1} e_{\bar{R}_2}$ (kNmm)	(12) $N_{\bar{R}_2} e_{\bar{R}_2}$ (kNmm)
AB	2000	0.005	−5.0	−0.7071	0	0.0177	0	0.0025	0	0	0
AC	2000	0.005	0	−0.7071	−1.0	0	0	0.0025	0.00354	0.00354	0.005
AD	2828.4	0.02357	0	1.0	0	0	0	0.0236	0	0	0
BC	2828.4	0.02357	+7.071	1.0	0	0.1667	0	0.0236	0	0	0
BD	2000	0.005	−5.0	−0.7071	0	0.0177	0	0.0025	0	0	0
CD	2000	0.005	−10.0	−0.7071	0	0.0354	0	0.0025	0	0	0
CE	2000	0.005	0	0	−1.0	0	0	0	0	0	0.005
CF	2828.4	0.02357	+7.071	0	0	0	0	0	0	0	0
DF	2000	0.005	−5.0	0	0	0	0	0	0	0	0
EF	2000	0.005	−5.0	0	0	0	0	0	0	0	0
				Displacements (mm)		$u_{10} = 0.2375$	$u_{20} = 0.0$	$u_{11} = 0.0572$	$u_{21} = 0.00354$	$u_{12} = 0.00354$	$u_{22} = 0.01$

Note: $e_0 = N_0 L/EA$; $e_{\bar{R}_1} = N_{\bar{R}_1} L/EA$; $e_{\bar{R}_2} = N_{\bar{R}_2} L/EA$.

of Node A in the direction of H_A is also zero, i.e. $u_A = u_2 = 0$ (where the subscript "2" refers to the direction of redundant R_2). However, in the primary truss, the deformation in each member caused by each set of member forces may result in non-zero values of these displacements.

The displacement caused by the external loads at the release at D (release 1) is u_{10} and represents the sum of the gaps between D and D′ in the primary truss caused by the extension e_0 (= $N_0 L / EA$) in each member. This gap between D and D′ caused by the extensions e_0 of each member of the truss may be determined by the work equation $W_{ext} = W_{int}$ (see Sections 7.2 through 7.4), where $W_{ext} = 1 \times u_{10}$ and $W_{int} = \sum_{i=1}^{n}(N_{\bar{R}_1} e_0)_i$. For clarity, the product $(N_{\bar{R}_1} e_0)_i$ has been included for each bar in column 7 of Table 8.1. The corresponding displacement u_{10} is obtained by solving the following equation derived with the principle of virtual work:

$$1 \times u_{10} = \sum_{i=1}^{n}(N_{\bar{R}_1} e_0)_i \tag{8.1}$$

from which:

$$u_{10} = \frac{\sum_{i=1}^{n}(N_{\bar{R}_1} e_0)_i}{1} \tag{8.2}$$

where the unit force has been maintained in the denominator of Equation 8.2 to clarify the dimensions of all terms: u_{10} is a displacement (i.e. a length), $N_{\bar{R}_1}$ is a force and e_0 represents an elongation (i.e. a length). Performing the summation of all terms $(N_{\bar{R}_1} e_0)_i$, as specified in Equation 8.2, we can get u_{10}, which is given at the bottom of column 7 in Table 8.1. Similarly, the horizontal displacement u_{20} of the primary truss at A (release 2) caused by the external loads can be determined from $1 \times u_{20} = \sum_{i=1}^{n}(N_{\bar{R}_2} e_0)_i$ and is given at the bottom of column 8 in Table 8.1. The displacements of the primary truss at releases 1 and 2 caused by a unit force applied in member AD between D and D′ (u_{11} and u_{21}, respectively) are given in columns 9 and 10. The corresponding displacements caused by a unit horizontal force at A (u_{12} and u_{22}) are calculated in columns 11 and 12 in Table 8.1.

The actual values of the redundant forces N_{AD} and H_A are determined by enforcing the requirement that the displacement at each release is zero. Equation 8.3a states that the sum of the gaps between D and D′ owing to the external loads and the two unknown redundant forces is zero (i.e. $u_1 = 0$), while Equation 8.3b enforces the requirement that the horizontal displacement at A must also be zero (i.e. $u_2 = 0$):

$$u_{10} + f_{11}R_1 + f_{12}R_2 = 0.2375 + 0.0572 N_{AD} + 0.00354 H_A = 0 \tag{8.3a}$$

$$u_{20} + f_{21}R_1 + f_{22}R_2 = 0.0 + 0.00354 N_{AD} + 0.01 H_A = 0 \tag{8.3b}$$

where f_{ij} are the flexibility coefficients of the structure that specify the displacement at release i produced by a unit action $\bar{R}_j$ applied along release j. We have already calculated these values in Table 8.1. For example, f_{11} describes the displacement at release 1, i.e. relative movement between D and D′, produced by a unit force applied along release

1 (Figure 8.1c), and is equal to u_{11}. Despite this, these two terms are dimensionally different, as u_{11} is a displacement (i.e. a length) and f_{11} represents a displacement caused by a unit load (i.e. length/force). In a similar manner, f_{21} is the displacement along release 2, i.e. horizontal displacement at A, caused by a unit force applied along freedom 1 (Figure 8.1d) and is equal (in direction and magnitude) to u_{21} previously calculated in Table 8.1.

Solving these two simultaneous equations (Equations 8.3a and b) gives the two unknown redundant forces:

$$R_1 = N_{AD} = -4.247 \text{ kN} \quad \text{and} \quad R_2 = H_A = +1.502 \text{ kN}$$

We can now determine the force in each bar N by adding the forces in the bar caused by each of the three loadings on the primary truss (i.e. the external loads, the redundant force N_{AD} and the redundant reaction H_A). For any truss member:

$$N = N_0 + C_{\bar{R}_1} N_{AD} + C_{\bar{R}_2} H_A \tag{8.4}$$

where $C_{\bar{R}_1}$ and $C_{\bar{R}_2}$ are the transfer coefficients (see Section 7.4.3) determined from:

$$C_{\bar{R}_1} = \frac{N_{\bar{R}_1}}{R_1(=1)} \qquad C_{\bar{R}_2} = \frac{N_{\bar{R}_2}}{R_2(=1)} \tag{8.5a,b}$$

where the values for $N_{\bar{R}_1}$ and $N_{\bar{R}_2}$ were specified in columns 5 and 6 of Table 8.1.

The forces in each of the members of the indeterminate truss calculated using Equation 8.4 are shown in Figure 8.2 and are also provided in Table 8.2.

It can be deduced from this example that a truss member that is released (such as member AD in this example) remains unloaded after the application of any loads other than its own redundant force.

It is also true that the displacement at the release of one redundant force (force 1) caused by a second redundant force (force 2) is the same as the displacement at the release of force 2 caused by force 1. For example, $f_{21} = f_{12}$ in Table 8.1 or, in general, $f_{ij} = f_{ji}$.

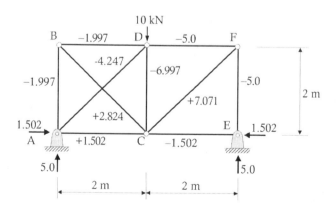

Figure 8.2 Loads, reactions and member forces (kN).

Table 8.2 Determination of member forces

Member	N_0 (kN)	$C_{\bar{R}_1}N_{AD}$ (kN)	$C_{\bar{R}_2}H_A$ (kN)	N (kN)
AB	−5.0	3.003	0.0	−1.997
AC	0.0	3.003	−1.502	1.502
AD	0.0	−4.247	0.0	−4.247
BC	+7.071	−4.247	0.0	2.824
BD	−5.0	3.003	0.0	−1.997
CD	−10.0	3.003	0.0	−6.997
CE	0.0	0.0	−1.502	−1.502
CF	+7.071	0.0	0.0	7.071
DF	−5.0	0.0	0.0	−5.0
EF	−5.0	0.0	0.0	−5.0

SUMMARY OF STEPS 8.1: Calculation of member forces in an indeterminate truss using the force method

1. Determine the degree of indeterminacy n, where $n = r + b - 2j$ (see Chapter 4).

2. Reduce the truss to a stable, statically determinate *primary* truss by making n appropriate releases.

3. Calculate the forces in each member of the primary truss caused by the external loads (N_0) and caused by n unit loads applied at each of the n releases (i.e. $\bar{R}_1$ to $\bar{R}_n$).

4. Determine the deformation of each truss member caused by each of the load systems considered in step 3. The deformation of each member caused by the external loads is $e_0 = fN_0$ (with f being the axial flexibility: $f = L/EA$), while the deformation caused by the ith unit load is $e_{\bar{R}_i} = fN_{\bar{R}_i}$.

5. Calculate the displacement or discontinuity at each release caused by the external loads and by each of the n unit loads. The displacement at release i caused by the external loads is:

$$u_{i0} = \sum \frac{N_{\bar{R}_i}}{1} e_0 \tag{8.6a}$$

and the displacement at release i caused by the k-th unit load is:

$$u_{ik} = \sum \frac{N_{\bar{R}_i}}{1} e_{\bar{R}_k} \text{ (m or mm)} \tag{8.6b}$$

6. By enforcing compatibility and equating the sum of the displacements at each release to zero, write the n simultaneous equations in terms of the n unknown redundant forces, R_1 to R_n. For example, at the i-th release:

$$u_{i0} + f_{i1}R_1 + f_{i2}R_2 + \dots + f_{in}R_n = 0 \tag{8.7}$$

where the flexibility coefficients f_{ij} represent the displacement produced at release i by a unit action applied at release j. The flexibility coefficient f_{ij} is equal in magnitude and direction to u_{ij} (because it is produced by a unit action), as calculated at step 5, but different in dimension and given by:

$$f_{ik} = u_{ik}/1 \quad \text{(length/force)} \tag{8.8}$$

7. Solve the n simultaneous equations of Equation 8.7 to determine the n unknown redundant forces, R_1 to R_n.

8. Calculate the force in each member of the truss as the sum of the forces in each member caused by the external loads and the n redundant forces. For example, the force in the i-th bar is:

$$N_i = N_0 + (C_{\bar{R}_1})_i R_1 + (C_{\bar{R}_2})_i R_2 + \ldots + (C_{\bar{R}_n})_i R_n \tag{8.9}$$

where $(C_{\bar{R}_j})_i$ are the transfer coefficients calculated from:

$$(C_{\bar{R}_j})_i = \frac{(N_{\bar{R}_j})_i}{R_j(=1)} \tag{8.10}$$

WORKED EXAMPLE 8.1

Determine the forces in each member of the truss shown in Figure 8.3. The cross-sectional area of members AD, BC, CF and DE is 1000 mm², and for all other truss members, A = 3000 mm². Take E = 200 kN/mm².

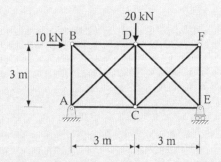

Figure 8.3 Truss for Worked Example 8.1.

(1) There are 6 nodes (j = 6), 11 members (b = 11) and 3 reactions (r = 3). Therefore, $(r + b) - 2j = 2$ and the truss is 2-fold redundant.

(2) Two members must be released to form a stable primary truss. In this example, the diagonal members AD and DE are released at D as shown in Figure 8.4.

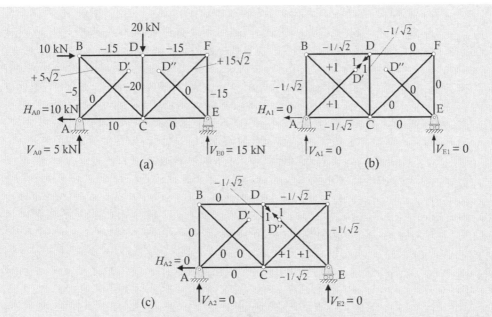

Figure 8.4 Forces and transfer coefficients in members of primary truss. (a) Forces in primary truss carrying external loads (F_0 in kN). (b) Member forces associated with unit force $(\bar{R}_1)$ in member AD (release 1). (c) Member forces associated with unit force $(\bar{R}_2)$ in member DE (release 2).

(3) The forces in each member of the primary truss (N_0) caused by the external loads, the forces in each member associated with unit loads $\bar{R}_1$ applied at the release in AD and $\bar{R}_2$ applied at the release in DE are shown in Figure 8.4a, b and c, respectively, and are also provided in columns 2, 3 and 4 of Table 8.3.

(4 and 5) The axial flexibility coefficients for the diagonal members are:

$$f_{AD} = f_{BC} = f_{CF} = f_{DE} = \frac{L}{EA} = \frac{3000\sqrt{2}}{200 \times 1000} = 0.0212 \text{ mm/kN}$$

and for the horizontal and vertical members:

$$f_{AB} = f_{AC} = f_{BD} = f_{CD} = f_{CE} = f_{DF} = f_{EF} = \frac{L}{EA} = \frac{3000}{200 \times 3000} = 0.005 \text{ mm/kN}$$

The deformations of each truss member caused by the load systems shown in Figure 8.4a, b and c are now used to find the discontinuity at each member release. The displacements at release 1 (DD′ in member AD) caused by the external loads and by the unit loads in the two redundant members are determined in columns 5, 7 and 9 of Table 8.3 and are respectively:

$$f_{10} = 0.256 \quad f_{11} = 0.0524 \quad f_{12} = 0.0025$$

while the corresponding displacements at release 2 (DD″ in member DE) are shown in columns 6, 8 and 10 of Table 8.3 as

$$f_{20} = 0.6266 \quad f_{21} = 0.0025 \quad f_{22} = 0.0524$$

Table 8.3 Determination of displacements for the truss of Figure 8.3

(1) Member	(2) N_0 (kN)	(3) $N_{\bar{R}_1}$ (kN)	(4) $N_{\bar{R}_2}$ (kN)	(5) $N_{\bar{R}_1} e_0$ (kNmm)	(6) $N_{\bar{R}_2} e_0$ (kNmm)	(7) $N_{\bar{R}_1} e_{\bar{R}_1}$ (kNmm)	(8) $N_{\bar{R}_2} e_{\bar{R}_1}$ (kNmm)	(9) $N_{\bar{R}_1} e_{\bar{R}_2}$ (kNmm)	(10) $N_{\bar{R}_2} e_{\bar{R}_2}$ (kNmm)
AB	−5.0	−0.707	0.0	0.0177	0.0	0.0025	0.0	0.0	0.0
AC	10.0	−0.707	0.0	−0.0354	0.0	0.0025	0.0	0.0	0.0
AD	0.0	1.0	0.0	0.0	0.0	0.0212	0.0	0.0	0.0
BC	+7.07	1.0	0.0	0.15	0.0	0.0212	0.0	0.0	0.0
BD	−15.0	−0.707	0.0	0.053	0.0	0.0025	0.0	0.0	0.0
CD	−20.0	−0.707	−0.707	0.0707	0.0707	0.0025	0.0025	0.0025	0.0025
CE	0.0	0.0	−0.707	0.0	0.0	0.0	0.0	0.0	0.0025
CF	+21.21	0.0	1.0	0.0	0.4499	0.0	0.0	0.0	0.0212
DF	−15.0	0.0	−0.707	0.0	0.053	0.0	0.0	0.0	0.0025
DE	0.0	0.0	1.0	0.0	0	0.0	0.0	0.0	0.0212
EF	−15.0	0.0	−0.707	0.0	0.053	0.0	0.0	0.0	0.0025
			Sum	0.256	0.6266	0.0524	0.0025	0.0025	0.0524

(6) From Equation 8.7, the two simultaneous compatibility equations enforcing the requirement that the displacement at each release is zero are:

$$0.256 + 0.0524R_1 + 0.0025R_2 = 0 \tag{1}$$

$$0.6266 + 0.0025R_1 + 0.0524R_2 = 0 \tag{2}$$

(7) Solving for the unknown redundant forces gives:

$$R_1 = -4.325 \text{ kN} \quad \text{and} \quad R_2 = -11.752 \text{ kN}$$

(8) The force in each member of the indeterminate truss is obtained from Equation 8.9, where the transfer coefficient $C_{\bar{R}_j}$ is equal in magnitude to $N_{\bar{R}_j}$ (given in columns 3 and 4 of Table 8.3). For example, in member AB:

$$N_{AB} = -5.0 + [-0.707 \times (-4.325)] + [0.0 \times (-11.752)] = -1.942 \text{ kN}$$

The calculation of each bar force is given in Table 8.4 and the forces are shown in Figure 8.5.

Table 8.4 Determination of member forces

Member	N_0 (kN)	$C_{\bar{R}_1}R_1$ (kN)	$C_{\bar{R}_2}R_2$ (kN)	N (kN)
AB	−5.0	3.058	0.0	−1.942
AC	10.0	3.058	0.0	13.058
AD	0.0	−4.325	0.0	−4.325
BC	+7.07	−4.325	0.0	2.745
BD	−15.0	3.058	0.0	−11.942
CD	−20.0	3.058	8.308	−8.634
CE	0.0	0.0	8.308	8.308
CF	+21.21	0.0	−11.752	9.458
DF	−15.0	0.0	8.308	−6.692
DE	0.0	0.0	−11.752	−11.752
EF	−15.0	0.0	8.308	−6.692

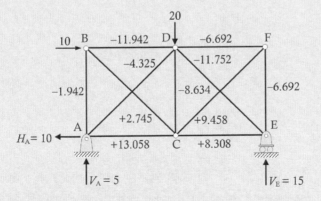

Figure 8.5 Loads, reactions and member forces (kN).

WORKED EXAMPLE 8.2

For the truss shown in Figure 8.3, calculate the forces in each member caused by a temperature rise of $\Delta T = 30°C$ in the top chord members BD and DF. As in the previous example, the cross-sectional area of all truss members is $A = 3000$ mm², except for members AD, BC, CF and DE, where $A = 1000$ mm². Take $E = 200$ kN/mm² and the coefficient of thermal expansion is $\alpha = 10 \times 10^{-6}/°C$.

(1 and 2) As in Worked Example 8.1, the diagonal members AD and DE are released at D to form the statically determinate primary truss.

(3) In this example, there are no external loads to consider (so $F_0 = 0$). However, members BD and DF in the primary truss undergo a thermal strain of:

$$\varepsilon_T = \Delta T + = +30 \times 10 \times 10^{-6} = +300 \times 10^{-6}$$

and a resultant elongation of:

$$(e_T)_{BD} = (e_T)_{DF} = \varepsilon_T L = 300 \times 10^{-6} \times 3000 = +0.9 \text{ mm}$$

These member elongations caused by the rise in temperature are treated in exactly the same way as the member deformations caused by the external loads (N_0) in the previous examples. The member forces $N_{\bar{R}_1}$ and $N_{\bar{R}_2}$ associated with unit loads applied at the release in AD and the release in DE, respectively, are as shown in Figure 8.4b and c.

(4 and 5) As in Worked Example 8.1, the axial flexibility for each of the diagonal members is $f = 0.0212$ mm/kN and that for the horizontal and vertical members is $f = 0.005$ mm/kN. The deformation of each truss member caused by the temperature change and the unit load systems shown in Figure 8.4b and c is now used to find the discontinuity at each member release. The displacements at release 1 (DD′ in member AD) caused by the temperature change in BD and DF and by the unit loads in the two redundant members are determined in columns 5, 7 and 9 of Table 8.5 and are respectively:

$$f_{10} = -0.6363 \quad f_{11} = 0.0524 \quad f_{12} = 0.0025$$

while the corresponding displacements at release 2 (DD″ in member DE) are shown in columns 6, 8 and 10 of Table 8.5:

$$f_{20} = -0.6363 \quad f_{21} = 0.0025 \quad f_{22} = 0.0524$$

(6) From Equation 8.7, the two simultaneous compatibility equations enforcing the requirement that the displacement at each release is zero are:

$$-0.6363 + 0.0524R_1 + 0.0025R_2 = 0 \tag{1}$$

$$-0.6363 + 0.0025R_1 + 0.0524R_2 = 0 \tag{2}$$

(7) Solving for the unknown redundant forces gives:

$$R_1 = +11.592 \text{ kN} \quad \text{and} \quad R_2 = +11.592 \text{ kN}$$

Table 8.5 Determination of displacements of the truss for Worked Example 8.2

(1) Member	(2) e_T (mm)	(3) $N_{\bar{R}_1}$ (kN)	(4) $N_{\bar{R}_2}$ (kN)	(5) $N_{\bar{R}_1} e_T$ (kNmm)	(6) $N_{\bar{R}_2} e_T$ (kNmm)	(7) $N_{\bar{R}_1} e_{\bar{R}_1}$ (kNmm)	(8) $N_{\bar{R}_2} e_{\bar{R}_1}$ (kNmm)	(9) $N_{\bar{R}_1} e_{\bar{R}_2}$ (kNmm)	(10) $N_{\bar{R}_2} e_{\bar{R}_2}$ (kNmm)
AB	0.0	−0.707	0.0	0.0	0.0	0.0025	0.0	0.0	0.0
AC	0.0	−0.707	0.0	0.0	0.0	0.0025	0.0	0.0	0.0
AD	0.0	1.0	0.0	0.0	0.0	0.0212	0.0	0.0	0.0
BC	0.0	1.0	0.0	0.0	0.0	0.0212	0.0	0.0	0.0
BD	+0.9	−0.707	0.0	−0.6363	0.0	0.0025	0.0	0.0	0.0
CD	0.0	−0.707	−0.707	0.0	0.0	0.0025	0.0025	0.0025	0.0025
CE	0.0	0.0	−0.707	0.0	0.0	0.0	0.0	0.0	0.0025
CF	0.0	0.0	1.0	0.0	0.0	0.0	0.0	0.0	0.0212
DF	+0.9	0.0	−0.707	0.0	−0.6363	0.0	0.0	0.0	0.0025
DE	0.0	0.0	1.0	0.0	0.0	0.0	0.0	0.0	0.0212
EF	0.0	0.0	−0.707	0.0	0.0	0.0	0.0	0.0	0.0025
			Sum	−0.6363	−0.6363	0.0524	0.0025	0.0025	0.0524

(8) The force in each member of the indeterminate truss is obtained from Equation 8.9, except that in this example, N_0 is zero for all members. The transfer coefficient $C_{\bar{R}_i}$ is equal in magnitude to $N_{\bar{R}_i}$ (given in columns 3 and 4 of Table 8.5). For example, in member AB:

$$N_{AB} = (-0.707 \times 11.592) + (0.0 \times 11.592) = -8.196 \text{ kN}$$

The calculation of each member force is given in Table 8.6 and the forces are shown in Figure 8.6.

Table 8.6 Determination of member forces caused by a temperature rise in BD and DF

Member	N_0 (kN)	$C_{\bar{R}_1}R_1$ (kN)	$C_{\bar{R}_2}R_2$ (kN)	N (kN)
AB	0	−8.196	0	−8.196
AC	0	−8.196	0	−8.196
AD	0	11.592	0	11.592
BC	0	11.592	0	11.592
BD	0	−8.196	0	−8.196
CD	0	−8.196	−8.196	−16.392
CE	0	0	−8.196	−8.196
CF	0	0	11.592	11.592
DF	0	0	−8.196	−8.196
DE	0	0	11.592	11.592
EF	0	0	−8.196	−8.196

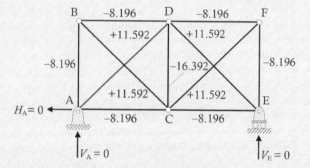

Figure 8.6 Member forces (kN) due to a 30°C temperature rise in BD and DF.

8.2.2 Determination of joint displacements

When all the member forces have been determined, the elongation of each truss member can be readily determined. For example, the elongation of members BD and CD in Worked Example 8.2 is:

$$e_{BD} = fN_{BD} + (e_T)_{BD} = 0.005 \times (-8.196) + 0.9 = 0.859 \text{ mm}$$

$$e_{CD} = fN_{CD} = 0.005 \times (-16.392) = -0.082 \text{ mm}$$

The method of virtual forces was discussed in Section 7.4.1 and was used to calculate joint displacements in the statically determinate trusses of Worked Examples 7.3, 7.4 and 7.5. The method can be applied equally well to statically indeterminate trusses. A unit virtual force is applied to the truss at the joint and in the direction of the required joint displacement. The internal virtual work associated with any internal force field that is in equilibrium with the external unit virtual force is required. In a statically indeterminate truss, there are an infinite number of such force fields. By selecting an appropriate number of member releases (or reaction releases), we can analyse any stable primary truss to determine an appropriate internal force field.

WORKED EXAMPLE 8.3

Determine the vertical displacement at joint C of the truss shown in Figure 8.3 caused by the loadings shown.

The forces in each member of the truss caused by applied loads were calculated in Worked Example 8.1, together with the axial flexibility for each bar and are provided here in columns 2 and 3 of Table 8.7, and the corresponding extensions of each bar e are shown in column 4.

In Worked Example 8.1, a stable primary truss was established by releasing the diagonal members AD and DE. In Figure 8.7, the numbered displacement directions at each node of the truss are shown, together with the virtual member forces $\bar{N}_6$ in this primary truss, caused by a vertical unit virtual force applied at C in direction 6 (the numbering of two directions at nodes in plane trusses was introduced in Section 4.5). These virtual member forces are shown in column 5 of Table 8.7. The external virtual work caused by the unit vertical load at C is $\bar{W}_{ext} = 1 \times v_C$ and the internal virtual work is $\bar{W}_{int} = \Sigma \bar{N}_6 e$ as calculated in column 6 of Table 8.7.

Table 8.7 Determination of the vertical displacement at C in Worked Example 8.3

(1) Member	(2) f (mm/kN)	(3) N (kN)	(4) e = fN (mm)	(5) $\bar{N}_6$ (kN)	(6) $\bar{N}_6 e$ (kNmm)
AB	0.005	−1.942	−0.0097	0.5	−0.0049
AC	0.005	13.058	0.0653	0.0	0.0
AD	0.0212	−4.325	−0.0917	0.0	0.0
BC	0.0212	2.745	0.0582	−0.7071	−0.0412
BD	0.005	−11.942	−0.0597	0.5	−0.0299
CD	0.005	−8.634	−0.0432	0.0	0.0
CE	0.005	8.308	0.0415	0.0	0.0
CF	0.0212	9.458	0.2006	−0.707	−0.1418
DF	0.005	−6.692	−0.0335	0.5	−0.0168
DE	0.0212	−11.752	−0.2493	0.0	0.0
EF	0.005	−6.692	−0.0335	0.5	−0.0168
				Sum	−0.2514

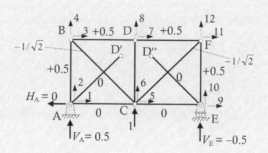

Figure 8.7 Truss for Worked Example 8.3.

Equating the external virtual work to the internal virtual work gives the vertical displacement at C:

$$v_C = -0.251 \text{ mm (i.e. downwards)}$$

WORKED EXAMPLE 8.4

Consider the truss analysed in Worked Example 8.2 and determine the vertical displacement at joint C caused by a temperature rise of 30°C in members BD and DF.

The forces in each member of the truss caused by a 30°C temperature rise in BD and DF were calculated in Worked Example 8.2 and are provided here in column 3 of Table 8.8. The corresponding extensions of each bar $e \ (= fN + e_T)$ are shown in column 4 of Table 8.8.

Table 8.8 Determination of the vertical displacement at C in Worked Example 8.4

(1) Member	(2) f (mm/kN)	(3) N (kN)	(4) e = fN + e_T (mm)	(5) $\bar{N}_6$ (kN)	(6) $\bar{N}_6 e$ (kNmm)
AB	0.005	−8.196	−0.041	0.5	−0.0205
AC	0.005	−8.196	−0.041	0.0	0.0
AD	0.0212	11.592	0.2459	0.0	0.0
BC	0.0212	11.592	0.2459	−0.7071	−0.1739
BD	0.005	−8.196	0.859	0.5	0.4295
CD	0.005	−16.392	−0.082	0.0	0.0
CE	0.005	−8.196	−0.041	0.0	0.0
CF	0.0212	11.592	0.2459	−0.7071	−0.1739
DF	0.005	−8.196	0.859	0.5	0.4295
DE	0.0212	11.592	0.2459	0.0	0.0
EF	0.005	−8.196	−0.041	0.5	−0.0205
				Sum	0.4702

The analysis of the stable primary truss obtained by releasing the diagonal members AD and DE is again used, with the virtual internal forces $\bar{N}_6$ associated with the unit vertical virtual force at C as shown in Figure 8.7 and repeated here in column 5 of Table 8.8.

Equating the external virtual work to the internal virtual work gives the vertical displacement at C (rounded to 3 significant figures):

$v_C = +0.470$ mm (i.e. upwards)

8.3 THE FORCE METHOD APPLIED TO BEAMS AND FRAMES

8.3.1 Determination of internal actions

The force method can also be used to find the redundant reactions in a statically indeterminate beam or frame. If the beam or frame is n-fold indeterminate, n reactions or internal actions at particular locations are selected as redundants and releases are made so that a statically determinate *primary structure* is established. Consider the 1-fold indeterminate frame shown in Figure 8.8a. If the horizontal reaction at D is selected as the redundant reaction, the appropriate release is to convert the pin joint at D to a roller, thereby permitting horizontal displacement. The resulting primary frame is shown in Figure 8.8b. Alternatively, we could select the bending moment at a particular point to be zero and the appropriate release involves the introduction of an internal hinge at that point. If we introduce an internal hinge at B, the primary frame is shown in Figure 8.8c and can now be analysed using only the principles of statics. With this release, the moment at B becomes the redundant internal action. The hinge could in fact be introduced at any location provided the resulting primary frame is stable.

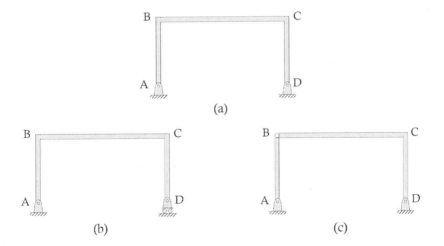

Figure 8.8 Example of primary frames for a onefold indeterminate frame. (a) Indeterminate frame. (b) Primary frame 1. (c) Primary frame 2.

For a 1-fold indeterminate beam or frame, when the redundant reaction or internal action is selected, the primary structure is analysed under the action of:
i. the external loads (and any environmental actions such as temperature changes), and
ii. the unknown redundant R

The sum of the displacements at the realease caused by i and ii is equated to zero (since this displacement is zero in the indeterminate structure) and the resulting equation is solved for the unknown R.

WORKED EXAMPLE 8.5

For the two-span beam shown in Figure 8.9, determine the reactions at the three supports using the force method. Consider only the flexural deformation of the beam and assume linear–elastic material behaviour. Take $EI = 200 \times 10^3$ kNm².

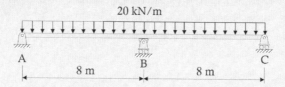

Figure 8.9 Beam for Worked Example 8.5.

---

(1) The beam is 1-fold indeterminate and the reaction at B is selected as the redundant. The corresponding release is the removal of the roller support at B. A free-body diagram of the resulting primary beam under the action of the external loads is shown in Figure 8.10a, together with the corresponding bending moment diagram. A free-body diagram of the primary beam under the action of a unit vertical force at B (in the direction of the redundant reaction at B) and the corresponding bending moment diagram are shown in Figure 8.10b.

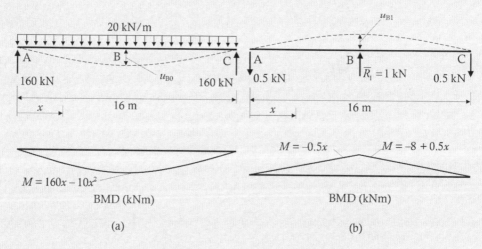

Figure 8.10 Free-body diagrams and bending moment diagrams of the primary beam. (a) Subjected to external loads. (b) Subjected to a unit vertical load at B.

(2) The principle of virtual forces is first used to determine the displacement of the primary beam at B caused by the uniformly distributed external load (u_{B0} in Figure 8.10a). The unit load shown in Figure 8.10b is treated as the virtual force.

The internal work is determined from Equation 7.14, with the curvature diagram expressed as $\kappa = M/EI = (160x - 10x^2)/(200 \times 10^3)$ and the virtual moment is $\bar{M} = -0.5x$ in AB. Both the virtual moment diagram and the actual curvature diagram are symmetrical about the mid-span and, therefore, the internal virtual work for the entire beam is twice the internal virtual work in the length of beam from A to B:

$$\bar{W}_{int} = 2 \int_0^8 \bar{M}\left(\frac{M}{EI}\right) dx = 2 \int_0^8 -0.5x\left[\frac{1}{200 \times 10^3}(160x - 10x^2)\right] dx$$

$$= \frac{2}{200 \times 10^3} \int_0^8 (-80x^2 + 5x^3) dx = \frac{2}{200 \times 10^3}\left[-\frac{80}{3}x^3 + \frac{5}{4}x^4\right]_0^8 = -0.0853 \text{ kNm}$$

The external work is $\bar{W}_{ext} = 1 \times u_{B0}$ and the principle of virtual work gives $u_{B0} = -0.0853$ m.

(3) The principle of virtual forces is then used to determine the displacement of the primary beam at B caused by a unit load at B (u_{B1} in Figure 8.10b). The curvature diagram in the half-span from A to B owing to the unit load is expressed as $\kappa = M/EI = (-0.5x)/(200 \times 10^3)$ and the virtual moment $\bar{M} = -0.5x$. The internal work is:

$$\bar{W}_{int} = 2 \int_0^8 \bar{M}\left(\frac{M}{EI}\right) dx = 2 \int_0^8 -0.5x\left(\frac{1}{200 \times 10^3}(-0.5x)\right) dx$$

$$= \frac{2}{200 \times 10^3} \int_0^8 \left(\frac{1}{4}x^2\right) dx = \frac{2}{200 \times 10^3}\left[\frac{1}{12}x^3\right]_0^8 = 0.427 \times 10^{-3} \text{ kNm}$$

The external work is $\bar{W}_{ext} = 1 \times u_{B1}$ and the work theorem gives $u_{B1} = 0.427 \times 10^{-3}$ m.

Following the same notation as used for trusses, the flexibility coefficient associated with member release 1 is $f_{B1} = \dfrac{u_{B1}}{1} = \dfrac{0.427 \times 10^{-3}}{1} = 0.427 \times 10^{-3}$ m/kN.

(4) Enforcing compatibility at point B, we have:

$$u_{B0} + R_B f_{B1} = -0.0853 + 0.427 \times 10^{-3} R_B = 0 \quad \text{and therefore} \quad R_B = 200 \text{ kN}$$

(5) With the redundant reaction at B determined, the other vertical reactions at A and C can now be obtained from statics. Taking moments about A in the free-body diagram of Figure 8.11 gives:

$$R_C \times 16 + 200 \times 8 - 20 \times 16 \times 8 = 0 \quad \therefore R_C = 60 \text{ kN}$$

and taking moments about C, or simply summing the vertical forces, gives $R_A = 60$ kN.

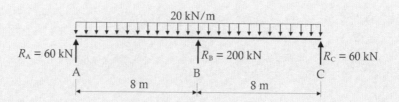

Figure 8.11 Free-body diagram for Worked Example 8.5.

WORKED EXAMPLE 8.6

The portal frame shown in Figure 8.12a is fixed at support A and pinned at support D. Considering only the flexural deformation of the frame and assuming linear–elastic material behaviour, determine the five reactions shown in the free-body diagram of Figure 8.12b using the force method.

Take $E = 200$ kN/mm^2 and $I = 250 \times 10^6$ mm^4. Therefore, $EI = 50 \times 10^3$ kNm^2.

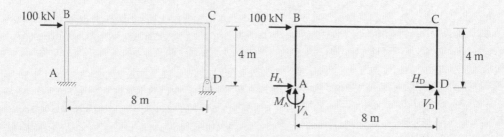

Figure 8.12 Frame for Worked Example 8.6. (a) 2-fold indeterminate frame. (b) Free-body diagram.

(1) The frame is 2-fold indeterminate and the two reactions at D are selected as the two redundants: R_1 (= V_D) and R_2 (= H_D). The corresponding release is the removal of the pin support at D. A free-body diagram of the resulting primary frame under the action of the external loads is shown in Figure 8.13a, together with the corresponding bending moment diagram.

Free-body diagrams of the primary frame under the action of a unit vertical force $\bar{R}_1$ at D (in the direction of the redundant vertical reaction at D) and a unit horizontal force $\bar{R}_2$ at D (in the direction of the redundant horizontal reaction at D), together with the corresponding bending moment diagrams, are shown in Figure 8.13b and c, respectively.

(2) Displacements at release 1: The vertical displacement (u_{10}) at D (release 1) caused by the external load is calculated using the principle of virtual work. The unit vertical load shown in Figure 8.13b is treated as the virtual force. Clearly, with no bending in BC and CD, internal work due to the external load occurs only in the vertical leg AB, where the bending moment may be

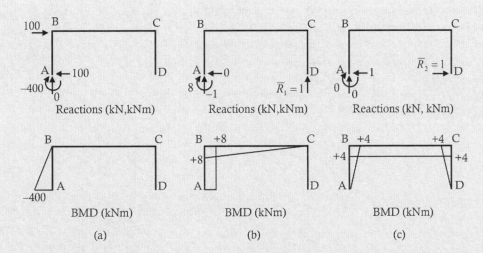

Figure 8.13 Free-body diagrams and bending moment diagrams of the primary frame. (a) Due to external load. (b) Due to unit value of redundant 1 $\left(\bar{R}_1\right)$. (c) Due to unit value of redundant 2 $\left(\bar{R}_2\right)$.

expressed as $M = -400 + 100x$ kNm, where x is measured upwards from A. The virtual moment in AB is uniform at $\bar{M} = 8$ kNm. The internal work is determined from Equation 7.14:

$$\bar{W}_{int} = (\bar{W}_{int})_{AB} = \int_0^4 \bar{M}\left(\frac{M}{EI}\right)dx = \int_0^4 8\left(\frac{-400+100x}{50\times10^3}\right)dx$$

$$= \frac{1}{50\times10^3}\int_0^4(-3200+800x)\,dx = \frac{1}{50\times10^3}\left[-3200x+\frac{800}{2}x^2\right]_0^4$$

$$= -0.128 \text{ kNm}$$

The external work is $\bar{W}_{ext} = 1\times u_{10}$ and the principle of virtual work gives $u_{10} = -0.128$ m. The vertical displacement (u_{11}) of the primary frame at D (release 1) caused by the unit vertical load at D is next calculated. In AB, $\kappa = M/EI = (+8)/(50\times10^3)$ m^{-1} and the virtual moment $\bar{M} = +8$ kNm. In BC, $\kappa = M/EI = (+8-x)/(50\times10^3)$ m^{-1} and the virtual moment $\bar{M} = +8-x$ kNm, where x is measured from B. In CD, the curvature is everywhere zero. The internal work is:

$$\bar{W}_{int} = (\bar{W}_{int})_{AB} + (\bar{W}_{int})_{BC} = \int_0^4 8\left(\frac{8}{50\times10^3}\right)dx + \int_0^8 (8-x)\left(\frac{8-x}{50\times10^3}\right)dx$$

$$= \frac{1}{50\times10^3}[+64x]_0^4 + \frac{1}{50\times10^3}[+64x-8x^2+x^3/3]_0^8$$

$$= 8.53\times10^{-3} \text{ kNm}$$

The external work is $\bar{W}_{ext} = 1\times u_{11}$ and the principle of virtual work gives $u_{11} = 8.53\times10^{-3}$ m, from which $f_{11} = u_{11}/1 = 8.53\times10^{-3}$ m/kN.

The vertical displacement (u_{12}) of the primary frame at D (release 1) caused by the unit horizontal load at D is then calculated. In AB, $\kappa = M/EI = (+x)/(50 \times 10^3)$ and the virtual moment $\overline{M} = +8$. In BC, $\kappa = M/EI = +4/(50 \times 10^3)$ and the virtual moment $\overline{M} = +8 - x$, where x is measured from B. In CD, the virtual moment is everywhere zero. The internal work is:

$$\overline{W}_{int} = (\overline{W}_{int})_{AB} + (\overline{W}_{int})_{BC} = \int_0^4 8\left(\frac{x}{50 \times 10^3}\right)dx + \int_0^8 (8-x)\left(\frac{4}{50 \times 10^3}\right)dx$$

$$= \frac{1}{50 \times 10^3}[+4x^2]_0^4 + \frac{1}{50 \times 10^3}[+32x - 2x^2]_0^8 = 3.84 \times 10^{-3} \text{ kNm}$$

The external work is $\overline{W}_{ext} = 1 \times u_{12}$ and the work theorem gives $u_{12} = 3.84 \times 10^{-3}$ m, based on which $f_{12} = u_{12}/1 = 3.84 \times 10^{-3}$ m/kN.

(3) Displacements at release 2: The horizontal displacement (u_{20}) at D (release 2) caused by the external load is calculated using the unit horizontal load shown in Figure 8.13c as the virtual force. With no bending in BC and CD, internal work due to the external load occurs only in the vertical leg AB, where the moment may be expressed as $M = -400 + 100x$ and virtual moment is $\overline{M} = x$ kNm. The internal work is determined from Equation 7.14:

$$\overline{W}_{int} = (\overline{W}_{int})_{AB} = \int_0^4 x\left(\frac{-400 + 100x}{50 \times 10^3}\right)dx$$

$$= \frac{1}{50 \times 10^3}\left[-200x^2 + \frac{100}{3}x^3\right]_0^4 = -0.0213 \text{ kNm}$$

The external work is $\overline{W}_{ext} = 1 \times u_{20}$ and with the principle of virtual work: $u_{20} = -0.0213$ m. The horizontal displacement (u_{21}) of the primary frame at D (release 2) caused by the unit vertical load at D is next calculated. In AB, $\kappa = M/EI = (+8)/(50 \times 10^3)$ and the virtual moment $\overline{M} = +x$. In BC, $\kappa = M/EI = (+8 - x)/(50 \times 10^3)$ and the virtual moment $\overline{M} = 4$. In CD, the curvature is everywhere zero. The internal work is:

$$\overline{W}_{int} = (\overline{W}_{int})_{AB} + (\overline{W}_{int})_{BC} = \int_0^4 x\left(\frac{8}{50 \times 10^3}\right)dx + \int_0^8 4\left(\frac{8-x}{50 \times 10^3}\right)dx$$

$$= \frac{1}{50 \times 10^3}[+4x^2]_0^4 + \frac{1}{50 \times 10^3}[+32x - 2x^2]_0^8 = 3.84 \times 10^{-3} \text{ kNm}$$

The external work is $\overline{W}_{ext} = 1 \times u_{21}$ and with the principle of virtual work: $u_{21} = 3.84 \times 10^{-3}$ m, from which $f_{21} = u_{21}/1 = 3.84 \times 10^{-3}$ m/kN.

The horizontal displacement (u_{22}) of the primary frame at D (release 2) caused by the unit horizontal load at D is now calculated. In AB, $\kappa = M/EI = (+x)/(50 \times 10^3)$ and the virtual moment $\overline{M} = +x$. In BC, $\kappa = M/EI = +4/(50 \times 10^3)$ and the virtual moment $\overline{M} = +4$. In CD, $\kappa = M/EI = +x/$

(50×10^3) and the virtual moment $\bar{M} = +x$, where x is measured upward from D. The internal work is:

$$\bar{W}_{int} = (\bar{W}_{int})_{AB} + (\bar{W}_{int})_{BC} + (\bar{W}_{int})_{BC}$$

$$= \int_0^4 x\left(\frac{x}{50 \times 10^3}\right)dx + \int_0^8 4\left(\frac{4}{50 \times 10^3}\right)dx + \int_0^4 x\left(\frac{x}{50 \times 10^3}\right)dx$$

$$= \frac{1}{50 \times 10^3}\left[+\frac{x^3}{3}\right]_0^4 + \frac{1}{50 \times 10^3}[+16x]_0^8 + \frac{1}{50 \times 10^3}\left[+\frac{x^3}{3}\right]_0^4$$

$$= 3.41 \times 10^{-3} \text{ kNm}$$

The external work is $\bar{W}_{ext} = 1 \times u_{22}$ and from the principle of virtual work: $u_{22} = 3.41 \times 10^{-3}$ m, from which $f_{22} = u_{22}/1 = 3.41 \times 10^{-3}$ m/kN.

(4) Solving for redundant reactions: By enforcing compatibility at releases 1 and 2, we get two simultaneous equations in terms of the two unknown redundant reactions $R_1 = V_D$ and $R_2 = H_D$:

$$u_{10} + R_1 f_{11} + R_2 f_{12} = -0.128 + 0.00853 R_1 + 0.00384 R_2 = 0$$

$$u_{20} + R_1 f_{21} + R_2 f_{22} = -0.0213 + 0.00384 R_1 + 0.00341 R_2 = 0$$

Solving gives $R_1 = V_D = 24.59$ kN and $R_2 = H_D = -21.36$ kN.

(5) Find remaining reactions using statics: With the two reactions at D now known, the three reactions at A can be determined by enforcing the three equations of statics. Referring to the free-body diagram in Figure 8.12b:

Vertical equilibrium: $V_A + V_D = 0$ $\therefore V_A = -24.59$ kN

Horizontal equilibrium: $100 + H_A + H_D = 0$ $\therefore H_A = -78.64$ kN

Moment equilibrium about A: $M_A + 100 \times 4 - V_D \times 8 = 0$ $\therefore M_A = -203.3$ kNm

WORKED EXAMPLE 8.7

For the two-span beam shown in Figure 8.9, determine the change in the vertical reactions at A, B and C if, during a hot day, exposure to the sun results in a uniform temperature-induced curvature of -0.5×10^{-6} mm^{-1} along the beam. Take $EI = 200 \times 10^3$ kNm^2.

(1) As in Worked Example 8.5, for this 1-fold indeterminate beam, the reaction at B is selected as the redundant (release 1). For the statically determinate primary beam, the uniform temperature-induced curvature causes deformation but no reactions or internal actions as shown in Figure 8.14a. The free-body diagram of the primary beam and the corresponding bending moment diagram under the action of a unit vertical force at B was shown in Figure 8.10b and is repeated here in Figure 8.14b.

(2) The principle of virtual forces is first used to determine the displacement of the primary beam at B caused by the uniform temperature-induced curvature (u_{10} in Figure 8.14a). Treating the unit load shown in Figure 8.14b as the virtual force, the internal work is determined from

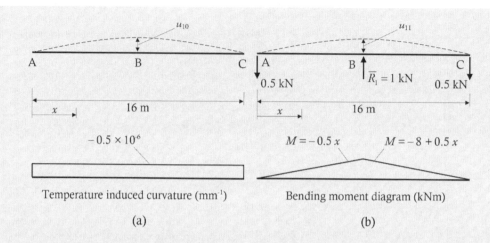

Figure 8.14 Free-body diagrams and bending moment diagrams of the primary beam. (a) Due to temperature gradient. (b) Subjected to a unit vertical load at B.

Equation 7.14. The change in curvature owing to the temperature rise is uniform at $\Delta\kappa_T = -0.5 \times 10^{-3}$ m^{-1} and the virtual moment $\bar{M} = -0.5x$ for the span AB. As in Worked Example 8.3, the virtual moment diagram and the actual curvature diagram are symmetrical about point B:

$$\bar{W}_{int} = 2\int_0^8 \bar{M}\Delta\kappa_T \, dx = 2\int_0^8 -0.5x \times (-0.5\times10^{-3}) \, dx = [0.25\times10^{-3}x^2]_0^8 = 0.016 \text{ kNm}$$

The external work is $\bar{W}_{ext} = 1\times u_{10}$ and the principle of virtual work gives $u_{10} = 0.016$ m.

(3) As calculated in Worked Example 8.3, the displacement of the primary beam at B (release I) caused by a unit load at B (u_{11} in Figure 8.14b) is $u_{11} = 0.427 \times 10^{-3}$ m and $f_{11} = 0.427 \times 10^{-3}$ m/kN.

(4) By enforcing compatibility at point B, the change in the reaction at B is:

$$u_{10} + \Delta R_1 f_{11} = 0.016 + 0.427 \times 10^{-3}\Delta R_1 = 0$$

and therefore:

$$\Delta R_1 = \Delta R_B = -37.5 \text{ kN (i.e. downwards)}$$

(5) With the change in the reaction at B determined, the changes in the other vertical reactions at A and C can be obtained from statics. By inspection:

$$\Delta R_A = \Delta R_C = +18.75 \text{ kN (i.e. upwards)}$$

8.3.2 Flexibility coefficients and transfer functions

We have seen that the deflection (or rotation) at any point in a beam or frame caused by flexural deformation can be determined by applying a unit force (or couple) at that point and then equating external and internal virtual work using an equation of the form $\bar{1}u = \int\kappa\bar{M}dx$. Dividing both sides by the unit virtual force, we get:

$$u = \int \kappa(\bar{M}/\bar{1})dx = \int \kappa\bar{m}dx \qquad (8.11)$$

When the deformation is caused by bending and material behaviour is linear–elastic, $\kappa = M/EI$. The term $\bar{m}$ is known as the *transfer function*, so called because it transfers a unit force or couple applied at a release to a bending moment at a particular section. At any point in the beam or frame, the transfer function $\bar{m}$ is numerically equal to the bending moment at that point caused by the unit load or couple. If a unit force is performing the external work, $\bar{m}$ has the dimension of length and it is dimensionless if a unit couple has been applied to the beam or frame.

When determining the deformation of a beam or frame, deformations caused by axial force and shear force are usually very small compared to deformation caused by bending and are usually ignored. However, if deformation caused by axial force needed to be considered, an equation of the form $u = \int \varepsilon(\bar{N}/\bar{1})\mathrm{d}x = \int \varepsilon\bar{n}\,\mathrm{d}x$ may be used, where $\bar{n}$ is the *transfer function* that transfers a unit force or couple applied at a release to an axial force at a particular section.

The transfer functions may be used to calculate the *flexibility coefficients* associated with each redundant reaction or internal action. For example, reconsidering Worked Example 8.6, the two compatibility equations for the 2-fold indeterminate frame were:

$$u_{10} + R_1 f_{11} + R_2 f_{12} = 0 \quad \text{and} \quad u_{20} + R_1 f_{21} + R_2 f_{22} = 0 \qquad (8.12\text{a,b})$$

The displacement at release 1 caused by external loads (or by environmental actions, such as temperature changes) is given by (Equation 8.11):

$$u_{10} = \int \frac{M_0}{EI}\,\bar{m}_1\,\mathrm{d}x \qquad (8.13)$$

The term $R_1 u_{11}$ is the displacement at release 1 caused by the redundant reaction R_1 and may be expressed as:

$$R_1 u_{11} = \int \frac{M_{R_1}}{EI}\,\bar{m}_1\,\mathrm{d}x \qquad (8.14)$$

where M_{R_1} is the moment caused by the redundant reaction R_1 applied to the primary beam or frame, and $\bar{m}_1$ is the transfer function associated with a unit redundant action applied along release 1. With this notation, we assume that u_{11} represents the deflection along release 1 produced by a unit action of R_1.

Dividing both sides of Equation 8.14 by R_1 gives:

$$u_{11} = \int \frac{M_{R_1}}{EIR_1}\,\bar{m}_1\,\mathrm{d}x = \int \frac{m_1 \bar{m}_1}{EI}\,\mathrm{d}x \qquad (8.15\text{a})$$

where m_1 and $\bar{m}_1$ are identical, with the only conceptual difference that m_1 and $\bar{m}_1$ transfer a real and a virtual force or couple, respectively.

In a similar manner, we can calculate the deflection at a release i caused by a unit action applied at release j as follows:

$$u_{ij} = \int \frac{M_{R_j}}{EIR_j}\,\bar{m}_i\,\mathrm{d}x = \int \frac{m_j \bar{m}_i}{EI}\,\mathrm{d}x \qquad (8.15\text{b})$$

and this can be used to determine the flexibility coefficients f_{ij} required in Equations 8.12. In fact, f_{ij} is equal to the displacement produced by a unit action and is equal in magnitude to u_{ij}, but different in dimension. f_{ij} is defined as a length per unit action, while u_{ij} has dimension of length. On the basis of this notation, the unknown redundant actions can be calculated from Equations 8.12a and b, which are summarised in matrix form as:

$$\begin{bmatrix} u_{10} \\ u_{20} \end{bmatrix} + \begin{bmatrix} f_{11} & f_{12} \\ f_{21} & f_{22} \end{bmatrix} \begin{bmatrix} R_1 \\ R_2 \end{bmatrix} = 0 \qquad (8.16)$$

When determining the moment at any point in the redundant frame of Worked Example 8.6, we can simply add the moments caused by the external loads on the primary structure to those caused by the two redundants (R_1 and R_2) using the transfer functions:

$$M = M_0 + m_1 R_1 + m_2 R_2 \qquad (8.17)$$

WORKED EXAMPLE 8.8

Re-solve Worked Example 8.6 using transfer functions and flexibility coefficients.

As in Worked Example 8.6, the two reactions at D are selected as the two redundants, R_1 ($= V_D$) and R_2 ($= H_D$), and the frame is reduced to a primary frame by removing the pin support at D. The bending moment diagrams for the primary frame caused by the external loads (M_o), by a unit vertical force at D (m_1) and unit horizontal forces at D (m_2) are shown in Figures 8.15 a, b and c, respectively.

In AB, $m_1 = +8$ and $m_2 = +x$, where x is measured upward from A. In BC, $m_1 = +8 - x$ and $m_2 = +4$, where x is measured from B. In CD, $m_1 = 0$ and $m_2 = +x$, where x is measured upward from D.

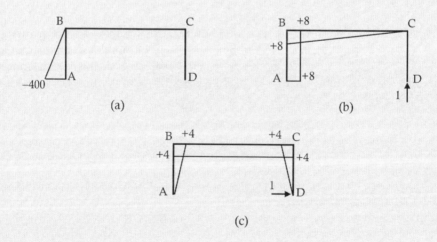

Figure 8.15 Bending moment diagrams. (a) M_0. (b) m_1. (c) m_2.

Because of the external loads, in AB, $M_0 = -400 + 100x$ and, in BC and CD, M_0 is everywhere zero. The vertical displacement u_{10} at D (release 1) and the horizontal displacement u_{20} at D (release 2) caused by the external load are calculated from:

$$u_{10} = \int_0^4 \frac{M_0}{EI}\,\bar{m}_1\,dx = \frac{1}{50\times 10^3}\int_0^4 (-400+100x)8\,dx = \frac{1}{50\times 10^3}[-3200x+400x^2]_0^4$$

$$= -0.128\,m$$

and

$$u_{20} = \int_0^4 \frac{M_0}{EI}\,\bar{m}_2\,dx = \frac{1}{50\times 10^3}\int_0^4 (-400+100x)x\,dx = \frac{1}{50\times 10^3}[-200x^2+100x^3/3]_0^4$$

$$= -0.0213\,m$$

The flexibility coefficients are:

$$f_{11} = \left[\int \frac{m_1\bar{m}_1}{EI}\,dx\right]_{AB} + \left[\int \frac{m_1\bar{m}_1}{EI}\,dx\right]_{BC} + \left[\int \frac{m_1\bar{m}_1}{EI}\,dx\right]_{CD}$$

$$= \frac{1}{50\times 10^3}\left[\int_0^4 64\,dx + \int_0^8 (64-16x+x^2)\,dx + \int_0^4 0\,dx\right]$$

$$= \frac{1}{50\times 10^3}\left[[64x]_0^4 + [64x-8x^2+x^3/3]_0^8\right]$$

$$= 8.53\times 10^{-3}\,m/kN$$

$$f_{12} = \left[\int \frac{\bar{m}_1 m_2}{EI}\,dx\right]_{AB} + \left[\int \frac{\bar{m}_1 m_2}{EI}\,dx\right]_{BC} + \left[\int \frac{\bar{m}_1 m_2}{EI}\,dx\right]_{CD}$$

$$= \frac{1}{50\times 10^3}\left[\int_0^4 8x\,dx + \int_0^8 (32-4x)\,dx + \int_0^4 0\,dx\right]$$

$$= \frac{1}{50\times 10^3}\left[[4x^2]_0^4 + [32x-2x^2]_0^8\right]$$

$$= 3.84\times 10^{-3}\,m/kN$$

$$f_{21} = f_{12} = 3.84\times 10^{-3}\,m/kN$$

and

$$f_{22} = \left[\int \frac{m_2 \bar{m}_2}{EI} dx\right]_{AB} + \left[\int \frac{m_2 \bar{m}_2}{EI} dx\right]_{BC} + \left[\int \frac{m_2 \bar{m}_2}{EI} dx\right]_{CD}$$

$$= \frac{1}{50 \times 10^3}\left[\int_0^4 x^2 dx + \int_0^8 16 dx + \int_0^4 x^2 dx\right]$$

$$= \frac{1}{50 \times 10^3}\left[[x^3/3]_0^4 + [16x]_0^8 + [x^3/3]_0^4\right]$$

$$= 3.41 \times 10^{-3} \, \text{m/kN}$$

For compatibility at release 1 and 2, we have:

$$u_{10} + R_1 f_{11} + R_2 f_{12} = -0.128 + 0.00853 R_1 + 0.00384 R_2 = 0$$

$$u_{20} + R_1 f_{21} + R_2 f_{22} = -0.0213 + 0.00384 R_1 + 0.00341 R_2 = 0$$

and solving these two equations gives:

$$R_1 = 24.59 \, \text{kN} \quad \text{and} \quad R_2 - 21.36 \, \text{kN}$$

As in Worked Example 8.6, the other three reactions are obtained using the equations of statics, giving:

$$V_A = 24.59 \, \text{kN} \, (\downarrow), \, H_A = 78.64 \, \text{kN} \, (\leftarrow), \, M_A = 203.3 \, \text{kNm} \, (\curvearrowleft)$$

$$V_D = 24.59 \, \text{kN} \, (\uparrow) \text{ and } H_D = 21.36 \, \text{kN} \, (\leftarrow)$$

We can now calculate the moments at A, B, C and D using Equation 8.17:

$$M_A = M_{0A} + m_{1A} R_1 + m_{2A} R_2 = -400 + 8 \times 24.59 + 0 \times (-21.36) = -202.6 \, \text{kNm}$$

$$M_B = M_{0B} + m_{1B} R_1 + m_{2B} R_2 = 0 + 8 \times 24.59 + 4 \times (-21.36) = +111.4 \, \text{kNm}$$

$$M_C = M_{0C} + m_{1C} R_1 + m_{2C} R_2 = 0 + 0 \times 24.59 + 4 \times (-21.36) = -86.1 \, \text{kNm}$$

and the final reactions and bending moment diagram for the indeterminate frame are shown in Figure 8.16.

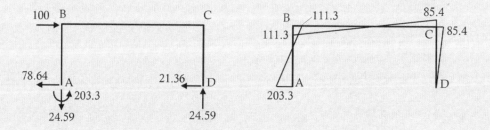

Figure 8.16 Reactions and bending moment diagram (drawn on tension side).

For a n-fold indeterminate beam or frame, after n redundant reactions (or internal actions) are selected and the corresponding n releases are made to form a stable primary structure, the n compatibility equations may be expressed as:

$$\begin{bmatrix} u_{10} \\ u_{20} \\ \cdot \\ \cdot \\ \cdot \\ u_{n0} \end{bmatrix} + \begin{bmatrix} f_{11} & f_{12} & \cdot & \cdot & \cdot & \cdot & f_{1n} \\ f_{21} & f_{22} & \cdot & \cdot & \cdot & \cdot & f_{2n} \\ \cdot & \cdot & & & & & \cdot \\ \cdot & \cdot & & & & & \cdot \\ \cdot & \cdot & & & & & \cdot \\ f_{n1} & f_{n2} & \cdot & \cdot & \cdot & \cdot & f_{nn} \end{bmatrix} \begin{bmatrix} R_1 \\ R_2 \\ \cdot \\ \cdot \\ \cdot \\ R_n \end{bmatrix} = 0 \qquad (8.18)$$

When axial deformations are significant, these need to be included in the determination of the release displacements caused by external loads and in the determination of the flexibility coefficients. This can be done using the more general forms of Equations 8.13 and 8.15:

$$u_{i0} = \int \frac{M_0}{EI} \bar{m}_i \, dx + \int \frac{N_0}{EA} \bar{n}_i \, dx \qquad (8.19)$$

$$f_{ij} = \int \frac{m_j \bar{m}_i}{EI} \, dx + \int \frac{n_j \bar{n}_i}{EA} \, dx \qquad (8.20)$$

8.3.3 Deformations of statically indeterminate beams and frames

After the reactions have been determined and then used to calculate the internal actions in a beam or frame (M, N, etc.), the displacement at any point may be readily determined using a unit virtual force applied at that point and then applying the principle of virtual work. The curvature and axial strain on every cross-section represent the real displacement field, while any distribution of bending moments and axial forces that is in equilibrium with the unit virtual force may be adopted as the virtual force field. Any stable primary beam or frame can be selected to establish the virtual force field.

WORKED EXAMPLE 8.9

For the frame analysed in Worked Examples 8.6 and 8.8, determine the horizontal displacement of the frame at point B caused by bending deformation. As before, $EI = 50 \times 10^3$ kNm^2.

For the virtual force method, the frame may be reduced to a statically determinate primary frame by removing the pin support at D. We will call this primary frame 1. Another possibility is to create the primary frame by setting M_A and H_A to zero (i.e. by replacing the fixed support at A with a roller support) — primary frame 2. Figure 8.17 shows the reactions and bending

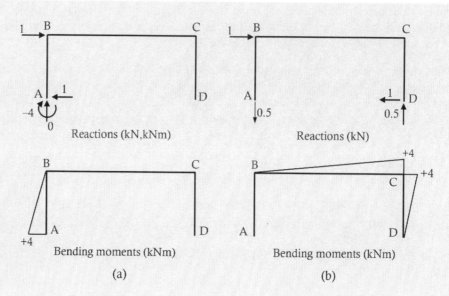

Figure 8.17 Reactions and bending moment diagrams in primary frames. (a) Primary frame 1. (b) Primary frame 2.

moment diagrams caused by a unit horizontal virtual force applied at B for each of these two primary frames.

Flexural deformations: The bending moment diagram and reactions caused by the external load on the frame are shown in Figure 8.16. In Member AB, $(M_0)_{AB} = -203.3 + 78.64x$, where x is measured upward from A. In BC, $(M_0)_{BC} = +111.3 - 24.59x$, where x is measured from B. In CD, $(M_0)_{CD} = -21.36x$, where x is measured upward from D.

Considering the virtual force field for primary frame 1, internal work only occurs in member AB, the virtual moment caused by the unit virtual force at B is $(\bar{m})_{AB} = -4 + x$. Therefore, $\bar{W}_{ext} = \bar{1}\,u_B$ and:

$$\bar{W}_{int} = \int_0^4 \frac{M_0}{EI}(\bar{m})_{AB}\,dx = \int_0^4 \frac{-203.3 + 78.64x}{50\times 10^3} \times (-4 + x)\,dx$$

$$= \frac{1}{50\times 10^3}[813.2x - 258.9x^2 + 26.21x^3]_0^4 = 0.0157$$

and the horizontal displacement at B caused by bending is $u_B = 0.0157$ m.

The same results are obtained by considering the virtual force field for primary frame **2**. In this case, internal work only occurs in members BC and CD, with $(\bar{m})_{BC} = -0.5x$ and $(\bar{m})_{CD} = -x$. With $\bar{W}_{ext} = \bar{1}u_B$ and:

$$\bar{W}_{int} = \int_0^8 \frac{(M_0)_{BC}}{EI}(\bar{m})_{BC}\,dx + \int_0^4 \frac{(M_0)_{CD}}{EI}(\bar{m})_{CD}\,dx$$

$$= \int_0^8 \frac{111.3 - 24.59x}{50\times10^3}\times(-0.5x)\,dx + \int_0^4 \frac{-21.36x}{50\times10^3}\times(-x)\,dx$$

$$= \frac{1}{50\times10^3}[-27.82x^2 + 4.10x^3]_0^8 + \frac{1}{50\times10^3}[7.12x^3]_0^4$$

$$= 0.0157$$

The horizontal displacement at B caused by bending is $u_B = 0.0157$ m (as previously calculated using primary frame I).

PROBLEMS

8.1   Determine the forces in each member of the truss shown using the force method. Assume EA is the same for all members.

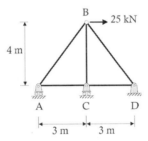

8.2   Determine the horizontal displacement at B for the truss of Problem 8.1.

8.3   Determine the forces in each member of the truss shown using the force method. Assume all members have the same EA.

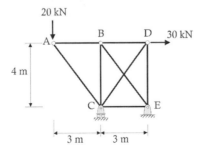

8.4   Determine (i) the vertical displacement at A and (ii) the horizontal displacement at D for the truss of Problem 8.3.

8.5   For the truss of Problem 8.3, determine the change in member forces caused by a temperature rise of 30°C in the top chord ABD. Take $EA = 200 \times 10^3$ kN and the coefficient of thermal expansion is $\alpha_T = 10 \times 10^{-6}/°C$ for all members. Use the force method.

8.6   Determine the forces in each member of the truss shown using the force method, if $A_{AB} = A_{AC} = A_{BD} = A_{CE} = A_{DF} = A_{EG} = A_{FH} = A_{GH} = 10,000$ mm², $A_{AD} = A_{CD} = A_{CF} = A_{DE} = A_{EF} = A_{EH} = 3000$ mm² and $E = 20$ kN/mm².

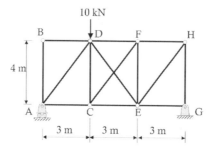

8.7   Determine (i) the vertical displacement at E and (i) the horizontal displacement at A for the truss of Problem 8.6.

8.8   Determine the forces in each member of the truss shown using the force method. Assume all members have the same elastic modulus and cross-sectional area.

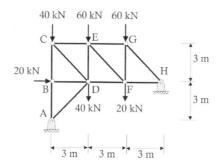

8.9   Determine the vertical displacement at D for the truss of Problem 8.8.

8.10 For the truss of Problem 8.8, determine the change in member forces caused by a temperature rise of 30°C in members AB, BC, CE, EG and GH. Take $EA = 150 \times 10^3$ kN and the coefficient of thermal expansion is $\alpha_T = 10 \times 10^{-6}/°C$ for all members.

8.11 For the truss shown, the top chord members AB = BD = DF = FG = 5.5 m. For the loading shown, determine the reactions and the forces in each member using the force method. Assume all members have the same elastic modulus and cross-sectional area.

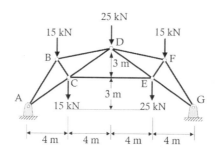

8.12 If the members CD and EG are selected as the redundants in the truss shown, the resulting primary truss is the same as that of Problem 4.7. The three applied loads are at right angles to the top chord ABDG. If all members have the same EA, determine the forces in each member of the truss using the force method.

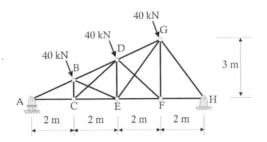

8.13 Determine the forces in each member of the truss shown using the force method. Also calculate the vertical deflections at C. Assume $EA = 100 \times 10^3$ kN for all members.

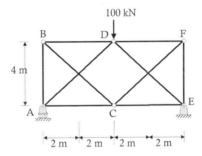

8.14 If the flexural rigidity EI of each of the propped cantilevers shown is constant, using the force method determine (i) the reactions at A and B, and (ii) the deflection at the mid-span of each beam in terms of the load (w or F), the span L and flexural rigidity EI.

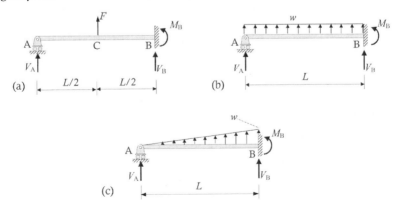

8.15 For the three propped cantilever beams of Problem 8.14, determine the rotation at the roller support at A using the force method.

8.16 If the flexural rigidity EI of each of the fixed-ended beams shown below is constant, determine (i) the reactions at A and B, and (ii) the deflection at the mid-span of each beam in terms of the load (w or F), the span L and flexural rigidity EI. Use the force method.

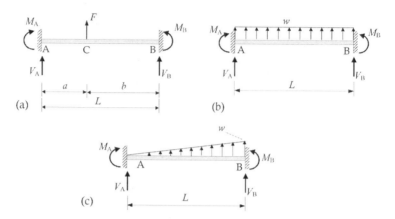

(a)

(b)

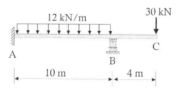

(c)

8.17 If EI is constant throughout the beam shown, determine (i) the reactions at A and B induced by the couple M_A applied at the support A, and (ii) the rotation at support A. Use the force method.

8.18 For the beam shown, determine the reactions at A and B using the force method and draw the bending moment diagram. Assume EI is constant along the beam.

8.19 If $EI = 40 \times 10^3 \text{ kNm}^2$, determine (i) the vertical deflection at C and (ii) the slope of the beam at C for the beam of Problem 8.18.

8.20 The beam of Problem 8.18 has a uniform temperature of 20°C at 6:00 a.m. During the morning, the top surface of the beam is exposed to the sun, and at midday, the temperature varies from 40°C at the top surface of the beam to 20°C on the bottom surface. Calculate the change in reactions at A and B using the force method. Take $EI = 40 \times 10^3 \text{ kNm}^2$ and the coefficient of thermal expansion is $\alpha_T = 10 \times 10^{-6}/°C$. The height of the cross-section is $h = 400$ mm.

8.21 For the beam shown, determine the reactions at A, B and C using the force method and draw the bending moment diagram. Assume EI is constant along the beam.

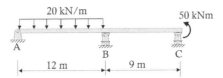

8.22 For the beam of Problem 8.21, if $EI = 50 \times 10^3$ kNm², determine the vertical deflection at the midpoint of span AB and the slope of the beam at A.

8.23 If EI is constant throughout the frame shown below, determine the reactions at A and C using the force method, and draw the bending moment diagram for the frame.

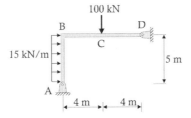

8.24 If EI is constant throughout the frame shown, determine the reactions at A and D using the force method, and draw the bending moment diagram for the frame.

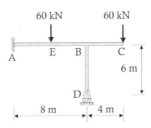

8.25 If $EI = 40 \times 10^3$ kNm², determine the vertical deflection at C and the horizontal displacement at D for the frame of Problem 8.24.

8.26 If EI is constant throughout the frame shown, determine the reactions at A and D using the force method, and draw the bending moment diagram for the frame.

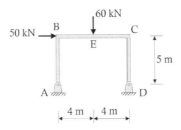

8.27 If $EI = 20 \times 10^3$ kNm² throughout the frame of Problem 8.26, determine the horizontal displacement at B and the vertical deflection at E.

8.28 If EI is constant throughout the frame shown, determine the reactions at A and D using the force method, and draw the bending moment diagram for the frame.

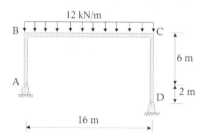

8.29 Determine the deflection at the mid-span of member BC in the frame of Problem 8.28, if $EI = 40 \times 10^3$ kNm² throughout.

8.30 If EI is constant throughout the frame shown below, determine the reactions at A, C and D using the force method, and draw the bending moment diagram for the frame.

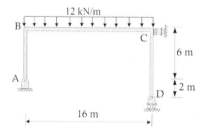

8.31 Determine the deflection at the mid-span of member BC in the frame of Problem 8.30, if $EI = 40 \times 10^3$ kNm² throughout. Also determine the horizontal displacement at D.

8.32 The frame shown has a pinned support at A and a roller support at F. A pin-ended tie restrains the frame at B and E. If the flexural rigidity of members AC, CD and DF is $EI = 20 \times 10^3$ kNm² and the axial rigidity of the tie BE is $EA = 1.2 \times 10^6$ kN, determine the frame reactions and the axial force in member BE using the force method.

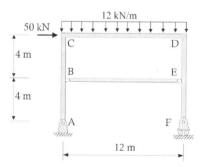

8.33 For the frame analysed in Problem 8.32, calculate the horizontal displacements at points D and F.

Chapter 9

Moment distribution

9.1 INTRODUCTION

In Chapter 6, we discussed the slope-deflection equations that prove useful for the analysis of relatively simple statically indeterminate beams and frames. In Chapter 8, we discussed the *force method* (or *flexibility method*) and demonstrated that it too proved to be a convenient approach for the analysis of simple indeterminate members with one or two redundancies. In this chapter, we will consider a method of analysis that falls under the category of the *displacement (or stiffness) method*. The particularity of the displacement method is that the problem is formulated in terms of unknown displacements that, once determined, enable the entire structural model to be established. Displacement methods are usually the most convenient approaches for the analysis of indeterminate structures with more than one or two redundant reactions.

In this chapter, a form of the displacement method known as *moment distribution* is presented. Moment distribution is suitable for the analysis of statically determinate beams and frames by manual calculation and is a useful introduction to the more general stiffness methods of analysis discussed subsequently in Chapters 10 through 13.

Moment distribution was developed by Hardy Cross in the early 1930s. At that time, it represented a very significant advancement in structural analysis and, in the days before computers, it was widely and routinely used for the analysis of continuous beams and frames. The method begins by assuming that all members in a continuous beam are fixed at the internal supports or that all members in a frame are fixed at the joints. The fixed-end moments (see Appendix B) for each member are then determined and the *unbalanced moment* at each joint is calculated as the sum of the fixed-end moments of all the members entering the joint. Each joint is then released one at a time, by applying a *balancing moment* to the joint, equal and opposite to the *unbalanced moment* at that joint. This balancing moment is then distributed to the members framing into the joint according to their relative rotational stiffness. When a balancing moment M_{bal} is applied at the end of a member that is fixed at the far end, a carry-over moment of magnitude $M_{bal}/2$ is induced at the far end, as observed in the derivation of the equations used in the slope-deflection method (see Equations 6.17a and b). This results in new unbalanced moments at adjacent joints. The joint is then re-locked and an adjacent joint is unlocked and balanced. This process of *moment distribution* continues until the unbalanced moment at each joint is small enough to ignore and the final end moments and end rotations of each member are determined. It is an iterative process that approaches the final solution by successive approximation. As we will see, the method is relatively simple, repetitive and suitable for manual calculation.

9.2 BASIC CONCEPTS

Sign Convention for External Moment and Rotation: The sign convention adopted for fixed-end moments and joint rotations in Section 6.5 is adopted here. Namely, *external moments and rotations are positive when they act in an anti-clockwise sense*. With this sign convention, the direction of positive moment is not affected by the orientation of the members (horizontal, vertical or inclined) and the unbalanced moment at any joint is simply the algebraic sum of the end moments of the members framing into the joint. After the end moments of each member are determined, the signs of the internal actions (bending moments, shear forces and axial forces) are determined using the statics sign convention adopted throughout the book (refer Section 3.3).

Fixed-End Moments: In Section 6.5, we saw how to determine the moment reactions at each end of a fixed-ended beam (i.e. the fixed-end moments). For some common loading conditions, the fixed-end moments are given in Appendix B. For other loading conditions, the fixed-end moments can be readily calculated using the moment-area method, the conjugate beam methods or other suitable approaches outlined in previous chapters.

Rotational Stiffness: We saw in Section 6.5 (in particular Equations 6.12 through 6.14) that when a moment M_A is applied to the pin end A of a beam AB that is fixed at its far end, such as that shown in Figure 9.1, the moment reaction at B is $M_A/2$ and the rotation at A is given by:

$$\theta_A = \frac{L}{4EI} M_A \tag{9.1}$$

The rotational stiffness (k_{AB}) of a span AB, pinned at A and fixed at B, such as that shown in Figure 9.1, is defined here as the moment that must be applied at end A to cause a unit rotation at end A. From Equation 9.1:

$$k_{AB} = \frac{4EI}{L} \tag{9.2}$$

Joint Stiffness and Distribution Factors: When n members are connected at a joint, the joint stiffness K_J is the sum of the rotational stiffness of each member at the joint and is calculated as:

$$K_J = \sum_{i=1}^{n} \frac{4E_i I_i}{L_i} \tag{9.3}$$

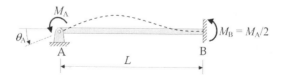

Figure 9.1 Propped cantilever subjected to an end moment.

The joint stiffness K_J is equal to the moment required to rotate the joint through an angle of 1 radian. For example, the stiffness of joint C in the frame shown in Figure 9.2 ($K_{J.C}$) is obtained as:

$$K_{J.C} = k_{AC} + k_{BC} + k_{CD} + k_{CF} = 120 + 130 + 150 + 100 = 500 \text{ kNm}$$

The fraction of a balancing moment at a joint that is distributed to a particular member is called the distribution factor (DF) for that member and is the ratio of the rotational stiffness of the member to the joint stiffness. For a joint with n members framing into it, the DF for the i-th member is:

$$\text{DF}_i = \frac{k_i}{K_J} = \frac{(4E_iI_i/L_i)}{\sum\limits_{i=1}^{n}(4E_iI_i/L_i)} \qquad (9.4)$$

For example, the distribution factors for the four members framing into joint C of the frame in Figure 9.2 are:

$$\text{DF}_{AC} = \frac{k_{AC}}{K_{J.C}} = \frac{120}{500} = 0.24 \qquad\qquad \text{DF}_{BC} = \frac{130}{500} = 0.26$$

$$\text{DF}_{CD} = \frac{150}{500} = 0.30 \qquad\qquad \text{DF}_{CF} = \frac{100}{500} = 0.20$$

and the sum of all the distribution factors at the joint is 1.0.

When all members of a beam or frame are made from the same material, i.e. E is the same for each member, the distribution factor of Equation 9.4 reduces to:

$$\text{DF}_i = \frac{k_i}{K_J} = \frac{(I_i/L_i)}{\sum\limits_{i=1}^{n}(I_i/L_i)} \qquad (9.5)$$

and the term I_i/L_i is known as the relative stiffness factor for the i-th member.

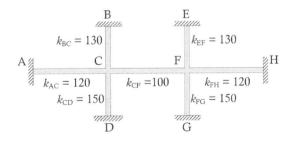

Figure 9.2 Rotational stiffness (4EI/L) of frame members (kNm).

Carry-over Factor: When a balancing moment $M_{\text{bal.A}}$ is applied to member AB at joint A, a *carry-over moment* is induced at the far end of the member B. We have seen that if end B is fixed, this carry-over moment is $M_{\text{bal.A}}/2$. The carry-over factor (COF) is the ratio of the carry-over moment at B to the balancing moment at A and is therefore:

$$\text{COF} = 0.5 \qquad\qquad (9.6)$$

9.3 CONTINUOUS BEAMS

9.3.1 Basic approach

We will demonstrate the application of moment distribution by analysing the two-span beam ABC shown in Figure 9.3. Let us assume that EI is constant throughout.

Step 1 — Fixed-End Moments: Assuming all members are fixed-ended, the fixed-end moments for each span are determined. From Appendix B:

For span AB:

$$(M_{\text{FE.A}})_{\text{AB}} = -\frac{wL^2}{12} = -\frac{-24 \times 20^2}{12} = 800 \text{ kNm}$$

$$(M_{\text{FE.B}})_{\text{AB}} = \frac{wL^2}{12} = \frac{-24 \times 20^2}{12} = -800 \text{ kNm}$$

For span CB:

$$(M_{\text{FE.B}})_{\text{BC}} = -\frac{Pa_1b_1^2}{L^2} - \frac{Pa_2b_2^2}{L^2}$$

$$= -\frac{-150 \times 5 \times 10^2}{15^2} - \frac{-150 \times 10 \times 5^2}{15^2} = 500 \text{ kNm}$$

$$(M_{\text{FE.C}})_{\text{BC}} = \frac{Pa_1b_1^2}{L^2} + \frac{Pa_2b_2^2}{L^2}$$

$$= \frac{-150 \times 5 \times 10^2}{15^2} + \frac{-150 \times 10 \times 5^2}{15^2} = -500 \text{ kNm}$$

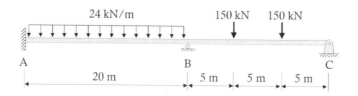

Figure 9.3 Example of continuous beam.

Step 2 — Relative Stiffness: Determine the relative stiffness of each member.

For span AB: $(I/L)_{AB} = I/20$

For span BC: $(I/L)_{BC} = I/15$

Step 3 — Distribution Factors: Determine the distribution factors at each joint.

At A: Since the rotation at the fixed support at A is zero, its stiffness is infinite. From Equation 9.5, the distribution factor for member AB at A is:

$$DF_{AB} = \frac{(I/20)}{\sum\limits_{i=1}^{n}(I_i/L_i)} = \frac{(I/20)}{\infty + (I/20)} = 0$$

At B: From Equation 9.5, the distribution factors for span AB and span BC at B are:

$$DF_{AB} = \frac{(I/20)}{(I/20)+(I/15)} = 0.429 \quad \text{and} \quad DF_{BC} = \frac{(I/15)}{(I/20)+(I/15)} = 0.571$$

At C: Since the support at C is a roller with zero rotational stiffness, the distribution factor for member BC at C is:

$$DF_{BC} = \frac{(I/15)}{(I/15)+0} = 1.0$$

Step 4 — Moment Distribution: Carry out the moment distribution by releasing one joint at a time, distributing the balancing moments to the relevant spans and applying the carry-over moments. Figure 9.4 shows a free-body diagram of the beam and a suitable sequence of calculations.

In the first step of Figure 9.4, joint B is unlocked. The unbalanced moment is the sum of the fixed-end moments at B from member BA and members BC, i.e. −800 + 500 = −300 kNm. A balancing moment of +300 kNm is applied to the joint and distributed to members BA and BC in accordance with the distribution factors:

0.429 × 300 = +128.7 kNm is applied to member BA at joint B

0.571 × 300 = +171.3 kNm is applied to member BC at joint B

With these moments applied at the unlocked end B of each member, carry-over moments are induced at the far (fixed) end of each member. The carry-over moment induced at joint A is 0.5 × 128.7 = +64.3 kNm and that at joint C is 0.5 × 171.3 = +85.7 kNm.

The next step is to lock joint B and unlock joint C (step 2 in Figure 9.4). The unbalanced moment at C is the sum of the fixed-end moment at C and the carry-over moment calculated above, i.e. −500 + 85.7 = −414.3 kNm. A balancing moment of +414.3 kNm is applied to support C and, as the distribution factor for member CB is 1.0, the entire balancing moment is applied to member CB, thereby inducing a carry-over moment at B of 0.5 × 414.3 = +207.2 kNm.

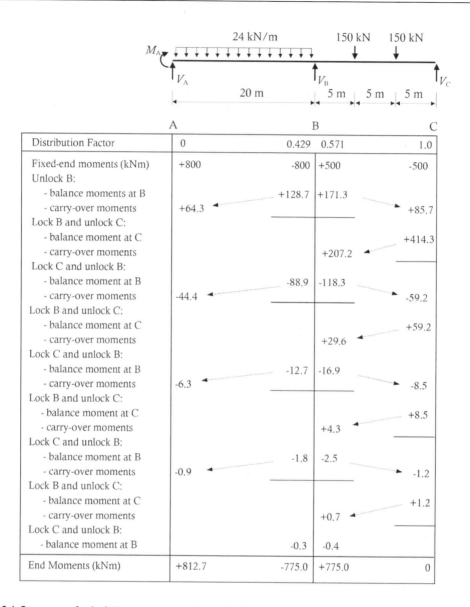

	A	B		C
Distribution Factor	0	0.429	0.571	1.0
Fixed-end moments (kNm)	+800	−800	+500	−500
Unlock B:				
- balance moments at B		+128.7	+171.3	
- carry-over moments	+64.3			+85.7
Lock B and unlock C:				
- balance moment at C				+414.3
- carry-over moments			+207.2	
Lock C and unlock B:				
- balance moment at B		−88.9	−118.3	
- carry-over moments	−44.4			−59.2
Lock B and unlock C:				
- balance moment at C				+59.2
- carry-over moments			+29.6	
Lock C and unlock B:				
- balance moment at B		−12.7	−16.9	
- carry-over moments	−6.3			−8.5
Lock B and unlock C:				
- balance moment at C				+8.5
- carry-over moments			+4.3	
Lock C and unlock B:				
- balance moment at B		−1.8	−2.5	
- carry-over moments	−0.9			−1.2
Lock B and unlock C:				
- balance moment at C				+1.2
- carry-over moments			+0.7	
Lock C and unlock B:				
- balance moment at B		−0.3	−0.4	
End Moments (kNm)	+812.7	−775.0	+775.0	0

Figure 9.4 Summary of calculations — consecutive balancing.

Now we lock C and release B (step 3 in Figure 9.4). The unbalanced moment at B is now the carry-over moment of +207.2 kNm determined in the previous step. A balancing moment of −207.2 kNm must be applied at B distributed to BA and BC in accordance with their distribution factors as done previously.

We continue releasing and relocking joints B and C in turn, balancing the joint and carrying over moments until the unbalanced moments become small enough to ignore. In this example, support A is fixed, the distribution factor is therefore zero and the carry-over moments from joint B simply accumulate at the fixed support.

When the moments being balanced and the carry-over moments are suitably small, the end moments on each span are established by summing the fixed-end moments and all the balancing and carry-over moments associated with each member at each end. These are shown in the bottom row of Figure 9.4.

It is noted that because the support at A is fixed with a distribution factor of 0, there is no need to unlock A and the carry-over moments from B simply accumulate at A. Although for this simple example, there is limited choice in the order in which the supports are released, it can be easily demonstrated that the final solution does not depend on the sequence selected.

In Figure 9.4, we applied the carry-over moments after each unlocked support was balanced and before relocking the support. As an alternative to this consecutive balancing approach, we can balance all supports at the same step and only carry-over moments after all the supports have been unlocked and relocked. In Figure 9.5, the beam is re-analysed using such a simultaneous balancing approach.

In both Figures 9.4 and 9.5, further iterations would refine the answers to any desired degree of accuracy, but for most practical purposes, the approximations arrived at here are sufficiently accurate.

Step 5 — Reactions and Internal Actions: With the end moments for each span determined, the vertical reactions and internal actions can be determined using the principles of statics. The free-body diagrams for each span of the beam of Figure 9.3 are shown in Figure 9.6, with the end moments and reactions shown in their actual directions. The reaction V_B has been subdivided into V_{BA} and V_{BC}, which represent the reaction components from members AB and BC, respectively.

	A		B	C
Distribution Factor	0	0.429	0.571	1.0
Fixed-end moments (kNm)	+800		-800	+500	-500
Balance at all supports:		+128.7	+171.3	+500
Carry-over moments:	+64.3		+250.0	+85.7
Balance at all supports:		-107.1	-142.9	-85.7
Carry-over moments:	-53.6		-42.9	-71.4
Balance at all supports:		+18.4	+24.5	+71.4
Carry-over moments:	+9.2		+35.7	+12.2
Balance at all supports:		-15.3	-20.4	-12.2
Carry-over moments:	-7.7		-6.1	-10.2
Balance at all supports:		+2.6	+3.5	+10.2
Carry-over moments:	+1.3		+5.1	+1.7
Balance at all supports:		-2.2	-2.9	-1.7
Carry-over moments:	-1.1		-0.9	-1.5
Balance at all supports:		+0.4	+0.5	+1.5
Carry-over moments:	+0.2		+0.7	+0.2
Balance at all supports:		-0.3	-0.4	-0.2
End moments (kNm)	+812.6		-775.0	+775.0	0

Figure 9.5 Summary of calculations — simultaneous balancing.

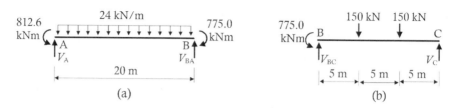

Figure 9.6 Free-body diagrams of each span. (a) Free-body diagram of span AB. (b) Free-body diagram of each span BC.

Moment equilibrium about support B in Figure 9.6a gives: $V_A = 241.9$ kN (↑)
Moment equilibrium about support A in Figure 9.6a gives: $V_{BA} = 238.1$ kN (↑)
Moment equilibrium about support B in Figure 9.6b gives: $V_C = 98.3$ kN (↑)
Moment equilibrium about support C in Figure 9.6b gives: $V_{BC} = 201.7$ kN (↑)
The vertical reaction at the support at B is: $V_B = V_{BA} + V_{BC} = 439.8$ kN (↑)

A quick check of vertical equilibrium of the whole structure indicates that the above reactions are correct.

With the reactions for each span determined, the bending moment and shear force diagrams can be determined using the principles outlined in Chapter 3 and are shown in Figure 9.7.

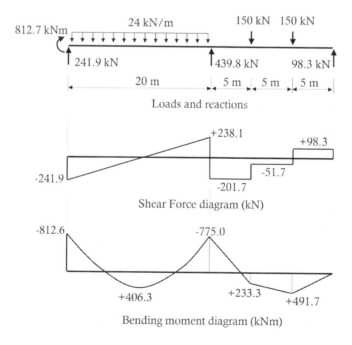

Figure 9.7 Shear force and bending moment diagrams.

9.3.2 Modification for an end span with a pinned support

The convergence of the moment distribution process can be speeded up if a pinned end (such as support C in Figure 9.3) is treated as a pin, instead of being locked with a fixed-end moment and a distribution factor of 1.0.

First, the fixed-end moment at the pinned end is zero and the fixed-end moment at the continuous end may be obtained for common loading cases from the expressions given for propped cantilevers in the right column of Appendix B.1. The relationship between the fixed-end moments of a beam fixed at both ends to an identical beam with one end pinned is shown in Figure 9.8. The fixed-end moment at A in the beam pinned at B, $M_{\mathrm{FE^*.A}}$ can be obtained from the fixed-end moments of the beam fixed at both ends using $M_{\mathrm{FE^*.A}} = M_{\mathrm{FE.A}} - 0.5M_{\mathrm{FE.B}}$, as derived in Equation 6.20 when deriving the expressions for the slope-deflection method. The term $-0.5M_{\mathrm{FE.B}}$ is the carry-over moment induced at A when B is unlocked and changed to a pin.

In this case, the rotational stiffness of the end span must also be adjusted. It can be readily shown from Equation 6.19 that, if a beam is pinned at end B (instead of fixed) and continuous at end A, the rotation at A causes a moment M_A of:

$$M_A = \frac{3EI}{L}\theta_A \tag{9.7}$$

and the rotational stiffness (k_{AB}) can be calculated as:

$$k_{\mathrm{AB}} = \frac{3EI}{L} \tag{9.8}$$

which corresponds to 3/4 of the rotational rigidity of an internal span (Equation 9.2). Because of this, the relative stiffness factor used in the calculation of the joint stiffness and distribution factor, when all members are made from the same material, is $0.75I/L$.

We will now re-analyse the beam of Figure 9.3 making these adjustments for the end span BC. From Appendix B, the fixed-end moments for span BC are now $M_{\mathrm{FE^*.B}} = +750\,\mathrm{kNm}$ and $M_{\mathrm{FE^*.C}} = 0$.

The relative stiffness of the each span is now:

Span AB: $(I/L)_{\mathrm{AB}} = I/20$ Span BC: $0.75(I/L)_{\mathrm{BC}} = I/20$

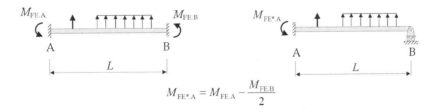

Figure 9.8 End moments for fixed-ended and propped cantilever beams subjected to member loads.

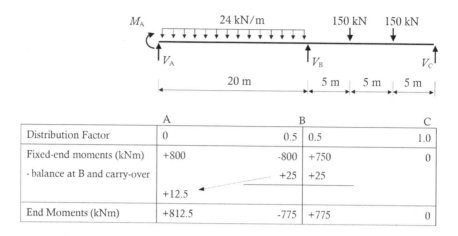

	A		B		C
Distribution Factor	0		0.5	0.5	1.0
Fixed-end moments (kNm)	+800		-800	+750	0
- balance at B and carry-over			+25	+25	
	+12.5				
End Moments (kNm)	+812.5		-775	+775	0

Figure 9.9 Summary of calculations — modification for end span BC.

and the distribution factors at support B become:

$$DF_{AB} = \frac{(I/20)}{(I/20)+(I/20)} = 0.5 \quad \text{and} \quad DF_{BC} = \frac{(I/20)}{(I/20)+(I/20)} = 0.5$$

In Figure 9.9, the moment distribution carried out in Figure 9.4 is repeated with a much more rapid convergence.

WORKED EXAMPLE 9.1

For the beam in Figure 9.10, calculate the reactions at the supports, and draw the bending moment and shear force diagrams. Assume EI is constant throughout.

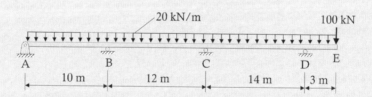

Figure 9.10 Beam for Worked Example 9.1.

Fixed-end moments

Span AB:

$$(M_{FE*.A})_{AB} = 0 \quad \text{and} \quad (M_{FE*.B})_{AB} = \frac{wL^2}{8} = \frac{-20 \times 10^2}{8} = -250 \text{ kNm}$$

Span BC:

$$(M_{FE.B})_{BC} = -\frac{wL^2}{12} = -\frac{-20 \times 12^2}{12} = 240 \text{ kNm} \quad \text{and} \quad (M_{FE.C})_{BC} = +\frac{wL^2}{12} = -240 \text{ kNm}$$

Span CD:

$$(M_{FE.C})_{CD} = -\frac{wL^2}{12} = -\frac{-20 \times 14^2}{12} = +326.7 \text{ kNm} \quad \text{and} \quad (M_{FE.D})_{CD} = \frac{wL^2}{12} = -326.7 \text{ kNm}$$

Span DE: DE is a cantilever with the value of moment at D calculated from statics:

$$(M_{FE.D})_{DE} = -\frac{wL^2}{2} - PL = -\frac{-20 \times 3^2}{2} - (-100 \times 3) = +390 \text{ kNm}$$

Relative stiffness

Span AB: $(0.75I/L_{AB}) = 0.075I$
Span BC: $(I/L_{BC}) = 0.0833I$
Span CD: $(I/L_{CD}) = 0.0714I$
and since DE is a cantilever free to rotate at D, and with no restraint at all at E, it is assigned zero rotational stiffness.

Distribution factors

At support A: $DF_{AB} = 1.0$

At support B:

$$DF_{AB} = \frac{(0.075I)}{(0.075I) + (0.0833I)} = 0.474$$

$$DF_{BC} = \frac{(0.0833I)}{(0.075I) + (0.0833I)} = 0.526$$

At support C:

$$DF_{BC} = \frac{(0.0833I)}{(0.0833I) + (0.0714I)} = 0.539$$

$$DF_{CD} = \frac{(0.0714I)}{(0.0833I) + (0.0714I)} = 0.461$$

At support D:

$$DF_{CD} = \frac{(0.0714I)}{(0.0714I) + 0} = 1.0$$

We can now proceed with the moment distribution.

Moment distribution

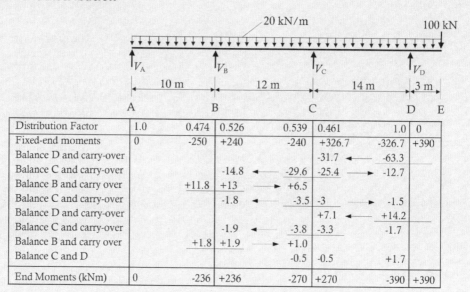

Distribution Factor	1.0	0.474	0.526		0.539	0.461	1.0	0
Fixed-end moments	0	-250	+240		-240	+326.7	-326.7	+390
Balance D and carry-over						-31.7 ←	-63.3	
Balance C and carry-over			-14.8 ←		-29.6	-25.4 →	-12.7	
Balance B and carry over		+11.8	+13	→	+6.5			
Balance C and carry-over			-1.8 ←		-3.5	-3 →	-1.5	
Balance D and carry-over						+7.1 ←	+14.2	
Balance C and carry-over			-1.9 ←		-3.8	-3.3	-1.7	
Balance B and carry over		+1.8	+1.9	→	+1.0			
Balance C and D					-0.5	-0.5	+1.7	
End Moments (kNm)	0	-236	+236		-270	+270	-390	+390

Reactions and internal actions

Taking moments about B in the free-body diagram of span AB, shown in Figure 9.11a, gives:

$$V_A \times 10 + 236 - 20 \times 10 \times 5 = 0 \qquad \therefore V_A = 76.4 \text{ kN}$$

Taking moments about C in the free-body diagram of ABC, shown in Figure 9.11b:

$$76.4 \times 22 + V_B \times 12 + 270 - 20 \times 22 \times 11 = 0 \qquad \therefore V_B = 240.8 \text{ kN}$$

Taking moments about C in the free-body diagram of CDE, shown in Figure 9.11c:

$$-V_D \times 14 - 270 + 20 \times 17 \times 8.5 + 100 \times 17 = 0 \qquad \therefore V_D = 308.6 \text{ kN}$$

Summing the forces vertically, we get:

$$76.4 + 240.8 + V_C + 308.6 \text{ kN} - 20 \times 39 - 100 = 0 \qquad \therefore V_C = 254.2 \text{ kN}$$

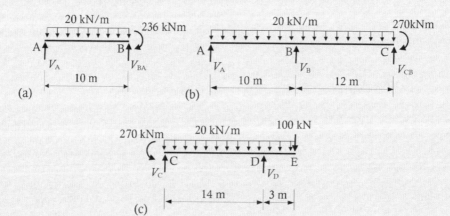

Figure 9.11 Free-body diagrams for Worked Example 9.1.

These reactions can be checked by taking moments about any point on a free-body of the entire beam.

Using the procedures discussed in Chapter 3, the shear force and bending moment diagrams are shown in Figure 9.12.

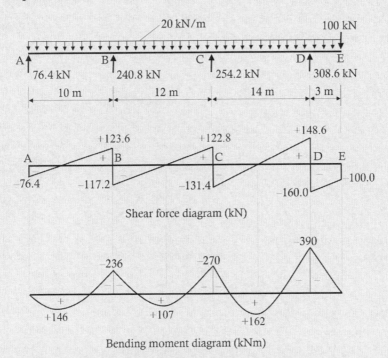

Figure 9.12 Shear force and bending moment diagrams for Worked Example 9.1.

WORKED EXAMPLE 9.2

For the beam in Figure 9.10, calculate the change in reactions at the support and draw the bending moment and shear force diagrams induced by a (downwards) settlement of support C by 80 mm. Assume $EI = 50 \times 10^3$ kNm² throughout.

Fixed-end moments

Because of the settlement $\Delta_C = -80$ mm, the fixed-end moments in spans BC and CD are as follows:

Span BC: $(M_{FE.B})_{BC} = (M_{FE.C})_{BC} = -\dfrac{6EI\Delta_C}{L^2} = -\dfrac{6 \times 50 \times 10^3 \times (-0.08)}{12^2} = 166.7$ kNm

Span CD: $(M_{FE.C})_{CD} = (M_{FE.D})_{CD} = \dfrac{6EI\Delta_C}{L^2} = \dfrac{6 \times 50 \times 10^3 \times (-0.08)}{14^2} = -122.4$ kNm

Relative stiffness and distribution factors

The relative stiffness and distribution factors are the same as calculated in Worked Example 9.1.

Moment distribution

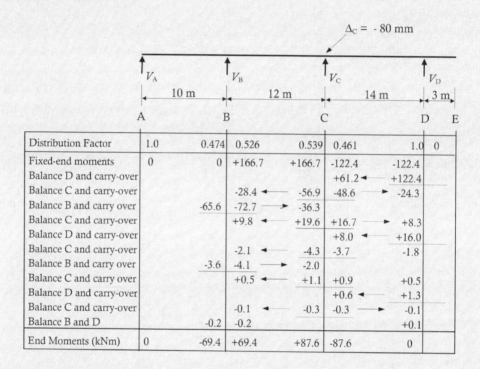

Distribution Factor	1.0	0.474	0.526	0.539	0.461	1.0	0
Fixed-end moments	0	0	+166.7	+166.7	-122.4	-122.4	
Balance D and carry-over					+61.2 ◄	+122.4	
Balance C and carry-over			-28.4 ◄	-56.9	-48.6	► -24.3	
Balance B and carry over		-65.6	-72.7	► -36.3			
Balance C and carry-over			+9.8 ◄	+19.6	+16.7	► +8.3	
Balance D and carry-over					+8.0 ◄	+16.0	
Balance C and carry-over			-2.1 ◄	-4.3	-3.7	-1.8	
Balance B and carry over		-3.6	-4.1	► -2.0			
Balance C and carry over			+0.5 ◄	+1.1	+0.9	+0.5	
Balance D and carry-over					+0.6 ◄	+1.3	
Balance C and carry-over			-0.1 ◄	-0.3	-0.3	► -0.1	
Balance B and D		-0.2	-0.2			+0.1	
End Moments (kNm)	0	-69.4	+69.4	+87.6	-87.6	0	

Reactions and internal actions

The reactions, shear forces and bending moments are shown in Figure 9.13.

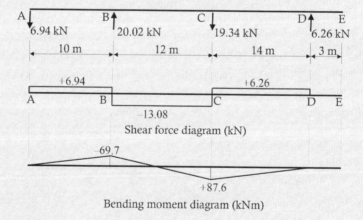

Figure 9.13 Shear force and bending moment diagrams for Worked Example 9.2.

9.4 FRAMES WITHOUT SIDESWAY

Frames in which joint translation is prevented may be analysed using moment distribution in the same way as described for continuous beams. The same is true for frames that are symmetrical and are symmetrically loaded since such frames will not sway. We are of course assuming that displacements are small compared to the frame geometry and the effect of axial forces on the bending in frame members is negligible. Similar considerations were provided in Section 6.5.7 when analysing frames without sidesway with the slope-deflection method.

WORKED EXAMPLE 9.3

The frame shown in Figure 9.14 was analysed previously in Worked Example 6.11 and is prevented from lateral movement by the roller support at C. The joints of the frame are therefore prevented from translation. Analyse the frame using moment distribution and calculate the support reactions caused by the applied loads.

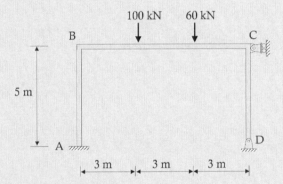

Figure 9.14 Frame for Worked Example 9.3.

Fixed-end moments

Member AB: $(M_{FE.A})_{AB} = (M_{FE.B})_{AB} = 0$

Member BC: From Appendix B, we get:

$$(M_{FE.B})_{BC} = -\frac{-100 \times 6^2 \times 3}{9^2} - \frac{-60 \times 3^2 \times 6}{9^2} = +173.3 \text{ kNm}$$

$$(M_{FE.C})_{BC} = \frac{-100 \times 3^2 \times 6}{9^2} + \frac{-60 \times 6^2 \times 3}{9^2} = -146.6 \text{ kNm}$$

Member CD: $(M_{FE.C})_{CD} = (M_{FE.D})_{CD} = 0$

Relative stiffness

Member AB: $(I/L_{AB}) = 0.2I$
Member BC: $(I/L_{BC}) = 0.111I$
Member CD: $(0.75I/L_{CD}) = 0.15I$

Distribution factors

At support A: $DF_{AB} = 0$

At joint B:

$$DF_{AB} = \frac{(0.2I)}{(0.2I)+(0.111I)} = 0.643$$

$$DF_{BC} = \frac{(0.111I)}{(0.2I)+(0.111I)} = 0.357$$

At joint C:

$$DF_{BC} = \frac{(0.111I)}{(0.111I)+(0.15I)} = 0.426$$

$$DF_{CD} = \frac{(0.15I)}{(0.111I)+(0.15I)} = 0.574$$

At support D:

$$DF_{CD} = \frac{(0.15I)}{(0.15I)+0} = 1.0$$

Moment distribution

	A		B		C	D
Distribution Factor	0	0.643	0.357	0.426	0.574	1.0
Fixed-end moments	0	0	+173.3	-146.7	0	0
Balance B and carry-over	-55.7 ←	-111.4	-61.9 ⟶	-30.9		
Balance C and carry-over			+37.8 ←	+75.6	+102.0	
Balance B and carry over	-12.1 ←	-24.3	-13.5 ⟶	-6.8		
Balance C and carry-over			+1.5 ←	+2.9	+3.9	
Balance B and carry-over	-0.5 ←	-1.0	-0.5 ⟶	-0.3		
Balance C				+0.1	+0.2	
End Moments (kNm)	-68.3	-136.7	+136.7	-106.1	+106.1	0

As expected, these end moments are exactly the same as those determined in Worked Example 6.11. The support reactions are determined from statics as done previously in Worked Example

6.11 and shown on the free-body diagram of the frame in Figure 9.15. The axial force, shear force, and bending moment diagrams for the frame were plotted in Figure 6.29.

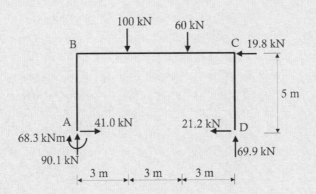

Figure 9.15 Free-body diagram for Worked Example 9.3.

9.5 FRAMES WITH SIDESWAY

In a frame where joint translation is not prevented by lateral restraints, sidesway will occur if the frame is not symmetric, if the vertical loading is not symmetric or if lateral loads are applied to the frame. In reality, construction inaccuracies, dimensional tolerances and variations in material properties will ensure that all frames without lateral restraints will sway.

Let us consider the single bay frame shown in Figure 9.16a. If we ignore the axial deformation of the members, it is clear that the lateral displacement (sway) of B and C will be the same and the only possible joint translation is shown in Figure 9.16b. This frame has *one degree of freedom with respect to joint translation*. The magnitude of this translation, referred to as Δ, depends on the stiffness of the frame members and it will affect the final distribution of moments in the frame.

A superposition approach may be used to include the effects of joint translation whereby sway is initially ignored by introducing a fictitious lateral support and the moment distribution M_0 caused by the applied loads is determined. The reaction that develops

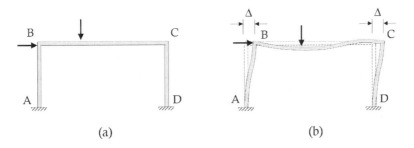

Figure 9.16 Single bay frame with sidesway.

at the fictitious support R_0 is calculated. Next, the frame with the fictitious support is subjected to an arbitrary sway (support settlement) and the moments caused by this sway M_Δ are determined. The reaction at the fictitious support R_Δ caused by the arbitrary sway is determined from statics. In the real frame, the reaction at the fictitious support must be zero, and so:

$$R_0 + f_\Delta R_\Delta = 0 \tag{9.9}$$

where f_Δ is the ratio of the actual sway to the arbitrary sway. The magnitude of the sway required to satisfy Equation 9.9 is determined and the corresponding sway moments $f_\Delta M_\Delta$ are established. The moment at any point in the sway frame M is obtained by adding the moment calculated at that point without sway to the moment calculated due to sway:

$$M = M_0 + f_\Delta M_\Delta \tag{9.10}$$

WORKED EXAMPLE 9.4

Analyse the frame shown in Figure 9.17 using moment distribution and calculate the support reactions. Assume $EI = 50 \times 10^3$ kNm² throughout.

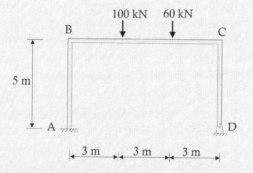

Figure 9.17 Frame for Worked Example 9.4.

Introduce fictitious support at C

In Worked Example 9.3, the moments in this frame with a lateral support at C were determined and the horizontal reaction at C was calculated as $H_C = 19.8$ kN ($\leftarrow$) $= R_0$.
The reactions and the bending moment diagram assuming lateral restraint at C are shown in Figure 9.18.

Introduce arbitrary sway Δ

We now calculate the moments in the frame caused by an arbitrary sway Δ (see Figure 9.19).

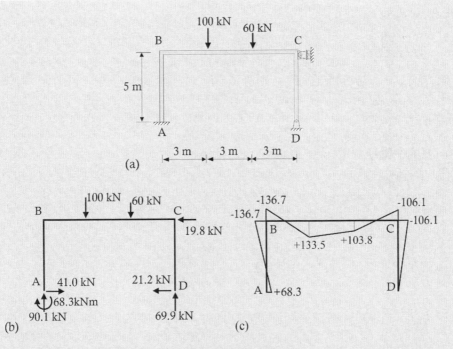

Figure 9.18 Frame with fictitious support at C. (a) Frame with restraint at C. (b) Free-body with reactions, (R_o). (c) Bending moment diagram, M_o (kNm).

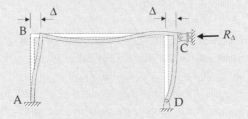

Figure 9.19 Frame with arbitrary sway.

Fixed-end moments

We will now induce an arbitrary sway of say $\Delta = -0.01$ m (moving BC to the right).
In Member AB:

$$(M_{FE.A})_{AB} = (M_{FE.B})_{AB} = -\frac{6EI\Delta}{L^2} = -\frac{6 \times 50,000 \times (-0.01)}{5^2} = 120 \text{ kNm}$$

In Member DC:

$$(M_{FE*.C})_{DC} = -\frac{3EI}{L^2} = -\frac{3 \times 50,000 \times (-0.01)}{5^2} = 60 \text{ kNm}$$

$$(M_{FE*D})_{DC} = 0$$

Moment distribution

The relative stiffness of each member and the distribution coefficients are as determined in Worked Example 9.3.

	A		B			C		D
Distribution Factor	0	0.643	0.375			0.426	0.574	1.0
Fixed-end moments (kNm)	+120	+120	0			0	+60	0
Balance at B and carry-over	-38.6 ←	-77.2	-42.8		→	-21.4		
Balance at C and carry-over			-8.2 ←			-16.4	-22.2	
Balance at B and carry-over	+2.6 ←	+5.3	+2.9		→	+1.5		
Balance at C and carry-over			-0.3 ←			-0.6	-0.9	
Balance at B		+0.2	+0.1					
End Moments (kNm)	+84.0	48.3	-48.3			-37.0	37.0	0

The support reactions caused by the arbitrary sway of −0.01 m are determined from statics and shown on the free-body diagram of the frame in Figure 9.20a. The bending moment diagram caused by the arbitrary sway is shown in Figure 9.20b.

The reaction at the fictitious support at C is R_Δ = 33.9 kN (→). From Equation 9.9, and taking forces acting to the right (→) as positive, we get:

$$R_o + f_\Delta R_\Delta = -19.8 + f_\Delta \times 33.9 = 0 \quad \therefore f_\Delta = 0.584$$

The actual sway is f_Δ multiplied by the arbitrarily selected sway of −0.01 m, i.e. −0.00584 m, and the end moments caused by sway are the end moments calculated in the moment distribution above multiplied by f_Δ.

The final reactions for the frame are shown in the free-body in Figure 9.21a and are the sum of reactions shown in Figure 9.18b (R_o) and those shown in Figure 9.20a multiplied by f_Δ (i.e. $f_\Delta R_\Delta$).

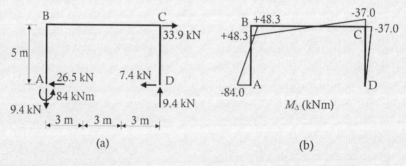

(a)   (b)

Figure 9.20 Reactions and bending moments caused by Δ = −0.01 m. (a) Reactions. (b) Bending moment diagram, M_Δ (kNm).

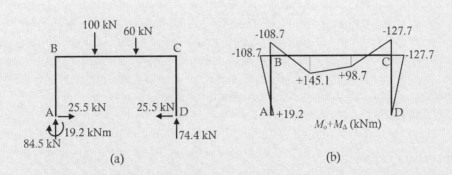

Figure 9.21 Frame for Worked Example 9.4. (a) Free-body. (b) Bending moment diagram, $M_o + M_\Delta$ (kNm).

The final bending moments diagram for the frame is shown in Figure 9.21b and is obtained from Equation 9.10 as the sum of the bending moments in Figure 9.18c (M_o) and those shown in Figure 9.20b multiplied by f_Δ (i.e. $f_\Delta M_\Delta$).

If a frame has n degrees of freedom with respect to joint translation, a fictitious support may be introduced to prevent each possible displacement and the frame is first analysed under the applied loads. There will be a reaction caused by the loads at each of the n fictitious supports (R_{oi} where $i = 1$ to n). An arbitrary displacement (sway) is then introduced at each fictitious support (one support at a time) and the frame analysed, i.e. n arbitrary displacements and n separate analyses. For each analysis, the moment distribution is performed, and from the end moments $M_{\Delta i}$, the n fictitious reactions are determined. The requirement that, in the real structure, the reaction at each fictitious support must be zero (Equation 9.9) results in n simultaneous equations:

$$R_{o1} + f_{\Delta 1}R_{\Delta 1}(1) + f_{\Delta 2}R_{\Delta 2}(1) + \quad + f_{\Delta n}R_{\Delta n}(1) = 0$$
$$R_{o2} + f_{\Delta 1}R_{\Delta 1}(2) + f_{\Delta 2}R_{\Delta 2}(2) + \quad + f_{\Delta n}R_{\Delta n}(2) = 0$$
$$\cdot$$
$$\cdot \qquad\qquad\qquad\qquad\qquad\qquad\qquad (9.11)$$
$$\cdot$$
$$R_{on} + f_{\Delta 1}R_{\Delta 1}(n) + f_{\Delta 2}R_{\Delta 2}(n) + \quad + f_{\Delta n}R_{\Delta n}(n) = 0$$

where $R_{\Delta j}(i)$ is the reaction at the i-th fictitious support owing to the j-th arbitrary displacement. Solving these equations gives the n correction factors (ratios of actual to arbitrary sway) $f_{\Delta 1}$ to $f_{\Delta n}$. The actual moment in the frame at any point is then given by:

$$M = M_o + f_{\Delta 1}M_{\Delta 1} + f_{\Delta 2}M_{\Delta 2} + \quad + f_{\Delta n}M_{\Delta n} \qquad\qquad (9.12)$$

WORKED EXAMPLE 9.5

Analyse the frame shown in Figure 9.22 using moment distribution and calculate the support reactions. Assume $EI = 50 \times 10^3$ kNm^2 throughout.

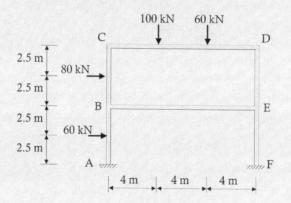

Figure 9.22 Fictitious restraints and arbitrary displacements.

The frame has two degrees of freedom with regard to joint translation as indicated in Figure 9.23, where fictitious supports are introduced at D and E.

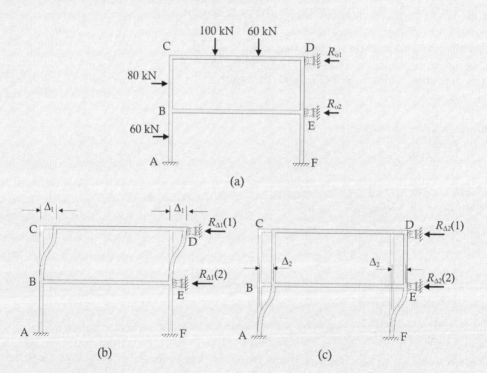

Figure 9.23 Fictitious supports and reactions at D and E. (a) Due to applied loads. (b) Due to arbitrary displacement Δ_1. (c) Due to arbitrary displacement Δ_2.

(1) Analysis with Zero Sidesway (with fictitious supports at D and E)

Fixed-end moments: From Appendix B, we get:

Member AB:

$$(M_{FE.A})_{AB} = -\frac{-60 \times 5}{8} = +37.5 \text{ kNm} \qquad (M_{FE.B})_{AB} = \frac{-60 \times 5}{8} = -37.5 \text{ kNm}$$

Member BC:

$$(M_{FE.B})_{BC} = -\frac{-80 \times 5}{8} = +50 \text{ kNm} \qquad (M_{FE.C})_{BC} = \frac{-80 \times 5}{8} = -50 \text{ kNm}$$

Member CD:

$$(M_{FE.C})_{CD} = -\frac{-100 \times 8^2 \times 4}{12^2} - \frac{-60 \times 4^2 \times 8}{12^2} = +231.1 \text{ kNm}$$

$$(M_{FE.D})_{CD} = \frac{-100 \times 4^2 \times 8}{12^2} + \frac{-60 \times 8^2 \times 4}{12^2} = -195.5 \text{ kNm}$$

Members BE, DE and EF:

$$(M_{FE.B})_{BE} = (M_{FE.E})_{BE} = 0 \quad (M_{FE.D})_{DE} = (M_{FE.E})_{DE} = 0 \quad (M_{FE.F})_{EF} = (M_{FE.F})_{EF} = 0$$

Relative stiffness:

Members AB, BC, DE and EF: $(I/L) = 0.2I$
Members BE and CD: $(I/L) = 0.0833I$

Distribution factors:

At supports A and F:

$$DF_{AB} = 0 \text{ and } DF_{FE} = 0$$

At joint B:

$$DF_{BA} = DF_{BC} = \frac{(0.2I)}{(0.2I) + (0.2I) + (0.0833I)} = 0.414$$

$$DF_{BE} = \frac{(0.0833I)}{(0.2I) + (0.2I) + (0.0833I)} = 0.172$$

Similarly at joint E:

$$DF_{EB} = 0.172 \quad \text{and} \quad DF_{ED} = DF_{EF} = 0.414$$

At joint C:

$$DF_{CB} = \frac{(0.2l)}{(0.2l)+(0.0833l)} = 0.706$$

$$DF_{CD} = \frac{(0.0833l)}{(0.2l)+(0.0833l)} = 0.294$$

Similarly at joint D:

$$DF_{DC} = 0.294 \quad \text{and} \quad DF_{DE} = 0.706$$

Moment distribution:

Joint	A	B			C		D		E			F
Member	AB	BA	BE	BC	CB	CD	DC	DE	ED	EB	EF	FE
DF	0	.414	.172	.414	.706	.294	.294	.706	.414	.172	.414	0
FEM (kNm)	+37.5	-37.5	0	+50	-50	+231.1	-195.6	0	0	0	0	0
Balance B	-2.6	-5.2	-2.2	-5.2	-2.6					-1.1		
Balance C				-63.0	-126.0	-52.5	-26.3					
Balance D						+32.6	+65.2	+156.6	+78.3			
Balance E			-6.7					-16.0	-31.9	-13.3	-31.9	-16.0
Balance D						+2.3	+4.7	+11.3	+5.6			
Balance C				-12.3	-24.7	-10.3	-5.1					
Balance B	+17.0	+33.9	+14.1	+33.9	+17.0					+7.1		
Balance C				-6.0	-12.0	-5.0	-2.5					
Balance D						+1.1	+2.2	+5.4	+2.7			
Balance E			-1.3					-3.2	-6.4	-2.7	-6.4	-3.2
Balance D						+0.5	+0.9	+2.2	+1.1			
Balance C				-0.5	-1.1	-0.5	-0.2					
Balance B	+1.6	+3.3	+1.4	+3.3	+1.6					+0.7		
Balance C				-0.6	-1.2	-0.5	-0.2					
Balance D							+0.1	+0.3	+0.2			
Balance E+B	+0.1	+0.2	+0.1	+0.2					-0.8	-0.3	-0.8	-0.4
M_0 (kNm)	+53.6	-5.2	+5.5	-0.2	-198.9	+198.9	-156.6	+156.6	+48.8	-9.6	-39.1	-19.6

With the end moments now determined, the reactions of the frame with the two fictitious supports are determined from statics and are shown in Figure 9.24.

(2) Analysis with arbitrary sidesway Δ_1

Fixed-end moments: From Appendix B, let us assume that Δ_1 is such that:

$$(M_{FE.B})_{BC} = (M_{FE.C})_{BC} = (M_{FE.E})_{ED} = (M_{FE.D})_{ED} = -\frac{6EI\Delta_1}{L^2} = +100 \text{ kNm}$$

This is equivalent to assuming $\Delta_1 = -1/120$ m.

All other fixed-end moments are zero.

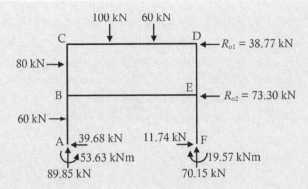

Figure 9.24 Reactions caused by applied load with fictitious restraints I and 2.

Moment distribution:

Joint	A	B			C		D		E			F
Member	AB	BA	BE	BC	CB	CD	DC	DE	ED	EB	EF	FE
DF	0	.414	.172	.414	.706	.294	.294	.706	.414	.172	.414	0
FEM (kNm)	0	0	0	+100	+100	0	0	+100	+100	0	0	0
Balance B	-20.7	-41.4	-17.2	-41.4	-20.7					-8.6		
Balance C				-28.0	-56.0	-23.3	-11.7					
Balance D						-13.0	-26.0	-62.4	-31.2			
Balance E			-5.2					-12.5	-24.9	-10.4	-24.9	-12.5
Balance D						+1.8	+3.7	+8.8	+4.4			
Balance C				+3.9	+7.9	+3.3	+1.6					
Balance B	+6.1	+12.1	+5.0	+12.1	+6.1					+2.5		
Balance C				-2.1	-4.3	-1.8	-0.9					
Balance D						-0.1	-0.2	-0.5	-0.3			
Balance E			-0.6					-1.4	-2.8	-1.1	-2.8	-1.4
Balance D						+0.2	+0.4	+1.0	+0.5			
Balance C				0	-0.1	0.0	0					
Balance B	+0.6	+1.1	+0.5	+1.1	+0.6					+0.2		
Balance C				-0.2	-0.4	-0.2	-0.1					
Balance B,D,E	+0.1	-0.1	0	+0.1				+0.1	-0.3	-0.1	-0.3	-0.1
$M_{\Delta 1}$ (kNm)	-14.0	-28.0	-17.5	+45.5	+33.1	-33.1	-33.1	+33.1	+45.5	-17.5	-28.0	-14.0

With the end moments now determined, the reactions of the frame with the two fictitious supports due to the arbitrary displacement Δ_1 are determined from statics and are shown in Figure 9.25a.

(3) Analysis with Arbitrary Sidesway Δ_2

Fixed-end moments: From Appendix B, let us assume that Δ_2 is such that:

$$(M_{FE.A})_{AB} = (M_{FE.B})_{AB} = (M_{FE.F})_{EF} = (M_{FE.E})_{EF} = -\frac{6EI\Delta_2}{L^2} = +100 \text{ kNm}$$

This is equivalent to assuming $\Delta_2 = -1/120$ m.

All other fixed-end moments are zero.

Moment distribution:

Joint	A	B			C		D		E			F
Member	AB	BA	BE	BC	CB	CD	DC	DE	ED	EB	EF	FE
DF	0	.414	.172	.414	.706	.294	.294	.706	.414	.172	.414	0
FEM (kNm)	+100	+100	0	0	0	0	0	0	0	0	+100	+100
Balance B	-20.7	-41.4	-17.2	-41.4	-20.7					-8.6		
Balance C				+7.3	+14.6	+6.1	+3.0					
Balance D						-0.4	-0.9	-2.1	-1.1			
Balance E			-7.8					-18.7	-37.4	-15.6	-37.4	-18.7
Balance D						+2.7	+5.5	+13.2	+6.6			
Balance C				-0.8	-1.6	-0.7	-0.3					
Balance B	+0.3	+0.5	+0.2	+0.5	+0.3					+0.1		
Balance C				-0.1	-0.2	-0.1	-0.1					
Balance D						+0.1	+0.1	+0.3	+0.1			
Balance E			-0.6					-1.4	-2.8	-1.2	-2.8	-1.4
Balance D						+0.2	+0.4	+1.0	+0.5			
Balance B,C,E	+0.2	+0.3	+0.1	+0.3	-0.1	0			-0.2	-0.1	-0.2	-0.1
$M_{\Delta2}$ (kNm)	+79.8	+59.5	-25.3	-34.2	-7.8	+7.8	+7.8	-7.8	-34.2	-25.3	+59.5	+79.8

With the end moments now determined, the reactions of the frame with the two fictitious supports due to the arbitrary displacement Δ_2 are determined from statics and are shown in Figure 9.25b.

Setting Fictitious Reactions at D and E to zero by using Equations 9.11, the reactions at D and E from the above three analyses are combined to give:

$$R_{o1} + f_{\Delta1}R_{\Delta1}(1) + f_{\Delta2}R_{\Delta2}(1) = 0 \quad \therefore -38.77 + 31.44\,f_{\Delta1} - 16.8\,f_{\Delta2} = 0$$

$$R_{o2} + f_{\Delta1}R_{\Delta1}(2) + f_{\Delta2}R_{\Delta2}(2) = 0 \quad \therefore -73.30 - 48.24\,f_{\Delta1} + 72.52\,f_{\Delta2} = 0$$

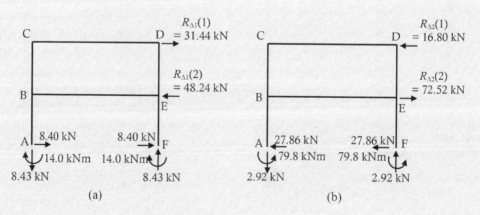

Figure 9.25 Reactions caused by arbitrary displacements Δ_1 and Δ_2. (a) Due to Δ_1. (b) Due to Δ_2.

and, solving these two simultaneous equations, we get:

$$f_{\Delta 1} = 2.751 \quad \text{and} \quad f_{\Delta 2} = 2.841$$

Final end moments, reactions and bending moment diagram:

The final end moments are obtained from Equation 9.12 (i.e. $M = M_o + f_{\Delta 1}M_{\Delta 1} + f_{\Delta 2}M_{\Delta 2}$) as follows:

Joint	A	B			C		D		E			F
Member	AB	BA	BE	BC	CB	CD	DC	DE	ED	EB	EF	FE
M_o (kNm)	+53.6	-5.2	+5.5	-0.2	-198.9	+198.9	-156.6	+156.6	+48.8	-9.6	-39.1	-19.6
$f_{\Delta 1}M_{\Delta 1}$ (kNm)	-38.5	-77.0	-48.1	+125.1	+91.1	-91.1	-91.1	+91.1	+125.1	-48.1	-77.0	-38.5
$f_{\Delta 2}M_{\Delta 2}$ (kNm)	+226.7	+169.0	-71.9	-97.2	-22.2	+22.2	+22.2	-22.2	-97.2	-71.9	+169.0	+226.7
M(kNm)	+241.8	+86.8	-114.5	+27.7	-130.0	+130.0	-225.5	+225.5	+76.7	-129.6	+52.9	+168.6

With these end moments, the final reactions are calculated from statics and are shown in Figure 9.26a. The bending moment diagram for the frame is shown in Figure 9.26b.

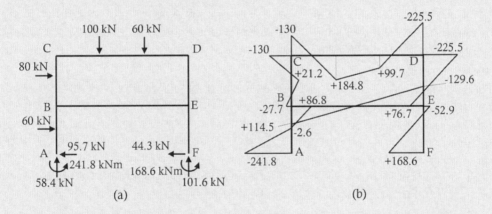

Figure 9.26 Reactions and bending moment diagram. (a) Loads and reactions. (b) Bending moment diagram (kNm).

For sway frames with more than two degrees of freedom, moment distribution by manual calculation becomes tedious and other methods of analysis might be preferred, such as the stiffness method presented in Chapter 12. Techniques are available to extend the moment distribution approach to cover the analysis of frames with inclined members or split level frames. Even frames containing non-prismatic members can be analysed using moment distribution. However, the complications associated with such problems are handled much more conveniently using the stiffness method and, with computer software readily available to practicing engineers, such problems are rarely solved using moment distribution today and we will not take the method any further here.

PROBLEMS

9.1   For the beam shown, determine the reactions at A, B and C using moment distribution and draw the shear force and bending moment diagrams. Assume EI is constant along the beam.

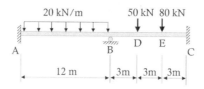

9.2   For the beam shown, determine the reactions at A and B using moment distribution and draw the shear force and bending moment diagrams. Assume EI is constant along the beam.

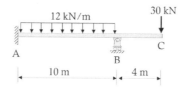

9.3   For the beam shown, determine the reactions at A, B and C using moment distribution and draw the shear force and bending moment diagrams. Assume EI is constant along the beam.

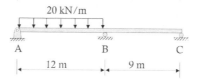

9.4   For the beam of Problem 9.1, if $EI = 50 \times 10^3$ kNm², determine the change in reactions if support B settles (downwards) by 60 mm.

9.5   For the beam of Problem 9.1, if $EI = 50 \times 10^3$ kNm², determine the change in reactions if support C settles (downwards) by 60 mm.

9.6   For the beam of Problem 9.3, if $EI = 50 \times 10^3$ kNm², determine the change in reactions if support B settles (downwards) by 60 mm.

9.7   For the beam shown, the 200 kN concentrated load is at the mid-span of BC. Determine the reactions at A, B, C and D using moment distribution, and draw the shear force and bending moment diagrams. Assume EI is constant along the beam.

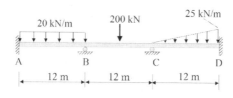

9.8   For the beam shown, calculate the reactions at supports A, B, C and D using moment distribution, and draw the shear force and bending moment diagrams. Assume *EI* is constant throughout.

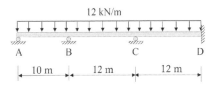

9.9   Re-analyse the beam of Problem 9.8 with the load on the span BC removed, while keeping all other spans loaded with a uniform load of 12 kN/m. Use moment distribution for the analysis. Comment on the effect of this change in loading on the magnitude and position of the maximum positive and negative moments in the beam.

9.10 Re-analyse the beam of Problem 9.8 with the load on the spans AB and CD removed, while keeping span BC loaded with a uniform load of 12 kN/m. Use moment distribution for the analysis. Comment on the effect on the magnitude and position of the maximum positive and negative moments in the beam.

9.11 Consider the wall of a water tank fixed at the base and supported as shown. Determine the reactions at the supports per 1 m width of wall under the linearly varying water pressure. Use moment distribution and assume *EI* is constant throughout.

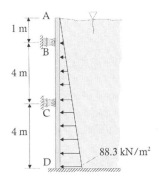

9.12 For the frame shown, determine the reactions at supports A and D using moment distribution and plot the axial force, shear force and bending moment diagrams.

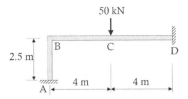

9.13 If *EI* is constant throughout the frame illustrated below, determine the reactions at A and C using moment distribution, and draw the shear force and bending moment diagrams for the frame.

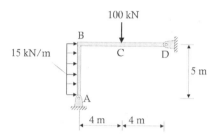

9.14 Determine the reactions at A and D for the frame shown using moment distribution. Assume $EI = 50 \times 10^3$ kNm² constant throughout, and draw the shear force and bending moment diagrams for the frame.

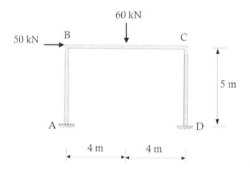

9.15 If the supports at A and D of Problem 9.14 are both pinned (rather than fixed), re-analyse the frame under the loading shown and plot the bending moment diagram.

9.16 Determine the reactions at A and D for the frame illustrated below using moment distribution. Assume $EI = 60 \times 10^3$ kNm² constant throughout, and draw the shear force and bending moment diagrams for the frame.

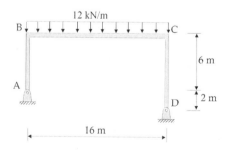

9.17 If *EI* is constant throughout the frame illustrated below, determine the reactions at A and F using moment distribution, and draw the shear force and bending moment diagrams for the frame.

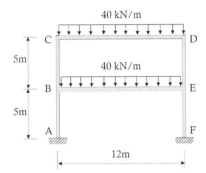

9.18 Determine the reactions at A and F for the frame shown using moment distribution. Assume $EI = 70 \times 10^3$ kNm^2 constant throughout, and draw the shear force and bending moment diagrams for the frame.

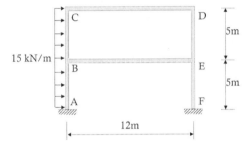

Chapter 10

Truss analysis using the stiffness method

10.1 OVERVIEW OF THE STIFFNESS METHOD

The stiffness method is a powerful approach for the analysis of statically determinate and indeterminate structures. Its popularity is mainly attributed to its ability to be easily programmed for computer calculations thanks to its well-defined solution procedure.

The first step to be carried out when using the stiffness method is to subdivide the structure into its constituent members, referred to as *stiffness elements* (or *elements*), and the joints connecting these members are called *nodes*. A stiffness relationship is written for each element, which relates the displacements at its nodes to the actions at its nodes and, because of this, it is referred to as the *load–displacement relationship*. These expressions are then combined to obtain the governing system of equations by enforcing equilibrium at each node expressed in terms of the unknown nodal displacements. Once these displacements are determined, the other variables describing the structural response, for example, internal actions and reactions, are calculated in the post-processing of the results. This approach is also known as the *displacement method* to emphasise the role played by the displacements in the solution process.

In this chapter, the simplest of the available stiffness elements, referred to as the *truss element*, is introduced. This element is assumed to be able to resist only axial forces and it is to be used in the framework of *small displacements*. In the first part of the chapter, we deal with the stiffness analysis of plane trusses using truss elements and this is followed by the stiffness analysis of three-dimensional space trusses.

10.2 SIGN CONVENTION, NOTATION, COORDINATE SYSTEMS AND DEGREES OF FREEDOM

10.2.1 Sign convention and notation

The sign convention and notation for trusses introduced in Chapter 4 are used here. Tensile (compressive) forces are assumed to be positive (negative). Nodes are identified with either alphabetic (uppercase) characters or numbers as shown in Figure 10.1, while members are denoted using the alphabetic identifications of their end nodes or simply with numbers (e.g. member DC or member 5 in Figure 10.1).

10.2.2 Local and global coordinate systems

The stiffness method requires the definition of two coordinate systems, one related to each individual stiffness element, denoted as the *local* (or *member*) *coordinate system*, and one applicable to the entire truss, referred to as the *global* (or *structure*) *coordinate system*.

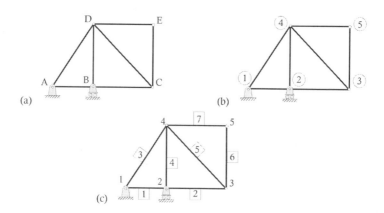

Figure 10.1 Subdivision of a structure in stiffness elements and nodes.

We will assume these coordinate systems to be *orthogonal* (i.e. all axes are perpendicular to each other) and to satisfy the *right-hand rule* (see Section 2.2).

For each element in a two-dimensional truss, we assign a member coordinate system in which the *x*-axis is parallel to the member axis and the *y*-axis is perpendicular to it in the plane of the truss. The positive *z*-axis comes out of the page. A location vector is introduced to specify the positive direction of *x* as shown, for example, in Figure 10.2 for member 3 from the truss in Figure 10.1. Location vectors were introduced earlier in Section 4.5. The positive direction of a location vector is arbitrary, but once assigned, it cannot be changed during the solution process. The origin of the local *x*-axis identifies the first node of the element (e.g. node 1 for element 3 in Figure 10.2) with the other node referred to as the second element node (node 4 for element 3 in Figure 10.2).

As the elements of a truss do not all possess the same inclination, we cannot specify one single set of local coordinates for all of them. For this purpose, we need to establish a global (or structures) reference system applicable to the entire truss.

Lowercase and uppercase characters are here used to distinguish between the local coordinate system (*x,y,z*) and the global coordinate system (*X,Y,Z*). For example, Figure 10.3 illustrates one possible set of local and global coordinate axes for the analysis of the truss of Figure 10.1. Figure 10.3a shows the selected global coordinate axes *X* and *Y* for the whole truss, while the member (or local) coordinates assigned on the basis of the assumed location vectors are illustrated for each element in Figure 10.3b.

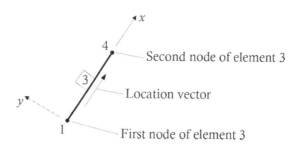

Figure 10.2 Local coordinate system and location vector for member 3 (in Figure 10.1c).

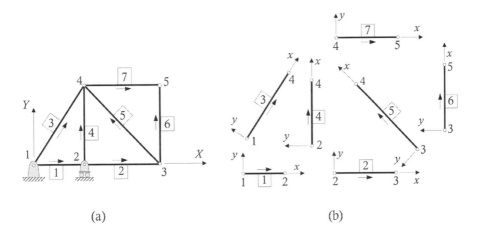

Figure 10.3 Example of global and local coordinate systems. (a) Global coordinate system. (b) Local coordinate systems.

10.2.3 Degrees of freedom of the structure

The movement of each node is defined by two independent displacements, which are usually (and conveniently) assumed to be parallel to the global axes. Similarly, equilibrium is enforced at each node along two independent directions, usually taken as the global coordinate directions. These independent directions (along which equilibrium is enforced and along which node displacements are described) are referred to as *degrees of freedom (dof)*, or simply *freedoms*, which are represented graphically by arrows defining their positive directions and are labelled with numbers. Degrees of freedom were introduced earlier in Section 4.5 for the method of joints in matrix form.

It is convenient for hand calculations (as outlined later) to first number the freedoms along which displacements are not restrained, followed by the remaining (restrained) freedoms. For example, let us consider the truss of Figure 10.4a. After introducing a global coordinate system, we can assign to each node two freedoms, depicted by arrows parallel to the global axes (Figure 10.4b). By inspection, we can see that dofs at nodes 3, 4 and 5, together with the horizontal freedom at node 2, are unrestrained (i.e. free to move) and are numbered first from 1 to 7 (Figure 10.4c). Remaining dofs related to restrained freedoms are then considered and numbered from 8 to 10 (Figure 10.4d).

10.3 DERIVATION OF THE STIFFNESS MATRIX IN LOCAL COORDINATES

The basis of the stiffness method relies on the *load–displacement relationship* applied along the member axis (i.e. along the local *x*-axis) at its two end nodes, which, for ease of reference, are denoted nodes 1 and 2 (Figure 10.5). For a truss element, let us denote the nodal displacements in the direction of the element as d_1 and d_2 and the nodal forces as q_1 and q_2, as shown in Figure 10.5. The nodal displacements and nodal forces are related by the following relationship:

$$\begin{bmatrix} q_1 \\ q_2 \end{bmatrix} = \begin{bmatrix} k_{11} & k_{12} \\ k_{21} & k_{22} \end{bmatrix} \begin{bmatrix} d_1 \\ d_2 \end{bmatrix} \tag{10.1}$$

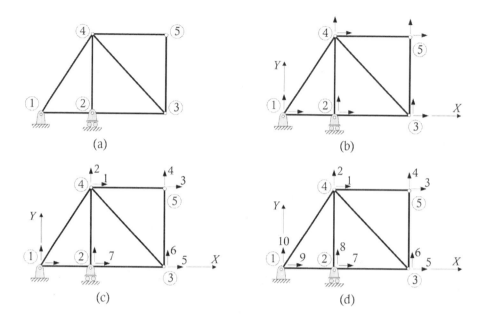

Figure 10.4 Numbering the degrees of freedom of the structure. (a) Layout of the truss. (b) Global freedoms. (c) Numbering of understrained freedoms. (d) Numbering of restrained freedoms.

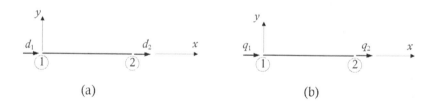

Figure 10.5 Isolated stiffness element in local coordinates. (a) Local nodal displacements. (b) Local nodal actions.

where nodal actions and displacements are assumed to be positive in the positive direction of the location vector or x-axis (Figure 10.5). The matrix terms k_{ij} (with $i,j = 1,2$) are referred to as the *member stiffness influence coefficients* and k_{ij} represents the reaction induced at node i in the direction of the element axis by a unit displacement in the direction of the element axis at node j. The physical representation of these coefficients is described graphically in Figure 10.6.

The member stiffness influence coefficients can be determined recalling that the behaviour of a truss element is similar to the one of a common spring for which we know that its elongation e is related to the internal force N by means of the spring stiffness k as (Figure 10.7)

$$N = ke \tag{10.2a}$$

The elongation of a linear–elastic truss element (of length L, cross-sectional area A and elastic modulus E) caused by an axial force N is (see Section 4.8):

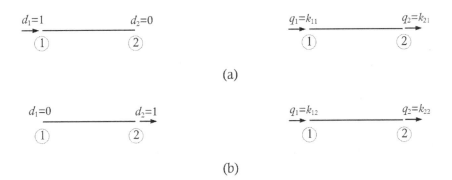

Figure 10.6 Physical representation of the member stiffness influence coefficients. (a) Unit displacement along local freedom 1. (b) Unit displacement along local freedom 2.

Figure 10.7 Nodal actions and internal force of a spring. (a) Nodal actions of the spring. (b) Free-body diagrams of the end spring nodes.

$$e = \varepsilon L = \frac{\sigma}{E} L = \frac{N}{EA} L \tag{10.2b}$$

and substituting into Equation 10.2a gives:

$$k = EA/L \tag{10.2c}$$

Referring back to the spring in Figure 10.7, a tensile force in the spring (positive N) produces an elongation (positive e) and, vice-versa, a compression force (negative N) produces a shortening of the spring (negative e).

Let us now make use of Equations 10.2 to define the load–displacement relationship. Enforcing equilibrium at the two end nodes (Figure 10.7b) gives:

$$q_1 = -N \qquad q_2 = N \tag{10.3a,b}$$

From a kinematic viewpoint, the contribution of the nodal displacements d_1 and d_2 to the elongation e of the spring is outlined in Figure 10.8. In particular, an elongation of the spring (positive e) is produced by a positive displacement of d_2 (Figure 10.8a) and a contraction of the spring (negative e) is produced by a positive displacement of d_1. On the basis of the principle of superposition, the total elongation that can originate from the displacements d_1 and d_2 is therefore:

(a) Influence of d_2 on e (b) Influence of d_1 on e

Figure 10.8 Relationship between end displacements and truss extensions.

$$e = d_2 - d_1 \tag{10.4a}$$

from which the internal axial force N can be calculated (recalling Equation 10.2a) as:

$$N = ke = k(d_2 - d_1) \tag{10.4b}$$

Substituting Equations 10.4 into Equations 10.3 enables the calculation of the end nodal forces of the spring in terms of its nodal end displacements:

$$q_1 = -N = -k(d_2 - d_1) \qquad q_2 = N = k(d_2 - d_1) \tag{10.5a,b}$$

These can be re-written in more compact form as:

$$\begin{bmatrix} q_1 \\ q_2 \end{bmatrix} = k \begin{bmatrix} 1 & -1 \\ -1 & 1 \end{bmatrix} \begin{bmatrix} d_1 \\ d_2 \end{bmatrix} \tag{10.6}$$

which represents the *load–displacement relationship* of a spring (or, in our context, the *load–displacement relationship* of a truss element) expressed in local coordinates.

REFLECTION ACTIVITY 10.1

On the basis of the stiffness relationship obtained in Equation 10.6, determine the corresponding expression for a truss element knowing that the element is equivalent to a spring having rigidity EA/L (Equation 10.2c), with A and L being the cross-sectional area and length of the truss member, and E being the elastic modulus of its material.

The response of a truss element is identical to the one of a spring. Substituting $k = EA/L$ into Equation 10.6 gives:

$$\begin{bmatrix} q_1 \\ q_2 \end{bmatrix} = \frac{EA}{L} \begin{bmatrix} 1 & -1 \\ -1 & 1 \end{bmatrix} \begin{bmatrix} d_1 \\ d_2 \end{bmatrix} \tag{10.7}$$

This expression is usually written in more compact form as:

$$\mathbf{q} = \mathbf{kd} \tag{10.8}$$

where **k** is the *stiffness matrix of the truss in local coordinates*, **q** is the vector of nodal forces and **d** is the vector of nodal displacements (Figure 10.5). That is:

$$\mathbf{q} = \begin{bmatrix} q_1 \\ q_2 \end{bmatrix} \qquad \mathbf{k} = \frac{EA}{L} \begin{bmatrix} 1 & -1 \\ -1 & 1 \end{bmatrix} \qquad \mathbf{d} = \begin{bmatrix} d_1 \\ d_2 \end{bmatrix} \tag{10.9a–c}$$

REFLECTION ACTIVITY 10.2

In the writing of the load–displacement relationship of the spring, we have only considered local nodal freedoms (and corresponding displacements and forces) parallel to the member axis (Figure 10.5). When a member is part of a truss, its nodal displacements can move in directions also perpendicular to the element (i.e. parallel to the y-axis). Determine whether we should include the influence of nodal displacements perpendicular to the member axis in the load–displacement relationship of Equation 10.8 and, if so, in what manner.

To address the question, we apply transverse displacements at the two end nodes of the truss element, one at a time, and see whether they induce any change in length, and any corresponding change in the axial force in the member.

In Figure 10.9a, the only non-zero displacement is the transverse displacement at node 1 denoted as d_{1y}, with subscripts '1' and 'y' identifying the first node of the element and the movement taking place in the direction of the y-axis. From trigonometry, the elongation induced in the truss element is:

$$e = L_f - L = L_f - L_f \cos \theta \approx L_f - L_f \times 1 = 0$$

where L and L_f denote the original and displaced lengths, respectively, while $\cos \theta$ is approximated by 1 because we are working under the assumptions of small displacements (and rotations are assumed to remain small). Because e is nil, the transverse displacement d_{1y} does not

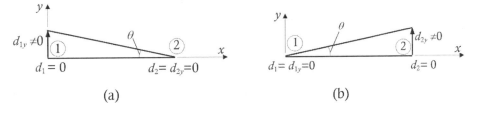

Figure 10.9 Transverse displacements at the end nodes of the truss element. (a) Transverse displacement at node 1. (b) Transverse displacement at node 2.

induce any axial force in the truss element. It follows that the effects of nodal displacements transverse to a truss element can be ignored in the calculation of the load–displacement relationship for the element, because its axial rigidity does not contribute to resisting this deformation.

Similarly, we can also show that the transverse displacement at node 2 does not induce any elongation and, as such, does not need to be included in the calculation of the load–displacement relationship of a truss member. In particular, the nil elongation at node 2 owing to d_{2y} is calculated as (Figure 10.9b):

$$e = L_f - L = L_f - L_f \cos \theta \approx L_f - L_f \times 1 = 0$$

10.4 TRANSFORMATION BETWEEN LOCAL AND GLOBAL COORDINATE SYSTEMS

10.4.1 Transformation matrix for vectors

Considering that the variables we are dealing with consist of vectors (in our case, displacement and force vectors), we need to establish a procedure to relate vector components in local and global coordinates.

Let us consider the coordinate systems shown in Figure 10.10a, one representing the local coordinates (x,y) and one the global ones (X,Y). Any vector **V** possesses different components when defined in these two systems. In global coordinates, these are expressed as R_X and R_Y (as shown in Figure 10.10b), while in the local system, these are denoted with r_x and r_y (as shown in Figure 10.10c).

We will now assume that a relationship between the local (r_x, r_y) and global (R_X, R_Y) components of the same vector can be expressed in terms of a 2×2 matrix **H** as:

$$\begin{bmatrix} r_x \\ r_y \end{bmatrix} = \begin{bmatrix} H_{11} & H_{12} \\ H_{21} & H_{22} \end{bmatrix} \begin{bmatrix} R_X \\ R_Y \end{bmatrix} \tag{10.10}$$

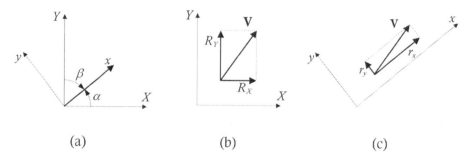

(a)   (b)   (c)

Figure 10.10 Example of a vector defined in global and local coordinates. (a) Global and local coordinates. (b) Vector and global coordinates. (c) Vector and local coordinates.

which could be used to determine (r_x, r_y) for known values of (R_X, R_Y). We will now evaluate the coefficients of **H** based on trigonometry considering two particular vectors: $(R_X, R_Y) = (1,0)$ and $(R_X, R_Y) = (0,1)$.

We start by substituting $(R_X, R_Y) = (1,0)$ in Equation 10.10:

$$\begin{bmatrix} r_x \\ r_y \end{bmatrix} = \begin{bmatrix} H_{11} & H_{12} \\ H_{21} & H_{22} \end{bmatrix} \begin{bmatrix} R_X \\ R_Y \end{bmatrix} = \begin{bmatrix} H_{11} & H_{12} \\ H_{21} & H_{22} \end{bmatrix} \begin{bmatrix} 1 \\ 0 \end{bmatrix} = \begin{bmatrix} H_{11} \\ H_{21} \end{bmatrix} \tag{10.11}$$

from which $r_x = H_{11}$ and $r_y = H_{21}$. It is possible to determine the values H_{11} and H_{21} from trigonometric considerations. It can be seen in Figure 10.11a that for the vector $(R_X, R_Y) = (1,0)$, the components in the member axes are $r_x = 1 \cos \alpha$ and $r_y = -1 \cos \beta$, i.e.:

$$r_x = l \quad \text{and} \quad r_y = -m \tag{10.12a,b}$$

where l and m are the direction cosines of the local x axis:

$$l = \cos \alpha \quad \text{and} \quad m = \cos \beta \tag{10.13a,b}$$

Substituting Equations 10.12 into Equation 10.11, the coefficients of the first column of **H**, i.e. H_{11} and H_{21}, are:

$$H_{11} = l \quad \text{and} \quad H_{21} = -m \tag{10.14a,b}$$

In a similar manner, the coefficients included in the second column of **H** can be obtained from the local components of vector $(R_X, R_Y) = (0,1)$:

$$\begin{bmatrix} r_x \\ r_y \end{bmatrix} = \begin{bmatrix} H_{11} & H_{12} \\ H_{21} & H_{22} \end{bmatrix} \begin{bmatrix} R_X \\ R_Y \end{bmatrix} = \begin{bmatrix} H_{11} & H_{12} \\ H_{21} & H_{22} \end{bmatrix} \begin{bmatrix} 0 \\ 1 \end{bmatrix} = \begin{bmatrix} H_{12} \\ H_{22} \end{bmatrix} \tag{10.15}$$

The corresponding values for r_x and r_y can be evaluated based on Figure 10.11b as:

$$r_x = m \quad \text{and} \quad r_y = l \tag{10.16a,b}$$

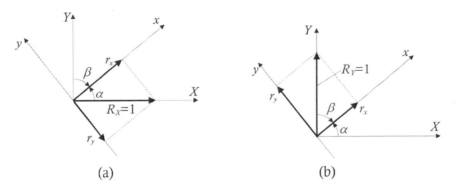

(a)　　　　　　　　　　　　(b)

Figure 10.11 Unit vectors expressed in the global coordinate system. (a) Vector with $(R_X, R_Y) = (1,0)$. (b) Vector with $(R_X, R_Y) = (0,1)$.

which, based on Equation 10.15, equal coefficients H_{12} and H_{22}:

$$H_{12} = m \quad \text{and} \quad H_{22} = l \tag{10.17a,b}$$

The matrix **H** is therefore defined as:

$$\mathbf{H} = \begin{bmatrix} H_{11} & H_{12} \\ H_{21} & H_{22} \end{bmatrix} = \begin{bmatrix} l & m \\ -m & l \end{bmatrix} \tag{10.18}$$

Considering its role in Equation 10.10, **H** is referred to as the *transformation matrix to convert vector components from global to local coordinates*.

REFLECTION ACTIVITY 10.3

Determine the transformation matrix $\tilde{\mathbf{H}}$ required to convert the components of a vector from local to global coordinates following the procedure previously adopted to derive the coefficients for **H** (which is the *transformation matrix to convert vector components from global to local coordinates*).

The global and local components of a vector are (R_X, R_Y) and (r_x, r_y), respectively. With the use of $\tilde{\mathbf{H}}$, we can calculate the global coordinates (R_X, R_Y) corresponding to a vector defined in the local system with (r_x, r_y) as:

$$\begin{bmatrix} R_X \\ R_Y \end{bmatrix} = \begin{bmatrix} \tilde{H}_{11} & \tilde{H}_{12} \\ \tilde{H}_{21} & \tilde{H}_{22} \end{bmatrix} \begin{bmatrix} r_x \\ r_y \end{bmatrix} \tag{10.19}$$

The coefficients of $\tilde{\mathbf{H}}$ are obtained by considering two particular vectors: $(r_x, r_y) = (1,0)$ and $(r_x, r_y) = (0,1)$. The terms in its first column, i.e. $\tilde{H}_{11}$ and $\tilde{H}_{21}$, are determined considering that their values equal those of the global components corresponding to the vector $(r_x, r_y) = (1,0)$:

$$\begin{bmatrix} R_X \\ R_Y \end{bmatrix} = \begin{bmatrix} \tilde{H}_{11} & \tilde{H}_{12} \\ \tilde{H}_{21} & \tilde{H}_{22} \end{bmatrix} \begin{bmatrix} r_x \\ r_y \end{bmatrix} = \begin{bmatrix} \tilde{H}_{11} & \tilde{H}_{12} \\ \tilde{H}_{21} & \tilde{H}_{22} \end{bmatrix} \begin{bmatrix} 1 \\ 0 \end{bmatrix} = \begin{bmatrix} \tilde{H}_{11} \\ \tilde{H}_{21} \end{bmatrix} \tag{10.20}$$

From Figure 10.12a, we can see that $R_X = 1 \cos \alpha = l$ and $R_Y = 1 \cos \beta = m$ and, substituting into Equation 10.20, we get:

$$\tilde{H}_{11} = l \quad \text{and} \quad \tilde{H}_{21} = m$$

With a similar approach, the second column of $\tilde{\mathbf{H}}$ can be obtained considering the global components of $(r_x, r_y) = (0,1)$:

$$\begin{bmatrix} R_X \\ R_Y \end{bmatrix} = \begin{bmatrix} \tilde{H}_{11} & \tilde{H}_{12} \\ \tilde{H}_{21} & \tilde{H}_{22} \end{bmatrix} \begin{bmatrix} r_x \\ r_y \end{bmatrix} = \begin{bmatrix} \tilde{H}_{11} & \tilde{H}_{12} \\ \tilde{H}_{21} & \tilde{H}_{22} \end{bmatrix} \begin{bmatrix} 0 \\ 1 \end{bmatrix} = \begin{bmatrix} \tilde{H}_{12} \\ \tilde{H}_{22} \end{bmatrix} \tag{10.21}$$

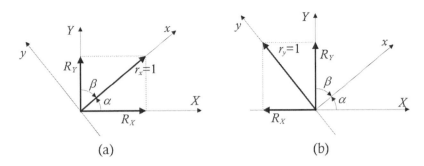

Figure 10.12 Unit vectors expressed in the local coordinate system. (a) Vector with $(r_x, r_y) = (1,0)$. (b) Vector with $(r_x, r_y) = (0,1)$.

Referring to Figure 10.12b, we can see that $R_X = -m$ and $R_Y = l$ and therefore $\tilde{H}_{12} = -m$ and $\tilde{H}_{22} = l$.

In summary, the *transformation matrix* $\tilde{\mathbf{H}}$ *to convert vector components from local to global coordinates* is:

$$\tilde{\mathbf{H}} = \begin{bmatrix} \tilde{H}_{11} & \tilde{H}_{12} \\ \tilde{H}_{21} & \tilde{H}_{22} \end{bmatrix} = \begin{bmatrix} l & -m \\ m & l \end{bmatrix} \tag{10.22}$$

It should be noted that the two matrices $\mathbf{H}$ and $\tilde{\mathbf{H}}$ are the transpose of each other:

$$\mathbf{H} = \tilde{\mathbf{H}}^{\mathrm{T}} \quad \text{or} \quad \tilde{\mathbf{H}} = \mathbf{H}^{\mathrm{T}} \tag{10.23a,b}$$

WORKED EXAMPLE 10.1

Consider the local and global coordinate systems shown in Figure 10.13 with $\alpha = 30°$ and $\beta = 60°$. Determine (i) the local components of vector $\mathbf{R} = (R_X, R_Y) = (10,5)$, where $\mathbf{R}$ is defined in global coordinates; (ii) the global components of vector $\mathbf{r} = (r_x, r_y) = (8,6)$, where $\mathbf{r}$ is specified in the local system.

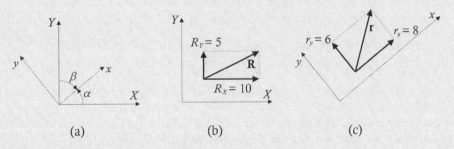

Figure 10.13 Vector components for Worked Example 10.1. (a) Global and local coordinates. (b) Vector and global coordinates. (c) Vector and local coordinates.

(i) Local components of **R** are calculated using Equation 10.10, determining matrix **H** from Equation 10.18. The direction cosines l and m are calculated from angles α and β as $l = \cos \alpha = \cos 30° = 0.866$ and $m = \cos \beta = \cos 60° = 0.5$. Therefore:

$$\mathbf{H} = \begin{bmatrix} l & m \\ -m & l \end{bmatrix} = \begin{bmatrix} 0.866 & 0.5 \\ -0.5 & 0.866 \end{bmatrix}$$

and the local components of vector **R** are (Equation 10.10):

$$\begin{bmatrix} r_x \\ r_y \end{bmatrix} = \begin{bmatrix} 0.866 & 0.5 \\ -0.5 & 0.866 \end{bmatrix} \begin{bmatrix} 10 \\ 5 \end{bmatrix} = \begin{bmatrix} 11.16 \\ -0.67 \end{bmatrix}$$

(ii) The global components of **r** are calculated from Equations 10.19 and 10.22:

$$l = \cos 30° = 0.8660; \quad m = \cos 60° = 0.5; \quad \therefore \tilde{\mathbf{H}} = \begin{bmatrix} l & -m \\ m & l \end{bmatrix} = \begin{bmatrix} 0.866 & -0.5 \\ 0.5 & 0.866 \end{bmatrix} \quad \text{and}$$

$$\begin{bmatrix} R_X \\ R_Y \end{bmatrix} = \tilde{\mathbf{H}}\mathbf{r} = \mathbf{H}^T\mathbf{r} = \begin{bmatrix} 0.866 & -0.5 \\ 0.5 & 0.866 \end{bmatrix} \begin{bmatrix} 8 \\ 6 \end{bmatrix} = \begin{bmatrix} 3.928 \\ 9.196 \end{bmatrix}$$

10.4.2 Transformation matrix for the truss element

The load–displacement expression introduced in Equation 10.8 relates the nodal displacements **d** to the nodal forces **q**. Consider the truss member shown in Figure 10.14.

For clarity, we will use local freedoms 1 and 1y at node 1, and 2 and 2y at node 2 (see Figure 10.14a), while the global freedoms are assumed to be 1 and 2 at node 1, and 3 and 4 at node 2 (see Figure 10.14b). The freedom numbering is based on the direction of the location vector (which defines the first node as the origin of the local coordinate system).

We will now consider the two displacement vectors at nodes 1 and 2 ($\mathbf{d}_1$ and $\mathbf{d}_2$), illustrated in Figure 10.15, and see how their components can be transformed into global coordinates. For each vector, we can apply Equations 10.10 and 10.18 as follows:

$$\begin{bmatrix} d_1 \\ d_{1y} \end{bmatrix} = \begin{bmatrix} l & m \\ -m & l \end{bmatrix} \begin{bmatrix} D_{e1} \\ D_{e2} \end{bmatrix} \quad \text{and} \quad \begin{bmatrix} d_2 \\ d_{2y} \end{bmatrix} = \begin{bmatrix} l & m \\ -m & l \end{bmatrix} \begin{bmatrix} D_{e3} \\ D_{e4} \end{bmatrix} \tag{10.24a,b}$$

where l and m represent the direction cosines related to α and β, respectively, which are the angles between the local and global coordinates at node 1 (as illustrated in Figure 10.14c). D_{e1} and D_{e2} are the components of the displacement vector at node 1 in global coordinates in the direction of global freedoms 1 and 2, respectively, and D_{e3} and D_{e4} are the components of the displacement vector at node 2 in global coordinates in the direction of global freedoms 3 and 4, respectively.

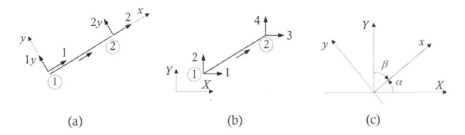

Figure 10.14 Local and global degrees of freedom. (a) Local freedoms. (b) Global freedoms. (c) Angles for the direction cosines.

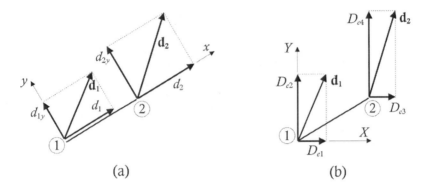

Figure 10.15 Nodal displacements in local and global coordinates. (a) Nodal displacements in local coordinates. (b) Nodal displacements in global coordinates.

Considering that the local displacements perpendicular to the truss element are not relevant to its structural response (see Reflection Activity 10.2), Equations 10.24 can be simplified to:

$$[d_1] = [\; l \;\; m \;]\begin{bmatrix} D_{e1} \\ D_{e2} \end{bmatrix} \quad \text{and} \quad [d_2] = [\; l \;\; m \;]\begin{bmatrix} D_{e3} \\ D_{e4} \end{bmatrix} \tag{10.25a,b}$$

These expressions can be combined in compact form as:

$$\begin{bmatrix} d_1 \\ d_2 \end{bmatrix} = \begin{bmatrix} l & m & 0 & 0 \\ 0 & 0 & l & m \end{bmatrix}\begin{bmatrix} D_{e1} \\ D_{e2} \\ D_{e3} \\ D_{e4} \end{bmatrix} \quad \text{or} \quad \mathbf{d} = \mathbf{T}\mathbf{D}_e \tag{10.26a,b}$$

where:

$$\mathbf{d} = \begin{bmatrix} d_1 \\ d_2 \end{bmatrix} \qquad \mathbf{T} = \begin{bmatrix} l & m & 0 & 0 \\ 0 & 0 & l & m \end{bmatrix} \qquad \mathbf{D}_e = \begin{bmatrix} D_{e1} \\ D_{e2} \\ D_{e3} \\ D_{e4} \end{bmatrix} \tag{10.27a–c}$$

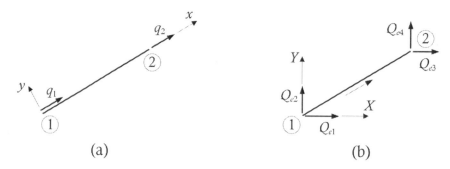

Figure 10.16 Nodal forces in local and global coordinates. (a) Nodal forces in local coordinates. (b) Nodal forces in global coordinates.

and **T** is referred to as the *transformation matrix* for the truss element from global to local coordinates. This can be also used to relate global and local components of the nodal forces:

$$
\begin{bmatrix} q_1 \\ q_2 \end{bmatrix} = \begin{bmatrix} l & m & 0 & 0 \\ 0 & 0 & l & m \end{bmatrix} \begin{bmatrix} Q_{e1} \\ Q_{e2} \\ Q_{e3} \\ Q_{e4} \end{bmatrix} \quad \text{or} \quad \mathbf{q} = \mathbf{T}\mathbf{Q}_e \tag{10.28a,b}
$$

where the nodal forces expressed in local and global coordinates are depicted in Figure 10.16 and are collected in vectors **q** and **Q**$_e$ as:

$$
\mathbf{q} = \begin{bmatrix} q_1 \\ q_2 \end{bmatrix} \qquad \mathbf{Q}_e = \begin{bmatrix} Q_{e1} \\ Q_{e2} \\ Q_{e3} \\ Q_{e4} \end{bmatrix} \tag{10.29a,b}
$$

REFLECTION ACTIVITY 10.4

Determine the transformation matrix to be used to convert the local components of the nodal displacements of a truss element into its corresponding global ones.

The global components of the nodal displacements can be calculated from the local coordinates using Equations 10.19, 10.22 and 10.23:

$$
\begin{bmatrix} D_{e1} \\ D_{e2} \end{bmatrix} = \begin{bmatrix} l & -m \\ m & l \end{bmatrix} \begin{bmatrix} d_1 \\ d_{1y} \end{bmatrix} \quad \text{and} \quad \begin{bmatrix} D_{e3} \\ D_{e4} \end{bmatrix} = \begin{bmatrix} l & -m \\ m & l \end{bmatrix} \begin{bmatrix} d_2 \\ d_{2y} \end{bmatrix}
$$

and, neglecting the local displacements perpendicular to the truss element (see Reflection Activity 10.2), the above expressions can be simplified to:

$$\begin{bmatrix} D_{e1} \\ D_{e2} \end{bmatrix} = \begin{bmatrix} l \\ m \end{bmatrix}[d_1] \qquad \begin{bmatrix} D_{e3} \\ D_{e4} \end{bmatrix} = \begin{bmatrix} l \\ m \end{bmatrix}[d_2]$$

Combining these in compact form leads to:

$$\begin{bmatrix} D_{e1} \\ D_{e2} \\ D_{e3} \\ D_{e4} \end{bmatrix} = \begin{bmatrix} l & 0 \\ m & 0 \\ 0 & l \\ 0 & m \end{bmatrix}\begin{bmatrix} d_1 \\ d_2 \end{bmatrix} = \begin{bmatrix} l & m & 0 & 0 \\ 0 & 0 & l & m \end{bmatrix}^{\mathsf{T}}\begin{bmatrix} d_1 \\ d_2 \end{bmatrix} \quad \text{or} \quad \mathbf{D}_e = \mathbf{T}^{\mathsf{T}}\mathbf{d} \qquad (10.30\text{a,b})$$

where $\mathbf{T}^{\mathsf{T}}$ is referred to as the *transformation matrix* for the truss element from local to global coordinates. It could also be used to transform components of the nodal forces:

$$\begin{bmatrix} Q_{e1} \\ Q_{e2} \\ Q_{e3} \\ Q_{e4} \end{bmatrix} = \begin{bmatrix} l & 0 \\ m & 0 \\ 0 & l \\ 0 & m \end{bmatrix}\begin{bmatrix} q_1 \\ q_2 \end{bmatrix} = \begin{bmatrix} l & m & 0 & 0 \\ 0 & 0 & l & m \end{bmatrix}^{\mathsf{T}}\begin{bmatrix} q_1 \\ q_2 \end{bmatrix} \quad \text{or} \quad \mathbf{Q}_e = \mathbf{T}^{\mathsf{T}}\mathbf{q} \qquad (10.31\text{a,b})$$

10.5 TRUSS ELEMENT IN GLOBAL COORDINATES

The load–displacement relationship previously derived in local coordinates for a truss element is reproduced below for ease of reference:

$$\mathbf{q} = \mathbf{kd} \tag{10.8}$$

with:

$$\mathbf{q} = \begin{bmatrix} q_1 \\ q_2 \end{bmatrix} \qquad \mathbf{k} = \frac{AE}{L}\begin{bmatrix} 1 & -1 \\ -1 & 1 \end{bmatrix} \qquad \mathbf{d} = \begin{bmatrix} d_1 \\ d_2 \end{bmatrix} \tag{10.9a–c}$$

This expression is now transformed into global coordinates based on the following steps:

i. Substitute $\mathbf{d} = \mathbf{TD}_e$ (Equation 10.26b) into Equation 10.8:

$$\mathbf{q} = \mathbf{kTD}_e \tag{10.32a}$$

ii. Pre-multiply both sides of Equation 10.32a by the transpose of $\mathbf{T}$:

$$\mathbf{T}^{\mathsf{T}}\mathbf{q} = \mathbf{T}^{\mathsf{T}}\mathbf{kTD}_e \tag{10.32b}$$

iii. Recalling Equation 10.31b, the left-hand side of Equation 10.32b represents the nodal forces expressed in global coordinates $\mathbf{Q}_e \, (= \mathbf{T}^{\mathsf{T}}\mathbf{q})$. Therefore:

$$\mathbf{Q}_e = \mathbf{T}^{\mathsf{T}}\mathbf{kTD}_e \tag{10.32c}$$

iv. Equation 10.32c can be re-arranged to highlight the stiffness matrix of an isolated truss element:

$$\mathbf{Q}_e = \mathbf{K}_e \mathbf{D}_e \tag{10.33}$$

where $\mathbf{Q}_e$ and $\mathbf{D}_e$ are vectors that collect the nodal forces and displacements, respectively, expressed in global coordinates (Equations 10.29b and 10.27c), and $\mathbf{K}_e$ represents the element stiffness matrix:

$$\mathbf{K}_e = \mathbf{T}^\mathsf{T} \mathbf{k} \mathbf{T} \tag{10.34}$$

WORKED EXAMPLE 10.2

Calculate all coefficients of the stiffness matrix in global coordinates for an isolated truss element $\mathbf{K}_e$ based on Equation 10.34.

The terms of $\mathbf{K}_e$ are evaluated recalling the definitions of $\mathbf{T}$ (Equation 10.27b) and $\mathbf{k}$ (Equation 10.9b) as:

$$\mathbf{K}_e = \mathbf{T}^\mathsf{T} \mathbf{k} \mathbf{T} = \begin{bmatrix} l & 0 \\ m & 0 \\ 0 & l \\ 0 & m \end{bmatrix} \frac{EA}{L} \begin{bmatrix} 1 & -1 \\ -1 & 1 \end{bmatrix} \begin{bmatrix} l & m & 0 & 0 \\ 0 & 0 & l & m \end{bmatrix} = \frac{EA}{L} \begin{bmatrix} l & -l \\ m & -m \\ -l & l \\ -m & m \end{bmatrix} \begin{bmatrix} l & m & 0 & 0 \\ 0 & 0 & l & m \end{bmatrix}$$

$$= \frac{EA}{L} \begin{bmatrix} l^2 & lm & -l^2 & -lm \\ lm & m^2 & -lm & -m^2 \\ -l^2 & -lm & l^2 & lm \\ -lm & -m^2 & lm & m^2 \end{bmatrix} \tag{10.35}$$

The stiffness relationship of an isolated truss element relates its nodal displacements $\mathbf{D}_e$ to its nodal forces $\mathbf{Q}_e$ specified along the global freedoms numbered from 1 to 4 (Figures 10.14b, 10.15b and 10.16b). In particular, the i-th components of vectors $\mathbf{Q}_e$ and $\mathbf{D}_e$ represent the nodal force and nodal displacement, respectively, assigned along freedom number i. Similarly, the rows and columns of $\mathbf{K}_e$ are associated with the global freedoms as follows:

$$\mathbf{K}_e = \frac{EA}{L} \begin{bmatrix} l^2 & lm & -l^2 & -lm \\ lm & m^2 & -lm & -m^2 \\ -l^2 & -lm & l^2 & lm \\ -lm & -m^2 & lm & m^2 \end{bmatrix} \begin{matrix} 1 \\ 2 \\ 3 \\ 4 \end{matrix} \tag{10.36}$$

This numbering will be useful for the *assembling* procedure described in the next section.

10.6 ASSEMBLING

The stiffness relationship of a truss relates nodal displacements **D** and nodal forces **Q** along all of its global freedoms. In particular, the size of both **D** and **Q** is equal to the number of global freedoms N_{dof}, and these are related by the $N_{dof} \times N_{dof}$ stiffness matrix **K** as:

$$Q = KD \tag{10.37}$$

where **K** represents the stiffness matrix of the entire truss.

The assembling process combines the stiffness matrix of each element considered in isolation $K_{e(n)}$ (where the additional subscript 'n' identifies the element number) into the stiffness matrix of the entire truss **K**, being careful to relate correctly the freedom numbers of each element to the appropriate global freedoms (or structure freedoms).

Let us reconsider the truss of Figure 10.4 (reproduced for ease of reference in Figure 10.17a) to outline the main aspects involved in the assembling procedure. The truss possesses 10 freedoms (i.e. $N_{dof} = 10$) and, because of this, **K** is a 10 × 10 matrix. At the beginning of the assembling, **K** is a nil matrix and it is populated by adding, one at a time, the stiffness coefficients of each truss member $K_{e(n)}$ based on the truss global freedoms. For example, element 2 has freedoms 1, 2, 3 and 4 when considered in isolation (Figure 10.17b), but these correspond to structure dof 7, 8, 5 and 6 based on the assembled truss numbering (Figure 10.17c).

This implies that rows and columns 1, 2, 3 and 4 of the isolated element stiffness matrix $K_{e(2)}$, as described in Equation 10.36 and Figure 10.18a, need to be mapped to rows and columns 7, 8, 5 and 6 of the structure stiffness matrix **K** (Figure 10.18b). We can then include each coefficient from the element stiffness matrix into the structure stiffness matrix making sure to match row and column numbering. An example illustrating the assembly

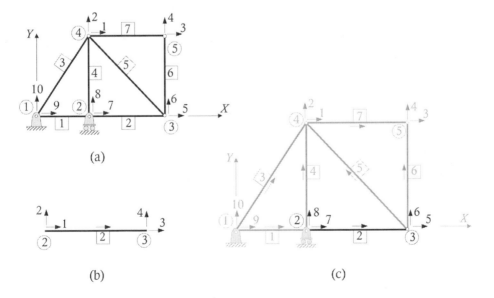

Figure 10.17 Relationship between isolated and assembled dof for member 2. (a) Truss of Figure 10.4. (b) Dof of isolated truss element 2. (c) Dof of truss element 2 in the assembled structure.

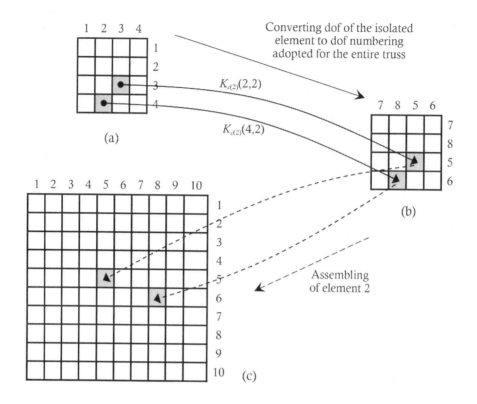

Figure 10.18 Assembling of two coefficients from $\mathbf{K}_{e2}$ into $\mathbf{K}$. (a) Dof of $\mathbf{K}_{e2}$ considered in isolation. (b) Dof of $\mathbf{K}_{e2}$ based on the dof numbering of the entire truss. (c) Dof of $\mathbf{K}$.

of coefficients $K_{e(2)}(3,3)$ and $K_{e(2)}(4,2)$ (of the isolated element) into $K(5,5)$ and $K(6,8)$ of the structure matrix $\mathbf{K}$ is shown in Figure 10.18c.

The same procedure must be followed for each of the remaining elements of the truss. For example, for element 3, the dofs of the isolated truss 1 to 4 correspond to global structure freedoms 9, 10, 1 and 2 (Figure 10.19). The assembling procedure is illustrated for a simple truss in Worked Example 10.3.

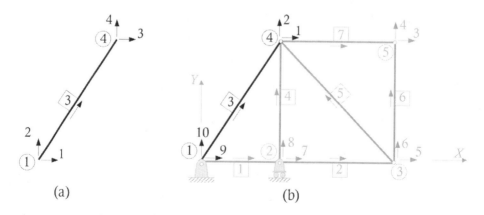

Figure 10.19 Relationship between isolated and assembled dof for member 3. (a) Dof of isolated truss element 3. (b) Dof of truss element 3 in the assembled structure.

WORKED EXAMPLE 10.3

Determine the structure stiffness matrix **K** for the truss shown in Figure 10.20 considering the location vectors assigned to each member. Assume *EA* to be the same for each truss element.

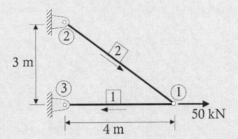

Figure 10.20 Truss for Worked Example 10.3.

In the first part of the solution, we introduce a global coordinate system and assign two global freedoms at each node (ensuring to use low numbers for unrestrained freedoms, followed by the restrained ones) as shown in Figure 10.21.

Because there are six dofs (i.e. $N_{dof} = 6$), **K** is a 6×6 matrix, and before assembling the element stiffness coefficients, all matrix terms are set to zero. That is:

$$\mathbf{K} = \begin{bmatrix} 0 & 0 & 0 & 0 & 0 & 0 \\ 0 & 0 & 0 & 0 & 0 & 0 \\ 0 & 0 & 0 & 0 & 0 & 0 \\ 0 & 0 & 0 & 0 & 0 & 0 \\ 0 & 0 & 0 & 0 & 0 & 0 \\ 0 & 0 & 0 & 0 & 0 & 0 \end{bmatrix}$$

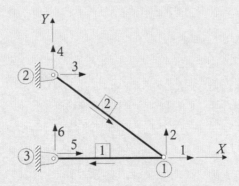

Figure 10.21 Global coordinate system, global freedoms, and location vectors.

ELEMENT 1

Direction cosines for element 1:

$$l_1 = \cos\alpha_{(1)} = \frac{x_3 - x_1}{L_1} = \frac{0-4}{4} = -1 \quad \text{and} \quad m_1 = \cos\beta_{(1)} = \frac{y_3 - y_1}{L_1} = \frac{0-0}{4} = 0$$

$\mathbf{K}_{e(1)}$ can then be calculated substituting $l_1 = -1$, $m_1 = 0$ and $L = 4$ m into Equation 10.36 as:

$$\mathbf{K}_{e(1)} = EA \begin{bmatrix} 0.25 & 0 & -0.25 & 0 \\ 0 & 0 & 0 & 0 \\ -0.25 & 0 & 0.25 & 0 \\ 0 & 0 & 0 & 0 \end{bmatrix}$$

The freedoms 1 to 4 of the isolated element correspond to structure freedoms 1, 2, 5 and 6 of the truss and the member coefficients of element 1 are mapped into the nil **K** matrix as follows:

$$\mathbf{K}_{e(1)} = EA \begin{bmatrix} & 1 & 2 & 5 & 6 & \\ 0.25 & 0 & -0.25 & 0 & 1 \\ 0 & 0 & 0 & 0 & 2 \\ -0.25 & 0 & 0.25 & 0 & 5 \\ 0 & 0 & 0 & 0 & 6 \end{bmatrix}$$

Assembling (adding)
$\mathbf{K}_{e(1)}$ to $\mathbf{K}$

$$\mathbf{K} = EA \begin{bmatrix} & 1 & 2 & 3 & 4 & 5 & 6 & \\ 0.25 & 0 & 0 & 0 & -0.25 & 0 & 1 \\ 0 & 0 & 0 & 0 & 0 & 0 & 2 \\ 0 & 0 & 0 & 0 & 0 & 0 & 3 \\ 0 & 0 & 0 & 0 & 0 & 0 & 4 \\ -0.25 & 0 & 0 & 0 & 0.25 & 0 & 5 \\ 0 & 0 & 0 & 0 & 0 & 0 & 6 \end{bmatrix}$$

ELEMENT 2

Direction cosines for element 2 are:

$$l_2 = \cos\alpha_{(2)} = \frac{x_1 - x_2}{L_2} = \frac{4-0}{5} = 0.8 \quad \text{and} \quad m_2 = \cos\beta_{(2)} = \frac{y_1 - y_2}{L_2} = \frac{0-3}{5} = -0.6$$

and with $L = 5$ m, $\mathbf{K}_{e(2)}$ can be calculated as:

$$\mathbf{K}_{e(2)} = EA \begin{bmatrix} 0.128 & -0.96 & -0.128 & 0.096 \\ -0.096 & 0.072 & 0.096 & -0.072 \\ -0.128 & 0.096 & 0.128 & -0.096 \\ 0.096 & -0.072 & -0.096 & 0.072 \end{bmatrix}$$

The element freedoms 1 to 4 are mapped to the global freedoms 3, 4, 1 and 2. This is carried out below adding the coefficients of element 2 to **K** (already populated with the contribution of element 1):

$$
\mathbf{K}_{e(2)} = EA
\begin{array}{cccc}
3 & 4 & 1 & 2 \\
\begin{bmatrix}
0.128 & -0.096 & -0.128 & 0.096 \\
-0.096 & 0.072 & 0.096 & -0.072 \\
-0.128 & 0.096 & 0.128 & -0.096 \\
0.096 & -0.072 & -0.096 & 0.072
\end{bmatrix}
\begin{array}{c} 3 \\ 4 \\ 1 \\ 2 \end{array}
\end{array}
$$

Assembling (adding) $\mathbf{K}_{e(2)}$ to $\mathbf{K}$

$$
\mathbf{K} = EA
\begin{array}{cccccc}
1 & 2 & 3 & 4 & 5 & 6 \\
\begin{bmatrix}
0.25+0.128 & -0.096 & -0.128 & 0.096 & -0.25 & 0 \\
-0.096 & 0.072 & 0.096 & -0.072 & 0 & 0 \\
-0.128 & 0.096 & 0.128 & -0.096 & 0 & 0 \\
0.096 & -0.072 & -0.096 & 0.072 & 0 & 0 \\
-0.25 & 0 & 0 & 0 & 0.25 & 0 \\
0 & 0 & 0 & 0 & 0 & 0
\end{bmatrix}
\begin{array}{c} 1 \\ 2 \\ 3 \\ 4 \\ 5 \\ 6 \end{array}
\end{array}
$$

With the contribution of both elements 1 and 2, **K** becomes:

$$
\mathbf{K} = EA
\begin{bmatrix}
0.378 & -0.096 & -0.128 & 0.096 & -0.25 & 0 \\
-0.096 & 0.072 & 0.096 & -0.072 & 0 & 0 \\
-0.128 & 0.096 & 0.128 & -0.096 & 0 & 0 \\
0.096 & -0.072 & -0.096 & 0.072 & 0 & 0 \\
-0.25 & 0 & 0 & 0 & 0.25 & 0 \\
0 & 0 & 0 & 0 & 0 & 0
\end{bmatrix}
$$

10.7 SOLUTION PROCEDURE

The analysis of a truss using the stiffness method is performed by solving the system of N_{dof} equilibrium equations expressed in Equation 10.37. The displacement vector **D** collects both known and unknown displacements and, for clarity, it is partitioned to differentiate between them:

$$
\mathbf{D} =
\begin{bmatrix}
\mathbf{D}_u \\
\mathbf{D}_k
\end{bmatrix}
\tag{10.38}
$$

where $\mathbf{D}_u$ and $\mathbf{D}_k$ depict unknown and known displacements, respectively. This partition reflects the dof numbering previously introduced, in which unrestrained freedoms are considered first (with unknown displacements listed in $\mathbf{D}_u$) followed by the restrained freedoms (with known, usually zero, displacements included in $\mathbf{D}_k$).

Forces applied along each freedom are collected in **Q**. Those associated with an unrestrained freedom, i.e. external loads, are included in the vector of known forces $\mathbf{Q}_k$ and the remaining ones in the vector of unknown forces $\mathbf{Q}_u$ (which correspond to the reactions required to provide the specified restraints):

$$
\mathbf{Q} =
\begin{bmatrix}
\mathbf{Q}_k \\
\mathbf{Q}_u
\end{bmatrix}
\tag{10.39}
$$

Based on Equations 10.38 and 10.39, Equation 10.37 can be re-expressed as follows:

$$\begin{bmatrix} \mathbf{Q}_k \\ \mathbf{Q}_u \end{bmatrix} = \begin{bmatrix} \mathbf{K}_{11} & \mathbf{K}_{12} \\ \mathbf{K}_{21} & \mathbf{K}_{22} \end{bmatrix} \begin{bmatrix} \mathbf{D}_u \\ \mathbf{D}_k \end{bmatrix} \tag{10.40a}$$

or

$$\mathbf{Q}_k = \mathbf{K}_{11}\mathbf{D}_u + \mathbf{K}_{12}\mathbf{D}_k \tag{10.40b}$$

$$\mathbf{Q}_u = \mathbf{K}_{21}\mathbf{D}_u + \mathbf{K}_{22}\mathbf{D}_k \tag{10.40c}$$

where $\mathbf{K}_{ij}$ represents partitions of $\mathbf{K}$. Introducing $N_{dof.u}$ and $N_{dof.k}$ to depict the number of unrestrained and restrained freedoms, respectively, the partition of $\mathbf{K}$ have the following sizes: $\mathbf{K}_{11}$ is a $N_{dof.u} \times N_{dof.u}$ matrix, $\mathbf{K}_{12}$ is a $N_{dof.u} \times N_{dof.k}$ matrix, $\mathbf{K}_{21}$ is a $N_{dof.k} \times N_{dof.u}$ matrix and $\mathbf{K}_{22}$ is a $N_{dof.k} \times N_{dof.k}$ matrix. Similarly, vectors $\mathbf{D}_u$ and $\mathbf{Q}_k$ have a length of $N_{dof.u}$, while $\mathbf{D}_k$ and $\mathbf{Q}_u$ have a length of $N_{dof.k}$.

The unknown displacements $\mathbf{D}_u$ can be obtained by applying one of the solution procedures presented in Appendix C. For example, Equation 10.40b can be rearranged as:

$$\mathbf{D}_u = \mathbf{K}_{11}^{-1}(\mathbf{Q}_k - \mathbf{K}_{12}\mathbf{D}_k) \tag{10.41}$$

and this can be substituted into Equation 10.40c to determine the unknown forces $\mathbf{Q}_u$.

10.8 CALCULATION OF INTERNAL ACTIONS

The internal forces resisted by the truss elements can be calculated, once the unknown displacements are determined, recalling Equation 10.32a (reproduced here for ease of reference)

$$\mathbf{q} = \mathbf{k}\mathbf{T}\mathbf{D}_e \tag{10.42}$$

where $\mathbf{D}_e$ corresponds to the nodal displacements of the element considered. The axial force can then be evaluated from equilibrium considerations at the nodes of the truss as (Figure 10.22):

$$N = -q_1 \quad \text{and} \quad N = q_2 \tag{10.43a,b}$$

The overall procedure required for the analysis of a truss based on the stiffness method is summarised below followed by a number of worked examples.

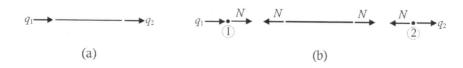

(a)　　　　　　　　　　　　　　　　　　(b)

Figure 10.22 Nodal actions and internal force of a truss element. (a) Nodal actions. (b) Free-body diagrams of the end nodes.

SUMMARY OF STEPS 10.1: Stiffness method — Solution procedure

The main steps required for the analysis of a truss with the stiffness method are as follows:

1. Specify a global reference system, and number the nodes and elements.

2. Assign an arbitrary location vector to each element.

3. Introduce two global freedoms at each node, preferably pointing in the same positive directions as the global coordinate axes and number them. Make sure to number unrestrained freedoms first, followed by the restrained ones.

4. For each element, determine its direction cosines and calculate its stiffness matrix based on Equation 10.36.

5. Assemble the contribution of each element to the structure stiffness matrix $\mathbf{K}$, making sure to relate correctly freedoms of the isolated element to those adopted for the entire truss.

6. Write vectors of known displacements $\mathbf{D}_k$ and external forces $\mathbf{Q}_k$.

7. Partition the stiffness matrix $\mathbf{K}$ into $\mathbf{K}_{11}$, $\mathbf{K}_{12}$, $\mathbf{K}_{21}$ and $\mathbf{K}_{22}$, as carried out in Equations 10.40.

8. Determine the unknown displacements $\mathbf{D}_u$ with Equation 10.41.

9. Calculate the unknown reactions $\mathbf{Q}_u$ (Equation 10.40c) and axial forces for each member of the truss $\mathbf{q}$ (Equation 10.42).

WORKED EXAMPLE 10.4

Reconsider the truss of Worked Example 10.3 and determine its unknown displacements if a 50 kN horizontal force is applied at node 1. Calculate also the unknown reactions and the axial forces in each member. Assume EA is constant throughout.

The solution is carried out following the steps detailed in Summary of Steps 10.1. Units used for lengths and forces are m and kN, respectively.

(1–5) Steps 1 to 5 have already been covered in Worked Example 10.3 and the results are reproduced for ease of reference in Figure 10.23.

From Worked Example 10.3, the structure stiffness matrix is:

$$\mathbf{K} = EA \begin{bmatrix} 0.378 & -0.096 & -0.128 & 0.096 & -0.25 & 0 \\ -0.096 & 0.072 & 0.096 & -0.072 & 0 & 0 \\ -0.128 & 0.096 & 0.128 & -0.096 & 0 & 0 \\ 0.096 & -0.072 & -0.096 & 0.072 & 0 & 0 \\ -0.25 & 0 & 0 & 0 & 0.25 & 0 \\ 0 & 0 & 0 & 0 & 0 & 0 \end{bmatrix}$$

(6) Based on the boundary conditions, unrestrained freedoms are 1 and 2 with the remaining freedoms from 3 to 6 restrained. Considering that no displacements are permitted along

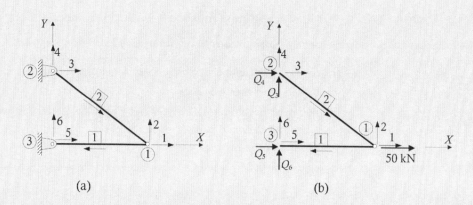

Figure 10.23 Global coordinate system and global freedoms. (a) Global freedoms and location vectors. (b) Free-body diagram of the truss.

freedoms 3 to 6 and that the only external load is a 50 kN force applied along freedom 1, the vectors of known displacements and forces can be written as:

$$\mathbf{D}_k = \begin{bmatrix} 0 \\ 0 \\ 0 \\ 0 \end{bmatrix} \qquad \mathbf{Q}_k = \begin{bmatrix} 50 \\ 0 \end{bmatrix}$$

where the positive sign of the force reflects the fact that it is acting in the positive direction of freedom 1.

(7) The structure stiffness matrix **K** is partitioned as follows based on the fact that freedoms 1 and 2 are unrestrained and freedoms 3 to 6 are restrained:

$$\mathbf{K} = EA \begin{bmatrix} \begin{array}{cc|cccc} & 1 & 2 & 3 & 4 & 5 & 6 \\ 0.378 & -0.096 & -0.128 & 0.096 & -0.25 & 0 \\ -0.096 & 0.072 & 0.096 & -0.072 & 0 & 0 \\ \hline -0.128 & 0.096 & 0.128 & -0.096 & 0 & 0 \\ 0.096 & -0.072 & -0.096 & 0.072 & 0 & 0 \\ -0.25 & 0 & 0 & 0 & 0.25 & 0 \\ 0 & 0 & 0 & 0 & 0 & 0 \end{array} \end{bmatrix} \begin{matrix} 1 \\ 2 \\ 3 \\ 4 \\ 5 \\ 6 \end{matrix} \quad \text{or} \quad \mathbf{K} = \begin{bmatrix} \mathbf{K}_{11} & \mathbf{K}_{12} \\ \hline \mathbf{K}_{21} & \mathbf{K}_{22} \end{bmatrix}$$

with individual partitions being:

$$\mathbf{K}_{11} = EA \begin{bmatrix} 0.378 & -0.096 \\ -0.096 & 0.072 \end{bmatrix} \qquad \mathbf{K}_{12} = EA \begin{bmatrix} -0.128 & 0.096 & -0.25 & 0 \\ 0.096 & -0.072 & 0 & 0 \end{bmatrix}$$

$$\mathbf{K}_{21} = EA \begin{bmatrix} -0.128 & 0.096 \\ 0.096 & -0.072 \\ -0.25 & 0 \\ 0 & 0 \end{bmatrix} \qquad \mathbf{K}_{22} = EA \begin{bmatrix} 0.128 & -0.096 & 0 & 0 \\ -0.096 & 0.072 & 0 & 0 \\ 0 & 0 & 0.25 & 0 \\ 0 & 0 & 0 & 0 \end{bmatrix}$$

(8) Unknown displacements $\mathbf{D}_u$ are determined using Equation 10.41.

$$\mathbf{D}_u = \mathbf{K}_{11}^{-1}(\mathbf{Q}_k - \mathbf{K}_{12}\mathbf{D}_k) = \frac{1}{EA}\begin{bmatrix} 4 & 5.33 \\ 5.33 & 21 \end{bmatrix}\left(\begin{bmatrix} 50 \\ 0 \end{bmatrix} - EA\begin{bmatrix} -0.128 & 0.096 & -0.25 & 0 \\ 0.096 & -0.072 & 0 & 0 \end{bmatrix}\begin{bmatrix} 0 \\ 0 \\ 0 \\ 0 \end{bmatrix}\right)$$

$$= \frac{1}{EA}\begin{bmatrix} 200 \\ 266.6 \end{bmatrix}$$

where the units of the displacements are in metres and the inverse of $\mathbf{K}_{11}$ is obtained from Appendix C as:

$$\mathbf{K}_{11}^{-1} = \frac{1}{EA}\begin{bmatrix} 4 & 5.33 \\ 5.33 & 21 \end{bmatrix}$$

(9) With the unknown displacements established, the unknown reactions and member forces can be readily calculated. In particular, the reactions (in kN) are obtained using Equation 10.40c:

$$\mathbf{Q}_u = \mathbf{K}_{21}\mathbf{D}_u + \mathbf{K}_{22}\mathbf{D}_k = EA\begin{bmatrix} -0.128 & 0.096 \\ 0.096 & -0.072 \\ -0.25 & 0 \\ 0 & 0 \end{bmatrix}\frac{1}{EA}\begin{bmatrix} 200 \\ 266.6 \end{bmatrix} + EA\begin{bmatrix} 0.128 & -0.096 & 0 & 0 \\ -0.096 & 0.072 & 0 & 0 \\ 0 & 0 & 0.25 & 0 \\ 0 & 0 & 0 & 0 \end{bmatrix}\begin{bmatrix} 0 \\ 0 \\ 0 \\ 0 \end{bmatrix}$$

$$= \begin{bmatrix} 0 \\ 0 \\ -50 \\ 0 \end{bmatrix}$$

Recalling that the member freedoms from 1 to 4 of isolated element 1 are mapped against structure freedoms 1, 2, 5 and 6, the nodal displacements of element 1 collected in $\mathbf{D}_{e(1)}$ include global displacements D_1, D_2, D_5 and D_6:

$$\mathbf{D}_{e(1)} = \begin{bmatrix} D_{e1} \\ D_{e2} \\ D_{e3} \\ D_{e4} \end{bmatrix} = \begin{bmatrix} D_1 \\ D_2 \\ D_5 \\ D_6 \end{bmatrix} = \begin{bmatrix} 200/EA \\ 266.6/EA \\ 0 \\ 0 \end{bmatrix}$$

The member forces of element 1 are determined substituting $\mathbf{D}_{e(1)}$ into Equation 10.42:

$$\mathbf{q}_{(1)} = \mathbf{kTD}_{e(1)} = \frac{EA}{4}\begin{bmatrix} 1 & -1 \\ -1 & 1 \end{bmatrix}\begin{bmatrix} -1 & 0 & 0 & 0 \\ 0 & 0 & -1 & 0 \end{bmatrix}\begin{bmatrix} 200/EA \\ 266.6/EA \\ 0 \\ 0 \end{bmatrix} = \begin{bmatrix} -50 \\ 50 \end{bmatrix}$$

from which $N_1 = 50$ kN.

Similarly, freedoms from 1 to 4 of isolated element 2 correspond to structure freedoms 3, 4, 1 and 2, and nodal displacements and member forces for element 2 are:

$$\mathbf{D}_{e(2)} = \begin{bmatrix} D_{e1} \\ D_{e2} \\ D_{e3} \\ D_{e4} \end{bmatrix} = \begin{bmatrix} D_3 \\ D_4 \\ D_1 \\ D_2 \end{bmatrix} = \begin{bmatrix} 0 \\ 0 \\ 200/EA \\ 266.6/EA \end{bmatrix}$$

$$\mathbf{q}_{(2)} = \mathbf{kTD}_{e(2)} = \frac{EA}{5}\begin{bmatrix} 1 & -1 \\ -1 & 1 \end{bmatrix}\begin{bmatrix} 0.8 & -0.6 & 0 & 0 \\ 0 & 0 & 0.8 & -0.6 \end{bmatrix}\begin{bmatrix} 0 \\ 0 \\ 200/EA \\ 266.6/EA \end{bmatrix} = \begin{bmatrix} 0 \\ 0 \end{bmatrix}$$

from which $N_2 = 0$ kN.

The correctness of the solution can be checked applying the method of joints to node 1. From equilibrium along freedom 2, the axial force in element 2 is nil as it is the only force included in the equilibrium equation (see Reflection Activity 4.1). The force in element 1 can then be calculated to be 50 kN based on equilibrium along freedom 1.

10.9 NODAL COORDINATES

It is always convenient, in the analysis of a truss, to assume all global freedoms to be parallel to the global X- and Y-axes. There are situations, such as those involving inclined restraints, where this freedom arrangement might not be suitable, in which case we need to introduce *nodal coordinates* and *freedoms*. Let us consider the truss shown in Figure 10.24a that possesses an inclined roller support at node 3. The unrestrained and restrained directions of the roller are not parallel to the global axes and, because of this, it is convenient to adopt nodal coordinates at this node as shown in Figure 10.24b where axes X_1 and Y_1 are introduced,

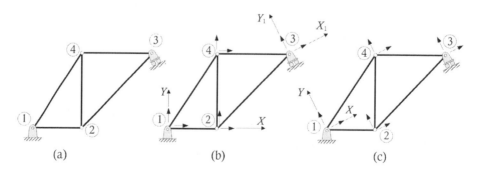

Figure 10.24 Use of nodal coordinates. (a) Truss layout. (b) Nodal coordinates. (c) Global freedoms.

with X_1 parallel to the unrestrained freedom and Y_1 placed along the restrained dof. The term *nodal coordinates* simply highlights that these axes are assigned to a particular node. Obviously, in this particular case, it would have been possible to specify global axes as depicted in Figure 10.24c, even if such an arrangement is not convenient for the calculation of the various nodal coordinates. Where we have two or more inclined roller supports not parallel to each other, it is necessary to introduce the nodal coordinates at one or more nodes.

With the stiffness method, the use of the nodal coordinates only affects the terms included in the transformation matrices **T** of the elements connected to the nodes with an inclined support. Let us consider the truss element of Figure 10.25, in which the local freedoms 1 and 2 are related to two sets of global freedoms parallel to global axes (X_1,Y_1) at node 1 and axes (X_2,Y_2) at node 2.

Local vector components can be related to the nodal components (as illustrated in Figure 10.26) following the same procedure previously adopted for the description of the transformation matrices (Equations 10.25 to 10.27) as:

$$[d_1]=[\begin{array}{cc} l_1 & m_1 \end{array}]\begin{bmatrix} D_{e1} \\ D_{e2} \end{bmatrix} \qquad [d_2]=[\begin{array}{cc} l_2 & m_2 \end{array}]\begin{bmatrix} D_{e3} \\ D_{e4} \end{bmatrix} \qquad (10.44a,b)$$

where l_1 and m_1 represent the direction cosines related to angles θ_{1x} and θ_{1y}, respectively, and θ_{1x} and θ_{1y} are the angles between the global axes X_1 and Y_1, at node 1 and the local

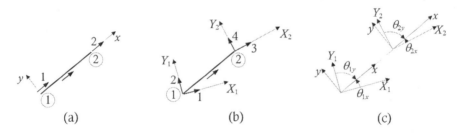

Figure 10.25 Local and nodal degrees of freedom. (a) Local freedoms. (b) Nodal freedoms. (c) Angles for the direction cosines.

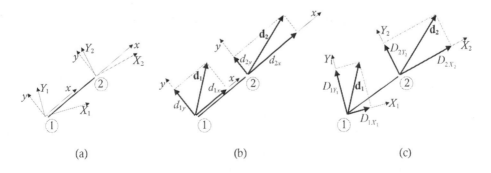

Figure 10.26 Nodal displacements in local and nodal coordinates. (a) Local and global coordinates. (b) Nodal displacements in local coordinates. (c) Nodal displacements in nodal coordinates.

x-axis, as shown in Figure 10.25c. Similarly, l_2 and m_2 are the direction cosines of θ_{2x} and θ_{2y} at node 2 (with θ_{2x} and θ_{2y} defined in Figure 10.25c). Equations 10.44 reflect the fact that the local displacements perpendicular to the truss element are not relevant to its structural response (see Reflection Activity 10.2).

The relationships of Equations 10.44 can be simplified to:

$$
\begin{bmatrix} d_1 \\ d_2 \end{bmatrix} = \begin{bmatrix} l_1 & m_1 & 0 & 0 \\ 0 & 0 & l_2 & m_2 \end{bmatrix} \begin{bmatrix} D_{e1} \\ D_{e2} \\ D_{e3} \\ D_{e4} \end{bmatrix} \quad \text{or} \quad \mathbf{d} = \mathbf{T}\mathbf{D}_e \tag{10.45a,b}
$$

in which case $\mathbf{T}$ is defined as:

$$
\mathbf{T} = \begin{bmatrix} l_1 & m_1 & 0 & 0 \\ 0 & 0 & l_2 & m_2 \end{bmatrix} \tag{10.46}
$$

The same matrix can be used to relate nodal and local components of the forces applied at the member end nodes.

Obviously, when the direction cosines at the two end nodes of an element coincide (i.e. the angles between local and global coordinates are the same at the two end nodes of an element), $\mathbf{T}$ simplifies to the expression already provided in Equation 10.27b, which can be obtained by substituting $l_1 = l_2 = l$ and $m_1 = m_2 = m$ in Equation 10.46.

The stiffness matrix of an element with a different set of nodal coordinates at each end can be determined by recalling Equation 10.34:

$$
\mathbf{K}_e = \mathbf{T}^\mathrm{T}\mathbf{k}\mathbf{T} = \begin{bmatrix} l_1 & 0 \\ m_1 & 0 \\ 0 & l_2 \\ 0 & m_2 \end{bmatrix} \frac{EA}{L} \begin{bmatrix} 1 & -1 \\ -1 & 1 \end{bmatrix} \begin{bmatrix} l_1 & m_1 & 0 & 0 \\ 0 & 0 & l_2 & m_2 \end{bmatrix}
$$

$$
= \frac{EA}{L} \begin{bmatrix} l_1 & -l_1 \\ m_1 & -m_1 \\ -l_2 & l_2 \\ -m_2 & m_2 \end{bmatrix} \begin{bmatrix} l_1 & m_1 & 0 & 0 \\ 0 & 0 & l_2 & m_2 \end{bmatrix}
$$

$$
= \frac{EA}{L} \begin{bmatrix} l_1^2 & l_1 m_1 & -l_1 l_2 & -l_1 m_2 \\ l_1 m_1 & m_1^2 & -l_2 m_1 & -m_1 m_2 \\ -l_1 l_2 & -l_2 m_1 & l_2^2 & l_2 m_2 \\ -l_1 m_2 & -m_1 m_2 & l_2 m_2 & m_2^2 \end{bmatrix} \tag{10.47}
$$

Figure 10.27 Freedom numbering related to nodal coordinates.

The use of this matrix in the assembling and solution procedure is the same as previously described, except that the global displacements and forces related to inclined supports are now specified along the nodal freedoms. The stiffness relationship of an isolated truss element can be expressed by considering the use of nodal coordinates as:

$$
\begin{array}{c} 1 \\ 2 \\ 3 \\ 4 \end{array}
\begin{bmatrix} Q_{e1} \\ Q_{e2} \\ Q_{e3} \\ Q_{e4} \end{bmatrix}
= \frac{EA}{L}
\begin{matrix} 1 & 2 & 3 & 4 \end{matrix}
\begin{bmatrix}
l_1^2 & l_1 m_1 & -l_1 l_2 & -l_1 m_2 \\
l_1 m_1 & m_1^2 & -l_2 m_1 & -m_1 m_2 \\
-l_1 l_2 & -l_2 m_1 & l_2^2 & l_2 m_2 \\
-l_1 m_2 & -m_1 m_2 & l_2 m_2 & m_2^2
\end{bmatrix}
\begin{bmatrix} D_{e1} \\ D_{e2} \\ D_{e3} \\ D_{e4} \end{bmatrix}
\begin{array}{c} 1 \\ 2 \\ 3 \\ 4 \end{array}
\tag{10.48}
$$

where the definition of node 1 or 2 for each element is defined by the positive direction of the location vector, as illustrated in Figure 10.27.

WORKED EXAMPLE 10.5

Determine the unknown displacement, reactions, and member forces for the truss depicted in Figure 10.28. Assume *EA* is constant throughout.

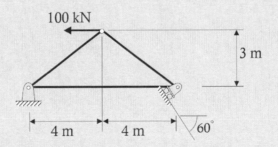

Figure 10.28 Truss for Worked Example 10.6.

The solution is outlined below following Summary of Steps 10.1.

(1–3) Elements and nodes are numbered as outlined in Figure 10.29a, and a global coordinate system is adopted with origin at node 1. In this system, the coordinates for each node are: node 1 (0,0), node 2 (8,0) and node 3 (4,3).

Arbitrary location vectors are introduced for each element, as outlined in Figure 10.29b. Two global freedoms are specified for each node parallel to the global coordinates (X,Y), except at the location of the inclined roller support, in which case these are parallel to the nodal coordinates placed at node 2 (X_2,Y_2) as depicted in Figure 10.29c.

(4) The lengths of the truss elements are as follows: $L_1 = 8$ m, $L_2 = L_3 = 5$ m.

The direction cosines for the three elements are calculated below, placing attention to differentiate between values for l and m calculated at the two end nodes of the elements connected to nodes with nodal coordinates (node 2 of the structure in our case).

Member 1: The location vector specifies that node 1 is the first node and node 2 is the second node: $l_1 = (8 - 0)/8 = 1$; $m_1 = (0 - 0)/8 = 0$; $l_2 = \cos \theta_{x2} = \cos 60 = 0.5$; $m_2 = \cos \theta_{y2} = \cos 30 = 0.866$ (Figure 10.29d).

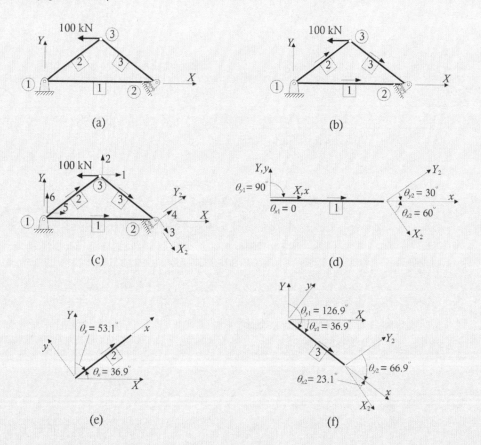

Figure 10.29 Global and nodal freedoms for Worked Example 10.6. (a) Global coordinates and numbering of nodes and elements. (b) Location vectors. (c) Global and nodal freedoms. (d) Element 1. (e) Element 2. (f) Element 3.

Member 2: $l = (4 - 0)/5 = 0.8$; $m = (3 - 0)/5 = 0.6$.

Member 3: The location vector specifies that node 3 is the first node and node 2 is the second node: $l_1 = (8 - 4)/5 = 0.8$; $m_1 = (0 - 3)/5 = -0.6$; $l_2 = \cos \theta_{x2} = \cos 23.1 = 0.92$; $m_2 = \cos \theta_{y2} = \cos 66.9 = 0.392$ (Figure 10.29f).

The stiffness matrices are then calculated by applying Equation 10.48:

$$\mathbf{K}_{e(1)} = EA \begin{bmatrix} 0.125 & 0 & -0.0625 & -0.1083 \\ 0 & 0 & 0 & 0 \\ -0.0625 & 0 & 0.0312 & 0.0541 \\ -0.1083 & 0 & 0.0541 & 0.0937 \end{bmatrix}$$

$$\mathbf{K}_{e(2)} = EA \begin{bmatrix} 0.128 & 0.096 & -0.128 & -0.096 \\ 0.096 & 0.072 & -0.096 & -0.072 \\ -0.128 & -0.096 & 0.128 & 0.096 \\ -0.096 & -0.072 & 0.096 & 0.072 \end{bmatrix}$$

$$\mathbf{K}_{e(3)} = EA \begin{bmatrix} 0.128 & -0.096 & -0.1472 & -0.0628 \\ -0.096 & 0.072 & 0.1104 & 0.0471 \\ -0.1472 & 0.1104 & 0.1692 & 0.0722 \\ -0.0628 & 0.0471 & 0.0722 & 0.0308 \end{bmatrix}$$

(5) The contribution of each element is included in the structural matrix $\mathbf{K}$, noting that freedoms 1 to 4 of each isolated element are mapped to structure freedoms 5, 6, 3 and 4 for element 1, and to structure freedoms 5, 6, 1 and 2 and structure freedoms 1, 2, 3 and 4 for elements 2 and 3, respectively. Based on this, $\mathbf{K}$ becomes:

$$\mathbf{K} = EA \begin{bmatrix} 0.256 & 0 & -0.1472 & -0.0628 & -0.128 & -0.096 \\ 0 & 0.144 & 0.1104 & 0.0471 & -0.096 & -0.072 \\ -0.1472 & 0.1104 & 0.2004 & 0.1263 & -0.0625 & 0 \\ -0.0628 & 0.0471 & 0.1263 & 0.1246 & -0.1083 & 0 \\ -0.128 & -0.096 & -0.0625 & -0.1083 & 0.253 & 0.096 \\ -0.096 & -0.072 & 0 & 0 & 0.096 & 0.072 \end{bmatrix}$$

(6) Considering that the structure freedoms from 4 to 6 are restrained, the vector of known displacements is $\mathbf{D}_k = [0\ 0\ 0]^T$. The external force applied at node 3 is included in the loading vector $\mathbf{Q}_k$ with a negative sign, because it is pointing in the negative direction of freedom 1:

$\mathbf{Q}_k = [-100\ 0\ 0]^T$

(7) The partitioning of $\mathbf{K}$ into $\mathbf{K}_{11}$, $\mathbf{K}_{12}$, $\mathbf{K}_{21}$ and $\mathbf{K}_{22}$ is carried out based on Equations 10.40:

$$\mathbf{K}_{11} = EA \begin{bmatrix} 0.256 & 0 & -0.1472 \\ 0 & 0.144 & 0.1104 \\ -0.1472 & 0.1104 & 0.2004 \end{bmatrix} \qquad \mathbf{K}_{12} = EA \begin{bmatrix} -0.0628 & -0.128 & -0.096 \\ 0.0471 & -0.096 & -0.072 \\ 0.1263 & -0.0625 & 0 \end{bmatrix}$$

$$\mathbf{K}_{21} = EA \begin{bmatrix} -0.0628 & 0.0471 & 0.1263 \\ -0.128 & -0.096 & -0.0625 \\ -0.096 & -0.072 & 0 \end{bmatrix} \qquad \mathbf{K}_{22} = EA \begin{bmatrix} 0.1246 & -0.1083 & 0 \\ -0.1083 & 0.253 & 0.096 \\ 0 & 0.096 & 0.072 \end{bmatrix}$$

(8) Unknown displacements (in m) are determined with Equation 10.41:

$$\mathbf{D}_u = \frac{1}{EA} \begin{bmatrix} -1448 \\ 1409 \\ -1839 \end{bmatrix}$$

(9) Equations 10.40c and 10.42 are used to evaluate the unknown reactions $\mathbf{Q}_u$ and member forces for each element, i.e. $\mathbf{q}_{(1)}$ $\mathbf{q}_{(2)}$ and $\mathbf{q}_{(3)}$ (in kN):

$$\mathbf{Q}_u = \begin{bmatrix} -75 \\ 165 \\ 37.5 \end{bmatrix}$$

$$\mathbf{q}_{(1)} = \mathbf{kTD}_{e(1)} = \frac{EA}{8} \begin{bmatrix} 1 & -1 \\ -1 & 1 \end{bmatrix} \begin{bmatrix} 1 & 0 & 0 & 0 \\ 0 & 0 & 0.5 & 0.866 \end{bmatrix} \begin{bmatrix} 0 \\ 0 \\ -1839/EA \\ 0 \end{bmatrix} = \begin{bmatrix} 115 \\ -115 \end{bmatrix}$$

Therefore, $N_1 = -115.0$ kN.

$$\mathbf{q}_{(2)} = \mathbf{kTD}_{e(1)} = \frac{EA}{5} \begin{bmatrix} 1 & -1 \\ -1 & 1 \end{bmatrix} \begin{bmatrix} 0.8 & 0.6 & 0 & 0 \\ 0 & 0 & 0.8 & 0.6 \end{bmatrix} \begin{bmatrix} 0 \\ 0 \\ -1448/EA \\ 1409/EA \end{bmatrix} = \begin{bmatrix} 62.5 \\ -62.5 \end{bmatrix}$$

Therefore, $N_2 = -62.5$ kN.

$$\mathbf{q}_{(3)} = \mathbf{kTD}_{e(1)} = \frac{EA}{5} \begin{bmatrix} 1 & -1 \\ -1 & 1 \end{bmatrix} \begin{bmatrix} 0.8 & -0.6 & 0 & 0 \\ 0 & 0 & 0.920 & 0.392 \end{bmatrix} \begin{bmatrix} -1448/EA \\ 1409/EA \\ -1839/EA \\ 0 \end{bmatrix} = \begin{bmatrix} -62.5 \\ 62.5 \end{bmatrix}$$

Therefore, $N_3 = 62.5$ kN.

10.10 SPACE TRUSS

The procedure required for the analysis of a space truss using the stiffness method is similar to that already described for plane trusses. The main difference being that each node can now have displacement components in three orthogonal directions. Because of this, three freedoms (instead of the two used for plane trusses) must be assigned at each node.

All the steps involved in the solution process are identical to those already detailed in Summary of Steps 10.1 for plane trusses, except that at point 3, three nodal freedoms need to be specified at each node and that the transformation matrix $\mathbf{T}$ needs to account for this change. Unknown displacements, reactions and member forces are determined following the steps specified in Equations 10.41, 10.40c and 10.42, respectively.

The transformation matrix **T** to be used with space trusses is evaluated in Reflection Activity 10.5, followed by the derivation of the corresponding stiffness matrix for the three-dimensional element.

REFLECTION ACTIVITY 10.5

Determine the transformation matrix **T** required to convert vector components of nodal displacements from global to local coordinates considering the freedoms of the isolated space truss element shown in Figure 10.30. Comment on how **T** needs to be modified to account for the possible presence of an inclined roller support at one of the element ends.

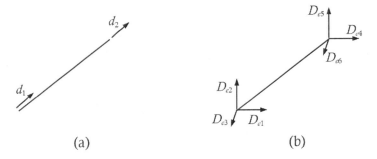

(a)   (b)

Figure 10.30 Nodal displacements in local and global coordinates for an isolated member of a space truss. (a) Displacements in local coordinates. (b) Displacements in global coordinates.

The procedure adopted to determine **T** follows the steps previously used to derive the transformation matrix for a plane truss, as outlined in Section 10.4.2.

Before considering the truss displacements, we determine how the global and local components of a vector relate to each other. Recalling that, for a truss element, the only relevant deformations are those parallel to the local x-axis (Reflection Activity 10.2), we only need to establish a relationship between a vector acting along the local x-axis and its global components. This significantly simplifies the calculation. Consider the vector $\mathbf{r} = [r_x\ 0\ 0]^\mathsf{T}$ in three-dimensional (X,Y,Z) space. The vector is pointing in the direction of the local x-axis (see Figure 10.31).

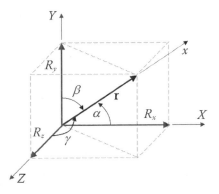

Figure 10.31 Global components of vector **r** acting along the local x-axis.

The relationship between the vector and its components in the X, Y and Z directions is:

$$[\mathbf{r}] = [r_x] = [\, l \ m \ n \,] \begin{bmatrix} R_x \\ R_y \\ R_z \end{bmatrix} \tag{10.49}$$

in which:

$$l = \cos \alpha \qquad m = \cos \beta \qquad n = \cos \gamma \tag{10.50a–c}$$

The transformation matrix for the nodal displacements of a truss at its two end nodes can be evaluated by applying Equation 10.49 to the two displacement vectors:

$$[\mathbf{d}_1] = [d_1] = [\, l \ m \ n \,] \begin{bmatrix} D_{x1} \\ D_{y1} \\ D_{z1} \end{bmatrix} \qquad [\mathbf{d}_2] = [d_2] = [\, l \ m \ n \,] \begin{bmatrix} D_{x2} \\ D_{y2} \\ D_{z2} \end{bmatrix} \tag{10.51a,b}$$

and combining these two relationships, we get:

$$\begin{bmatrix} d_1 \\ d_2 \end{bmatrix} = \begin{bmatrix} l & m & n & 0 & 0 & 0 \\ 0 & 0 & 0 & l & m & n \end{bmatrix} \begin{bmatrix} D_{x1} \\ D_{y1} \\ D_{z1} \\ D_{x2} \\ D_{y2} \\ D_{z2} \end{bmatrix} \tag{10.52}$$

The transformation matrix **T** is therefore:

$$\mathbf{T} = \begin{bmatrix} l & m & n & 0 & 0 & 0 \\ 0 & 0 & 0 & l & m & n \end{bmatrix} \tag{10.53}$$

The possibility of dealing with inclined roller supports can be easily accommodated by introducing nodal coordinates and differentiating between the direction cosines adopted for the two end nodes:

$$\mathbf{T} = \begin{bmatrix} l_1 & m_1 & n_1 & 0 & 0 & 0 \\ 0 & 0 & 0 & l_2 & m_2 & n_2 \end{bmatrix} \tag{10.54}$$

On the basis of the expression of **T** defined in Equation 10.53 (or Equation 10.54 when dealing with nodal coordinates), the stiffness matrix of an isolated element expressed in global coordinates is derived using Equation 10.34 as follows:

$$\mathbf{K}_e = \mathbf{T}^T \mathbf{k} \mathbf{T} = \begin{bmatrix} l & 0 \\ m & 0 \\ n & 0 \\ 0 & l \\ 0 & m \\ 0 & n \end{bmatrix} \frac{EA}{L} \begin{bmatrix} 1 & -1 \\ -1 & 1 \end{bmatrix} \begin{bmatrix} l & m & n & 0 & 0 & 0 \\ 0 & 0 & 0 & l & m & n \end{bmatrix}$$

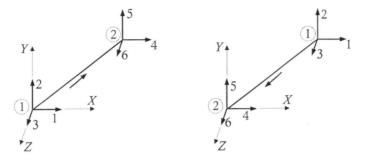

Figure 10.32 Freedom numbering of an isolated truss element in global coordinates.

$$= \frac{EA}{L} \begin{bmatrix} l^2 & lm & ln & -l^2 & -lm & -ln \\ lm & m^2 & mn & -lm & -m^2 & -mn \\ ln & mn & n^2 & -ln & -mn & -n^2 \\ -l^2 & -lm & -ln & l^2 & lm & ln \\ -lm & -m^2 & -mn & lm & m^2 & mn \\ -ln & -mn & -n^2 & ln & mn & n^2 \end{bmatrix} \qquad (10.55)$$

The contribution of the stiffness of an isolated element of a space truss to the structure stiffness matrix can be assembled when the global freedoms are assigned and when a location vector defining the positive direction of the x-axis is selected. For the element shown in Figure 10.32, the six nodal loads $\mathbf{Q}_e$ and six nodal displacements $\mathbf{D}_e$ are related as follows:

$$\mathbf{Q}_e = \mathbf{K}_e \mathbf{D}_e$$

$$\begin{matrix} & & & 1 & 2 & 3 & 4 & 5 & 6 \\ 1 \\ 2 \\ 3 \\ 4 \\ 5 \\ 6 \end{matrix} \begin{bmatrix} Q_{e1} \\ Q_{e2} \\ Q_{e3} \\ Q_{e4} \\ Q_{e5} \\ Q_{e6} \end{bmatrix} = \frac{EA}{L} \begin{bmatrix} l^2 & lm & ln & -l^2 & -lm & -ln \\ lm & m^2 & mn & -lm & -m^2 & -mn \\ ln & mn & n^2 & -ln & -mn & -n^2 \\ -l^2 & -lm & -ln & l^2 & lm & ln \\ -lm & -m^2 & -mn & lm & m^2 & mn \\ -ln & -mn & -n^2 & ln & mn & n^2 \end{bmatrix} \begin{bmatrix} D_{e1} \\ D_{e2} \\ D_{e3} \\ D_{e4} \\ D_{e5} \\ D_{e6} \end{bmatrix} \begin{matrix} 1 \\ 2 \\ 3 \\ 4 \\ 5 \\ 6 \end{matrix} \qquad (10.56)$$

PROBLEMS

10.1   Calculate the structural stiffness matrix of the truss shown.

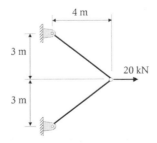

10.2 Reconsider the truss of Problem 10.1, and determine the unknown displacements and reactions using the stiffness method.

10.3 For the truss of Problem 10.1, evaluate the member forces of each element using the stiffness method and specify whether these are in compression or tension.

10.4 Determine the stiffness matrix for the truss shown.

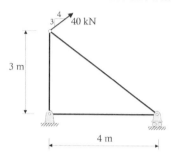

10.5 Calculate the unknown displacements and reactions for the truss of Problem 10.4 using the stiffness method.

10.6 Reconsider the truss of Problem 10.4 and determine, for each truss element, its member forces using the stiffness method. Clarify whether these are in compression or tension.

10.7 Calculate the stiffness matrix for the truss shown.

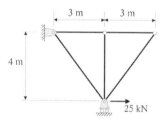

10.8 Evaluate the unknown nodal displacements and reactions for the truss of Problem 10.7 using the stiffness method.

10.9 For the truss of Problem 10.7, determine using the stiffness method the force in each truss element. Clarify whether these are in compression or tension.

10.10 Calculate the stiffness matrix for the truss shown.

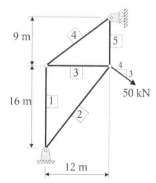

10.11 Reconsider the truss of Problem 10.10, and determine the unknown nodal displacements and reactions using the stiffness method.

10.12 Using the stiffness method, calculate member forces for elements 2, 3 and 4 for the truss of Problem 10.10 and specify whether these are compressive or tensile ones.

10.13 Calculate the stiffness matrix for the truss shown below.

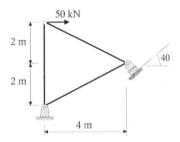

10.14 Reconsider the truss of Problem 10.13, and evaluate the unknown nodal displacements and reactions using the stiffness method.

10.15 Determine member forces for all elements of the truss of Problem 10.13 using the stiffness method. Specify whether they are compressive or tensile forces.

10.16 Calculate the stiffness matrix for the truss shown.

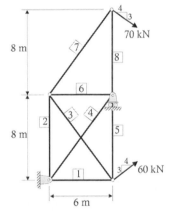

10.17 For the truss of Problem 10.16, determine the unknown nodal displacements and reactions using the stiffness method.

10.18 Calculate member forces for elements 3, 4, 5 and 7 for the truss of Problem 10.16.

10.19 Calculate the stiffness matrix of the truss shown and evaluate the unknown nodal displacements using the stiffness method.

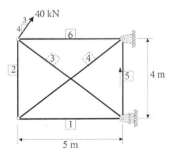

10.20 Using the stiffness method, determine member forces for elements 3, 4 and 6 for the truss of Problem 10.19.

10.21 Reconsider the truss of Problem 10.19 rotating the roller support as shown below. Determine the unknown displacements and unknown reactions using the stiffness method.

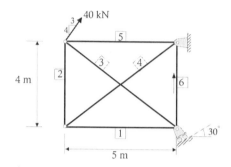

10.22 Using the stiffness method, calculate member forces for elements 3, 4 and 6 for the truss of Problem 10.21.

10.23 Determine the unknown nodal displacements for the space truss shown using the stiffness method.

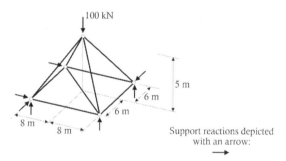

10.24 Using the stiffness method, calculate the unknown reactions for the truss of Problem 10.23.

10.25 Using the stiffness method, determine the member forces for the truss elements of Problem 10.23.

10.26 Calculate the unknown reactions for the truss shown using the stiffness method.

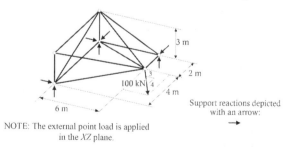

10.27 For the truss of Problem 10.26, evaluate reactions and member axial forces using the stiffness method.

Chapter 11

Beam analysis using the stiffness method

11.1 THE BEAM ELEMENT

We have seen that a beam is a structural member that resists loads transverse to its longitudinal axis (x-axis) and is supported at different locations along its length. In this chapter, we will present a particular stiffness element, called a *beam element*, that has been developed for the stiffness analysis of beams. The distinguishing feature of the beam element is its ability to resist moment and shear force. The member is assumed to possess a plane of symmetry and to resist forces applied within this plane of symmetry. It may also resist couples applied about an axis perpendicular to the plane of symmetry. Even if these conditions might appear to be restrictive, they are applicable to a large number of structural members commonly used in real structures.

The displacements at its nodal freedoms consist of transverse deflections and rotations, with nodal actions corresponding to transverse forces and moments. These are illustrated in Figure 11.1. For the purpose of this book, the use of the beam element is limited to the cases in which local and global coordinates coincide (Figure 11.1). This condition on the coordinate systems is enforced by assuming that the location vector of each beam element (which defines the positive direction of the local x-axis) points in the positive direction of the global X-axes. This simplifies the solution process because it avoids the need for a transformation matrix to relate global and local freedoms.

Stiffness analysis using the beam element is a very useful approach to the analysis of both statically determinate and indeterminate beams. For members resisting axial forces, shear forces and moments, the frame element presented in the next chapter should be used.

The solution procedure required for the analysis of beams is similar to that already presented in Chapter 10 for trusses, where the problem was expressed in terms of unknown nodal displacements. In the case of the beam element, the unknown variables consist of the *transverse displacements* and *rotations*. Once these are determined, all other variables, such as the support reactions and the internal actions, are calculated in the post-processing stage.

The *discretisation* of the beam requires nodes to be placed at supports and at free ends of the beam. When point loads are applied within a beam span, it is usually convenient to introduce a node at these locations, even though point loads can also be included in the analysis using the procedures outlined for member loads in Section 11.5. For the continuous beam illustrated in Figure 11.2a we introduce nodes at the left pinned support (node 1), at the two roller supports (nodes 3 and 4) and at the end of the beam (node 5). For convenience, we also specify a node at the location of the applied load P_1 (node 2), as shown in Figure 11.2b. There is no need to introduce a separate node for load P_2 as its point of application coincides with node 5. In this manner, the continuous beam is discretised into four beam elements.

We are now in a position to assign the nodal freedoms at each of the nodes of the structure, as illustrated in Figure 11.2c. Considering that for the beam element we adopt identical

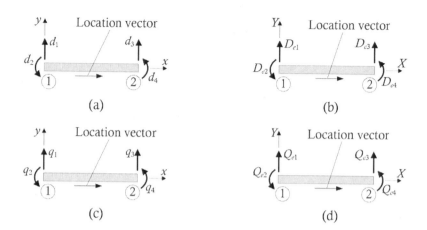

Figure 11.1 Nodal displacements and actions of the beam element. (a) Nodal displacements in local coordinates. (b) Nodal displacements in global coordinates. (c) Nodal actions in local coordinates. (d) Nodal actions in global coordinates.

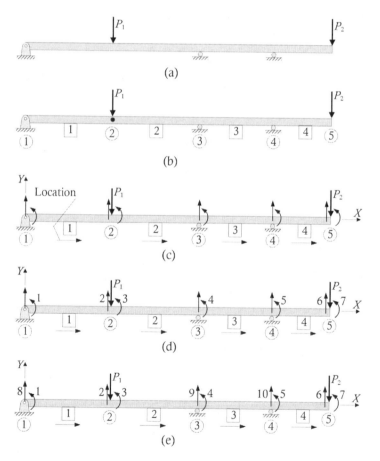

Figure 11.2 Beam discretisation and numbering of the local (global) freedoms. (a) Layout of the beam. (b) Beam discretisation (5 nodes, 4 elements). (c) Local (global) freedoms. (d) Numbering of unrestrained freedoms. (e) Numbering of restrained freedoms.

local and global coordinate systems, the location vectors of all elements are assumed to be pointing toward the right along the positive direction of the x-axis.

The numbering of the nodal freedoms follows the same approach used in the analysis of trusses in Chapter 10, in which unrestrained freedoms are numbered first (freedoms 1–7 in Figure 11.2d), followed by the restrained ones (freedoms 8–10 in Figure 11.2e). Although this separation between restrained and unrestrained freedoms is convenient when solutions are carried out by hand, it is not necessary when implementing the analysis in a computer program.

II.2 DERIVATION OF THE STIFFNESS MATRIX

The *load–displacement relationship* for the beam element relates the nodal displacements (d_1, d_2, d_3, d_4) and nodal actions (q_1, q_2, q_3, q_4), illustrated in Figure 11.1, with the following relationship:

$$\begin{bmatrix} q_1 \\ q_2 \\ q_3 \\ q_4 \end{bmatrix} = \begin{bmatrix} k_{11} & k_{12} & k_{13} & k_{14} \\ k_{21} & k_{22} & k_{23} & k_{24} \\ k_{31} & k_{32} & k_{33} & k_{34} \\ k_{41} & k_{42} & k_{43} & k_{44} \end{bmatrix} \begin{bmatrix} d_1 \\ d_2 \\ d_3 \\ d_4 \end{bmatrix} = \begin{bmatrix} \dfrac{12EI}{L^3} & \dfrac{6EI}{L^2} & \dfrac{-12EI}{L^3} & \dfrac{6EI}{L^2} \\[2ex] \dfrac{6EI}{L^2} & \dfrac{4EI}{L} & -\dfrac{6EI}{L^2} & \dfrac{2EI}{L} \\[2ex] -\dfrac{12EI}{L^3} & -\dfrac{6EI}{L^2} & \dfrac{12EI}{L^3} & -\dfrac{6EI}{L^2} \\[2ex] \dfrac{6EI}{L^2} & \dfrac{2EI}{L} & -\dfrac{6EI}{L^2} & \dfrac{4EI}{L} \end{bmatrix} \begin{bmatrix} d_1 \\ d_2 \\ d_3 \\ d_4 \end{bmatrix} \qquad (11.1a)$$

or in compact matrix form as:

$$\mathbf{q} = \mathbf{kd} \qquad (11.1b)$$

where nodal actions and displacements are assumed to be positive in the positive direction of the nodal freedoms. Each *member stiffness influence coefficient* k_{ij} (with $i,j = 1,...,4$) can be interpreted as the reaction produced along freedom i by a unit displacement enforced along freedom j. This definition is illustrated graphically in Figure 11.3 specifying a unit displacement along each freedom considered separately.

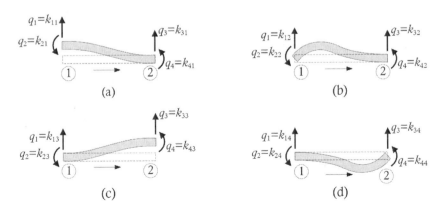

Figure 11.3 Physical representation of the member stiffness influence coefficients. (a) $d_1 = 1$ (with $d_2 = d_3 = d_4 = 0$). (b) $d_2 = $ (with $d_1 = d_3 = d_4 = 0$). (c) $d_3 = 1$ (with $d_1 = d_2 = d_4 = 0$). (d) $d_4 = 1$ (with $d_1 = d_2 = d_3 = 0$).

REFLECTION ACTIVITY 11.1

Suggest one possible procedure to derive the member stiffness influence coefficients k_{ij} (with $i,j = 1,...,4$) included in Equation 11.1a, relying on their physical interpretation provided in Figure 11.3.

The representation of the member stiffness influence coefficients in Figure 11.3 reflects the fact that the j-th column of the element stiffness matrix $\mathbf{k}$ (Equation 11.1a) contains the actions required to be applied to the ends of the beam element to produce a unit displacement along freedom j, while maintaining all other freedoms restrained. For example, the set of end displacements representing a unit displacement applied along freedom 1, while restraining all other freedoms, is:

$$\mathbf{d} = \begin{bmatrix} d_1 \\ d_2 \\ d_3 \\ d_4 \end{bmatrix} = \begin{bmatrix} 1 \\ 0 \\ 0 \\ 0 \end{bmatrix} \tag{11.2}$$

Substituting Equation 11.2 into Equation 11.1a enables the calculation of the corresponding end actions $\mathbf{q}$ to be applied at the supports to maintain the enforced displaced shape:

$$\begin{bmatrix} q_1 \\ q_2 \\ q_3 \\ q_4 \end{bmatrix} = \begin{bmatrix} k_{11} & k_{12} & k_{13} & k_{14} \\ k_{21} & k_{22} & k_{23} & k_{24} \\ k_{31} & k_{32} & k_{33} & k_{34} \\ k_{41} & k_{42} & k_{43} & k_{44} \end{bmatrix} \begin{bmatrix} 1 \\ 0 \\ 0 \\ 0 \end{bmatrix} = \begin{bmatrix} k_{11} \\ k_{21} \\ k_{31} \\ k_{41} \end{bmatrix} \tag{11.3}$$

Equation 11.3 shows how the set of end actions $\mathbf{q}$ related to the end displacements $\mathbf{d} = [1\ 0\ 0\ 0]^T$ equals the coefficients included in the first column of the element stiffness matrix $\mathbf{k}$.

The expression provided in Equation 11.3 is useful only if we are able to determine the end actions corresponding to the different sets of induced displacements so that we have numerical values for each influence coefficient (k_{ij}). Fortunately, we have already derived the expressions for these end actions in Worked Example 5.6, where we considered a fixed-ended beam with particular support displacements, such as those considered here. In particular, in Worked Example 5.6 part ii, we considered a unit transverse displacement at one support (equivalent to the displacements illustrated in Figures 11.3a and c), and in Worked Example 5.6 part iii, we considered a unit rotation at one support (equivalent to the displacements illustrated in Figures 11.3b and d). For ease of reference, the set of displacements and set of nodal reactions obtained in Worked Example 5.6 (and those included in the solution of Problem 5.18) are tabulated below for each of the four cases considered here.

Case	Set of displacements	Reactions
1	$\begin{bmatrix} d_1 \\ d_2 \\ d_3 \\ d_4 \end{bmatrix} = \begin{bmatrix} 1 \\ 0 \\ 0 \\ 0 \end{bmatrix}$	$\begin{bmatrix} q_1 \\ q_2 \\ q_3 \\ q_4 \end{bmatrix} = \begin{bmatrix} \dfrac{12EI}{L^3} \\ \dfrac{6EI}{L^2} \\ -\dfrac{12EI}{L^3} \\ \dfrac{6EI}{L^2} \end{bmatrix}$
2	$\begin{bmatrix} d_1 \\ d_2 \\ d_3 \\ d_4 \end{bmatrix} = \begin{bmatrix} 0 \\ 1 \\ 0 \\ 0 \end{bmatrix}$	$\begin{bmatrix} q_1 \\ q_2 \\ q_3 \\ q_4 \end{bmatrix} = \begin{bmatrix} \dfrac{6EI}{L^2} \\ \dfrac{4EI}{L} \\ -\dfrac{6EI}{L^2} \\ \dfrac{2EI}{L} \end{bmatrix}$
3	$\begin{bmatrix} d_1 \\ d_2 \\ d_3 \\ d_4 \end{bmatrix} = \begin{bmatrix} 0 \\ 0 \\ 1 \\ 0 \end{bmatrix}$	$\begin{bmatrix} q_1 \\ q_2 \\ q_3 \\ q_4 \end{bmatrix} = \begin{bmatrix} -\dfrac{12EI}{L^3} \\ -\dfrac{6EI}{L^2} \\ \dfrac{12EI}{L^3} \\ -\dfrac{6EI}{L^2} \end{bmatrix}$
4	$\begin{bmatrix} d_1 \\ d_2 \\ d_3 \\ d_4 \end{bmatrix} = \begin{bmatrix} 0 \\ 0 \\ 0 \\ 1 \end{bmatrix}$	$\begin{bmatrix} q_1 \\ q_2 \\ q_3 \\ q_4 \end{bmatrix} = \begin{bmatrix} \dfrac{6EI}{L^2} \\ \dfrac{2EI}{L} \\ -\dfrac{6EI}{L^2} \\ \dfrac{4EI}{L} \end{bmatrix}$

The stiffness matrix of the beam can then be obtained by combining these reaction vectors (which correspond to the end nodal actions of the free-body diagram of the beam element) as follows:

$$
\mathbf{k} = \begin{bmatrix} k_{11} & k_{12} & k_{13} & k_{14} \\ k_{21} & k_{22} & k_{23} & k_{24} \\ k_{31} & k_{32} & k_{33} & k_{34} \\ k_{41} & k_{42} & k_{43} & k_{44} \end{bmatrix} = \begin{bmatrix} \dfrac{12EI}{L^3} & \dfrac{6EI}{L^2} & -\dfrac{12EI}{L^3} & \dfrac{6EI}{L^2} \\ \dfrac{6EI}{L^2} & \dfrac{4EI}{L} & -\dfrac{6EI}{L^2} & \dfrac{2EI}{L} \\ -\dfrac{12EI}{L^3} & -\dfrac{6EI}{L^2} & \dfrac{12EI}{L^3} & -\dfrac{6EI}{L^2} \\ \dfrac{6EI}{L^2} & \dfrac{2EI}{L} & -\dfrac{6EI}{L^2} & \dfrac{4EI}{L} \end{bmatrix} \tag{11.4}
$$

which is the element stiffness matrix included in Equation 11.1a for the beam element, I is the second moment of area of the cross-section, E is the elastic modulus of the material and L is the length of the beam element.

The method described for deriving the stiffness matrix by assigning one unit displacement at a time to each of the available freedoms is usually referred to as the *direct stiffness method*.

11.3 BEAM ELEMENT IN GLOBAL COORDINATES

We have assumed that, for beam elements only, global and local coordinate systems coincide (Figure 11.1) and, as a consequence, there is no need to transform global and local displacements, forces and stiffness coefficients.

REFLECTION ACTIVITY 11.2

Under the assumptions that local and global coordinate systems coincide for beam elements, what would the transformation matrix **T** look like and how would **T** affect the relationships between local and global components of the displacements $\mathbf{d} = \mathbf{T}\mathbf{D}_e$, end actions $\mathbf{Q}_e = \mathbf{T}^T\mathbf{q}$ and stiffness matrix $\mathbf{K}_e = \mathbf{T}^T\mathbf{k}\mathbf{T}$ for an element unloaded along its length?

Because the local and global coordinate systems coincide, the transformation matrix **T** of the beam element is simply represented by an identity matrix:

$$
\mathbf{T} = \begin{bmatrix} 1 & 0 & 0 & 0 \\ 0 & 1 & 0 & 0 \\ 0 & 0 & 1 & 0 \\ 0 & 0 & 0 & 1 \end{bmatrix} \tag{11.5}
$$

Therefore, the local and global representations of both displacements and end actions coincide:

$$
\mathbf{d} = \mathbf{T}\mathbf{D}_e = \mathbf{D}_e \qquad \mathbf{Q}_e = \mathbf{T}^T\mathbf{q} = \mathbf{q} \tag{11.6a,b}
$$

with:

$$\mathbf{d} = \begin{bmatrix} d_1 \\ d_2 \\ d_3 \\ d_4 \end{bmatrix} \qquad \mathbf{D}_e = \begin{bmatrix} D_{e1} \\ D_{e2} \\ D_{e3} \\ D_{e4} \end{bmatrix} \qquad \mathbf{q} = \begin{bmatrix} q_1 \\ q_2 \\ q_3 \\ q_4 \end{bmatrix} \qquad \mathbf{Q}_e = \begin{bmatrix} Q_{e1} \\ Q_{e2} \\ Q_{e3} \\ Q_{e4} \end{bmatrix} \qquad (11.7a\text{–}d)$$

In a similar manner, the local stiffness matrix and the global stiffness matrix are identical:

$$\mathbf{K}_e = \mathbf{T}^\mathsf{T}\mathbf{k}\mathbf{T} = \mathbf{k} \qquad (11.8)$$

Note that global and local displacements, actions and stiffness matrices coincide, as stated in Equations 11.6 and 11.8, only under the assumption that local and global coordinates coincide.

11.4 ASSEMBLING OF THE STIFFNESS ELEMENTS

The assembling procedure follows the same steps already presented for truss elements in the previous chapter. The stiffness relationship for an isolated beam element can be expressed in global coordinates (which here coincide with the local coordinates) as:

$$\mathbf{Q}_e = \mathbf{K}_e\mathbf{D}_e \qquad (11.9)$$

where $\mathbf{Q}_e$ and $\mathbf{D}_e$ collect the nodal forces and displacements (Equations 11.6), and $\mathbf{K}_e$ represents the element stiffness matrix (Equation 11.8).

The stiffness relationship for the entire beam relates the nodal displacements $\mathbf{D}$ to the nodal actions $\mathbf{Q}$ defined along the global freedoms as:

$$\mathbf{Q} = \mathbf{K}\mathbf{D} \qquad (11.10)$$

where $\mathbf{K}$ is the stiffness matrix of the entire beam (called the structure stiffness matrix). Denoting with N_{dof} the number of global freedoms of the beam, the length of vectors $\mathbf{D}$ and $\mathbf{Q}$ is N_{dof}, while the size of $\mathbf{K}$ is $N_{dof} \times N_{dof}$. For example, the beam of Figure 11.2 possesses 10 freedoms and so $N_{dof} = 10$.

The structure stiffness matrix $\mathbf{K}$ is determined by combining the contribution of the stiffness matrix of each beam element considered in isolation $\mathbf{K}_{e(n)}$ (where 'n' identifies the element number). In this process, it is important to carefully relate the freedom numbers of the isolated element to the appropriate global freedoms.

11.5 MEMBER LOADS

The loading vector included in the governing system of equations for determining the unknown displacements (Equation 11.10) does not allow member loads to be specified directly. For this purpose, consideration of member loads is carried out in two stages, making use of the *principle of superposition* and introducing the concept of *equivalent nodal loads*. For example, consider the two-span beam shown in Figure 11.4a, where one of the spans is subjected to a uniformly distributed load w.

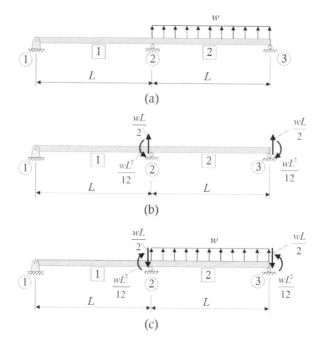

Figure 11.4 Equivalent nodal loads for a uniformly distributed load. (a) Beam A. (b) Beam B. (c) Beam C.

Relying on the principle of superposition, the behaviour of the beam shown in Figure 11.4a, referred to as beam A, is equivalent to the sum of the responses obtained for the beams in Figures 11.4b and c, denoted as beams B and C, respectively. This is the case because the sum of the external loads applied to beams B and C produces the same loading condition as beam A (because the applied vertical forces and moments applied at nodes 2 and 3 in beams B and C cancel each other out when added together). The set of nodal actions introduced at nodes 2 and 3 are the actions corresponding to the reactions of a fixed-ended beam with the same length as the loaded beam element, as highlighted in Figures 11.5 (and calculated in Worked Example 5.5).

If for the three beams and loading conditions shown in Figures 11.4a, b, and c, $w = 8$ kN/m, $L = 10$ m and EI is constant over the length of each beam, the vertical displacements, rotations and reactions calculated at the three nodes of beams A, B and C are reported in Table 11.1. These values can be readily confirmed using one or more of the analysis techniques already discussed in Chapters 7, 8 and 9.

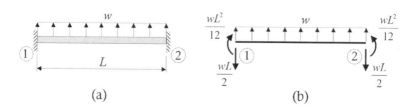

Figure 11.5 Reactions of a fixed-ended beam subjected to a uniformly distributed load. (a) Support and loading conditions. (b) Free-body diagram.

Table 11.1 Rotations and support reactions for two-span beams A, B and C of
Figure 11.4 (with *w* = 8 kN/m, *L* = 10 m and constant *EI*)

Beam	Rotations at nodes 1, 2 and 3 (rad)	Support reactions at nodes 1, 2 and 3 (kN)
A	D(rotation at node 1) = −83.33/*IE* D(rotation at node 2) = 166.67/*IE* D(rotation at node 3) = −250/*IE*	Q(reaction at node 1) = 5 Q(reaction at node 2) = −50 Q(reaction at node 3) = −35
B	D(rotation at node 1) = −83.33/*IE* D(rotation at node 2) = 166.67/*IE* D(rotation at node 3) = −250/*IE*	Q(reaction at node 1) = 5 Q(reaction at node 2) = −50 Q(reaction at node 3) = −35
C	D(rotation at node 1) = 0 D(rotation at node 2) = 0 D(rotation at node 3) = 0	Q(reaction at node 1) = 0 Q(reaction at node 2) = 0 Q(reaction at node 3) = 0

Based on these results, we can observe that the displacements and rotations undergone by
beam A (which is the beam that we want to analyse) are identical to those determined with
beam B, while zero nodal displacements or reactions are found for beam C. This is due to
the fact that the *equivalent nodal loads* specified for Beam B produce an *equivalent* effect to
that induced by the member loading, enabling the correct calculation of the nodal displacements and support reactions. The results for beam C reflect the fact that, specifying the
actual reactions of the fixed-ended beam as nodal loads produces the zero nodal displacements expected in the fixed-ended configuration. The zero reactions for Beam C indicate
that all actions are in equilibrium, as expected because they belong to the free-body diagram
of the fixed-ended beam shown in Figure 11.5b.

The shear force and bending moment diagrams obtained from the analyses of beams A,
B and C are plotted in Figure 11.6. These highlight the need to superimpose the solution of
beams B and C to obtain the values describing the behaviour of beam A.

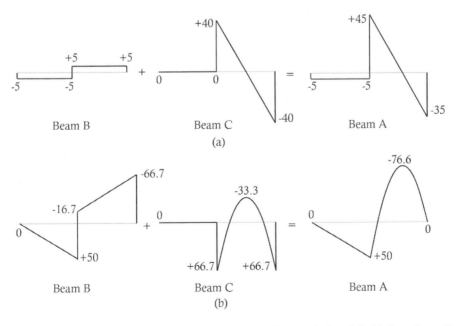

Figure 11.6 Distribution of internal actions along the lengths of beams A, B and C. (a) Shear force diagrams
(kN). (b) Bending moment diagrams (kNm).

Based on the above considerations, the effects of a member load may be included by performing one analysis where the equivalent nodal loads, referred to as q_M in the following, are applied to the entire structure. The vector q_M is equal and opposite in sign to the support reactions of a fixed-ended beam subjected to the same member load, denoted as q_F. This analysis enables the calculation of the global displacements and reactions of the structure (similar to the case of beam B in Figure 11.4b). A second step is required for the evaluation of the internal actions (similar to the case of beam C in Figure 11.4c) and this will be addressed in the next section when dealing with the post-processing of the solution.

As we saw earlier in Chapter 9 when discussing the moment distribution method, the use of the equivalent nodal loads is a very effective (and convenient) way of including member loads when using the stiffness method. Of course, we need to know the reactions induced by the member load on a fixed-ended beam of equivalent length. These can be readily remembered for common distributions of member loading, but for other sets of member loads (see Appendix B), they can be calculated based on the differential equations and boundary conditions presented in Chapter 5 (and discussed further in Chapter 12).

11.6 SOLUTION PROCEDURE AND POST-PROCESSING

The response of the beam can be evaluated by solving Equation 11.10 for the unknown nodal displacements and reactions. For clarity, displacements included in vector **D** are subdivided into known displacements D_k and unknown displacements D_u. In a similar manner, nodal actions can be grouped into known nodal actions Q_k (the external loads) and unknown nodal actions Q_u (consisting of the support reactions). With this notation, we can rewrite Equation 11.10 highlighting known and unknown terms as follows:

$$\begin{bmatrix} Q_k \\ Q_u \end{bmatrix} = \begin{bmatrix} K_{11} & K_{12} \\ K_{21} & K_{22} \end{bmatrix} \begin{bmatrix} D_u \\ D_k \end{bmatrix} \tag{11.11a}$$

or

$$Q_k = K_{11}D_u + K_{12}D_k \tag{11.11b}$$

$$Q_u = K_{21}D_u + K_{22}D_k \tag{11.11c}$$

where the beam stiffness matrix **K** is partitioned into K_{ij} (with $i,j = 1,2$), with **D** and **Q** being defined as:

$$D = \begin{bmatrix} D_u \\ D_k \end{bmatrix} \qquad Q = \begin{bmatrix} Q_k \\ Q_u \end{bmatrix} \tag{11.12a,b}$$

The unknown displacements D_u can be obtained by solving Equation 11.11b (see Appendix C for possible solution procedures of a system of simultaneous equations). For example:

$$D_u = K_{11}^{-1}(Q_k - K_{12}D_k) \tag{11.13}$$

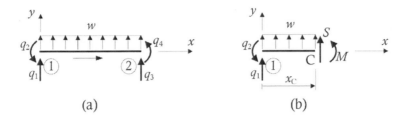

(a)   (b)

Figure 11.7 Evaluation of internal actions along the length of a beam element. (a) Free-body diagram of a beam element subjected to member loading. (b) Cut along the element length to calculate the internal shear force S and bending moment M.

When $\mathbf{D_u}$ is determined, the unknown reactions $\mathbf{Q_u}$ can be evaluated from Equation 11.11c.

The partitioning introduced in Equations 11.11 highlights that, when performing the solution by hand, it is convenient to number the unrestrained freedom first (i.e. those related to the unknown displacements $\mathbf{D_u}$), followed by the numbering of the restrained freedoms (i.e. those related to the unknown reactions $\mathbf{Q_u}$).

The internal actions resisted at the ends of each member can be calculated by post-processing the solution for $\mathbf{D_u}$. This is carried out by determining the nodal actions $\mathbf{q}$ applied to the free-body diagram of the element as follows:

$$\mathbf{q} = \mathbf{kd} + \mathbf{q_F} = \mathbf{kD}_e + \mathbf{q_F} \tag{11.14}$$

where $\mathbf{D}_e$ is the vector of the nodal displacements for the particular element considered and $\mathbf{q_F}$ is the vector of the reactions of the element subjected to the member load and fixed at both its ends (see Section 11.5). The distribution of the internal actions along the beam length can then be evaluated from statics by performing a cut at the locations of interest, as shown, for example, in Figure 11.7b at point C, located at a distance x_C from node 1. Another simple way to calculate the internal actions is to include a node at the location where internal actions are sought, so that these can be determined directly from the values of $\mathbf{q}$ without having to apply statics to a cut along the length of an element.

The steps required in the analysis of beams using the stiffness method are detailed below and are followed by the presentation of worked examples.

SUMMARY OF STEPS 11.1: Stiffness method — Solution procedure

The main steps to be followed when analysing a beam with the stiffness method are detailed below.

1. Specify a global reference system and then number the nodes and beam elements.

2. Assign two freedoms at each node, i.e. one transverse to the beam axis and one for the rotation taken positive when anti-clockwise. In the presence of end releases, such as hinges, additional freedoms need to be specified to account for the relative movements between adjacent elements. Ensure the local axes of each member have the same positive directions as the global axes. In the numbering of the freedoms, number the unrestrained freedoms first, followed by the restrained freedoms.

3. Calculate the stiffness matrix for each beam element based on Equation 11.4. When an element is subjected to member loads, determine the corresponding equivalent nodal loads $\mathbf{q}_M$ $(= -\mathbf{q}_F)$. The vector $\mathbf{q}_F$ collects, for each loaded element, the support reactions of a fixed-ended beam with identical geometry and properties to those of the element and subjected to the same member loads.

4. Assemble the stiffness matrix for the entire beam $\mathbf{K}$.

5. Determine vectors of known displacements $\mathbf{D}_k$ and external loads $\mathbf{Q}_k$. Partition the stiffness matrix $\mathbf{K}$ into $\mathbf{K}_{11}$, $\mathbf{K}_{12}$, $\mathbf{K}_{21}$ and $\mathbf{K}_{22}$ (Equations 11.11), as well as vectors $\mathbf{D}$ and $\mathbf{Q}$ into $[\mathbf{D}_u\ \mathbf{D}_k]^T$ and $[\mathbf{Q}_k\ \mathbf{Q}_u]^T$, respectively. Define the stiffness relationship $\mathbf{Q} = \mathbf{KD}$ (Equation 11.10 and Equation 11.11a) based on these partitions. For elements subjected to member loads, the contribution of their equivalent nodal actions $\mathbf{q}_M$ $(= -\mathbf{q}_F)$ needs to be mapped in the corresponding locations of the loading vector $\mathbf{Q}$.

6. Determine the unknown displacements $\mathbf{D}_u$ with Equation 11.13.

7. Evaluate the unknown reactions $\mathbf{Q}_u$ (Equation 11.11c) and calculate the end nodal actions $\mathbf{q}$ for each beam element (Equation 11.14).

WORKED EXAMPLE 11.1

Determine the nodal displacements and reactions for the beam shown in Figure 11.8. Assume *EI* is constant along the beam length.

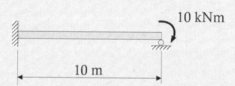

Figure 11.8 Beam for Worked Example 11.1.

The solution follows the steps outlined in Summary of Steps 11.1. Units used for length, rotation and force are m, rad and kN, respectively.

(1) We introduce a global coordinate system, and select two nodes and one element, numbered as shown.

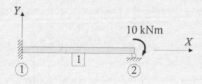

(2) At each node, we insert two freedoms, i.e. a transverse displacement and a rotation. The location vector is assigned to have local x-axis and global X-axis positive in the same direction. The unrestrained freedoms are numbered first. In this case, the only unrestrained freedom is the rotation at node 2 (numbered as freedom 1). Next, the restrained freedoms are numbered (freedoms 2 to 4 in the figure below).

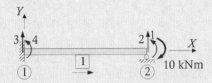

(3) The stiffness matrix for element 1 is calculated using Equation 11.4:

$$\mathbf{k}_{(1)} = \begin{bmatrix} \dfrac{12EI}{L^3} & \dfrac{6EI}{L^2} & \dfrac{-12EI}{L^3} & \dfrac{6EI}{L^2} \\[2mm] \dfrac{6EI}{L^2} & \dfrac{4EI}{L} & \dfrac{6EI}{L^2} & \dfrac{2EI}{L} \\[2mm] -\dfrac{12EI}{L^3} & -\dfrac{6EI}{L^2} & \dfrac{12EI}{L^3} & -\dfrac{6EI}{L^2} \\[2mm] \dfrac{6EI}{L^2} & \dfrac{2EI}{L} & \dfrac{6EI}{L^2} & \dfrac{4EI}{L} \end{bmatrix} = EI \begin{bmatrix} 0.012 & 0.06 & -0.012 & 0.06 \\ 0.06 & 0.4 & -0.06 & 0.2 \\ -0.012 & -0.06 & 0.012 & -0.06 \\ 0.06 & 0.2 & -0.06 & 0.4 \end{bmatrix}$$

where the stiffness coefficients are calculated as:

$$\frac{12EI}{L^3} = 0.012EI \qquad \frac{6EI}{L^2} = 0.06EI \qquad \frac{4EI}{L} = 0.4EI \qquad \frac{2EI}{L} = 0.2EI$$

Because local and global coordinates coincide, we have $\mathbf{K}_{e(1)} = \mathbf{k}_{(1)}$.

(4) In the assembling process, it is important to relate the freedom numbering adopted for the isolated beam element to the one of the entire structure. In this case, the entire beam is described by one element only, whose isolated dofs 1, 2, 3 and 4 correspond to global freedoms 3, 4, 2 and 1 adopted for the beam being analysed.

$$\mathbf{k}_{(1)} = \mathbf{K}_{e(1)} = EI \begin{array}{c} \\ \\ \\ \\ \end{array} \begin{matrix} 3 & 4 & 2 & 1 \\ \begin{bmatrix} 0.012 & 0.06 & -0.012 & 0.06 \\ 0.06 & 0.4 & -0.06 & 0.2 \\ -0.012 & -0.06 & 0.012 & -0.06 \\ 0.06 & 0.2 & -0.06 & 0.4 \end{bmatrix} & \begin{matrix} 3 \\ 4 \\ 2 \\ 1 \end{matrix} \end{matrix}$$

Assembling (adding)
$\mathbf{K}_{e(1)}$ to $\mathbf{K}$

$$\mathbf{K} = EI \begin{matrix} 1 & 2 & 3 & 4 \\ \begin{bmatrix} 0.4 & -0.06 & 0.06 & 0.2 \\ -0.06 & 0.012 & -0.012 & -0.06 \\ 0.06 & -0.012 & 0.012 & 0.06 \\ 0.2 & -0.06 & 0.06 & 0.4 \end{bmatrix} & \begin{matrix} 1 \\ 2 \\ 3 \\ 4 \end{matrix} \end{matrix}$$

(5) The only unrestrained freedom in this problem is 1 and, based on this, the vector of known actions is $\mathbf{Q}_k = [-10]$ to account for the clockwise couple applied at node 2. Its sign is negative because it acts in the opposite direction to freedom 1. The vector of known displacements $\mathbf{D}_k = [0\ 0\ 0]^T$ contains the zero displacements related to the restrained freedoms 2, 3 and 4.

The stiffness relationship is next partitioned in accordance with Equation 11.11a and the system of four equations in four unknowns is expressed as:

$$
\begin{bmatrix} -10 \\ Q_2 \\ Q_3 \\ Q_4 \end{bmatrix} = EI \begin{bmatrix} 0.4 & -0.06 & 0.06 & 0.2 \\ -0.06 & 0.012 & -0.012 & -0.06 \\ 0.06 & -0.012 & 0.012 & 0.06 \\ 0.2 & -0.06 & 0.06 & 0.4 \end{bmatrix} \begin{bmatrix} D_1 \\ 0 \\ 0 \\ 0 \end{bmatrix}
\begin{matrix} 1 \\ 2 \\ 3 \\ 4 \end{matrix}
\tag{11.15}
$$

$$
\text{or} \quad \begin{bmatrix} \mathbf{Q}_k \\ \mathbf{Q}_u \end{bmatrix} = \begin{bmatrix} \mathbf{K}_{11} & \mathbf{K}_{12} \\ \mathbf{K}_{21} & \mathbf{K}_{22} \end{bmatrix} \begin{bmatrix} \mathbf{D}_u \\ \mathbf{D}_k \end{bmatrix}
$$

with individual partitioned terms being:

$$
\mathbf{Q}_k = [-10] \quad \mathbf{Q}_u = \begin{bmatrix} Q_2 \\ Q_3 \\ Q_4 \end{bmatrix} \quad \mathbf{D}_u = [D_1] \quad \mathbf{D}_k = \begin{bmatrix} 0 \\ 0 \\ 0 \end{bmatrix}
$$

$$
\mathbf{K}_{11} = 0.4EI \quad \mathbf{K}_{12} = EI \begin{bmatrix} -0.06 & 0.06 & 0.2 \end{bmatrix}
$$

$$
\mathbf{K}_{21} = EI \begin{bmatrix} -0.06 \\ 0.06 \\ 0.2 \end{bmatrix} \quad \mathbf{K}_{22} = EI \begin{bmatrix} 0.012 & -0.012 & -0.06 \\ -0.012 & 0.012 & 0.06 \\ -0.06 & 0.06 & 0.4 \end{bmatrix}
$$

(6) The unknown displacement $\mathbf{D}_u(=[D_1])$ can be obtained solving the first equation expressed by the system of Equation 11.15 as follows:

$$
-10 = 0.4EI\ D_1 \quad \text{from which:} \quad D_1 = \frac{-10}{0.4EI} = -\frac{25}{EI}
$$

(7) Once the unknown displacements are determined, it is possible to calculate the unknown reactions (Equation 11.11c):

$$
\mathbf{Q}_u = \mathbf{K}_{21}\mathbf{D}_u + \mathbf{K}_{22}\mathbf{D}_k = EI \begin{bmatrix} -0.06 \\ 0.06 \\ 0.2 \end{bmatrix} \left(-\frac{25}{EI} \right) + EI \begin{bmatrix} 0.012 & -0.012 & -0.06 \\ -0.012 & 0.012 & 0.06 \\ -0.06 & 0.06 & 0.4 \end{bmatrix} \begin{bmatrix} 0 \\ 0 \\ 0 \end{bmatrix} = \begin{bmatrix} 1.5 \\ -1.5 \\ -5 \end{bmatrix}
$$

and these reactions are illustrated on the free-body diagram of the beam in Figure 11.9.

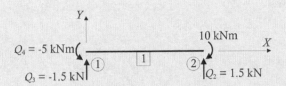

Figure 11.9 Reactions calculated for Worked Example 11.1.

WORKED EXAMPLE 11.2

Calculate nodal displacements and reactions for the beam shown in Figure 11.10 and evaluate the internal actions for each element specified. Assume EI is constant along the beam.

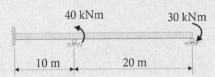

Figure 11.10 Beam for Worked Example 11.2.

(1) We start by specifying a global coordinate system and by numbering nodes and members. In this case, we select three nodes (one at each support) and two elements, as shown below.

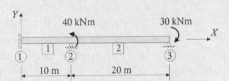

(2) Two freedoms (one transverse freedom and one rotational freedom) are assigned to each node and numbered starting from the unrestrained freedoms, followed by the restrained freedoms. In this case, unrestrained freedoms correspond to the rotations at nodes 2 and 3 and are labelled as freedoms 1 and 2. As all remaining freedoms are restrained, these are then numbered from 3 to 6 as shown. Location vectors are specified for each member so that the local x-axis and global X-axis are positive in the same direction.

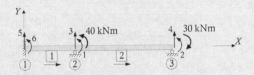

(3) With $L_{(1)} = 10$ m and $L_{(2)} = 20$ m, the stiffness matrices for elements 1 and 2 are determined with Equation 11.4 as:

$$\mathbf{k}_{(1)} = \begin{bmatrix} \dfrac{12EI}{L_{(1)}^3} & \dfrac{6EI}{L_{(1)}^2} & \dfrac{-12EI}{L_{(1)}^3} & \dfrac{6EI}{L_{(1)}^2} \\ \dfrac{6EI}{L_{(1)}^2} & \dfrac{4EI}{L_{(1)}} & -\dfrac{6EI}{L_{(1)}^2} & \dfrac{2EI}{L_{(1)}} \\ -\dfrac{12EI}{L_{(1)}^3} & -\dfrac{6EI}{L_{(1)}^2} & \dfrac{12EI}{L_{(1)}^3} & -\dfrac{6EI}{L_{(1)}^2} \\ \dfrac{6EI}{L_{(1)}^2} & \dfrac{2EI}{L_{(1)}} & -\dfrac{6EI}{L_{(1)}^2} & \dfrac{4EI}{L_{(1)}} \end{bmatrix} = EI \begin{bmatrix} 0.012 & 0.06 & -0.012 & 0.06 \\ 0.06 & 0.4 & -0.06 & 0.2 \\ -0.012 & -0.06 & 0.012 & -0.06 \\ 0.06 & 0.2 & -0.06 & 0.4 \end{bmatrix}$$

and

$$\mathbf{k}_{(2)} = \begin{bmatrix} \dfrac{12EI}{L_{(2)}^3} & \dfrac{6EI}{L_{(2)}^2} & \dfrac{-12EI}{L_{(2)}^3} & \dfrac{6EI}{L_{(2)}^2} \\ \dfrac{6EI}{L_{(2)}^2} & \dfrac{4EI}{L_{(2)}} & -\dfrac{6EI}{L_{(2)}^2} & \dfrac{2EI}{L_{(2)}} \\ -\dfrac{12EI}{L_{(2)}^3} & -\dfrac{6EI}{L_{(2)}^2} & \dfrac{12EI}{L_{(2)}^3} & -\dfrac{6EI}{L_{(2)}^2} \\ \dfrac{6EI}{L_{(2)}^2} & \dfrac{2EI}{L_{(2)}} & -\dfrac{6EI}{L_{(2)}^2} & \dfrac{4EI}{L_{(2)}} \end{bmatrix} = EI \begin{bmatrix} 0.0015 & 0.015 & -0.0015 & 0.015 \\ 0.015 & 0.2 & -0.015 & 0.1 \\ -0.0015 & -0.015 & 0.0015 & -0.015 \\ 0.015 & 0.1 & -0.015 & 0.2 \end{bmatrix}$$

Since the local and global coordinates coincide: $\mathbf{K}_{e(1)} = \mathbf{k}_{(1)}$ and $\mathbf{K}_{e(2)} = \mathbf{k}_{(2)}$.

(4) The stiffness matrices of the two beams are assembled into the structure stiffness matrix $\mathbf{K}$. This is carried out mapping the freedoms 1 to 4 of isolated element 1 to the global freedoms 5, 6, 3 and 1 and freedoms 1 to 4 of isolated element 2 to the global freedoms 3, 1, 4 and 2. This produces the following matrix $\mathbf{K}$:

$$\mathbf{K} = EI \begin{bmatrix} 0.6 & 0.1 & -0.045 & -0.015 & 0.06 & 0.2 \\ 0.1 & 0.2 & 0.015 & -0.015 & 0 & 0 \\ -0.045 & 0.015 & 0.0135 & -0.0015 & -0.012 & -0.06 \\ -0.015 & -0.015 & -0.0015 & 0.0015 & 0 & 0 \\ 0.06 & 0 & -0.012 & 0 & 0.012 & 0.06 \\ 0.2 & 0 & -0.06 & 0 & 0.06 & 0.4 \end{bmatrix}$$

(5) The vector of known actions is $\mathbf{Q}_k = [40 \ -30]^T$. The 40 kNm anti-clockwise moment applied at node 2 has been included with a positive sign because it is pointing in the same direction of freedom 1, while the clockwise moment of 30 kNm at node 3 is negative because it is applied in the opposite direction to freedom 2. The zero displacements related to the restrained freedoms are $\mathbf{D}_k = [0 \ 0 \ 0 \ 0]^T$.

The stiffness relationship for the entire structure (Equation 11.11a) can then be expressed as follows:

$$
\begin{bmatrix}
40 \\
-30 \\
Q_3 \\
Q_4 \\
Q_5 \\
Q_6
\end{bmatrix}
= EI
\begin{bmatrix}
0.6 & 0.1 & -0.045 & -0.015 & 0.06 & 0.2 \\
0.1 & 0.2 & 0.015 & -0.015 & 0 & 0 \\
-0.045 & 0.015 & 0.0135 & -0.0015 & -0.012 & -0.06 \\
-0.015 & -0.015 & -0.0015 & 0.0015 & 0 & 0 \\
0.06 & 0 & -0.012 & 0 & 0.012 & 0.06 \\
0.2 & 0 & -0.06 & 0 & 0.06 & 0.4
\end{bmatrix}
\begin{bmatrix}
D_1 \\
D_2 \\
0 \\
0 \\
0 \\
0
\end{bmatrix}
\tag{11.16}
$$

(6) The unknown displacements $\mathbf{D}_u$ are obtained by solving the first two of these equations, which can be rewritten as:

$$40 = EI(0.6D_1 + 0.1D_2) \tag{11.17a}$$

$$-30 = EI(0.1D_1 + 0.2D_2) \tag{11.17b}$$

Solving gives:

$$D_1 = \frac{100}{IE} \text{ rad} \quad \text{and} \quad D_2 = -\frac{200}{IE} \text{rad}$$

(7) The unknown reactions $\mathbf{Q}_u$ are calculated by solving the last four equations included in Equation 11.16:

$$Q_3 = EI(-0.045D_1 + 0.015D_2) = -7.5 \text{ kN}$$

$$Q_4 = EI(-0.015D_1 - 0.015D_2) = 1.5 \text{ kN}$$

$$Q_5 = EI\, 0.6D_1 = 6 \text{ kN}$$

$$Q_6 = EI\, 0.2D_2 = 20 \text{ kNm}$$

Values of $\mathbf{Q}_u$ are illustrated in the free-body diagram of the beam shown in Figure 11.11.

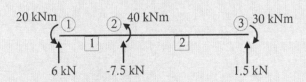

Figure 11.11 Free-body diagram for the two-span beam.

The nodal actions of elements 1 and 2 are calculated with Equation 11.14. For this purpose, the nodal displacements of member 1 are:

$$\mathbf{d}_{(1)} = \mathbf{D}_{e(1)} = \begin{bmatrix} D_{e1} \\ D_{e2} \\ D_{e3} \\ D_{e4} \end{bmatrix} = \begin{bmatrix} D_5 \\ D_6 \\ D_3 \\ D_1 \end{bmatrix} = \begin{bmatrix} 0 \\ 0 \\ 0 \\ 100/EI \end{bmatrix}$$

because member freedoms 1 to 4 of the isolated element 1 are mapped to the structure freedoms 5, 6, 3 and 1. The element actions are therefore:

$$\mathbf{q}_{(1)} = \mathbf{kd}_{(1)} = EI \begin{bmatrix} 0.012 & 0.06 & -0.012 & 0.06 \\ 0.06 & 0.4 & -0.06 & 0.2 \\ -0.012 & -0.06 & 0.012 & -0.06 \\ 0.06 & 0.2 & -0.06 & 0.4 \end{bmatrix} \begin{bmatrix} 0 \\ 0 \\ 0 \\ 100/EI \end{bmatrix} = \begin{bmatrix} 6 \\ 20 \\ -6 \\ 40 \end{bmatrix}$$

which are illustrated in the free-body diagram shown in Figure 11.12a.

A similar procedure is applied for element 2:

$$\mathbf{q}_{(2)} = \mathbf{kd}_{(2)} = EI \begin{bmatrix} 0.0015 & 0.015 & -0.0015 & 0.015 \\ 0.015 & 0.2 & -0.015 & 0.1 \\ -0.0015 & -0.015 & 0.0015 & -0.015 \\ 0.015 & 0.1 & -0.015 & 0.2 \end{bmatrix} \begin{bmatrix} 0 \\ 100/EI \\ 0 \\ -200/EI \end{bmatrix} = \begin{bmatrix} -1.5 \\ 0 \\ 1.5 \\ -30 \end{bmatrix}$$

and the results are shown in Figure 11.12b.

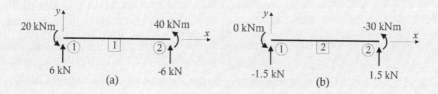

Figure 11.12 Free-body diagrams of the two elements. (a) Element 1. (b) Element 2.

WORKED EXAMPLE 11.3

Determine nodal displacements and reactions for the beam shown in Figure 11.13 and calculate the internal end actions for each element. Assume EI is constant along the beam.

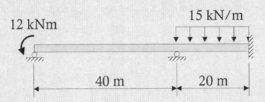

Figure 11.13 Beam for Worked Example 11.3.

---

(1) The global coordinate system and the node and member numbering are shown below.

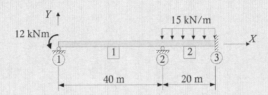

(2) Two freedoms are assigned at each node and numbered. The two unrestrained freedoms are numbered 1 and 2, and the restrained freedoms are numbered 3 to 6 as shown. The location vectors for each of the two elements are specified so that the local x-axis and the global X-axis coincide and are in the same direction.

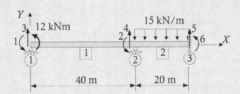

(3) Element 1: With $L_{(1)} = 40$ m:

$$\mathbf{k}_{(1)} = \begin{bmatrix} \dfrac{12EI}{L_{(1)}^3} & \dfrac{6EI}{L_{(1)}^2} & \dfrac{-12EI}{L_{(1)}^3} & \dfrac{6EI}{L_{(1)}^2} \\[2mm] \dfrac{6EI}{L_{(1)}^2} & \dfrac{4EI}{L_{(1)}} & -\dfrac{6EI}{L_{(1)}^2} & \dfrac{2EI}{L_{(1)}} \\[2mm] \dfrac{12EI}{L_{(1)}^3} & -\dfrac{6EI}{L_{(1)}^2} & \dfrac{12EI}{L_{(1)}^3} & -\dfrac{6EI}{L_{(1)}^2} \\[2mm] \dfrac{6EI}{L_{(1)}^2} & \dfrac{2EI}{L_{(1)}} & -\dfrac{6EI}{L_{(1)}^2} & \dfrac{4EI}{L_{(1)}} \end{bmatrix} = EI \begin{bmatrix} 1.875 \times 10^{-4} & 3.75 \times 10^{-3} & -1.875 \times 10^{-4} & 3.75 \times 10^{-3} \\ 3.75 \times 10^{-3} & 0.1 & -3.75 \times 10^{-3} & 0.05 \\ -1.875 \times 10^{-4} & -3.75 \times 10^{-3} & 1.875 \times 10^{-4} & -3.75 \times 10^{-3} \\ 3.75 \times 10^{-3} & 0.05 & -3.75 \times 10^{-3} & 0.1 \end{bmatrix}$$

Element 2: With $L_{(2)} = 20$ m:

$$
\mathbf{k}_{(2)} = \begin{bmatrix} \dfrac{12EI}{L_{(2)}^3} & \dfrac{6EI}{L_{(2)}^2} & -\dfrac{12EI}{L_{(2)}^3} & \dfrac{6EI}{L_{(2)}^2} \\[2ex] \dfrac{6EI}{L_{(2)}^2} & \dfrac{4EI}{L_{(2)}} & -\dfrac{6EI}{L_{(2)}^2} & \dfrac{2EI}{L_{(2)}} \\[2ex] -\dfrac{12EI}{L_{(2)}^3} & -\dfrac{6EI}{L_{(2)}^2} & \dfrac{12EI}{L_{(2)}^3} & -\dfrac{6EI}{L_{(2)}^2} \\[2ex] \dfrac{6EI}{L_{(2)}^2} & \dfrac{2EI}{L_{(2)}} & -\dfrac{6EI}{L_{(2)}^2} & \dfrac{4EI}{L_{(2)}} \end{bmatrix} = EI \begin{bmatrix} 0.0015 & 0.015 & -0.0015 & 0.015 \\ 0.015 & 0.2 & -0.015 & 0.1 \\ -0.0015 & -0.015 & 0.0015 & -0.015 \\ 0.015 & 0.1 & -0.015 & 0.2 \end{bmatrix}
$$

Because the local and global coordinates coincide: $\mathbf{K}_{e(1)} = \mathbf{k}_{(1)}$ and $\mathbf{K}_{e(2)} = \mathbf{k}_{(2)}$.

The equivalent nodal loads $\mathbf{q}_M$ resulting from the uniformly distributed load applied to element 2 are calculated based on Figure 11.5. These correspond to the opposite of the reactions of the fixed-ended beam $\mathbf{q}_F$ with identical geometry, properties, and member loads as the element considered (as shown in Figure 11.14) and are collected in vector $\mathbf{q}_{F(2)}$:

$$\mathbf{q}_{F(2)} = [\ 150\ 500\ 150\ -500\ ]^T$$

The equivalent nodal loads for element 2, $\mathbf{q}_{M(2)}$, to be applied to the structure are therefore:

$$\mathbf{q}_{M(2)} = -\mathbf{q}_{F(2)} = [\ -150\ -500\ -150\ 500\ ]^T$$

(4) The structure stiffness matrix $\mathbf{K}$ is assembled by mapping the freedoms 1, 2, 3 and 4 of isolated element 1 to the global freedoms 3, 1, 4 and 2, and freedoms 1, 2, 3 and 4 of isolated element 2 to the global freedoms 4, 2, 5 and 6.

$$
\mathbf{K} = EI \begin{bmatrix} 0.1 & 0.05 & 3.75 \times 10^{-3} & -3.75 \times 10^{-3} & 0 & 0 \\ 0.05 & 0.3 & 3.75 \times 10^{-3} & 0.01125 & -0.015 & 0.1 \\ 3.75 \times 10^{-3} & 3.75 \times 10^{-3} & 1.875 \times 10^{-4} & -1.875 \times 10^{-4} & 0 & 0 \\ -3.75 \times 10^{-3} & 0.01125 & -1.875 \times 10^{-4} & 1.6875 \times 10^{-3} & -0.0015 & 0.015 \\ 0 & -0.015 & 0 & -0.0015 & 0.0015 & -0.015 \\ 0 & 0.1 & 0 & 0.015 & -0.015 & 0.2 \end{bmatrix}
$$

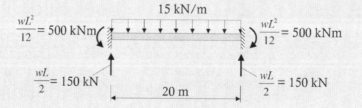

Figure 11.14 Reactions due to the member loads under fixed-end conditions.

(5) The loading vector **Q** includes the contribution of the anti-clockwise moment applied at node 1 and the equivalent nodal loads $\mathbf{q}_{M(2)}$, and is defined as:

$$\mathbf{Q} = \begin{bmatrix} 12 \\ -500 \\ Q_3 \\ Q_4 - 150 \\ Q_5 - 150 \\ Q_6 + 500 \end{bmatrix}$$

Each row of vector **Q** includes all actions related to the corresponding freedom, as shown in Figure 11.15. For example, row 1 includes the anti-clockwise moment of 12 kNm, taken as positive because it is pointing in the same direction of freedom 1, and at freedom 4 at node 2, we have the vertical reaction Q_4 and the download nodal load of 150 kN included in row 4 of **Q**. The zero displacements related to the restrained freedoms are depicted in $\mathbf{D}_k = [0\ 0\ 0\ 0]^T$.
The stiffness relationship for the entire structure is:

$$\begin{bmatrix} 12 \\ -500 \\ \hline Q_3 \\ Q_4 - 150 \\ Q_5 - 150 \\ Q_6 + 500 \end{bmatrix} = EI \begin{bmatrix} 0.1 & 0.05 & 3.75 \times 10^{-3} & -3.75 \times 10^{-3} & 0 & 0 \\ 0.05 & 0.3 & 3.75 \times 10^{-3} & 0.01125 & -0.015 & 0.1 \\ 3.75 \times 10^{-3} & 3.75 \times 10^{-3} & 1.875 \times 10^{-4} & -1.875 \times 10^{-4} & 0 & 0 \\ -3.75 \times 10^{-3} & 0.01125 & -1.875 \times 10^{-4} & 1.6875 \times 10^{-3} & -0.0015 & 0.015 \\ 0 & -0.015 & 0 & -0.0015 & 0.0015 & -0.015 \\ 0 & 0.1 & 0 & 0.015 & -0.015 & 0.2 \end{bmatrix} \begin{bmatrix} D_1 \\ D_2 \\ \hline 0 \\ 0 \\ 0 \\ 0 \end{bmatrix}$$

(11.18)

(6) The first two equations of Equation 11.18 are shown below and used to calculate D_1 and D_2.

$$12 = EI0.1D_1 + EI0.05D_2 \tag{11.19a}$$

$$-500 = EI0.05D_1 + EI0.3D_2 \tag{11.19b}$$

and solving gives:

$$D_1 = \frac{1040}{EI}\text{rad} \quad \text{and} \quad D_2 = -\frac{1840}{EI}\text{rad}$$

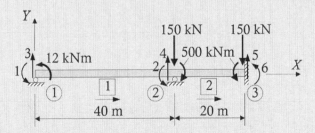

Figure 11.15 Freedoms and nodal loads.

(7) The unknown reactions are determined by solving the last four relationships in Equation 11.18:

$$Q_3 = EI(3.75 \times 10^{-3} D_1 + 3.75 \times 10^{-3} D_2) = -3 \text{ kN}$$

$$Q_4 = EI(-3.75 \times 10^{-3} D_1 + 0.01125 D_2) + 150 = 125.4 \text{ kN}$$

$$Q_5 = EI(-0.015 D_2) + 150 = 177.6 \text{ kN}$$

$$Q_6 = EI0.1 D_2 - 500 = -684 \text{ kNm}$$

These reactions are shown in the free-body diagram of Figure 11.16.

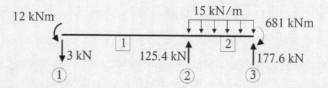

Figure 11.16 Free-body diagram for the two-span beam.

The nodal actions of elements 1 and 2 are calculated with Equation 11.14 (or by considering equilibrium of the free-body diagram of each element taken from Figure 11.16). Recalling that, for element 1, the isolated freedoms 1 to 4 are mapped against the global freedoms 3, 1, 4 and 2:

$$\mathbf{q}_{(1)} = \mathbf{kd}_{(1)} = EI \begin{bmatrix} 1.875 \times 10^{-4} & 3.75 \times 10^{-3} & -1.875 \times 10^{-4} & 3.75 \times 10^{-3} \\ 3.75 \times 10^{-3} & 0.1 & -3.75 \times 10^{-3} & 0.05 \\ -1.875 \times 10^{-4} & -3.75 \times 10^{-3} & 1.875 \times 10^{-4} & -3.75 \times 10^{-3} \\ 3.75 \times 10^{-3} & 0.05 & -3.75 \times 10^{-3} & 0.1 \end{bmatrix} \begin{bmatrix} 0 \\ 1040/EI \\ 0 \\ -1840/EI \end{bmatrix} = \begin{bmatrix} -3 \\ 12 \\ 3 \\ -132 \end{bmatrix}$$

which are shown in the free-body diagram of Figure 11.17.

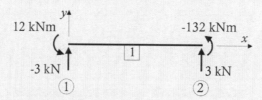

Figure 11.17 Free-body diagram for element 1.

Considering that element 2 is subjected to a member load, the calculation of $\mathbf{q}_{(2)}$ needs to account for the equivalent nodal actions equal and opposite to the actions included in $\mathbf{q}_{F(2)}$. For element 2, the isolated member freedoms 1 to 4 are mapped to the global freedoms 4, 2, 5 and 6.

$$\mathbf{q}_{(2)} = \mathbf{kd}_{(2)} + \mathbf{q}_{F(2)}$$

$$= EI \begin{bmatrix} 0.0015 & 0.015 & -0.0015 & 0.015 \\ 0.015 & 0.2 & -0.015 & 0.1 \\ -0.0015 & -0.015 & 0.0015 & -0.015 \\ 0.015 & 0.1 & -0.015 & 0.2 \end{bmatrix} \begin{bmatrix} 0 \\ -1840/EI \\ 0 \\ 0 \end{bmatrix} + \begin{bmatrix} 150 \\ 500 \\ 150 \\ -500 \end{bmatrix} = \begin{bmatrix} 122.4 \\ 132 \\ 177.6 \\ -684 \end{bmatrix}$$

These results are shown in Figure 11.18a. The contributions provided by the vector $\mathbf{kd}_{e(2)}$ are shown in Figure 11.18b and those coming from the equivalent uniformly loaded fixed-end element are shown in Figure 11.18c.

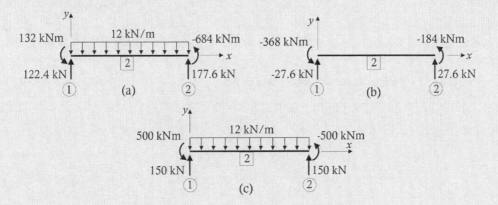

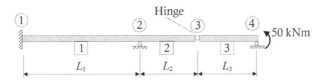

Figure 11.18 Free-body diagrams for element 2. (a) $\mathbf{q}_{(2)}$. (b) $\mathbf{kd}_{e(2)}$. (c) $\mathbf{q}_{F(2)}$.

REFLECTION ACTIVITY 11.3

Explain how you would account for the presence of the hinge at node 3 of the beam shown in Figure 11.19 when analysing the structure with the stiffness method.

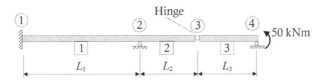

Figure 11.19 Beam for Reflection Activity 11.3.

The presence of the hinge enables the ends of elements 2 and 3 to rotate relative to each other. This can be easily handled with the stiffness method by introducing two different rotational freedoms at node 3, one for the right end rotation of element 2 and one for the left end rotation of element 3. This numbering is outlined in Figure 11.20, in which freedoms 1 to 5 are unrestrained and freedoms 6 to 9 are restrained.

In this manner, the isolated freedoms 1, 2, 3 and 4 of element 1 are mapped to the structure freedoms 8, 9, 6 and 1. Those of element 2 are mapped to the structure freedoms 6, 1, 2 and 3 and those of element 3 are mapped to freedoms 2, 4, 7 and 5.

The vectors of unknown and known displacements are then written as:

$$\mathbf{D}_u = [D_1 \ D_2 \ D_3 \ D_4 \ D_5]^\mathsf{T} \qquad \mathbf{D}_k = [0 \ 0 \ 0 \ 0]^\mathsf{T}$$

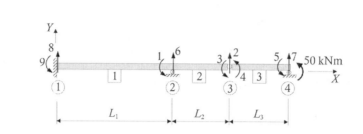

Figure 11.20 Freedom arrangements for Reflection Activity 11.3.

PROBLEMS

11.1   Evaluate the stiffness matrix of the structure shown. Assume $EI = 60 \times 10^3$ kNm².

11.2   Reconsider the beam of Problem 11.1 and calculate nodal displacements and support reactions using the stiffness method. Adopt $EI = 60 \times 10^3$ kNm².

11.3   Calculate the support reactions for the structure shown using the stiffness method. Assume $EI = 70 \times 10^3$ kNm².

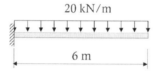

11.4   Reconsider the beam of Problem 11.3 and draw the free-body diagram of the stiffness element showing all actions applied to it. Adopt $EI = 70 \times 10^3$ kNm².

11.5   Consider the beam shown. Calculate the nodal displacements and support reactions using the stiffness method, and sketch the free-body diagram of element 2. Assume $EI = 50 \times 10^3$ kNm².

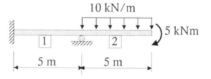

11.6   Evaluate the stiffness matrix of the structure shown. Assume EI to be constant.

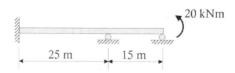

11.7   For the beam of Problem 11.6, calculate nodal displacements and support reactions using the stiffness method. Adopt $EI = 30 \times 10^3$ kNm².

11.8   Determine the support reactions for the structure shown using the stiffness method. Assume $EI = 40 \times 10^3$ kNm².

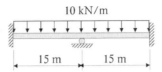

11.9   Reconsider the beam of Problem 11.8 and draw the free-body diagrams of the two elements showing all actions applied to them. Adopt $EI = 40 \times 10^3$ kNm².

11.10   For the beam shown, calculate the nodal displacements and support reactions using the stiffness method, and draw the free-body diagrams of elements 1 and 2. Assume $EI = 50 \times 10^3$ kNm².

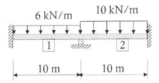

11.11   Consider the beam shown. Using the stiffness method, calculate the nodal displacements and reactions assuming node 2 settles (i.e. moves downwards) by 15 mm. Assume $EI = 48 \times 10^3$ kNm².

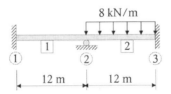

11.12   Sketch the free-body diagram of element 2 showing all actions applied to it for the beam analysed in Problem 11.11. Adopt $EI = 48 \times 10^3$ kNm².

11.13   The internal support of the beam shown settles (i.e. moves downward) by 10 mm. Calculate the nodal displacements and the reactions at the supports using the stiffness method. Assume $EI = 48 \times 10^3$ kNm².

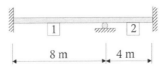

11.14   Draw the free-body diagrams of element 2 showing all actions applied to it for the beam of Problem 11.13. Adopt $EI = 48 \times 10^3$ kNm².

11.15   For the beam shown, determine the reactions at the supports and draw the free-body diagram of element 1. Use the stiffness method. Assume $EI = 60 \times 10^3$ kNm².

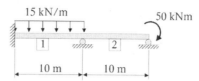

11.16   For the beam shown, calculate the nodal displacements and reactions at the supports using the stiffness method. Assume $EI = 72 \times 10^3$ kNm².

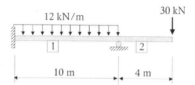

11.17   Sketch the free-body diagrams of elements 1 and 2 showing all actions applied to it for the beam of Problem 11.16.

11.18   Calculate the nodal displacements and support reactions for the structure shown using the stiffness method. Assume $EI = 60 \times 10^3$ kNm².

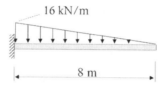

11.19   Plot the free-body diagram of element 1 highlighting all actions applied to it for the beam of Problem 11.18.

11.20   Determine the nodal displacements and support reactions for the structure illustrated below. Use the stiffness method and assume $EI = 70 \times 10^3$ kNm².

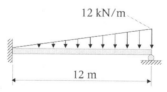

11.21   Reconsider the beam of Problem 11.20 and draw the free-body diagram of element 1 showing all actions applied to it.

11.22   For the beam shown, evaluate using the stiffness method the nodal displacements and support reactions, and sketch the free-body diagram of element 1. Assume $EI = 30 \times 10^3$ kNm².

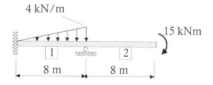

11.23 Consider the beam shown. Calculate the nodal displacements assuming node 3 settles (i.e. moves downwards) by 20 mm. Use the stiffness method and assume $EI = 80 \times 10^3$ kNm^2.

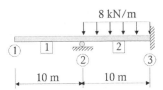

11.24 Reconsider the beam of Problem 11.23 and draw the free-body diagram of element 2 showing all actions applied to it. Adopt $EI = 80 \times 10^3$ kNm^2.

11.25 Using the stiffness method, calculate the nodal displacements and sketch the free-body diagram of element 2 illustrated below. Assume $EI = 50 \times 10^3$ kNm^2.

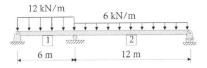

11.26 For the beam shown, determine the nodal displacements and reactions using the stiffness method, and plot the free-body diagrams of elements 1 and 2 showing all actions applied to them. Adopt $EI = 100 \times 10^3$ kNm^2.

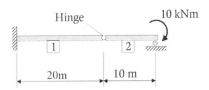

11.27 Calculate the nodal displacements and reactions for the beam illustrated below, with a hinge at node 3. Use the stiffness method and assume $EI = 60 \times 10^3$ kNm^2.

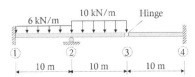

11.28 Evaluate the nodal displacements and reactions for the beam shown, with hinges at nodes 2 and 3. Use the stiffness method and assume $EI = 60 \times 10^3$ kNm^2.

11.29 For the beam of Problem 11.28, draw the free-body diagram of element 2 showing all actions applied to it.

11.30 Determine the reactions at the supports for the beam illustrated below and draw the free-body diagram of element 2. Use the stiffness method and assume $EI = 60 \times 10^3$ kNm^2.

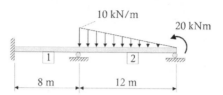

11.31 For the beam shown, determine reactions and sketch the free-body diagram of element 1. Use the stiffness method and assume $EI = 50 \times 10^3$ kNm^2.

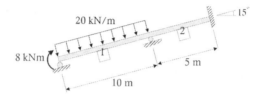

Chapter 12

Frame analysis using the stiffness method

12.1 THE FRAME ELEMENT

This chapter presents the *frame element* that is often used for the analysis of two-dimensional structures with the stiffness method. This element combines the characteristics of the truss and beam elements presented in previous chapters and is capable of resisting applied actions that induce axial force, shear force and bending moment. The development of the frame element is based on the assumptions of beam theory introduced in Chapter 5. The element is located in the x–y plane with the local x-axis in the direction of the member axis and with a cross-section that has an axis of symmetry in the direction of the local y-axis. Forces are applied in the x–y plane and couples are applied about an axis perpendicular to this plane (i.e. in the direction of the z-axis). Despite these limiting assumptions, the frame element is widely used in the analysis of many real structures.

The nodal freedoms of the element are illustrated in Figure 12.1, with displacements consisting of an *axial displacement* (in the direction of the x-axis), a *transverse displacement* (in the direction of the y-axis) and a *rotation* (about the z-axis) at the two ends of the element (i.e. six freedoms per element). These are shown in Figures 12.1a and b expressed in both the local coordinates (x,y,z) and the global coordinates (X,Y,Z). The corresponding nodal actions, which include the axial force, shear force and moment at the two nodes, are illustrated in Figures 12.1c and d considering both local and global coordinates.

When using the frame element, the *discretisation* of the structure needs to be carried out by inserting nodes at support locations and at free member ends. As for the beam element in Chapter 11, it is also preferable, even if not strictly necessary, to specify nodes where point loads are applied.

A stiffness relationship is written for each isolated frame element, based on the freedoms of Figure 12.1. In the analysis of a structure, the contribution of its members to the structural stiffness is combined together in the assembling procedure. The solution is then obtained in terms of displacements, from which all other variables describing the structural response are determined in the post-processing stage.

12.2 DERIVATION OF THE ELEMENT STIFFNESS MATRIX

The relationship between nodal actions and nodal displacements for the frame element of Figure 12.1 in local coordinates is as follows:

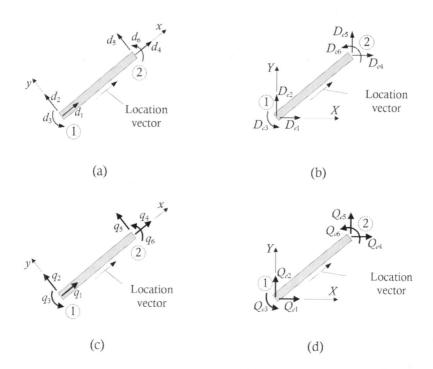

Figure 12.1 Nodal displacements and actions of the frame element. (a) Nodal displacements in local coordinates. (b) Nodal displacements in global coordinates. (c) Nodal actions in local coordinates. (d) Nodal actions in global coordinates.

$$
\begin{bmatrix} q_1 \\ q_2 \\ q_3 \\ q_4 \\ q_5 \\ q_6 \end{bmatrix} =
\begin{bmatrix}
\dfrac{EA}{L} & 0 & 0 & -\dfrac{EA}{L} & 0 & 0 \\[2mm]
0 & \dfrac{12EI}{L^3} & \dfrac{6EI}{L^2} & 0 & \dfrac{-12EI}{L^3} & \dfrac{6EI}{L^2} \\[2mm]
0 & \dfrac{6EI}{L^2} & \dfrac{4EI}{L} & 0 & -\dfrac{6EI}{L^2} & \dfrac{2EI}{L} \\[2mm]
-\dfrac{EA}{L} & 0 & 0 & \dfrac{EA}{L} & 0 & 0 \\[2mm]
0 & -\dfrac{12EI}{L^3} & -\dfrac{6EI}{L^2} & 0 & \dfrac{12EI}{L^3} & -\dfrac{6EI}{L^2} \\[2mm]
0 & \dfrac{6EI}{L^2} & \dfrac{2EI}{L} & 0 & -\dfrac{6EI}{L^2} & \dfrac{4EI}{L}
\end{bmatrix}
\begin{bmatrix} d_1 \\ d_2 \\ d_3 \\ d_4 \\ d_5 \\ d_6 \end{bmatrix}
\tag{12.1a}
$$

or in compact matrix form as:

$$
\mathbf{q} = \mathbf{kd} \tag{12.1b}
$$

where:

$$
\mathbf{q} = \begin{bmatrix} q_1 \\ q_2 \\ q_3 \\ q_4 \\ q_5 \\ q_6 \end{bmatrix} \qquad
\mathbf{k} = \begin{bmatrix}
\dfrac{EA}{L} & 0 & 0 & -\dfrac{EA}{L} & 0 & 0 \\[2ex]
0 & \dfrac{12EI}{L^3} & \dfrac{6EI}{L^2} & 0 & \dfrac{-12EI}{L^3} & \dfrac{6EI}{L^2} \\[2ex]
0 & \dfrac{6EI}{L^2} & \dfrac{4EI}{L} & 0 & -\dfrac{6EI}{L^2} & \dfrac{2EI}{L} \\[2ex]
-\dfrac{EA}{L} & 0 & 0 & \dfrac{EA}{L} & 0 & 0 \\[2ex]
0 & -\dfrac{12EI}{L^3} & -\dfrac{6EI}{L^2} & 0 & \dfrac{12EI}{L^3} & -\dfrac{6EI}{L^2} \\[2ex]
0 & \dfrac{6EI}{L^2} & \dfrac{2EI}{L} & 0 & -\dfrac{6EI}{L^2} & \dfrac{4EI}{L}
\end{bmatrix} \qquad
\mathbf{d} = \begin{bmatrix} d_1 \\ d_2 \\ d_3 \\ d_4 \\ d_5 \\ d_6 \end{bmatrix} \qquad (12.2\text{a--c})
$$

Nodal actions and displacements are assumed to be positive in the positive direction of the nodal freedoms. It can be seen that the 6×6 element stiffness matrix for a frame element (Equation 12.2b) can be obtained by appropriately combining the 2×2 stiffness matrix of the truss element (Equation 10.9b) and the 4×4 stiffness matrix of the beam element (Equation 11.4).

The derivation of the *member stiffness influence coefficient* k_{ij} (with $i,j = 1,...,6$) in the stiffness matrix $\mathbf{k}$ can be carried out by recalling that each k_{ij} can be interpreted, considering a free-body diagram of the frame element, to represent the reaction produced along freedom i by a unit displacement enforced along freedom j, as shown in Figure 12.2. This concept is at the basis of the *direct stiffness method* (introduced in Chapter 11), where the coefficients k_{ij} related to the j-th column of $\mathbf{k}$ are determined as the nodal actions required to enforce

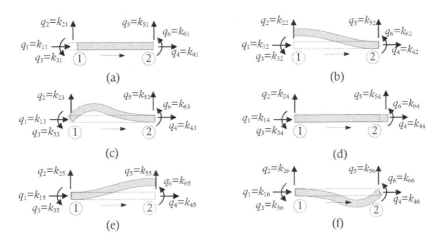

Figure 12.2 Physical representation of the member stiffness influence coefficients. (a) $d_1 = 1$ (with $d_2 = d_3 = d_4 = d_5 = d_6 = 0$). (b) $d_2 = 1$ (with $d_1 = d_3 = d_4 = d_5 = d_6 = 0$). (c) $d_3 = 1$ (with $d_1 = d_2 = d_4 = d_5 = d_6 = 0$). (d) $d_4 = 1$ (with $d_1 = d_2 = d_3 = d_5 = d_6 = 0$). (e) $d_5 = 1$ (with $d_1 = d_2 = d_3 = d_4 = d_6 = 0$). (f) $d_6 = 1$ (with $d_1 = d_2 = d_3 = d_4 = d_5 = 0$).

a unit displacement along freedom j while maintaining all other freedoms restrained. This relationship is useful and could be applied to the evaluation of the member stiffness influence coefficients, as long as we are able to calculate these nodal actions. We have already determined these actions in Worked Example 5.6, where we calculated the nodal actions (or reactions of the restrained member) related to a number of displaced configurations.

For example, the first column of $\mathbf{k}$ can be evaluated by assuming all freedoms to be restrained and enforcing a unit displacement along freedom 1, i.e. $\mathbf{d} = [1\ 0\ 0\ 0\ 0\ 0]^{\mathrm{T}}$. The required terms k_{i1} (with $i = 1,...,6$) are then equal to the nodal actions $\mathbf{q}$ required to keep this displaced shape, which is $\mathbf{q} = [AE/L\ 0\ 0\ -AE/L\ 0\ 0]^{\mathrm{T}}$ (see results of Worked Example 5.6). The terms included in the remaining columns of $\mathbf{k}$ can be obtained in a similar manner (see Reflection Activity 11.1 for more details).

12.3 TRANSFORMATION BETWEEN LOCAL AND GLOBAL COORDINATE SYSTEMS

12.3.1 Transformation matrix for vectors

The frame element possesses three freedoms at each node (two displacements and one rotation). Let us consider a vector with components in the local and global coordinate systems given by $\mathbf{r} = [r_1\ r_2\ r_3]^{\mathrm{T}}$ and $\mathbf{R} = [R_1\ R_2\ R_3]^{\mathrm{T}}$, respectively. The origins of the local and global reference systems are assumed to coincide. Under these assumptions, the expressions relating the local and global vector components are:

$$
\begin{bmatrix} r_1 \\ r_2 \\ r_3 \end{bmatrix} = \begin{bmatrix} l & m & 0 \\ -m & l & 0 \\ 0 & 0 & 1 \end{bmatrix} \begin{bmatrix} R_1 \\ R_2 \\ R_3 \end{bmatrix}
\qquad
\begin{bmatrix} R_1 \\ R_2 \\ R_3 \end{bmatrix} = \begin{bmatrix} l & -m & 0 \\ m & l & 0 \\ 0 & 0 & 1 \end{bmatrix} \begin{bmatrix} r_1 \\ r_2 \\ r_3 \end{bmatrix}
\tag{12.3a,b}
$$

or in more compact form as:

$$
\mathbf{r} = \mathbf{H}\mathbf{R} \qquad \mathbf{R} = \mathbf{H}^{\mathrm{T}}\mathbf{r}
\tag{12.4a,b}
$$

where:

$$
\mathbf{r} = \begin{bmatrix} r_1 \\ r_2 \\ r_3 \end{bmatrix}
\qquad
\mathbf{H} = \begin{bmatrix} l & m & 0 \\ -m & l & 0 \\ 0 & 0 & 1 \end{bmatrix}
\qquad
\mathbf{R} = \begin{bmatrix} R_1 \\ R_2 \\ R_3 \end{bmatrix}
\tag{12.5a–c}
$$

where l and m are the direction cosines related to α and β, respectively, representing the relative angles between the local and global coordinate systems, as shown in Figure 12.3, and are calculated using:

$$
l = \cos \alpha \qquad m = \cos \beta = \sin \alpha
\tag{12.6a,b}
$$

The relationships between local and global components specified in Equations 12.3 and 12.4 are obtained in terms of $\mathbf{H}$ following the procedure adopted in Chapter 10 for the derivation of the local and global representations of the truss element (see Section 10.4 and Reflection Activity 10.3 for more details). The only difference from the case of the truss element is that the vector now contains a rotation component, expressed by r_3 and R_3 in local and global coordinates, respectively. On the basis of the fact that the local and global

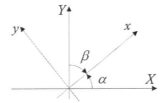

Figure 12.3 Angles relating local and global coordinate axes.

coordinate systems share the same origin, these rotations always coincide, i.e. $r_3 = R_3$, so that their representation does not change by varying the orientation of the reference system.

12.3.2 Transformation matrix for the frame element

In this section, we will develop a relationship between local and global components of the nodal displacements for the frame element. The vector of nodal displacements at node 1 expressed in local coordinates (Figure 12.1a) is $\mathbf{d}_1$ ($= [d_1 \, d_2 \, d_3]^T$), and the vector related to node 2 is $\mathbf{d}_2$ ($= [d_4 \, d_5 \, d_6]^T$). The corresponding global components of the displacements taking place at nodes 1 and 2 are $\mathbf{D}_{e1} = [D_{e1} \, D_{e2} \, D_{e3}]^T$ and $\mathbf{D}_{e2} = [D_{e4} \, D_{e5} \, D_{e6}]^T$, respectively (Figure 12.1b).

Applying the relationships of Equations 12.3a and 12.4a to the sets of local and global components of the displacements for nodes 1 and 2, we obtain the following relationships:

$$\begin{bmatrix} d_1 \\ d_2 \\ d_3 \end{bmatrix} = \begin{bmatrix} l & m & 0 \\ -m & l & 0 \\ 0 & 0 & 1 \end{bmatrix} \begin{bmatrix} D_{e1} \\ D_{e2} \\ D_{e3} \end{bmatrix} \qquad \begin{bmatrix} d_4 \\ d_5 \\ d_6 \end{bmatrix} = \begin{bmatrix} l & m & 0 \\ -m & l & 0 \\ 0 & 0 & 1 \end{bmatrix} \begin{bmatrix} D_{e4} \\ D_{e5} \\ D_{e6} \end{bmatrix} \qquad (12.7a,b)$$

or $\qquad \mathbf{d}_1 = \mathbf{H} \mathbf{D}_{e1} \qquad \mathbf{d}_2 = \mathbf{H} \mathbf{D}_{e2}$ $\qquad\qquad\qquad$ (12.8a,b)

These relationships are now combined in one expression as follows:

$$\begin{bmatrix} d_1 \\ d_2 \\ d_3 \\ d_4 \\ d_5 \\ d_6 \end{bmatrix} = \begin{bmatrix} l & m & 0 & 0 & 0 & 0 \\ -m & l & 0 & 0 & 0 & 0 \\ 0 & 0 & 1 & 0 & 0 & 0 \\ 0 & 0 & 0 & l & m & 0 \\ 0 & 0 & 0 & -m & l & 0 \\ 0 & 0 & 0 & 0 & 0 & 1 \end{bmatrix} \begin{bmatrix} D_{e1} \\ D_{e2} \\ D_{e3} \\ D_{e4} \\ D_{e5} \\ D_{e6} \end{bmatrix} \quad \text{or} \quad \mathbf{d} = \mathbf{T} \mathbf{D}_e \qquad (12.9a,b)$$

where:

$$\mathbf{d} = \begin{bmatrix} d_1 \\ d_2 \\ d_3 \\ d_4 \\ d_5 \\ d_6 \end{bmatrix} \qquad \mathbf{T} = \begin{bmatrix} l & m & 0 & 0 & 0 & 0 \\ -m & l & 0 & 0 & 0 & 0 \\ 0 & 0 & 1 & 0 & 0 & 0 \\ 0 & 0 & 0 & l & m & 0 \\ 0 & 0 & 0 & -m & l & 0 \\ 0 & 0 & 0 & 0 & 0 & 1 \end{bmatrix} \qquad \mathbf{D}_e = \begin{bmatrix} D_{e1} \\ D_{e2} \\ D_{e3} \\ D_{e4} \\ D_{e5} \\ D_{e6} \end{bmatrix} \qquad (12.10a\text{--}c)$$

Equations 12.9 allow us to determine the local components of $\mathbf{d}$ once $\mathbf{D}_e$ is known. For this reason, $\mathbf{T}$ is referred to as the *transformation matrix* for the frame element from global to local coordinates.

The transformation required to convert the local components of the displacements $\mathbf{d}$ to the global components $\mathbf{D}_e$ is performed by applying Equations 12.3b and 12.4b:

$$
\begin{bmatrix} D_{e1} \\ D_{e2} \\ D_{e3} \\ D_{e4} \\ D_{e5} \\ D_{e6} \end{bmatrix} =
\begin{bmatrix}
l & -m & 0 & 0 & 0 & 0 \\
m & l & 0 & 0 & 0 & 0 \\
0 & 0 & 1 & 0 & 0 & 0 \\
0 & 0 & 0 & l & -m & 0 \\
0 & 0 & 0 & m & l & 0 \\
0 & 0 & 0 & 0 & 0 & 1
\end{bmatrix}
\begin{bmatrix} d_1 \\ d_2 \\ d_3 \\ d_4 \\ d_5 \\ d_6 \end{bmatrix}
\quad \text{or} \quad \mathbf{D}_e = \mathbf{T}^\mathsf{T} \mathbf{d}
\tag{12.11a,b}
$$

where $\mathbf{T}^\mathsf{T}$ is referred to as the *transformation matrix* for the frame element from local to global coordinates.

Relationships similar to Equations 12.9 and 12.11 are also applicable to relate the local and global components of the nodal actions. That is:

$$
\begin{bmatrix} q_1 \\ q_2 \\ q_3 \\ q_4 \\ q_5 \\ q_6 \end{bmatrix} =
\begin{bmatrix}
l & m & 0 & 0 & 0 & 0 \\
-m & l & 0 & 0 & 0 & 0 \\
0 & 0 & 1 & 0 & 0 & 0 \\
0 & 0 & 0 & l & m & 0 \\
0 & 0 & 0 & -m & l & 0 \\
0 & 0 & 0 & 0 & 0 & 1
\end{bmatrix}
\begin{bmatrix} Q_{e1} \\ Q_{e2} \\ Q_{e3} \\ Q_{e4} \\ Q_{e5} \\ Q_{e6} \end{bmatrix}
\quad \text{or} \quad \mathbf{q} = \mathbf{T} \mathbf{Q}_e
\tag{12.12a,b}
$$

and

$$
\begin{bmatrix} Q_{e1} \\ Q_{e2} \\ Q_{e3} \\ Q_{e4} \\ Q_{e5} \\ Q_{e6} \end{bmatrix} =
\begin{bmatrix}
l & -m & 0 & 0 & 0 & 0 \\
m & l & 0 & 0 & 0 & 0 \\
0 & 0 & 1 & 0 & 0 & 0 \\
0 & 0 & 0 & l & -m & 0 \\
0 & 0 & 0 & m & l & 0 \\
0 & 0 & 0 & 0 & 0 & 1
\end{bmatrix}
\begin{bmatrix} q_1 \\ q_2 \\ q_3 \\ q_4 \\ q_5 \\ q_6 \end{bmatrix}
\quad \text{or} \quad \mathbf{Q}_e = \mathbf{T}^\mathsf{T} \mathbf{q}
\tag{12.13a,b}
$$

where the nodal actions expressed in local and global coordinates are shown in Figure 12.1 and given by:

$$
\mathbf{q} = \begin{bmatrix} q_1 \\ q_2 \\ q_3 \\ q_4 \\ q_5 \\ q_6 \end{bmatrix}
\qquad
\mathbf{Q}_e = \begin{bmatrix} Q_{e1} \\ Q_{e2} \\ Q_{e3} \\ Q_{e4} \\ Q_{e5} \\ Q_{e6} \end{bmatrix}
\tag{12.13c,d}
$$

12.4 FRAME ELEMENT IN GLOBAL COORDINATES

The load–displacement relationship $\mathbf{q} = \mathbf{kd}$, introduced in Equation 12.1b to describe the response of the frame element in local coordinates, is now rewritten in terms of the global components of the nodal displacements and actions. This is carried out by substituting $\mathbf{d} = \mathbf{TD}_e$ (Equation 12.9b) and $\mathbf{Q}_e = \mathbf{T}^\mathrm{T}\mathbf{q}$ (Equation 12.13b) into Equation 12.1b, which produces the global load–displacement relationship defined as:

$$\mathbf{Q}_e = \mathbf{K}_e\mathbf{D}_e \tag{12.14}$$

and $\mathbf{K}_e$ is the element stiffness matrix calculated as follows:

$$\mathbf{K}_e = \mathbf{T}^\mathrm{T}\mathbf{kT} =$$

$$
\begin{array}{ccccccc}
 & 1 & 2 & 3 & 4 & 5 & 6 \\
\end{array}
$$

$$
=\begin{bmatrix}
\left(\dfrac{EA}{L}l^2+\dfrac{12EI}{L^3}m^2\right) & \left(\dfrac{EA}{L}-\dfrac{12EI}{L^3}\right)lm & -\dfrac{6EI}{L^2}m & -\left(\dfrac{EA}{L}l^2+\dfrac{12EI}{L^3}m^2\right) & -\left(\dfrac{EA}{L}-\dfrac{12EI}{L^3}\right)lm & -\dfrac{6EI}{L^2}m \\[3mm]
\left(\dfrac{EA}{L}-\dfrac{12EI}{L^3}\right)lm & \left(\dfrac{EA}{L}m^2+\dfrac{12EI}{L^3}l^2\right) & \dfrac{6EI}{L^2}l & -\left(\dfrac{EA}{L}-\dfrac{12EI}{L^3}\right)lm & -\left(\dfrac{EA}{L}m^2+\dfrac{12EI}{L^3}l^2\right) & \dfrac{6EI}{L^2}l \\[3mm]
-\dfrac{6EI}{L^2}m & \dfrac{6EI}{L^2}l & \dfrac{4EI}{L} & \dfrac{6EI}{L^2}m & -\dfrac{6EI}{L^2}l & \dfrac{2EI}{L} \\[3mm]
-\left(\dfrac{EA}{L}l^2+\dfrac{12EI}{L^3}m^2\right) & -\left(\dfrac{EA}{L}-\dfrac{12EI}{L^3}\right)lm & \dfrac{6EI}{L^2}m & \left(\dfrac{EA}{L}l^2+\dfrac{12EI}{L^3}m^2\right) & \left(\dfrac{EA}{L}-\dfrac{12EI}{L^3}\right)lm & \dfrac{6EI}{L^2}m \\[3mm]
-\left(\dfrac{EA}{L}-\dfrac{12EI}{L^3}\right)lm & -\left(\dfrac{EA}{L}m^2+\dfrac{12EI}{L^3}l^2\right) & -\dfrac{6EI}{L^2}l & \left(\dfrac{EA}{L}-\dfrac{12EI}{L^3}\right)lm & \left(\dfrac{EA}{L}m^2+\dfrac{12EI}{L^3}l^2\right) & -\dfrac{6EI}{L^2}l \\[3mm]
-\dfrac{6EI}{L^2}m & \dfrac{6EI}{L^2}l & \dfrac{2EI}{L} & \dfrac{6EI}{L^2}m & -\dfrac{6EI}{L^2}l & \dfrac{4EI}{L}
\end{bmatrix}
\begin{matrix}1\\[3mm]2\\[3mm]3\\[3mm]4\\[3mm]5\\[3mm]6\end{matrix}
$$

$$\tag{12.15}$$

This stiffness relationship is applicable to an isolated frame element whose local and global freedoms are numbered as shown in Figure 12.1. In particular, the i-th component of $\mathbf{Q}_e$ and $\mathbf{D}_e$ relates to the nodal actions and nodal displacements assigned along freedom i, respectively. Similarly, rows and columns of $\mathbf{K}_e$ are associated with the global freedoms of the isolated element as specified in Equation 12.15.

12.5 MEMBER LOADS

Member loads are included in the analysis using equivalent nodal loads, consisting of nodal actions capable of inducing an equivalent structural response to the one being analysed. The selection of the set of actions to be used was outlined in Section 11.5, where it was shown that the member loads are equivalent to a set of equivalent nodal loads equal and opposite to the reactions of a fixed-ended beam subjected to the same member loading and with the same geometry and properties as the loaded element. That is, the equivalent nodal loads $\mathbf{q}_\mathrm{M}$ (where the subscript 'M' stands for member load) are:

$$\mathbf{q}_\mathrm{M} = -\mathbf{q}_\mathrm{F} \tag{12.16}$$

where $\mathbf{q}_\mathrm{F}$ represents the support reactions of a fixed-ended beam (see Section 11.5 for more details). For a member carrying a uniformly distributed load w as shown in Figure 12.4:

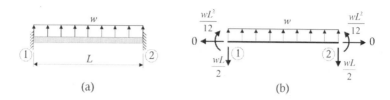

Figure 12.4 Support reactions for a uniformly loaded fixed-ended beam. (a) Support conditions of a fixed-ended member. (b) Free-body diagram.

$$\mathbf{q}_F = \left[\begin{array}{cccccc} 0 & -\dfrac{wL}{2} & -\dfrac{wL^2}{12} & 0 & -\dfrac{wL}{2} & \dfrac{wL^2}{12} \end{array}\right]^T \tag{12.17}$$

assuming w to be positive when pointing upward in accordance with the sign convention adopted throughout the book.

In the assembling process, it is necessary to convert the local actions included in $\mathbf{q}_F$ into global actions $\mathbf{Q}_{eF}$. This can be carried out either by treating each nodal action as an individual external action and, from trigonometry, evaluating its global components or by calculating the global components using the transformation matrix (Equations 12.13) with the direction cosines l and m determined based on the inclination of the loaded element.

REFLECTION ACTIVITY 12.1

Determine the expressions for the equivalent nodal loads to be used in the analysis of a frame to account for the situation where an element is subjected to a temperature gradient linearly varying over its cross-section and constant along its length (see Reflection Activity 5.2). The element is of uniform cross-section, with area A and second moment of area I, and with constant elastic modulus E and coefficient of thermal expansion α.

The equivalent nodal loads to be used in the stiffness method correspond to the support reactions of a fixed-ended beam subjected to the same temperature gradient. These can be calculated based on the differential equations and boundary conditions introduced in Chapter 5. For the thermal loading considered here (illustrated in Figure 12.5a), the reactions for a fixed-ended

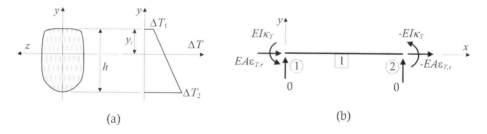

Figure 12.5 Fixed-ended beam reactions caused by temperature variations. (a) Cross-section and tempereature variation. (b) $\mathbf{q}_F$ collecting the support reactions of a fixed-ended team.

beam have already been evaluated in Worked Example 5.9 and are shown in Figure 12.5b. The vector of reactions of the fixed-ended beam is:

$$\mathbf{q}_F = [EA\varepsilon_{T,r} \quad 0 \quad EI\kappa_T \quad -EA\varepsilon_{T,r} \quad 0 \quad -EI\kappa_T]^T \tag{12.18a}$$

and (from Equations 5.45 and 5.46 reproduced here for ease of reference):

$$\varepsilon_T = \varepsilon_{T,r} - y\kappa_T \tag{5.45}$$

$$\varepsilon_{T,r} = \alpha\left[\Delta T_1 + (\Delta T_2 - \Delta T_1)\frac{y_t}{h}\right] \quad \kappa_T = \alpha\frac{\Delta T_2 - \Delta T_1}{h} \tag{5.46a,b}$$

where $\varepsilon_{T,r}$ and κ_T represent the thermal strain at the level of the reference axis and the thermal gradient, respectively, while ΔT_1 and ΔT_2 are the temperature changes at the top and bottom fibres of the cross-section, y_t depicts the distance from the centroid of the cross-section (because we are using a centroidal reference system) to the top fibre of the cross-section and h is the thickness of the cross-section.

Based on the above expressions and using Equation 12.16, the equivalent nodal loads $\mathbf{q}_M$ can be calculated as:

$$\mathbf{q}_M = -\mathbf{q}_F = [-EA\varepsilon_{T,r} \quad 0 \quad -EI\kappa_T \quad EA\varepsilon_{T,r} \quad 0 \quad EI\kappa_T]^T \tag{12.18b}$$

12.6 ASSEMBLING, SOLUTION AND POST-PROCESSING

The contribution of each frame element to the stiffness matrix of the structure $\mathbf{K}_e$ (from Equation 12.15) is combined in the assembling stage to form the stiffness relationship for the entire structure, written as:

$$\mathbf{Q} = \mathbf{KD} \tag{12.19}$$

where $\mathbf{Q}$ and $\mathbf{D}$ are actions and displacements specified along the structure freedoms, and $\mathbf{K}$ is the structure stiffness matrix. In this process, it is important to carefully relate the freedom numbers of the isolated element to those of the global freedoms of the structure.

Equation 12.19 can be rewritten to highlight the partitioning applied to $\mathbf{Q}$, $\mathbf{K}$ and $\mathbf{D}$ to separate freedoms related to the unrestrained and restrained freedoms, as:

$$\begin{bmatrix} \mathbf{Q}_k \\ \mathbf{Q}_u \end{bmatrix} = \begin{bmatrix} \mathbf{K}_{11} & \mathbf{K}_{12} \\ \mathbf{K}_{21} & \mathbf{K}_{22} \end{bmatrix} \begin{bmatrix} \mathbf{D}_u \\ \mathbf{D}_k \end{bmatrix} \tag{12.20a}$$

or

$$\mathbf{Q}_k = \mathbf{K}_{11}\mathbf{D}_u + \mathbf{K}_{12}\mathbf{D}_k \tag{12.20b}$$

$$\mathbf{Q}_u = \mathbf{K}_{21}\mathbf{D}_u + \mathbf{K}_{22}\mathbf{D}_k \tag{12.20c}$$

in which $\mathbf{D}_u$ and $\mathbf{D}_k$ represent unknown and known displacements, respectively, $\mathbf{Q}_k$ and $\mathbf{Q}_u$ describe the known (external) loads and unknown loads (support reactions), while matrix $\mathbf{K}$ is partitioned into $\mathbf{K}_{ij}$ (with $i,j = 1,2$).

Figure 12.6 Evaluation of internal actions along the length of a frame element. (a) Free-body diagram of a frame element subjected to member loading. (b) Cut along the element length to calculate the internal axial force N, shear force S and bending moment M.

The unknown displacements $\mathbf{D}_u$ can be obtained by solving the set of Equations 12.20b (see Appendix C for possible solution procedures for the system of equations). In matrix form:

$$\mathbf{D}_u = \mathbf{K}_{11}^{-1}(\mathbf{Q}_k - \mathbf{K}_{12}\mathbf{D}_k) \tag{12.21}$$

All other variables describing the structural response can be determined based on $\mathbf{D}_u$ in the post-processing phase. For example, $\mathbf{Q}_u$ is evaluated with Equation 12.20c and the nodal actions $\mathbf{q}$ applied to the free-body diagram of the element are calculated with:

$$\mathbf{q} = \mathbf{kd} + \mathbf{q}_F = \mathbf{kTD}_e + \mathbf{q}_F \tag{12.22}$$

where $\mathbf{D}_e$ depicts the nodal displacements of the particular element considered, and $\mathbf{q}_F$ represents the reactions of the element subjected to the member load and fixed at both its ends with the same properties and length of the actual element.

The distribution of the internal actions can be determined based on the nodal actions $\mathbf{q}$. This is carried out by performing a cut at the locations of interest and applying the equilibrium equations to the resulting free-body diagram to obtain the expressions for N, S and M along each element length. An example is provided in Figure 12.6, which shows the postprocessing of the internal actions for a frame element subjected to a uniformly distributed load w with a cut specified at a point C located at a distance x_C from node 1.

The internal actions at a particular location can also be evaluated directly from the vector $\mathbf{q}$ by inserting a node at the location of interest at the beginning of the solution process.

The main steps involved in the solution process are now summarised and illustrated by worked examples.

SUMMARY OF STEPS 12.1: Stiffness method — Solution procedure

The solution procedure for the analysis of a structure using the frame element with the stiffness method is based on the following steps:

1. Assign a global reference system. Select the number and locations of all nodes and members, and number them.

2. Specify three freedoms at each node, i.e. two displacements and one rotation (taken positive when anti-clockwise). In the presence of end releases specified at the element ends, such as hinges, additional freedoms need to be included to accommodate the different rotations of adjacent members. In the numbering of the freedoms, number the unrestrained freedoms first, followed by the restrained freedoms.

3. Determine the stiffness matrix $\mathbf{k}$ for each element (Equation 12.2b) and, if necessary, the set of equivalent nodal loads $\mathbf{q}_M$ (= $-\mathbf{q}_F$) for each member of the structure (Equation 12.17).

4. Assemble the stiffness matrix for the entire structure **K**.

5. Define, based on the boundary conditions of the problem, vectors of known displacements $\mathbf{D}_k$ and external loads $\mathbf{Q}_k$. Partition the stiffness matrix **K** into $\mathbf{K}_{11}$, $\mathbf{K}_{12}$, $\mathbf{K}_{21}$ and $\mathbf{K}_{22}$ (Equations 12.20). Also partition the vectors **D** and **Q** into $[\mathbf{D}_u\ \mathbf{D}_k]^T$ and $[\mathbf{Q}_k\ \mathbf{Q}_u]^T$, respectively. Write the governing system of equations $\mathbf{Q} = \mathbf{KD}$ (Equations 12.19 and 12.20a). For elements subjected to member loads, the contribution of their equivalent nodal actions $\mathbf{q}_M$ ($= -\mathbf{q}_F$) needs to be mapped in the corresponding locations of the loading vector **Q**.

6. Calculate the unknown displacements $\mathbf{D}_u$ with Equation 12.21.

7. Determine the unknown reactions $\mathbf{Q}_u$ (Equation 12.20c) and the end nodal actions **q** for each frame element forming the structure (Equation 12.22).

WORKED EXAMPLE 12.1

Calculate the nodal displacements and support reactions for the frame illustrated in Figure 12.7. Assume $A = 5000$ mm², $I = 240 \times 10^6$ mm⁴ and $E = 200$ GPa for all members of the frame.

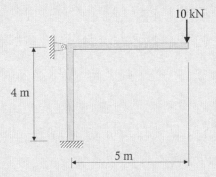

Figure 12.7   Frame for Worked Example 12.1.

The steps of the solution follow Summary of Steps 12.1.

(1) We adopt a global coordinate system and specify numbering for the nodes and members as shown below.

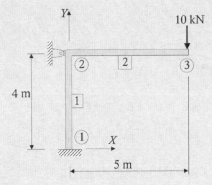

(2) Three freedoms are assigned at each node. These are numbered starting from the unrestrained ones (1 to 4), followed by the restrained ones (5 to 9). Location vectors are specified for each element to define the positive direction of the local x-axis, as shown.

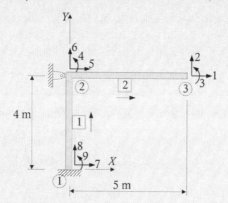

(3) The calculation of the stiffness matrices $\mathbf{K}_{e(1)}$ and $\mathbf{K}_{e(2)}$ requires the evaluation of the inclinations of the elements expressed in terms of the direction cosines l and m, evaluated based on the angles α and β.

Element 1

The determination of the direction cosines for element 1 is shown in Figure 12.8a, where $l = \cos \alpha = \cos 90° = 0$; $m = \sin \alpha = \sin 90° = 1$ (or $m = \cos \beta = \cos 0° = 1$). These could also be calculated from the coordinates of the nodes as:

$$l = \frac{x_2 - x_1}{L_{(1)}} = \frac{0-0}{4} = 0 \qquad m = \frac{y_2 - y_1}{L_{(1)}} = \frac{4-0}{4} = 1$$

where x_1 and y_1 represent the coordinates of the first node, while x_2 and y_2 are the coordinates of the second node (as defined by the location vector).

With $L_{(1)} = 4$ m, the relevant terms and stiffness coefficients for inclusion in $\mathbf{K}_{e(1)}$ are:

$$\frac{EA}{L_{(1)}} = 250 \times 10^3 \text{ kN/m} \qquad \frac{12EI}{L_{(1)}^3} = 9 \times 10^3 \text{ kN/m} \qquad \frac{6EI}{L_{(1)}^2} = 18 \times 10^3 \text{ kN}$$

$$\frac{4EI}{L_{(1)}} = 48 \times 10^3 \text{ kNm} \qquad \frac{2EI}{L_{(1)}} = 24 \times 10^3 \text{ kNm}$$

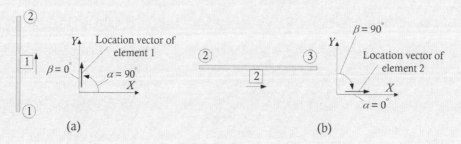

Figure 12.8 Loading vectors and orientations for elements 1 and 2. (a) Element 1. (b) Element 2.

Substituting these values into Equation 12.15 (and highlighting how freedoms of the structure relate to those of the isolated element $\mathbf{K}_{e(1)}$):

$$
\mathbf{K}_{e(1)} =
\begin{array}{cccccc}
7 & 8 & 9 & 5 & 6 & 4 \\
\left[\begin{array}{cccccc}
9\times10^3 & 0 & -18\times10^3 & -9\times10^3 & 0 & -18\times10^3 \\
0 & 250\times10^3 & 0 & 0 & -250\times10^3 & 0 \\
-18\times10^3 & 0 & 48\times10^3 & 18\times10^3 & 0 & 24\times10^3 \\
-9\times10^3 & 0 & 18\times10^3 & 9\times10^3 & 0 & 18\times10^3 \\
0 & -250\times10^3 & 0 & 0 & 250\times10^3 & 0 \\
-18\times10^3 & 0 & 24\times10^3 & 18\times10^3 & 0 & 48\times10^3
\end{array}\right] &
\begin{array}{c}
7 \\ 8 \\ 9 \\ 5 \\ 6 \\ 4
\end{array}
\end{array}
$$

Element 2

The direction cosines, length and stiffness coefficients for element 2 are $l = \cos \alpha = \cos 0° = 1$; $m = \sin \alpha = \sin 0° = 0$ (or $m = \cos \beta = \cos 90° = 0$) as shown in Figure 12.8b. These could also be calculated from the coordinates of the nodes:

$$
l = \frac{x_2 - x_1}{L_2} = \frac{5-0}{5} = 1 \qquad m = \frac{y_2 - y_1}{L_{(2)}} = \frac{0-0}{5} = 0
$$

With $L_{(2)} = 5$ m, $\dfrac{EA}{L_{(2)}} = 200\times10^3$ kN/m, $\dfrac{12EI}{L_{(2)}^3} = 4608$ kN/m, $\dfrac{6EI}{L_{(2)}^2} = 11{,}520$ kN, $\dfrac{4EI}{L_{(2)}} = 38.4\times10^3$ kNm,

and $\dfrac{2EI}{L_{(2)}} = 19.2\times10^3$ kN/m the element stiffness matrix $\mathbf{K}_{e(2)}$ is (Equation 12.15):

$$
\mathbf{K}_{e(2)} =
\begin{array}{cccccc}
5 & 6 & 4 & 1 & 2 & 3 \\
\left[\begin{array}{cccccc}
200\times10^3 & 0 & 0 & -200\times10^3 & 0 & 0 \\
0 & 4608 & 11{,}520 & 0 & -4608 & 11{,}520 \\
0 & 11{,}520 & 38.4\times10^3 & 0 & -11{,}520 & 19.2\times10^3 \\
-200\times10^3 & 0 & 0 & 200\times10^3 & 0 & 0 \\
0 & -4608 & -11{,}520 & 0 & 4608 & -11{,}520 \\
0 & 11{,}520 & 19.2\times10^3 & 0 & -11{,}520 & 38.4\times10^3
\end{array}\right] &
\begin{array}{c}
5 \\ 6 \\ 4 \\ 1 \\ 2 \\ 3
\end{array}
\end{array}
$$

(4) The structure stiffness matrix $\mathbf{K}$ is obtained by assembling the contribution of the two elements $\mathbf{K}_{e(1)}$ and $\mathbf{K}_{e(2)}$. For element 1, the freedoms of the isolated elements 1 to 6 are mapped against the structural freedoms 7, 8, 9, 5, 6 and 4, while the isolated freedoms 1 to 6 of element 2 are mapped against structural freedoms 5, 6, 4, 1, 2 and 3.

(5) Considering the restrained freedoms, the vector of known displacements $\mathbf{D}_k$ is defined as $\mathbf{D}_k = [0\ 0\ 0\ 0\ 0]^T$. The known term in the loading vector $\mathbf{Q}_k$ is the vertical load at node 1 (along freedom 1), i.e. $\mathbf{Q}_k = [0\ -10\ 0\ 0]^T$. The unknown displacements and nodal actions (reactions) are collected in vectors $\mathbf{D}_u$ and $\mathbf{Q}_u$: $\mathbf{D}_u = [D_1\ D_2\ D_3\ D_4]^T$ and $\mathbf{Q}_u = [Q_5\ Q_6\ Q_7\ Q_8\ Q_9]^T$.

After assembling, the stiffness relationship for the structure becomes:

$$
\begin{bmatrix} 0 \\ -10 \\ 0 \\ 0 \\ \hline Q_5 \\ Q_6 \\ Q_7 \\ Q_8 \\ Q_9 \end{bmatrix} =
\left[\begin{array}{cccc|ccccc}
200\times10^3 & 0 & 0 & 0 & -200\times10^3 & 0 & 0 & 0 & 0 \\
0 & 4608 & -11{,}520 & -11{,}520 & 0 & -4608 & 0 & 0 & 0 \\
0 & -11{,}520 & 38{,}400 & 19{,}200 & 0 & 11{,}520 & 0 & 0 & 0 \\
0 & -11{,}520 & 19{,}200 & 86{,}400 & 18\times10^3 & 11{,}520 & -18\times10^3 & 0 & 24\times10^3 \\
\hline
-200\times10^3 & 0 & 0 & 18{,}000 & 209\times10^3 & 0 & -9\times10^3 & 0 & 18\times10^3 \\
0 & -4608 & 11{,}520 & 11{,}520 & 0 & 254{,}608 & 0 & -250\times10^3 & 0 \\
0 & 0 & 0 & -18{,}000 & -9\times10^3 & 0 & 9\times10^3 & 0 & -18\times10^3 \\
0 & 0 & 0 & 0 & 0 & -250\times10^3 & 0 & 250\times10^3 & 0 \\
0 & 0 & 0 & 24\times10^3 & 18\times10^3 & 0 & -18\times10^3 & 0 & 48\times10^3
\end{array}\right]
\begin{bmatrix} D_1 \\ D_2 \\ D_3 \\ D_4 \\ \hline 0 \\ 0 \\ 0 \\ 0 \\ 0 \end{bmatrix}
\begin{matrix} 1 \\ 2 \\ 3 \\ 4 \\ 5 \\ 6 \\ 7 \\ 8 \\ 9 \end{matrix}
$$

$$(12.23)$$

(6) The unknown displacements $\mathbf{D}_u$ are determined from the first four equations in this set of simultaneous equations (Equations 12.23):

$$0 = 200 \times 10^3 D_1 \tag{12.24a}$$

$$-10 = 4608 D_2 - 11{,}520 D_3 - 11{,}520 D_4 \tag{12.24b}$$

$$0 = -11{,}520 D_2 + 38{,}400 D_3 + 19{,}200 D_4 \tag{12.24c}$$

$$0 = -11{,}520 D_2 + 19{,}200 D_3 + 86{,}400 D_4 \tag{12.24d}$$

and solving gives:

$$D_1 = 0\text{ m}, \quad D_2 = -13.89 \times 10^{-3}\text{ m}, \quad D_3 = -3.645 \times 10^{-3}\text{ rad}, \quad D_4 = -1.042 \times 10^{-3}\text{ rad}.$$

(7) Once the unknown displacements are determined, it is possible to calculate the unknown reactions from the bottom five equations of Equation 12.23:

$$Q_5 = -200 \times 10^3 D_1 + 18 \times 10^3 D_4 = -18.75\text{ kN}$$

$$Q_6 = -4608 D_2 + 11{,}520 D_3 + 11{,}520 D_4 = 10\text{ kN}$$

$$Q_7 = -18 \times 10^3 D_4 = 18.75\text{ kN}$$

$$Q_8 = 0\text{ kN}$$

$$Q_9 = 24 \times 10^3 D_4 = -25\text{ kNm}$$

and these reactions are shown on the free-body diagram of the frame in Figure 12.9.

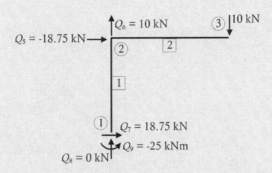

Figure 12.9 Free-body diagram with calculated reactions.

WORKED EXAMPLE 12.2

Determine the support reactions and the internal actions of the frame shown in Figure 12.10. Assume all members to have the following properties along their lengths: $A = 4000$ mm^2, $I = 160 \times 10^6$ mm^4 and $E = 200$ GPa.

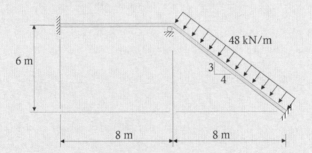

Figure 12.10 Frame for Worked Example 12.2.

(1 and 2) The global coordinate system and the numbering of nodes and members are shown below, together with the location vectors. The unrestrained freedoms (1 and 2) are numbered first, followed by the restrained ones (3–9).

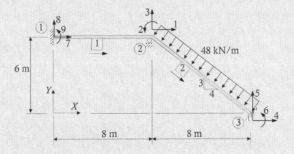

(3) Stiffness matrices for elements I and 2 (Equation 12.15):

$$
\mathbf{K}_{e(1)} =
\begin{bmatrix}
100\times10^3 & 0 & 0 & -100\times10^3 & 0 & 0 \\
0 & 750 & 3000 & 0 & -750 & 3000 \\
0 & 3000 & 16,000 & 0 & -3000 & 8000 \\
-100\times10^3 & 0 & 0 & 100\times10^3 & 0 & 0 \\
0 & -750 & -3000 & 0 & 750 & -3000 \\
0 & 3000 & 8000 & 0 & -3000 & 16,000
\end{bmatrix}
$$

where $L_{(1)} = 8$ m; $l_{(1)} = 1$, $m_{(1)} = 0$ ($\alpha = 0°$ and $\beta = 90°$); $EA/L_{(1)} = 100 \times 10^3$ kN/m; $12EI/L_{(1)}^3 = 750$ kN/m; $6EI/L_{(1)}^2 = 3000$ kN; $4EI/L_{(1)} = 16,000$ kNm; $2EI/L_{(1)} = 8000$ kNm; and

$$
\mathbf{K}_{e(2)} =
\begin{bmatrix}
51,338 & -38,215 & 1152 & -51,338 & 38,215 & 1152 \\
-38,215 & 29,045 & 1536 & 38,215 & -29,045 & 1536 \\
1152 & 1536 & 12,800 & -1152 & -1536 & 6400 \\
-51,338 & 38,215 & -1152 & 51,338 & -38,215 & -1152 \\
38,215 & -29,045 & -1536 & -38,125 & 29,045 & -1536 \\
1152 & 1536 & 6400 & -1152 & -1536 & 12,800
\end{bmatrix}
$$

where $L_{(2)} = 10$ m; $l_{(2)} = 0.8$; $m_{(2)} = -0.6$; $EA/L_{(2)}$ 80×10^3 kN/m; $12EI/L_{(2)}^3 = 384$ kN/m; $6EI/L_{(2)}^2 = 1920$ kN; $4EI/L_{(2)} = 12,800$ kNm; and $2EI/L_{(2)} = 6400$ kNm.

The equivalent nodal loads $\mathbf{q}_{M(2)}$ $(= -\mathbf{q}_{F(2)})$ in local coordinates related to the uniformly distributed load applied to element 2 are (Equation 12.17):

$$
\mathbf{q}_{M(2)} = \begin{bmatrix} 0 & \dfrac{wL}{2} & \dfrac{wL^2}{12} & 0 & \dfrac{wL}{2} & -\dfrac{wL^2}{12} \end{bmatrix}^{T} = \begin{bmatrix} 0 & -240 & -400 & 0 & -240 & 400 \end{bmatrix}^{T} \tag{12.25}
$$

Figure 12.11a highlights how the equivalent nodal loads $\mathbf{q}_{M(2)}$ are applied to the structure. The loading vector $\mathbf{Q}_{eM(2)}$ expressing the components of $\mathbf{q}_{M(2)}$ in global coordinates can be calculated from trigonometry based on the inclination of the equivalent nodal loads illustrated in Figure 12.11b:

$$
\mathbf{Q}_{eM(2)} = \begin{bmatrix} -144 & -192 & -400 & -144 & -192 & +400 \end{bmatrix}^{T}
$$

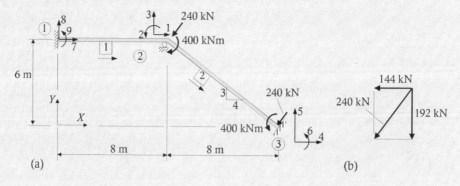

Figure 12.11 Structure subjected to equivalent nodal loads $\mathbf{q}_{M(2)}$.

The same loading vector could have been obtained by recalling that, based on Equations 12.13:

$$
\mathbf{Q}_{eM(2)} = \mathbf{T}_{(2)}^{T}\mathbf{q}_{M(2)} =
\begin{bmatrix}
0.8 & 0.6 & 0 & 0 & 0 & 0 \\
-0.6 & 0.8 & 0 & 0 & 0 & 0 \\
0 & 0 & 1 & 0 & 0 & 0 \\
0 & 0 & 0 & 0.8 & 0.6 & 0 \\
0 & 0 & 0 & -0.6 & 0.8 & 0 \\
0 & 0 & 0 & 0 & 0 & 1
\end{bmatrix}
\begin{bmatrix}
0 \\
-240 \\
-400 \\
0 \\
-240 \\
400
\end{bmatrix}
\tag{12.26}
$$

where $\mathbf{q}_{M(2)}$ is defined in Equation 12.25 and $\mathbf{T}_{(2)}$ represents the transformation matrix for element 2 calculated (with direction cosines of element 2: $l = 0.8$ and $m = -0.6$).

(4) Isolated freedoms (1–6) of elements 1 and 2 are mapped to the structure freedoms 7, 8, 9, 1, 3 and 2 and structure freedoms 1, 3, 2, 4, 5 and 6, respectively.

(5) The vectors of known actions and displacements are:

$$\mathbf{Q}_k = [-144\ -400]^T \quad \mathbf{D}_k = [0\ 0\ 0\ 0\ 0\ 0\ 0]^T$$

The stiffness relationship for the structure can now be expressed as:

$$
\begin{bmatrix}
-144 \\
-400 \\
Q_3 -192 \\
Q_4 -144 \\
Q_5 -192 \\
Q_6 +400 \\
Q_7 \\
Q_8 \\
Q_9
\end{bmatrix}
=
\begin{bmatrix}
151{,}338 & 1152 & -38{,}215 & -51{,}338 & 38{,}215 & 1152 & -100\times10^3 & 0 & 0 \\
1152 & 28{,}800 & -1464 & -1152 & -1536 & 6400 & 0 & 3000 & 8000 \\
-38{,}215 & -1464 & 29{,}796 & 38{,}216 & -29{,}046 & 1536 & 0 & -750 & -3000 \\
-51{,}338 & -1152 & 38{,}216 & 51{,}338 & -38{,}216 & -1152 & 0 & 0 & 0 \\
38{,}215 & -1536 & -29{,}046 & -38{,}216 & 29{,}046 & -1536 & 0 & 0 & 0 \\
1152 & 6400 & 1536 & -1152 & -1536 & 12{,}800 & 0 & 0 & 0 \\
-100\times10^3 & 0 & 0 & 0 & 0 & 0 & 100\times10^3 & 0 & 0 \\
0 & 3000 & -750 & 0 & 0 & 0 & 0 & 750 & 3000 \\
0 & 8000 & -3000 & 0 & 0 & 0 & 0 & 3000 & 16{,}000
\end{bmatrix}
\begin{bmatrix}
D_1 \\
D_2 \\
0 \\
0 \\
0 \\
0 \\
0 \\
0 \\
0
\end{bmatrix}
\tag{12.27}
$$

(6) The unknown displacements $\mathbf{D}_u$ are calculated from the first two rows of Equations 12.27 as:

$$D_1 = -0.846 \times 10^{-3}\ \text{m} \quad \text{and} \quad D_2 = -13.85 \times 10^{-3}\ \text{rad}$$

(7) The unknown reactions are then evaluated by substituting the values of D_1 and D_2 into the last seven rows of Equations 12.27, giving:

$$Q_3 = 244.6\ \text{kN} \qquad Q_4 = 203.4\ \text{kN} \qquad Q_5 = 180.9\ \text{kN} \qquad Q_6 = -489.6\ \text{kNm}$$

$$Q_7 = 84.60\ \text{kN} \qquad Q_8 = -41.5\ \text{kN} \qquad Q_9 = -110.8\ \text{kNm}$$

These are plotted on the free-body diagram of the structure in Figure 12.12.

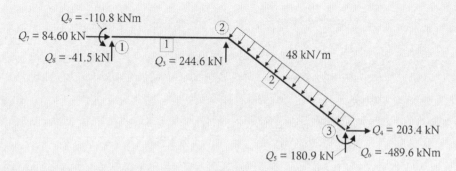

Figure 12.12 Free-body diagram with calculated reactions.

The nodal actions $\mathbf{q}_{(2)}$ related to element 2 are calculated with Equation 12.22:

$$\mathbf{q}_{(2)} = \mathbf{kTD}_{e(2)} + \mathbf{q}_{F(2)}$$

$$
= \begin{bmatrix}
80 \times 10^3 & 0 & 0 & -80 \times 10^3 & 0 & 0 \\
0 & 384 & 1920 & 0 & -384 & 1920 \\
0 & 1920 & 12{,}800 & 0 & -1920 & 6400 \\
-80 \times 10^3 & 0 & 0 & 80 \times 10^3 & 0 & 0 \\
0 & -384 & -1920 & 0 & 384 & -1920 \\
0 & 1920 & 6400 & 0 & -1920 & 12{,}800
\end{bmatrix}
\begin{bmatrix}
0.8 & -0.6 & 0 & 0 & 0 & 0 \\
0.6 & 0.8 & 0 & 0 & 0 & 0 \\
0 & 0 & 1 & 0 & 0 & 0 \\
0 & 0 & 0 & 0.8 & -0.6 & 0 \\
0 & 0 & 0 & 0.6 & 0.8 & 0 \\
0 & 0 & 0 & 0 & 0 & 1
\end{bmatrix}
\begin{bmatrix}
-0.846 \times 10^{-3} \\
0 \\
-13.85 \times 10^{-3} \\
0 \\
0 \\
0
\end{bmatrix}
$$

$$
+ \begin{bmatrix}
0 \\
240 \\
400 \\
0 \\
240 \\
-400
\end{bmatrix}
= \begin{bmatrix}
-54.15 \\
213.2 \\
221.7 \\
54.15 \\
266.8 \\
-489.6
\end{bmatrix}
$$

The terms collected in $\mathbf{q}_{(2)}$, including vectors $\mathbf{kd}_{e(2)}$ and $\mathbf{q}_{F(2)}$, are shown on the free-body diagram of element 2 in Figure 12.13.

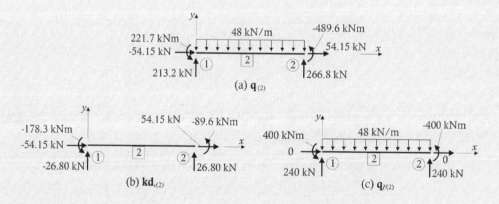

Figure 12.13 Free-body diagrams with internal actions for element 2.

WORKED EXAMPLE 12.3

Reconsider the frame of Worked Example 12.1 and determine the nodal displacements, support reactions and the internal actions of the frame elements induced by the following temperature changes (without the presence of external applied forces):

(i) vertical element subjected to a constant temperature change of 10°C

(ii) horizontal element subjected to a linear temperature change with top and bottom temperature changes of 20°C and 10°C, respectively.

The temperature changes are constant over the length of each element. Assume the cross-section to be doubly-symmetric with its depth equal to 0.3 m and adopt a coefficient of thermal expansion α of $11 \times 10^{-6}/°C$.

(1 and 2) From Worked Example 12.1:

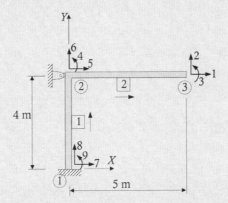

(3) The stiffness matrices for elements 1 and 2 have already been calculated in Worked Example 12.1 and these results are not repeated here.

The equivalent nodal loads related to the temperature changes in elements 1 and 2 are calculated using Equations 12.18:

$$\mathbf{q}_F = [EA\varepsilon_{T,r} \quad 0 \quad EI\kappa_T \quad -EA\varepsilon_{T,r} \quad 0 \quad -EI\kappa_T]^T$$

$$\mathbf{q}_M = -\mathbf{q}_F = [-EA\varepsilon_{T,r} \quad 0 \quad -EI\kappa_T \quad EA\varepsilon_{T,r} \quad 0 \quad EI\kappa_T]^T$$

where (see Reflection Activity 12.1):

$$\varepsilon_T = \varepsilon_{T,r} - y\kappa_T$$

$$\varepsilon_{T,r} = \alpha\left[\Delta T_1 + (\Delta T_2 - \Delta T_1)\frac{y_t}{h}\right] \qquad \kappa_T = \alpha\frac{\Delta T_2 - \Delta T_1}{h}$$

The equivalent nodal loads to account for the temperature changes are now calculated separately for elements 1 and 2 in the following.

Element I

From $\Delta T_{1(1)} = 10°C$, $\Delta T_{2(1)} = 10°C$, $h = 0.3$ m, $y_t = h/2 = 0.3/2 = 0.15$ m (because section is doubly-symmetric) and $\alpha = 11 \times 10^{-6}/°C$ (units in kN and m):

$$\varepsilon_{T,r(1)} = \alpha\left[\Delta T_{1(1)} + (\Delta T_{2(1)} - \Delta T_{1(1)})\frac{y_t}{h}\right] = 11\times10^{-6}\left[10 + (10-10)\frac{0.15}{0.3}\right] = 1.1\times10^{-4}$$

$$\kappa_{T(1)} = \alpha\frac{\Delta T_{2(1)} - \Delta T_{1(1)}}{h} = 11\times10^{-6}\frac{10-10}{0.3} = 0$$

$$\mathbf{q}_{F(1)} = [110 \quad 0 \quad 0 -110 \quad 0 \quad 0]^T$$

$$\mathbf{q}_{M(1)} = -\mathbf{q}_{F(1)} = [-110 \quad 0 \quad 0 \quad 110 \quad 0 \quad 0]^T$$

$$\mathbf{Q}_{eM(1)} = \mathbf{T}_{(1)}^T\mathbf{q}_{M(1)} = \begin{bmatrix} 0 & -1 & 0 & 0 & 0 & 0 \\ 1 & 0 & 0 & 0 & 0 & 0 \\ 0 & 0 & 1 & 0 & 0 & 0 \\ 0 & 0 & 0 & 0 & -1 & 0 \\ 0 & 0 & 0 & 1 & 0 & 0 \\ 0 & 0 & 0 & 0 & 0 & 1 \end{bmatrix}\begin{bmatrix} -110 \\ 0 \\ 0 \\ 110 \\ 0 \\ 0 \end{bmatrix} = \begin{bmatrix} 0 \\ -110 \\ 0 \\ 0 \\ 110 \\ 0 \end{bmatrix}$$

Element 2

From $\Delta T_{1(2)} = 20°C$, $\Delta T_{2(2)} = 10°C$, $h = 0.3$ m, $y_t = 0.15$ m and $\alpha = 11 \times 10^{-6}/°C$ (units in kN and m):

$$\varepsilon_{T,r(2)} = \alpha\left[\Delta T_{1(2)} + (\Delta T_{2(2)} - \Delta T_{1(2)})\frac{y_t}{h}\right] = 11\times10^{-6}\left[20 + (10-20)\frac{0.15}{0.3}\right] = 1.65\times10^{-4}$$

$$\kappa_{T(2)} = \alpha\frac{\Delta T_{2(2)} - \Delta T_{1(2)}}{h} = 11\times10^{-6}\frac{10-20}{0.3} = -3.666\times10^{-4} \text{ m}^{-1}$$

$$\mathbf{q}_{F(2)} = [165 \quad 0 \quad -17.6 \quad -165 \quad 0 \quad 17.6]^T$$

$$\mathbf{q}_{M(2)} = -\mathbf{q}_{F(2)} = [-165 \quad 0 \quad 17.6 \quad 165 \quad 0 \quad -17.6]^T$$

$$\mathbf{Q}_{eM(2)} = \mathbf{T}_{(2)}^T\mathbf{q}_{M(2)} = \begin{bmatrix} 1 & 0 & 0 & 0 & 0 & 0 \\ 0 & 1 & 0 & 0 & 0 & 0 \\ 0 & 0 & 1 & 0 & 0 & 0 \\ 0 & 0 & 0 & 1 & 0 & 0 \\ 0 & 0 & 0 & 0 & 1 & 0 \\ 0 & 0 & 0 & 0 & 0 & 1 \end{bmatrix}\begin{bmatrix} -165 \\ 0 \\ 17.6 \\ 165 \\ 0 \\ -17.6 \end{bmatrix} = \begin{bmatrix} -165 \\ 0 \\ 17.6 \\ 165 \\ 0 \\ -17.6 \end{bmatrix}$$

(4 and 5) The stiffness relationship for the structure is obtained by assembling the stiffness coefficients of elements 1 and 2 and considering the contributions of the member loads describing the temperature changes:

$$
\begin{bmatrix}
165 \\ 0 \\ -17.6 \\ 17.6 \\ \hline Q_5 - 165 \\ Q_6 + 110 \\ Q_7 \\ Q_8 - 110 \\ Q_9
\end{bmatrix}
=
\left[
\begin{array}{cccc:ccccc}
 & 1 & 2 & 3 & 4 & 5 & 6 & 7 & 8 & 9 \\
200 \times 10^3 & 0 & 0 & 0 & -200 \times 10^3 & 0 & 0 & 0 & 0 \\
0 & 4608 & -11,520 & -11,520 & 0 & -4608 & 0 & 0 & 0 \\
0 & -11,520 & 38,400 & 19,200 & 0 & 11,520 & 0 & 0 & 0 \\
0 & -11,520 & 19,200 & 86,400 & 18 \times 10^3 & 11,520 & -18 \times 10^3 & 0 & 24 \times 10^3 \\ \hdashline
-200 \times 10^3 & 0 & 0 & 18,000 & 209 \times 10^3 & 0 & -9 \times 10^3 & 0 & 18 \times 10^3 \\
0 & -4608 & 11,520 & 11,520 & 0 & 254,608 & 0 & -250 \times 10^3 & 0 \\
0 & 0 & 0 & -18,000 & -9 \times 10^3 & 0 & 9 \times 10^3 & 0 & -18 \times 10^3 \\
0 & 0 & 0 & 0 & 0 & -250 \times 10^3 & 0 & 250 \times 10^3 & 0 \\
0 & 0 & 0 & 24 \times 10^3 & 18 \times 10^3 & 0 & -18 \times 10^3 & 0 & 48 \times 10^3
\end{array}
\right]
\begin{bmatrix}
D_1 \\ D_2 \\ D_3 \\ D_4 \\ 0 \\ 0 \\ 0 \\ 0 \\ 0
\end{bmatrix}
\begin{matrix}
1 \\ 2 \\ 3 \\ 4 \\ 5 \\ 6 \\ 7 \\ 8 \\ 9
\end{matrix}
$$

$$(12.28)$$

(6) $\mathbf{D}_u$ is determined with the first four rows of Equations 12.28 as:

$$D_1 = 0.825 \times 10^{-3}\ \text{m} \qquad D_2 = -4.583 \times 10^{-3}\ \text{m} \qquad D_3 = -1.833 \times 10^{-3}\ \text{rad} \qquad D_4 = 0\ \text{rad}$$

(7) The unknown reactions are calculated from the last five rows of Equations 12.28:

$$Q_5 = 0\ \text{kN} \qquad Q_6 = -110\ \text{kN} \qquad Q_7 = 0\ \text{kN}$$

$$Q_8 = 110\ \text{kN} \qquad Q_9 = 0\ \text{kNm}$$

The support reactions are illustrated in Figure 12.14.
The nodal actions of elements 1 and 2 are calculated with Equation 12.22 and are plotted in the free-body diagrams in Figures 12.15 and 12.16, respectively.

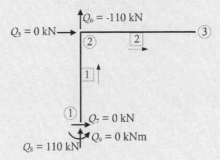

Figure 12.14 Free-body diagram with calculated reactions.

Element 1

$$\mathbf{q}_{(1)} = \mathbf{kTD}_{e(1)} + \mathbf{q}_{F(1)} =$$

$$= EI \begin{bmatrix} 250\times10^3 & 0 & 0 & -250\times10^3 & 0 & 0 \\ 0 & 9\times10^3 & 18\times10^3 & 0 & -9\times10^3 & 18\times10^3 \\ 0 & 18\times10^3 & 48\times10^3 & 0 & -18\times10^3 & 24\times10^3 \\ -250\times10^3 & 0 & 0 & 250\times10^3 & 0 & 0 \\ 0 & -9\times10^3 & -18\times10^3 & 0 & 9\times10^3 & -18\times10^3 \\ 0 & 18\times10^3 & 24\times10^3 & 0 & -18\times10^3 & 48\times10^3 \end{bmatrix}$$

$$\times \begin{bmatrix} 0 & 1 & 0 & 0 & 0 & 0 \\ -1 & 0 & 0 & 0 & 0 & 0 \\ 0 & 0 & 1 & 0 & 0 & 0 \\ 0 & 0 & 0 & 0 & 1 & 0 \\ 0 & 0 & 0 & -1 & 0 & 0 \\ 0 & 0 & 0 & 0 & 0 & 1 \end{bmatrix} \begin{bmatrix} 0 \\ 0 \\ 0 \\ 0 \\ 0 \\ 0 \end{bmatrix} + \begin{bmatrix} 110 \\ 0 \\ 0 \\ -110 \\ 0 \\ 0 \end{bmatrix} = \begin{bmatrix} 110 \\ 0 \\ 0 \\ -110 \\ 0 \\ 0 \end{bmatrix}$$

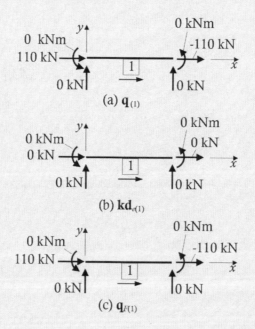

Figure 12.15 Element 1.

Element 2

$$\mathbf{q}_{(2)} = \mathbf{kTD}_{e(2)} + \mathbf{q}_{F(2)} =$$

$$= EI \begin{bmatrix} 200\times10^3 & 0 & 0 & -200\times10^3 & 0 & 0 \\ 0 & 4608 & 11,520 & 0 & -4608 & 11,520 \\ 0 & 11,520 & 38,400 & 0 & -11,520 & 19,200 \\ -200\times10^3 & 0 & 0 & 200\times10^3 & 0 & 0 \\ 0 & -4608 & -11,520 & 0 & 4608 & -11,520 \\ 0 & 11,520 & 19,200 & 0 & -11,520 & 38,400 \end{bmatrix}$$

$$\times \begin{bmatrix} 1 & 0 & 0 & 0 & 0 & 0 \\ 0 & 1 & 0 & 0 & 0 & 0 \\ 0 & 0 & 1 & 0 & 0 & 0 \\ 0 & 0 & 0 & 1 & 0 & 0 \\ 0 & 0 & 0 & 0 & 1 & 0 \\ 0 & 0 & 0 & 0 & 0 & 1 \end{bmatrix} \begin{bmatrix} 0 \\ 0 \\ 0 \\ 0.825\times10^{-3} \\ -4.583\times10^{-3} \\ -1.833\times10^{-3} \end{bmatrix} + \begin{bmatrix} 165 \\ 0 \\ -17.6 \\ -165 \\ 0 \\ 17.6 \end{bmatrix} = \begin{bmatrix} 0 \\ 0 \\ 0 \\ 0 \\ 0 \\ 0 \end{bmatrix}$$

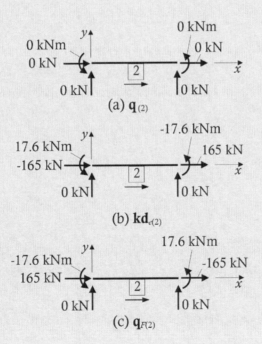

(a) $\mathbf{q}_{(2)}$

(b) $\mathbf{kd}_{e(2)}$

(c) $\mathbf{q}_{F(2)}$

Figure 12.16 Element 2.

PROBLEMS

12.1 Determine the reactions for the frame illustrated below using the stiffness method. Assume $I = 300 \times 10^6$ mm^4, $A = 10 \times 10^3$ mm^2 and $E = 200$ GPa.

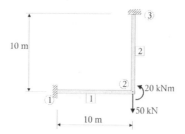

12.2 Consider the frame of Problem 12.1 and sketch the free-body diagram of element 1 highlighting all actions applied to it.

12.3 Evaluate the nodal displacements and reactions for the frame shown using the stiffness method. Adopt $I = 300 \times 10^6$ mm^4, $A = 10 \times 10^3$ mm^2 and $E = 200$ GPa.

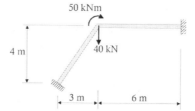

12.4 Consider the frame of Problem 12.3 and sketch the free-body diagram of element 1 highlighting all actions applied to it.

12.5 For the beam illustrated below, calculate the nodal displacements using the stiffness method and assuming $I = 240 \times 10^6$ mm^4, $A = 8 \times 10^3$ mm^2 and $E = 200$ GPa.

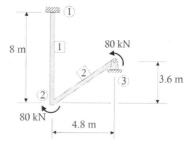

12.6 Draw the free-body diagram of element 2 of the beam of Problem 12.5 highlighting all actions applied to it.

12.7 Determine the nodal displacements for the structure illustrated below adopting $I = 240 \times 10^6$ mm^4, $A = 8 \times 10^3$ mm^2 and $E = 200$ GPa. Use the stiffness method.

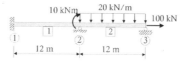

12.8 Sketch the free-body diagram of element 1 for the structure analysed in Problem 12.7 highlighting all actions applied to it.

12.9 Using the stiffness method, calculate the nodal displacements for the frame illustrated below with $I = 300 \times 10^6$ mm^4, $A = 10 \times 10^3$ mm^2 and $E = 200$ GPa.

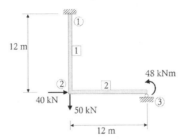

12.10 For the frame of Problem 12.9, draw the free-body diagram of element 2 highlighting all actions applied to it.

12.11 Consider the frame shown and calculate the reactions using the stiffness method, assuming $I = 300 \times 10^6$ mm^4, $A = 10 \times 10^3$ mm^2 and $E = 200$ GPa.

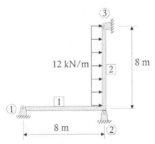

12.12 Plot the free-body diagram of element 1 of the frame analysed in Problem 12.11 highlighting all actions applied to it.

12.13 Evaluate the reactions for the frame illustrated below using the stiffness method. Assume $I = 320 \times 10^6$ mm^4, $A = 12 \times 10^3$ mm^2 and $E = 200$ GPa.

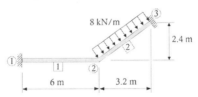

12.14 Consider the frame of Problem 12.13 and plot the free-body diagrams of elements 1 and 2 highlighting all actions applied to them.

12.15 Using the stiffness method, calculate the nodal displacements and reactions for the frame illustrated below. Assume $I = 240 \times 10^6$ mm^4, $A = 6 \times 10^3$ mm^2 and $E = 200$ GPa.

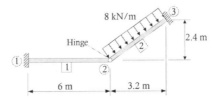

12.16 Sketch the free-body diagram of element 1 of the frame analysed in Problem 12.15 highlighting all actions applied to it.

12.17 Evaluate the reactions for the frame shown below using the stiffness method. Adopt $I = 240 \times 10^6$ mm^4, $A = 6 \times 10^3$ mm^2 and $E = 200$ GPa.

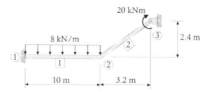

12.18 Consider the frame of Problem 12.17 and sketch the free-body diagrams of elements 1 and 2 highlighting all actions applied to them.

12.19 Consider the frame shown below and determine the support reactions using the stiffness method with $I = 300 \times 10^6$ mm^4, $A = 10 \times 10^3$ mm^2 and $E = 200$ GPa.

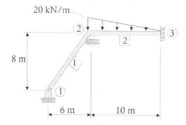

12.20 For the frame of Problem 12.19, draw the free-body diagram of element 1 highlighting all actions applied to it.

12.21 For the frame illustrated below, evaluate the support reactions using the stiffness method with $I = 320 \times 10^6$ mm^4, $A = 12 \times 10^3$ mm^2 and $E = 200$ GPa.

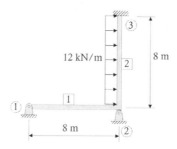

12.22 Draw the free-body diagram of element 2 of the frame analysed in Problem 12.11 highlighting all actions applied to it.

12.23 Using the stiffness method, determine the reactions for the frame illustrated below considering $I = 320 \times 10^6$ mm^4, $A = 12 \times 10^3$ mm^2 and $E = 200$ GPa.

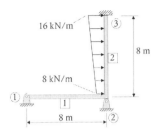

12.24 Consider the frame of Problem 12.23 and sketch the free-body diagram of element 2 highlighting all actions applied to it.

12.25 Using the stiffness method, evaluate the reactions induced by a temperature change for the frame illustrated below. Both elements 1 and 2 are assumed to be subjected to a linear temperature change over the cross-section with top and bottom variations equal to $\Delta T_a = 20$°C and $\Delta T_b = 10$°C. The temperature changes are constant over the length of each element. The cross-section is doubly-symmetric and its depth is equal to 0.4 m. Assume $I = 300 \times 10^6$ mm^4, $A = 10 \times 10^3$ mm^2, $E = 200$ GPa and $\alpha = 11 \times 10^{-6}$/°C.

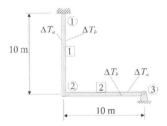

12.26 For the frame of Problem 12.25, draw the free-body diagram of element 2 highlighting all actions applied to it.

12.27 Using the stiffness method, determine the nodal displacements and reactions induced by a temperature change for the frame shown. A linear temperature change is assumed to be applied over the doubly-symmetric cross-sections of both elements 1 and 2 (with depth equal to 0.4 m), with top and bottom variations equal to $\Delta T_a = 20$°C and $\Delta T_b = 10$°C. The temperature changes are constant over the length of each element. Assume $I = 300 \times 10^6$ mm^4, $A = 10 \times 10^3$ mm^2, $E = 200$ GPa and $\alpha = 11 \times 10^{-6}$/°C.

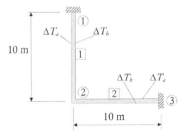

12.28 Sketch the free-body diagram of element 2 for the frame analysed in Problem 12.27 highlighting all actions applied to it.

12.29 Reconsider the frame of Problem 12.27 and draw the free-body diagrams of elements 1 and 2, highlighting all actions applied to them, assuming a hinge to be placed at node 2 as illustrated below.

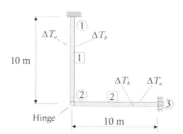

12.30 Using the stiffness method, evaluate the reactions for the structure illustrated below, which is subjected to a linear temperature change over the cross-sections defined by top and bottom variations of: $\Delta T_1 = 20°C$ and $\Delta T_2 = 10°C$ (as shown). The temperature changes are constant over the length of each element. The cross-section is doubly-symmetric and is 0.3 m deep. Adopt $I = 240 \times 10^6$ mm^4, $A = 8 \times 10^3$ mm^2, $E = 200$ GPa and $\alpha = 11 \times 10^{-6}/°C$.

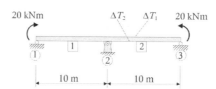

12.31 Reconsider the structure of Problem 12.30 and replace the end rollers with pinned supports as illustrated below. Calculate the reactions of the structure.

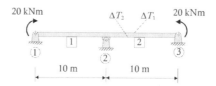

Chapter 13

Introduction to the finite element method

13.1 INTRODUCTION

The finite element method is a well-established method of analysis extensively used in all disciplines of engineering. This chapter provides an introduction to the method for the analysis of structures. The derivations presented deal with simple finite elements consisting of line elements describing two common beam models, namely the Euler–Bernoulli beam and the Timoshenko beam, readily available in some of the most common finite element analysis software. In this manner, we will cover some of the key aspects involved in a finite element derivation and analysis, while minimising the complexity of the numerical model. For the derivation of more refined and advanced elements, reference should be made to textbooks dedicated to the finite element method.

In Chapter 5, we derived the system of differential equations describing the behaviour of the Euler–Bernoulli beam model. It is common to refer to such set of governing differential equations, and its corresponding boundary conditions, as the *strong form* of the problem. The term *strong* intends to highlight the fact that the set of equations needs to be satisfied at any section of the beam and to distinguish it from its *weak form*, which represents an integral form of these equations. The weak form is very useful as it provides the basis for the derivation of finite elements. There are different approaches available in the literature to obtain the weak form of a problem. In this chapter, we will consider a simple approach for the finite element derivation of both Euler–Bernoulli and Timoshenko beam models. After the presentation of the kinematic model, which defines the possible displacements and deformations of the beam, the weak form is obtained by applying the principle of virtual work. The numerical solution is then derived based on the finite element method approximating the beam displacements with polynomial functions. Numerical results are provided to better outline some aspects related to finite element modelling.

13.2 EULER–BERNOULLI BEAM MODEL

The Euler–Bernoulli beam model was presented in Chapter 5 and is widely used for the analysis of structures. Structural designers use it routinely for the prediction of deformations and internal actions in beams and frames using commercial structural analysis software. Various closed-form solutions are also commonly used, such as figures and tables of elastic deflection coefficients and internal actions in standard beams and frames for common loading cases. The analytical formulation at the basis of the Euler–Bernoulli beam model is described in the following sections for a generic member, such as that shown in Figure 13.1.

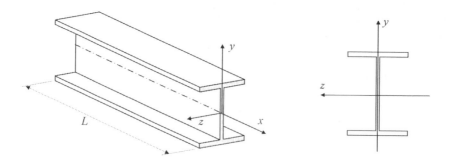

Figure 13.1 Typical structural member and cross-sections.

13.2.1 Kinematic model

In the undeformed state, the beam is assumed to be prismatic as shown in Figure 13.1. Plane sections are assumed to remain plane and perpendicular to the beam axis before and after deformations.

The formulation is derived for a beam segment of length L and the cross-section is assumed to be symmetric about the y-axis. Under these assumptions, no torsional and out-of-plane effects are considered. For generality, the level of the reference x-axis is taken as arbitrary.

The kinematic behaviour is illustrated in Figures 13.2a and b for a generic point P on the x-axis, highlighting both its axial displacement at the level of the reference axis $u(x_P)$ and its deflection $v(x_P)$, where x_P defines the position of P in the undeformed beam.

The final displacement of a point Q, not on the member axis, can be expressed in terms of $u(x_Q)$ and $v(x_Q)$ as well as the rotation $\theta(x_Q)$. In particular, the horizontal and vertical displacements, referred to as $d_x(x_Q, y_Q)$ and $d_y(x_Q, y_Q)$, are expressed as:

$$d_x(x_Q, y_Q) = u(x_Q) - y_Q \sin \theta(x_Q) \tag{13.1a}$$

$$d_y(x_Q, y_Q) = v(x_Q) - y_Q + y_Q \cos \theta(x_Q) \tag{13.1b}$$

and, for clarity, the kinematic response of point Q is illustrated in Figure 13.2c, where x_Q and y_Q represent the coordinates of point Q in the undeformed beam.

The expressions of Equations 13.1 describe all possible displacements that the points of the beam can undergo and, because of this, this set of displacements is usually referred to as the *displacement field* of the model. For structural engineering applications, it is usually sufficient and convenient to remain within the framework of small displacements. In this way, the cosine and sine of the angle $\theta(x)$ in Equations 13.1 can be approximated by $\cos \theta(x) \approx 1$ and $\sin \theta(x) \approx \theta(x)$. For ease of notation, $u(x)$, $v(x)$ and $\theta(x)$ will be referred to as u, v and θ, respectively, in the following.

The condition that plane sections remain plane and perpendicular to the beam axis before and after deformations is enforced by:

$$\theta = v' \tag{13.2}$$

where the prime represents differentiation with respect to x.

Based on these simplifications, the displacement field of Equations 13.1 is re-written as:

$$d_x(x, y) = u - y\theta = u - yv' \tag{13.3a}$$

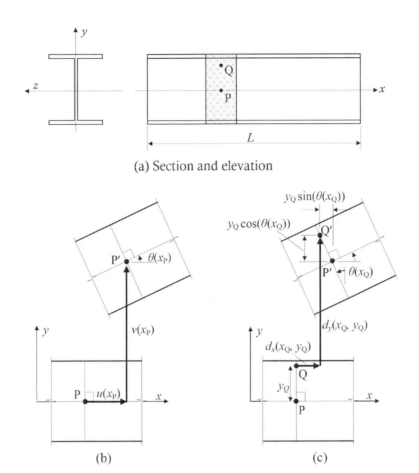

(a) Section and elevation

(b)   (c)

Figure 13.2 Displacement field for the Euler–Bernoulli beam model.

$$d_y(x, y) = v \tag{13.3b}$$

$$d_z(x, y) = 0 \tag{13.3c}$$

Equations 13.3 show how the kinematic response of a point in the beam can be determined once the displacements u and v are known. These independent variables (u and v) describing the displacement field are usually denoted as *generalised displacements*.

The corresponding strain field is calculated based on linear elasticity as:

$$\varepsilon_x = \frac{\partial d_x}{\partial x} = u' - yv'' \qquad \varepsilon_y = \frac{\partial d_y}{\partial y} = 0 \qquad \varepsilon_z = \frac{\partial d_z}{\partial z} = 0 \tag{13.4a–c}$$

$$\gamma_{xy} = \frac{\partial d_x}{\partial y} + \frac{\partial d_y}{\partial x} = 0 \qquad \gamma_{yz} = \frac{\partial d_y}{\partial z} + \frac{\partial d_z}{\partial y} = 0 \qquad \gamma_{xz} = \frac{\partial d_x}{\partial z} + \frac{\partial d_z}{\partial x} = 0 \tag{13.4d–f}$$

where strain normal to the cross-section in the x direction ε_x is the only non-zero strain. The expression for the curvature $\kappa\,(= v'')$ can be obtained by differentiating the deflection v twice with respect to the coordinate x, and a positive value of v'' indicates a positive curvature in sagging moment regions (i.e. where compressive and tensile strains occur in the top and bottom fibres of the section, respectively, in a horizontal beam in accordance with the sign convention adopted throughout this book).

13.2.2 Weak form

The weak form of the Euler–Bernoulli beam model is obtained using the principle of virtual work considering a beam segment of length L with the free-body diagram shown in Figure 13.3. The loads $w(x)$ and $n(x)$ represent the vertical and horizontal distributed member loads and, for ease of notation, will be referred to subsequently as w and n. The nodal actions at each end of the member represent external loads, internal actions or support reactions depending on the boundary conditions of the beam segment and have been referred to as N, S and M with the subscripts 'L' and 'R' specifying whether they relate to the left end (at $x = 0$) or the right end (at $x = L$), respectively, as shown in Figure 13.3.

We start the derivation by equating the work of internal stresses to the work of external actions for each virtual admissible variation of the displacements and corresponding strains (which, by definition, represent all variations of possible displacements satisfying the kinematic boundary conditions of the problem):

$$\int_L \int_A \sigma_x \hat{\varepsilon}_x \, dA \, dx = \int_L (w\hat{v} + n\hat{u}) dx + S_L \hat{v}_L + N_L \hat{u}_L + M_L \hat{\theta}_L + S_R \hat{v}_R + N_R \hat{u}_R + M_R \hat{\theta}_R \qquad (13.5)$$

where the variables with the hat '^' represent the virtual variations of displacements or strains. By substituting the expression for the axial strain ε_x of Equation 13.4a into Equation 13.5, the problem can be re-written as:

$$\int_L \int_A \sigma_x \left(\hat{u}' - y\hat{v}'' \right) dA \, dx = \int_L (w\hat{v} + n\hat{u}) dx + S_L \hat{v}_L + N_L \hat{u}_L + M_L \hat{\theta}_L + S_R \hat{v}_R + N_R \hat{u}_R + M_R \hat{\theta}_R$$

$$(13.6)$$

Recalling the definitions of internal axial force N and moment M about the z-axis (see Section 5.4):

$$N = \int_A \sigma_x \, dA \quad \text{and} \quad M = -\int_A y\sigma_x \, dA \qquad (13.7a,b)$$

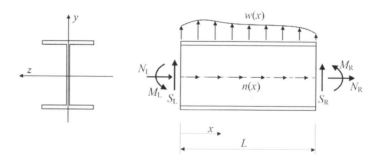

Figure 13.3 Member loads and nodal actions.

the integral at the cross-section (i.e. in dA) on the left-hand side of Equation 13.6 can be replaced by the internal actions N and M as:

$$\int_L (N\hat{u}' + M\hat{v}'')\,dx = \int_L (w\hat{v} + n\hat{u})\,dx + S_L\hat{v}_L + N_L\hat{u}_L + M_L\hat{\theta}_L + S_R\hat{v}_R + N_R\hat{u}_R + M_R\hat{\theta}_R \qquad (13.8)$$

This relationship can be further rearranged to isolate the terms related to N and M as follows:

$$\int_L \begin{bmatrix} N \\ M \end{bmatrix} \cdot \begin{bmatrix} \hat{u}' \\ \hat{v}'' \end{bmatrix} dx = \int_L \begin{bmatrix} n \\ w \end{bmatrix} \cdot \begin{bmatrix} \hat{u} \\ \hat{v} \end{bmatrix} dx + \begin{bmatrix} N_L \\ S_L \\ M_L \end{bmatrix} \cdot \begin{bmatrix} \hat{u}_L \\ \hat{v}_L \\ \hat{\theta}_L \end{bmatrix} + \begin{bmatrix} N_R \\ S_R \\ M_R \end{bmatrix} \cdot \begin{bmatrix} \hat{u}_R \\ \hat{v}_R \\ \hat{\theta}_R \end{bmatrix} \qquad (13.9)$$

It is possible to assume a nil variation of the virtual displacements at the ends of the beam segment, where $x = 0$ and $x = L$, without affecting the derivation of the stiffness matrix for the finite element and its loading vector accounting for n and w. In such a case, it is then possible to include the nodal actions in the finite element analysis during the assembly of the load vector. Based on these considerations, Equation 13.9 can be simplified to:

$$\int_L \begin{bmatrix} N \\ M \end{bmatrix} \cdot \begin{bmatrix} \hat{u}' \\ \hat{v}'' \end{bmatrix} dx = \int_L \begin{bmatrix} n \\ w \end{bmatrix} \cdot \begin{bmatrix} \hat{u} \\ \hat{v} \end{bmatrix} dx \qquad (13.10)$$

In this form, the constitutive models for the materials are not explicitly specified as they are included in the definitions of the internal actions N and M.

Under the assumptions of linear–elastic material properties (where the material follows Hooke's law $\sigma = E\varepsilon$), the internal actions N and M can be expressed as follows:

$$N = \int_A \sigma_x\,dA = \int_A E(\varepsilon_r - y\kappa)\,dA = R_A\varepsilon_r - R_B\kappa \qquad (13.11a)$$

$$M = -\int_A y\sigma_x\,dA = -\int_A Ey(\varepsilon_r - y\kappa)\,dA = -R_B\varepsilon_r + R_I\kappa \qquad (13.11b)$$

where ε_r and κ are the strain at the level of the reference axis and the curvature, respectively, while R_A, R_B and R_I represent the axial rigidity, the stiffness related to the first moment of area and the flexural rigidity of the cross-section, respectively, and are calculated as:

$$R_A = EA \quad R_B = EB \quad R_I = EI \qquad (13.12a\text{--}c)$$

Equations 13.11 can be re-written in more compact form as:

$$\mathbf{r} = \mathbf{D}\boldsymbol{\varepsilon} \qquad (13.13)$$

and the terms included in the matrix and vectors are:

$$\begin{bmatrix} N \\ M \end{bmatrix} = \begin{bmatrix} R_A & -R_B \\ -R_B & R_I \end{bmatrix} \begin{bmatrix} \varepsilon_r \\ \kappa \end{bmatrix} \qquad (13.14)$$

where $\mathbf{r}$ is the vector of internal actions N and M, $\mathbf{D}$ specifies the geometric and material properties of the cross-section and $\boldsymbol{\varepsilon}$ includes the strain calculated at the level of the reference axis ε_r and the curvature κ. These terms can be expressed in terms of the horizontal and vertical displacements (i.e. u and v) as detailed below:

$$\mathbf{r} = \begin{bmatrix} N \\ M \end{bmatrix} = \begin{bmatrix} \int_A \sigma_x \, dA \\ -\int_A y\sigma_x \, dA \end{bmatrix} \qquad \mathbf{D} = \begin{bmatrix} R_A & -R_B \\ -R_B & R_I \end{bmatrix}$$
(13.15a,b)

and

$$\boldsymbol{\varepsilon} = \begin{bmatrix} \varepsilon_r \\ \kappa \end{bmatrix} = \begin{bmatrix} u' \\ v'' \end{bmatrix} = \begin{bmatrix} \partial & 0 \\ 0 & \partial^2 \end{bmatrix} \begin{bmatrix} u \\ v \end{bmatrix} = \mathbf{Ae}$$
(13.16)

where $\mathbf{A}$ is a differential operator and the symbol ∂ defines the derivative with respect to the member coordinate x, and the generalised displacements u and v are collected in vector $\mathbf{e}$ as:

$$\mathbf{e} = \begin{bmatrix} u \\ v \end{bmatrix}$$
(13.17)

At this point, it is useful to re-write Equation 13.10 in terms of vectors $\mathbf{r}$ and $\boldsymbol{\varepsilon}$ (= $\mathbf{Ae}$):

$$\int_L \mathbf{r} \cdot \mathbf{A}\hat{\mathbf{e}} \, dx = \int_L \mathbf{p} \cdot \hat{\mathbf{e}} \, dx$$
(13.18)

where the member loads n and w have been collected in $\mathbf{p}$:

$$\mathbf{p} = \begin{bmatrix} n \\ w \end{bmatrix}$$
(13.19)

Substituting the constitutive properties defined in Equation 13.13 into Equation 13.18 produces the general expression for the weak formulation of the problem:

$$\int_L \mathbf{D}\boldsymbol{\varepsilon} \cdot \mathbf{A}\hat{\mathbf{e}} \, dx = \int_L \mathbf{p} \cdot \hat{\mathbf{e}} \, dx$$
(13.20)

13.2.3 Finite element formulation

The formulation described here is applicable to displacement-based finite elements, which are derived by approximating the generalised displacements of the model by means of polynomials or other selected functions. If necessary, the approach can be modified for the derivation of other elements, for example, force-based elements (where the approximation is applied to stress variables, such as the internal actions in the case of beams) or mixed elements (in which case, a combination of displacements, strains, and stresses is approximated).

The basis of the proposed displacement-based finite element formulation relies on the approximation of the generalised displacements u and v, previously collected in the vector $\mathbf{e}$ (Equation 13.17) by means of polynomial functions.

For example, if we approximate the axial displacement by means of a parabolic function and the deflection with a cubic polynomial, we have:

$$u = a_0 + a_1 x + a_2 x^2 \tag{13.21a}$$

$$v = b_0 + b_1 x + b_2 x^2 + b_3 x^3 \tag{13.21b}$$

where a_i and b_j (with $i = 0,1,2$ and $j = 0,1,2,3$) are unknown coefficients that need to be evaluated from the analysis. Equations 13.21 are used to approximate the displacements of each element. In this current form, these expressions are not easy to use as they do not enable a direct connection (assembling) among adjacent elements. For this purpose, we will try to replace these coefficients with more useful terms.

Let us start by considering the axial displacement (Equation 13.21a) and replace the three coefficients a_i with another three terms that describe, for example, the axial displacements at the left node u_L, in the middle of the element u_M, and at the right node u_R, as shown in Figure 13.4. In this manner, we could connect the axial displacements of adjacent elements at the element ends. This is carried out by enforcing the polynomial of Equation 13.21a to match the values for the selected nodal displacements where these are specified, and the equations describing these conditions are:

$$u(x = 0) = a_0 + a_1 \times 0 + a_2 \times 0^2 = u_L \tag{13.22a}$$

$$u(x = L/2) = a_0 + a_1 \times L/2 + a_2 \times (L/2)^2 = u_M \tag{13.22b}$$

$$u(x = L) = a_0 + a_1 \times L + a_2 \times L^2 = u_R \tag{13.22c}$$

based on which the coefficients a_i can be written in terms of u_L, u_M and u_R as:

$$a_0 = u_L \quad a_1 = -\frac{3u_L - 4u_M + u_R}{L} \quad a_2 = \frac{2(u_L - 2u_M + u_R)}{L^2} \tag{13.23a--c}$$

Substituting Equations 13.23 into Equations 13.21a produces:

$$u = u_L - \frac{3u_L - 4u_M + u_R}{L} x + \frac{2(u_L - 2u_M + u_R)}{L^2} x^2 \tag{13.24}$$

and collecting the nodal displacements, we get:

$$u = \left(1 - \frac{3x}{L} + \frac{2x^2}{L^2}\right) u_L + \left(\frac{4x}{L} - \frac{4x^2}{L^2}\right) u_M + \left(-\frac{x}{L} + \frac{2x^2}{L^2}\right) u_R \tag{13.25}$$

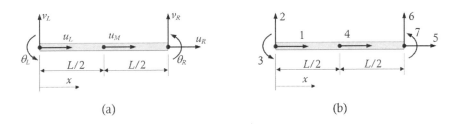

(a)   (b)

Figure 13.4 The 7-dof finite element (Euler–Bernoulli beam). (a) Nodal displacements. (b) Freedom numbering for the isolated finite element.

Equation 13.25 can be re-written in more compact form as:

$$u = N_{u1}u_L + N_{u2}u_M + N_{u3}u_R \qquad (13.26)$$

where:

$$N_{u1} = 1 - \frac{3x}{L} + \frac{2x^2}{L^2} \quad N_{u2} = \frac{4x}{L} - \frac{4x^2}{L^2} \quad N_{u3} = -\frac{x}{L} + \frac{2x^2}{L^2} \qquad (13.27a-c)$$

The expressions for N_{u1}, N_{u2} and N_{u3} are commonly referred to as shape functions because they describe the shapes associated with each nodal displacement, whose variations along the element axis are plotted in Figure 13.5. These shape functions are equal to unity only at the location (along the element axis) associated with their corresponding nodal freedom. For example, N_{u1} equals 1 at $x = 0$ (because it is associated with u_L) and is nil at the locations of the other nodes. N_{u2} and N_{u3} are equal to 1 at $x = L/2$ and $x = L$, respectively, and zero at other nodes.

In a similar manner, the coefficients b_j introduced in Equation 13.21b to describe the vertical displacement can be replaced by nodal displacements related to the deflections (v_L and v_R) and rotation (θ_L and θ_R), as shown in Figure 13.4a. Recalling that in the Euler–Bernoulli beam model $\theta = v'$ (Equation 13.2):

$$\theta = dv/dx = b_1 + 2b_2x + 3b_3x^2 \qquad (13.28)$$

the coefficients b_j are calculated from the following conditions:

$$v(x = 0) = b_0 + b_1 \times 0 + b_2 \times 0^2 + b_3 \times 0^3 = v_L \qquad (13.29a)$$

$$v(x = L) = b_0 + b_1 \times L + b_2 \times L^2 + b_3 \times L^3 = v_R \qquad (13.29b)$$

$$\theta(x = 0) = b_1 + 2b_2 \times 0 + 3b_3 \times 0^2 = \theta_L \qquad (13.29c)$$

$$\theta(x = L) = b_1 + 2b_2 \times L + 3b_3 \times L^2 = \theta_R \qquad (13.29d)$$

based on which:

$$b_0 = v_L \qquad\qquad b_1 = \theta_L \qquad (13.30a,b)$$

$$b_2 = -\frac{3v_L + 2\theta_L L - 3v_R + \theta_R L}{L^2} \qquad b_3 = \frac{\theta_L L + 2v_L - 2v_R + \theta_R L}{L^3} \qquad (13.30c,d)$$

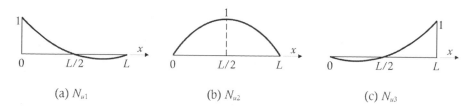

(a) N_{u1} (b) N_{u2} (c) N_{u3}

Figure 13.5 Shape functions used for the parabolic approximation of u.

The expression for v can then be re-arranged by collecting the nodal displacements as:

$$v = \left(1 - \frac{3x^2}{L^2} + \frac{2x^3}{L^3}\right)v_L + \left(x - \frac{2x^2}{L} + \frac{x^3}{L^2}\right)\theta_L + \left(\frac{3x^2}{L^2} - \frac{2x^3}{L^3}\right)v_R + \left(-\frac{x^2}{L} + \frac{x^3}{L^2}\right)\theta_R \qquad (13.31)$$

and, in compact form, as:

$$v = N_{v1}v_L + N_{v2}\theta_L + N_{v3}v_R + N_{v4}\theta_R \qquad (13.32)$$

where:

$$N_{v1} = 1 - \frac{3x^2}{L^2} + \frac{2x^3}{L^3}; \ N_{v2} = x - \frac{2x^2}{L} + \frac{x^3}{L^2}; \ N_{v3} = \frac{3x^2}{L^2} - \frac{2x^3}{L^3}; \ N_{v4} = -\frac{x^2}{L} + \frac{x^3}{L^2} \qquad (13.33a\text{–}d)$$

The variations of the shape functions N_{v1}, N_{v2}, N_{v3} and N_{v4} are plotted in Figure 13.6.

It is usually convenient to represent Equations 13.26 and 13.32 with the matrix of shape functions $\mathbf{N}_e$ and a vector of nodal displacements $\mathbf{d}_e$:

$$\begin{bmatrix} u \\ v \end{bmatrix} = \begin{bmatrix} N_{u1} & 0 & 0 & N_{u2} & N_{u3} & 0 & 0 \\ 0 & N_{v1} & N_{v2} & 0 & 0 & N_{v3} & N_{v4} \end{bmatrix} \begin{bmatrix} u_L \\ v_L \\ \theta_L \\ u_M \\ u_R \\ v_R \\ \theta_R \end{bmatrix} = \mathbf{N}_e\mathbf{d}_e \qquad (13.34)$$

Obviously, other polynomials could be used in the derivation and the associated element freedoms would need to be modified accordingly. In fact, the number of available freedoms is defined by the number of coefficients introduced in the approximating polynomials. For example, we introduced three and four coefficients for u and v, respectively, in Equations

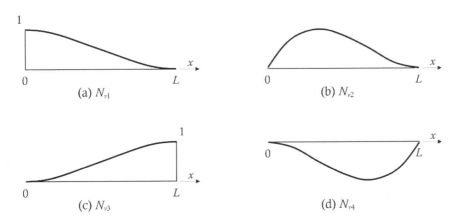

Figure 13.6 Shape functions used for the cubic approximation of v.

13.21. These have been replaced by three freedoms for u (u_L, u_M, u_R) and four freedoms for v (v_L, v_R, θ_L, θ_R), respectively, as shown in Figure 13.4.

The key step in the derivation of a displacement-based finite element is the approximation of the generalised displacements for the model considered by means of polynomial (or other) functions. On the basis of the notation introduced, this can be expressed as:

$$e \approx N_e d_e \tag{13.35}$$

from which the variables describing the strain distribution can be calculated by making use of Equation 13.16:

$$\varepsilon = AN_e d_e = Bd_e \tag{13.36}$$

with:

$$B = AN_e \tag{13.37}$$

By substituting the approximation of Equations 13.35 and 13.36 into Equation 13.20, the so-called weak form of the problem expressed in terms of nodal displacements is produced:

$$\int_L D(Bd_e) \cdot B\hat{d}_e \, dx = \int_L p \cdot N_e \hat{d}_e \, dx \tag{13.38}$$

This relationship can be re-arranged to isolate the virtual nodal displacements $\hat{d}_e$ on one side of the dot product. This is achieved recalling that $Aa \cdot Bb = B^T Aa \cdot b$:

$$\int_L B^T D(Bd_e) \, dx \cdot \hat{d}_e = \int_L N_e^T p \, dx \cdot \hat{d}_e \tag{13.39}$$

from which the stiffness relationship of the finite element can be obtained:

$$k_e d_e = q_e \tag{13.40}$$

where k_e is the finite element stiffness matrix and q_e represents the loading vector related to the member loads w and n. These are defined as:

$$k_e = \int_L B^T DB \, dx \tag{13.41a}$$

$$q_e = \int_L N_e^T p \, dx \tag{13.41b}$$

The derivation of the element stiffness matrix and the loading vector for the 7-dof element depicted in Figure 13.4 is presented below.

The stiffness matrix of the 7-dof finite element is calculated based on Equation 13.41a. This requires the calculation of matrix **B** and, recalling its definition in Equation 13.37, is obtained as:

$$
\mathbf{B} = \begin{bmatrix}
-\dfrac{3}{L}+\dfrac{4x}{L^2} & 0 & 0 & \dfrac{4}{L}-\dfrac{8x}{L^2} & -\dfrac{1}{L}+\dfrac{4x}{L^2} & 0 & 0 \\[2ex]
0 & \dfrac{12x}{L^3}-\dfrac{6}{L^2} & \dfrac{6x}{L^2}-\dfrac{4}{L} & 0 & 0 & \dfrac{6}{L^2}-\dfrac{12x}{L^3} & \dfrac{6x}{L^2}-\dfrac{2}{L}
\end{bmatrix}
\tag{13.42}
$$

Substituting the expression obtained for **B** into Equation 13.41a and carrying out the integration along the member length produces the element stiffness matrix:

$$
\mathbf{k}_e = \int_L \mathbf{B}^\mathsf{T}\mathbf{DB}\,\mathrm{d}x = \begin{bmatrix}
\dfrac{7R_A}{3L} & \dfrac{-4R_B}{L^2} & \dfrac{-3R_B}{L} & \dfrac{-8R_A}{3L} & \dfrac{R_A}{3L} & \dfrac{4R_B}{L^2} & \dfrac{-R_B}{L} \\[2ex]
\dfrac{-4R_B}{L^2} & \dfrac{12R_I}{L^3} & \dfrac{6R_I}{L^2} & \dfrac{8R_B}{L^2} & \dfrac{-4R_B}{L^2} & \dfrac{-12R_I}{L^3} & \dfrac{6R_I}{L^2} \\[2ex]
\dfrac{-3R_B}{L} & \dfrac{6R_I}{L^2} & \dfrac{4R_I}{L} & \dfrac{4R_B}{L} & \dfrac{-R_B}{L} & \dfrac{-6R_I}{L^2} & \dfrac{2R_I}{L} \\[2ex]
\dfrac{-8R_A}{3L} & \dfrac{8R_B}{L^2} & \dfrac{4R_B}{L} & \dfrac{16R_A}{3L} & \dfrac{-8R_A}{3L} & \dfrac{-8R_B}{L^2} & \dfrac{4R_B}{L} \\[2ex]
\dfrac{R_A}{3L} & \dfrac{-4R_B}{L^2} & \dfrac{-R_B}{L} & \dfrac{-8R_A}{3L} & \dfrac{7R_A}{3L} & \dfrac{4R_B}{L^2} & \dfrac{-3R_B}{L} \\[2ex]
\dfrac{4R_B}{L^2} & \dfrac{-12R_I}{L^3} & \dfrac{-6R_I}{L^2} & \dfrac{-8R_B}{L^2} & \dfrac{4R_B}{L^2} & \dfrac{12R_I}{L^3} & \dfrac{-6R_I}{L^2} \\[2ex]
\dfrac{-R_B}{L} & \dfrac{6R_I}{L^2} & \dfrac{2R_I}{L} & \dfrac{4R_B}{L} & \dfrac{-3R_B}{L} & \dfrac{-6R_I}{L^2} & \dfrac{4R_I}{L}
\end{bmatrix}
\tag{13.43}
$$

where R_A, R_B and R_I are the rigidities defined in Equations 13.12. The inclusion of the first moments of area becomes essential when the stiffness coefficients vary during the analysis, for example, to account for material nonlinearities.

The loading vector required to account for member loads n and w is calculated based on Equations 13.41b:

$$
\mathbf{q}_e = \int_L \mathbf{N}_e^\mathsf{T}\mathbf{p}\,\mathrm{d}x = \begin{bmatrix} \dfrac{L}{6}n & \dfrac{L}{2}w & \dfrac{L^2}{12}w & \dfrac{2L}{3}n & \dfrac{L}{6}n & \dfrac{L}{2}w & -\dfrac{L^2}{12}w \end{bmatrix}^\mathsf{T}
\tag{13.44}
$$

For illustrative purposes and to better outline all the steps involved in the solution process, the integrals in the expressions for $\mathbf{k}_e$ (Equation 13.41a) and $\mathbf{q}_e$ (Equation 13.41b) have been solved analytically in Equations 13.43 and 13.44. However, numerical integration could be easily implemented if preferred (see Chapter 15). Other procedures can be used to deal with internal freedoms, such as static condensation where the internal freedoms are expressed in terms of the other (boundary) freedoms and removed from the system of equations used in the solution process. Results related to the internal freedoms are then obtained in the post-processing phase of the analysis once the variables at the boundary freedoms are determined. These procedures will not be covered here.

13.2.4 Solution procedure

In a real structure, different members may have different orientations (with different local coordinate systems). It is therefore convenient to introduce a *global coordinate system* to be used for the whole structure, as already introduced in Chapter 12 for the frame analysis using the stiffness method. The load and displacement vectors for a particular element (defined in local coordinates in Equations 13.44 and 13.34, respectively) can be expressed in global coordinates by carrying out the following transformations:

$$\mathbf{d}_e = \mathbf{T}\mathbf{D}_e \quad \mathbf{Q}_e = \mathbf{T}^\mathsf{T}\mathbf{q} \qquad (13.45a,b)$$

where $\mathbf{Q}_e$ represents the load vector in global coordinates, $\mathbf{D}_e$ is the vector of nodal displacements in global coordinates and the transformation matrix $\mathbf{T}$ is given by (see Section 12.3):

$$\mathbf{T} = \begin{bmatrix} l & m & 0 & 0 & 0 & 0 & 0 \\ -m & l & 0 & 0 & 0 & 0 & 0 \\ 0 & 0 & 1 & 0 & 0 & 0 & 0 \\ 0 & 0 & 0 & 1 & 0 & 0 & 0 \\ 0 & 0 & 0 & 0 & l & m & 0 \\ 0 & 0 & 0 & 0 & -m & l & 0 \\ 0 & 0 & 0 & 0 & 0 & 0 & 1 \end{bmatrix} \qquad (13.46)$$

where l and m are respectively the cosine and the sine of the angle between the global and local coordinate systems (see Equations 12.6).

Substituting Equations 13.45 into Equation 13.40 gives the stiffness relationship of a particular element expressed in global coordinates:

$$\mathbf{Q}_e = \mathbf{K}_e\mathbf{D}_e \qquad (13.47)$$

where $\mathbf{K}_e$ is the stiffness matrix of the element in global coordinates given by:

$$\mathbf{K}_e = \mathbf{T}^\mathsf{T}\mathbf{k}_e\mathbf{T} \qquad (13.48)$$

The stiffness relationship for the whole structure is then obtained by assembling the contribution of each element, similarly to the procedure already outlined for the stiffness method in Chapter 12 (see Section 12.6) and can be expressed as:

$$\mathbf{Q} = \mathbf{K}\mathbf{D} \qquad (13.49)$$

where $\mathbf{K}$ is the structure stiffness matrix, while $\mathbf{Q}$ and $\mathbf{D}$ are, respectively, the vectors of nodal actions and displacements for the whole structure expressed in the global coordinate system.

Equation 13.49 is readily solved for the unknown displacements and reactions. In this process, it is convenient to partition Equation 13.49 to distinguish between known and unknown displacements ($\mathbf{D}_k$ and $\mathbf{D}_u$) and known and unknown actions ($\mathbf{Q}_k$ and $\mathbf{Q}_u$) as follows:

$$\begin{bmatrix} \mathbf{Q}_k \\ \mathbf{Q}_u \end{bmatrix} = \begin{bmatrix} \mathbf{K}_{11} & \mathbf{K}_{12} \\ \mathbf{K}_{21} & \mathbf{K}_{22} \end{bmatrix} \begin{bmatrix} \mathbf{D}_u \\ \mathbf{D}_k \end{bmatrix} \tag{13.50}$$

The determination of the unknown displacements $\mathbf{D}_u$ can be performed using one of the solution procedures presented in Appendix C. For example, these could be evaluated, together with the unknown actions, as:

$$\mathbf{D}_u = \mathbf{K}_{11}^{-1}(\mathbf{Q}_k - \mathbf{K}_{12}\mathbf{D}_k) \quad \text{and} \quad \mathbf{Q}_u = \mathbf{K}_{21}\mathbf{D}_u + \mathbf{K}_{22}\mathbf{D}_k \tag{13.51a,b}$$

13.2.5 Post-processing

When the analysis is completed, the solution is post-processed and, for each element, the different variables describing its structural response are determined on the basis of the calculated nodal displacements. For example, the variables defining the strain diagram are obtained from Equations 13.36 and 13.42:

$$u' = -\frac{3}{L}u_L + \frac{4}{L}u_M - \frac{1}{L}u_R + \left(\frac{4}{L^2}u_L - \frac{8}{L^2}u_M + \frac{4}{L^2}u_R\right)x \tag{13.52a}$$

$$v'' = -\frac{6}{L^2}v_L - \frac{4}{L}\theta_L + \frac{6}{L^2}v_R - \frac{2}{L}\theta_R + \left(\frac{12}{L^3}v_L + \frac{6}{L^2}\theta_L - \frac{12}{L^3}v_R + \frac{6}{L^2}\theta_R\right)x \tag{13.52b}$$

In a similar manner, the expressions for the internal axial force N and internal moment M can be calculated by substituting Equations 13.52 into Equations 13.13. The variation for the shear force along the member length needs to be evaluated from equilibrium considerations.

13.2.6 Remarks on the consistency requirements for finite elements

The proposed 7-dof finite element represents the simplest element that fulfils the consistency requirements by approximating the displacements by means of polynomials and thereby avoids potential locking problems that may arise when the member local x-axis does not pass through the centroid of the member cross-section. The ability to select a reference system with the origin not necessarily coincident with the centroid of the section is fundamental when dealing with material nonlinearities as the location of the actual centroid of a cross-section with nonlinear material behaviour varies depending on the level of applied loading or deformation.

From a practical viewpoint, the consistency requirement is satisfied when the independent displacements (or their derivatives) present in the expressions of the strains of the model possess the same order (i.e. u' and v'' have the same order). Adopting a cubic function for the deflection v leads to a linear contribution to the strain (i.e. due to v'', included in Equation 13.4a). Similarly, in order to produce the same (linear) contribution to the strain provided by

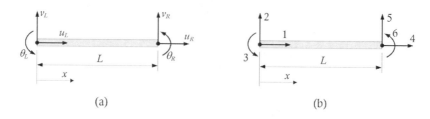

Figure 13.7 The 6-dof finite element (Euler–Bernoulli beam). (a) Nodal displacements. (b) Freedom numbering for the isolated finite element.

u' (specified in Equation 13.4a), it is necessary to have a parabolic function approximating the axial displacement u. A linear function for u, for example, would not be able to achieve this as its first derivative (i.e. u') is constant.

To better illustrate this behaviour, the results obtained using the 7-dof finite element (Figure 13.4) are compared to those calculated using a 6-dof element (Figure 13.7). The 6-dof element approximates u and v by means of linear and cubic polynomials, respectively. As previously discussed, this latter element does not satisfy the consistency requirements owing to the orders of its polynomials.

On the basis of the adopted approximated polynomials, the stiffness matrix of the 6-dof finite element is obtained based on Equation 13.41a as:

$$
\mathbf{k}_e =
\begin{bmatrix}
\dfrac{R_A}{L} & 0 & -\dfrac{R_B}{L} & -\dfrac{R_A}{L} & 0 & \dfrac{R_B}{L} \\[2ex]
0 & \dfrac{12R_I}{L^3} & \dfrac{6R_I}{L^2} & 0 & -\dfrac{12R_I}{L^3} & \dfrac{6R_I}{L^2} \\[2ex]
-\dfrac{R_B}{L} & \dfrac{6R_I}{L^2} & \dfrac{4R_I}{L} & \dfrac{R_B}{L} & -\dfrac{6R_I}{L^2} & \dfrac{2R_I}{L} \\[2ex]
-\dfrac{R_A}{L} & 0 & \dfrac{R_B}{L} & \dfrac{R_A}{L} & 0 & -\dfrac{R_B}{L} \\[2ex]
0 & -\dfrac{12R_I}{L^3} & -\dfrac{6R_I}{L^2} & 0 & \dfrac{12R_I}{L^3} & -\dfrac{6R_I}{L^2} \\[2ex]
\dfrac{R_B}{L} & \dfrac{6R_I}{L^2} & \dfrac{2R_I}{L} & -\dfrac{R_B}{L} & -\dfrac{6R_I}{L^2} & \dfrac{4R_I}{L}
\end{bmatrix}
\tag{13.53}
$$

The calculated matrix is equivalent to the stiffness matrix of the frame element presented in the previous chapter (Equation 12.1a), which was calculated assuming a centroidal reference system, i.e. a reference system for which $R_B = 0$.

For the case of a simply-supported beam with a prismatic rectangular section and subjected to a point load applied at mid-span, the mid-span deflections calculated using the 6-dof and 7-dof finite elements are shown in Figure 13.8. These results have been obtained by discretising the member with four elements to clearly emphasise the implications of the different sets of polynomials. The instantaneous mid-span deflection has been plotted for different positions of the reference axis (denoted by d_{ref} and measured from the top of the section) expressed as a function of the cross-section depth d. With this notation, the reference axis is located at the level of the centroid when $d_{ref}/d = 0.5$, in which case both elements produce the same mid-span deflection shown in Figure 13.8 where the ratio between the deflections calculated with the 6-dof and the 7-dof elements equals 1.

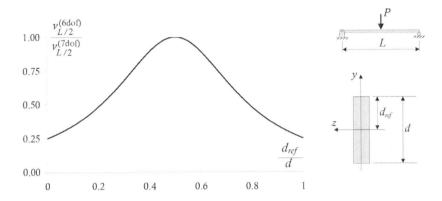

Figure 13.8 Comparisons between mid-span deflections calculated with 6-dof and 7-dof finite elements.

Based on Figure 13.8, it is apparent that, when using the 6-dof element with the origin of the reference system not coinciding with the cross-sectional centroid, a stiffer response than expected is obtained.

REFLECTION ACTIVITY 13.1

Consider the nodal displacements specified for the 6-dof finite element in Figure 13.7 and derive the stiffness matrix $\mathbf{k}_e$ and the loading vector $\mathbf{q}_e$ to account for constant member loads n and w applied over the member length.

The 6-dof finite element approximates displacements u and v by means of linear and cubic polynomials which can be expressed as:

$$u = a_0 + a_1x \tag{13.54a}$$

$$v = b_0 + b_1x + b_2x^2 + b_3x^3 \tag{13.54b}$$

$$\theta = dv/dx = b_1 + 2b_2x + 3b_3x^2 \tag{13.54c}$$

Following the procedure previously presented in Equations 13.21 through 13.34 for the 7-dof finite element, the coefficients a_i and b_j (with $i = 0,1$ and $j = 0,1,2,3$) can be re-written in terms of the nodal displacements (depicted in Figure 13.7) applying the following conditions to Equations 13.54:

$$u(x = 0) = u_L \quad u(x = L) = u_R \quad v(x = 0) = v_L$$

$$v(x = L) = v_R \quad \theta(x = 0) = \theta_L \quad \theta(x = L) = \theta_R$$

from which the expressions for a_i and b_j can be written as:

$$a_0 = u_L \quad a_1 = -\frac{u_L - u_R}{L} \quad\quad\quad b_0 = v_L$$

$$b_1 = \theta_L \quad b_2 = -\frac{3v_L + 2\theta_L L - 3v_R + \theta_R L}{L^2} \quad b_3 = \frac{\theta_L L + 2v_L - 2v_R + \theta_R L}{L^3}$$

Based on these, the functions describing the approximated generalised displacements become:

$$u = u_L + \left(-\frac{u_L - u_R}{L} \right) x \tag{13.55a}$$

$$v = v_L + \theta_L x - \frac{3v_L + 2\theta_L L - 3v_R + \theta_R L}{L^2} x^2 + \frac{\theta_L L + 2v_L - 2v_R + \theta_R L}{L^3} x^3 \tag{13.55b}$$

which can be re-arranged by collecting the nodal displacements as:

$$u = N_{u1} u_L + N_{u2} u_R \tag{13.56a}$$

$$v = N_{v1} v_L + N_{v2} \theta_L + N_{v3} v_R + N_{v4} \theta_R \tag{13.56b}$$

with the shape functions given by:

$$N_{u1} = 1 - \frac{x}{L} \qquad N_{u2} = \frac{x}{L}$$

$$N_{v1} = +1 - 3\frac{x^2}{L^2} + 2\frac{x^3}{L^3} \qquad N_{v2} = x - 2\frac{x^2}{L} + \frac{x^3}{L^2} \qquad N_{v3} = 3\frac{x^2}{L^2} - 2\frac{x^3}{L^3} \qquad N_{v4} = -\frac{x^2}{L} + \frac{x^3}{L^2}$$

Equations 13.56 can be re-written in more compact form as:

$$\begin{bmatrix} u \\ v \end{bmatrix} = \begin{bmatrix} N_{u1} & 0 & 0 & N_{u2} & 0 & 0 \\ 0 & N_{v1} & N_{v2} & 0 & N_{v3} & N_{v4} \end{bmatrix} \begin{bmatrix} u_L \\ v_L \\ \theta_L \\ u_R \\ v_R \\ \theta_R \end{bmatrix} = \mathbf{N_e d_e} \tag{13.57}$$

The stiffness matrix is then determined with Equation 13.41a:

$$\mathbf{k_e} = \int_L \mathbf{B^T D B}\, dx = \begin{bmatrix} \dfrac{R_A}{L} & 0 & -\dfrac{R_B}{L} & -\dfrac{R_A}{L} & 0 & \dfrac{R_B}{L} \\[2mm] 0 & \dfrac{12R_I}{L^3} & \dfrac{6R_I}{L^2} & 0 & -\dfrac{12R_I}{L^3} & \dfrac{6R_I}{L^2} \\[2mm] -\dfrac{R_B}{L} & \dfrac{6R_I}{L^2} & \dfrac{4R_I}{L} & \dfrac{R_B}{L} & -\dfrac{6R_I}{L^2} & \dfrac{2R_I}{L} \\[2mm] -\dfrac{R_A}{L} & 0 & \dfrac{R_B}{L} & \dfrac{R_A}{L} & 0 & -\dfrac{R_B}{L} \\[2mm] 0 & -\dfrac{12R_I}{L^3} & -\dfrac{6R_I}{L^2} & 0 & \dfrac{12R_I}{L^3} & -\dfrac{6R_I}{L^2} \\[2mm] \dfrac{R_B}{L} & \dfrac{6R_I}{L^2} & \dfrac{2R_I}{L} & -\dfrac{R_B}{L} & -\dfrac{6R_I}{L^2} & \dfrac{4R_I}{L} \end{bmatrix} \tag{13.58}$$

with matrix **D** defined in Equation 13.15b, and **B** is evaluated using Equation 13.37 as follows:

$$
\mathbf{B} = \begin{bmatrix} -\dfrac{1}{L} & 0 & 0 & \dfrac{1}{L} & 0 & 0 \\[2mm] 0 & 12\dfrac{x}{L^3} - \dfrac{6}{L^2} & \dfrac{6x}{L^2} - \dfrac{4}{L} & 0 & \dfrac{6}{L^2} - 12\dfrac{x}{L^3} & \dfrac{6x}{L^2} - \dfrac{2}{L} \end{bmatrix} \tag{13.59}
$$

The loading vector $\mathbf{q}_e$ required to account for constant member loads n and w is evaluated with Equation 13.41b as:

$$
\mathbf{q}_e = \int_L \mathbf{N}_e^T \mathbf{p}\, dx = \begin{bmatrix} \dfrac{nL}{2} & \dfrac{wL}{2} & \dfrac{wL^2}{12} & \dfrac{nL}{2} & \dfrac{wL}{2} & -\dfrac{wL^2}{12} \end{bmatrix}^T \tag{13.60}
$$

SUMMARY OF STEPS 13.1: Finite element analysis — Solution procedure

The main steps to be carried out for structural analysis using finite element line elements are detailed below.

1. Assign a global reference system.

2. Select the type of element to be used for the analysis and select the level of mesh refinement (i.e. number of elements to be used to discretise the structure). Usually, the selection of the number of nodes and elements requires an iterative process where the accuracy of the results is assessed for different discretisations. Assign location vectors for each element (this is usually carried out by specifying locations of start and end nodes for each element).

3. Assign freedoms at the nodes to accommodate nodal freedoms of the finite elements. Make sure to include additional freedoms, if necessary, to account for possible internal freedoms of the finite elements and possible end releases, such as hinges, which enable relative movements between adjacent elements. When performing the calculations by hand, number unrestrained freedoms first followed by restrained ones. In reality, because of the relatively large number of freedoms usually involved in the finite element modelling, calculations are performed with a computer where the managing of the unrestrained and restrained freedoms is independent from their numbering, and it is carried out by recording freedoms associated with the two unrestrained and restrained conditions separately.

4. Calculate stiffness matrices and loading vectors for each of the elements specified in the discretised structure.

5. Assemble all contributions from the individual finite elements to produce the structure stiffness matrix and vectors collecting loads and displacements.

6. Solve for the unknown nodal displacements.

7. Post-process the solution to evaluate all variables required to describe the structural response.

WORKED EXAMPLE 13.1

Consider the simply-supported beams shown in Figure 13.9. Beam 1 is subjected to a point load applied at mid-span and beam 2 carries a uniformly distributed load. Calculate and compare the mid-span deflections obtained using different levels of meshing (i.e. different number of elements). Use the 7-dof finite element of Figure 13.4, and assume the member to be prismatic and rectangular with the following cross-sectional and material properties: b (width) = 100 mm, d (depth) = 400 mm and E = 20 GPa.

Adopt a centroidal reference system (i.e. with $R_B = 0$).

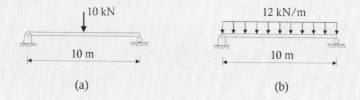

(a)   (b)

Figure 13.9 Beams for Worked Example 13.1. (a) Beam 1. (b) Beam 2.

The two beams are analysed separately following the Summary of Steps 13.1.

Beam 1

(1) Global coordinate system:

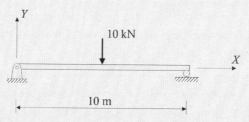

(2) The analyses are carried out using the 7-dof finite element of Figure 13.4.

The number of elements used in the analysis is varied from 2 (one on each side of the point load) to 100. Obviously, these simulations require the use of appropriate computer software to solve the system of equations for the unknown displacements following one of the solution procedures presented in Appendix C. To clarify the steps involved in the solution, the calculations when the beam is discretised into just two elements are shown here. The location vectors adopted in the analysis are shown below.

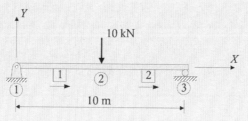

(3) The 7-dof element possesses three freedoms at end nodes and one internal freedom. These are illustrated below, where unrestrained freedoms have been numbered from 1 to 8, and restrained freedoms, from 9 to 11.

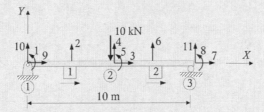

(4) The stiffness matrix of each element is calculated with Equation 13.43.

Element 1

$$k_{e(1)} = 10^3 \begin{bmatrix} 373.3 & 0 & 0 & -426.6 & 53.33 & 0 & 0 \\ 0 & 1.024 & 2.56 & 0 & 0 & -1.024 & 2.56 \\ 0 & 2.56 & 8.533 & 0 & 0 & -2.56 & 4.266 \\ -426.6 & 0 & 0 & 853.3 & -426.6 & 0 & 0 \\ 53.33 & 0 & 0 & -426.6 & 373.3 & 0 & 0 \\ 0 & -1.024 & -2.56 & 0 & 0 & 1.024 & -2.56 \\ 0 & 2.56 & 4.266 & 0 & 0 & -2.56 & 8.533 \end{bmatrix}$$

where:

$L_{(1)} = 5 \text{ m} \quad R_A = 0.4 \times 0.1 \times 20 \times 10^6 = 800 \times 10^3 \text{ kN} \quad R_B = 0$

$R_I = (0.1 \times 0.4^3/12) \times 20 \times 10^6 = 10.66 \times 10^3 \text{ kNm}^2$

Because of the element orientation, the transformation matrix **T** (Equation 13.46) is an identity matrix, with $l = 1$ and $m = 0$, and the local and global coordinates coincide. It is then possible to write the stiffness matrix $\mathbf{K}_{e(1)}$ of the isolated 7-dof element in global coordinates as follows:

$$\mathbf{K}_{e(1)} = \mathbf{T}_{(1)}^T \mathbf{k}_{e(1)} \mathbf{T}_{(1)} = \mathbf{k}_{e(1)}$$

Element 2

Taking advantage of the fact that elements 1 and 2 are identical:

$$\mathbf{k}_{e(2)} = \mathbf{k}_{e(1)} \text{ and } \mathbf{K}_{e(2)} = \mathbf{K}_{e(1)}$$

There is no need to calculate loading vectors as no member loads are applied.

(5) The stiffness matrix of the structure **K** is assembled by accounting for the contribution of both elements and making sure to correctly map their stiffness coefficients from the freedoms of the isolated finite elements to the structure freedoms. Freedoms 1 to 7 of the isolated elements 1 and 2 are mapped to the structure freedoms 9, 10, 1, 2, 3, 4 and 5 and to structure freedoms 3, 4, 5, 6, 7, 11 and 8, respectively.

$$\mathbf{K} = 10^3 \begin{bmatrix}
8.533 & 0 & 0 & -2.56 & 4.266 & 0 & 0 & 0 & 0 & 2.56 & 0 \\
0 & 853.3 & -426.6 & 0 & 0 & 0 & 0 & 0 & -426.6 & 0 & 0 \\
0 & -426.6 & 746.6 & 0 & 0 & -426.6 & 53.33 & 0 & 53.33 & 0 & 0 \\
-2.56 & 0 & 0 & 2.048 & 0 & 0 & 0 & 2.56 & 0 & -1.024 & -1.024 \\
4.266 & 0 & 0 & 0 & 17.06 & 0 & 0 & 4.266 & 0 & 2.56 & -2.56 \\
0 & 0 & -426.6 & 0 & 0 & 853.3 & -426.6 & 0 & 0 & 0 & 0 \\
0 & 0 & 53.33 & 0 & 0 & -426.6 & 373.3 & 0 & 0 & 0 & 0 \\
0 & 0 & 0 & 2.56 & 4.266 & 0 & 0 & 8.533 & 0 & 0 & -2.56 \\
0 & -426.6 & 53.33 & 0 & 0 & 0 & 0 & 0 & 373.3 & 0 & 0 \\
2.56 & 0 & 0 & -1.024 & 2.56 & 0 & 0 & 0 & 0 & 1.024 & 0 \\
0 & 0 & 0 & -1.024 & -2.56 & 0 & 0 & -2.56 & 0 & 0 & 1.024
\end{bmatrix}$$

Vectors related to known and unknown displacements and actions are:

$\mathbf{D}_k = [0\ 0\ 0]^T$ 　　　　　 $\mathbf{D}_u = [D_1\ D_2\ D_3\ D_4\ D_5\ D_6\ D_7\ D_8]^T$

$\mathbf{Q}_k = [0\ 0\ 0\ -10\ 0\ 0\ 0\ 0]^T$ 　 $\mathbf{Q}_u = [Q_9\ Q_{10}\ Q_{11}]^T$

(6) The unknown displacements are determined applying one of the solution procedures presented in Appendix C:

$\mathbf{D}_u = 10^{-3} \times [-5.859\ 0\ 0\ -19.53\ 0\ 0\ 0\ 5.859]^T$

(7) Based on $\mathbf{D}_k$ and $\mathbf{D}_u$, the nodal displacements of each element become:

$\mathbf{d}_{e(1)} = 10^{-3} \times [0\ 0\ -5.859\ 0\ 0\ -19.53\ 0]^T$

$\mathbf{d}_{e(2)} = 10^{-3} \times [0\ -19.53\ 0\ 0\ 0\ 5.859]^T$

The mid-span deflection corresponds to the displacement at global freedom number 4:

$v(x = 5\text{ m}) = -19.53 \times 10^{-3}\text{ m}$

Increasing the number of elements used for the discretisation of the beam does not change the results. This is due to the fact that the solutions of the differential equations describing the deflections of the beam are expressed in terms of a cubic function, when the beam is subjected to a point load (see Section 5.7). This polynomial is part of the approximations used for the generalised displacements and, for this reason, the solution is said to be *exact* on the basis of the assumptions of the adopted model.

Beam 2

The solution implemented for beam 2 is similar to the one already presented for beam 1 and its details are not repeated here.

The results for the mid-span deflections, calculated for different numbers of elements (i.e. different finite element meshes), are plotted in Figure 13.10. In this case, the mid-span deflections converge to the exact value at a very coarse discretisation, i.e. with just two elements (Figure 13.10). Despite this, finer meshes are required for an accurate evaluation of the mid-span curvature as illustrated in Figure 13.10.

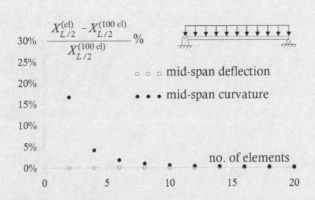

Figure 13.10 Calculated deflection of Beam 2 with different numbers of elements.

For this beam loaded with a uniformly distributed load, the solution is dependent on the specified discretisation because the polynomial adopted in the finite element formulation (i.e. cubic functions for the deflection) is one order smaller than the polynomial describing the actual deflection curve for the member that requires a polynomial of order 4 (see Section 5.7).

13.3 TIMOSHENKO BEAM MODEL

The Euler–Bernoulli beam is a very useful model for the representation of the behaviour of structural members. In particular, it works very well for beams whose length-to-depth ratio is greater than about 10. For shorter beams, also sometimes referred to as deep beams, the shearing deformations (not included in the Euler–Bernoulli beam formulation) need to be accounted for, as they may significantly affect the structural response. The Timoshenko beam represents the simplest analytical formulation capable of describing this behaviour.

In the following, the weak formulation describing the Timoshenko beam model is presented following the same procedure, based on the principle of virtual work, previously adopted for the Euler–Bernoulli beam. For this reason, the description for the Timoshenko beam will refer, where possible, to the previous section to avoid unnecessary repetitions. This formulation is then applied for the derivation of two finite elements. The possible occurrence of a problem known as *shear locking* in the analysis is discussed and simple recommendations to avoid this problem are provided.

13.3.1 Kinematic model

The analytical formulation of the Timoshenko Beam Model is also presented considering a prismatic member (Figure 13.1). Like the Euler–Bernoulli Model, the displacement field of the beam is described with Equations 13.1, reproduced here for ease of reference:

$$d_x(x, y) = u(x) - y \sin \theta(x) \tag{13.1a}$$

$$d_y(x, y) = v(x) - y + y \cos \theta(x) \tag{13.1b}$$

The graphical description of the displacement field is provided for a point P lying on the member axis in Figure 13.11a and for a point Q located away from the axis x in Figure 13.11b. From this representation, we can see that plane sections are assumed to remain

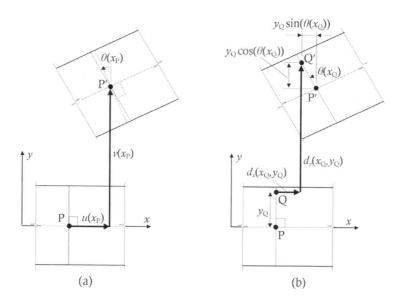

Figure 13.11 Displacement field for the Timoshenko beam model.

plane (as for the Euler–Bernoulli beam), but they may not remain orthogonal to the member axis after deformation (different to the Euler–Bernoulli beam). The main consequence of this latter condition is that the rotation of the beam is not always equal to v' because of the influence of shearing deformations.

Under these conditions, the kinematic behaviour of the Timoshenko beam is described by three variables: the axial displacement at the level of the reference axis $u(x)$, the vertical deflection $v(x)$ and the rotation $\theta(x)$. In the following, $u(x)$, $v(x)$ and $\theta(x)$ will be referred to as u, v and θ for ease of notation. These variables represent the generalised displacements and are collected in the vector **e** as:

$$\mathbf{e} = [u \quad v \quad \theta]^{\mathrm{T}} \tag{13.61}$$

The expressions for the curvature κ $(= \theta')$ along the beam can then be obtained by differentiating the rotation θ with respect to the coordinate x along the beam length.

In the framework of small displacements (i.e. $\cos\theta \approx 1$ and $\sin\theta \approx \theta$), the admissible displacement of a point in the beam is defined by the following expressions (Figure 13.11):

$$d_x(x, y) = u - y\theta \tag{13.62a}$$

$$d_y(x, y) = v \tag{13.62b}$$

The relevant strain field can then be obtained from the displacement field as:

$$\varepsilon_x = \frac{\partial d_x}{\partial x} = u' - y\theta' \qquad \varepsilon_y = \frac{\partial d_y}{\partial y} = 0 \qquad \varepsilon_z = \frac{\partial d_z}{\partial z} = 0 \tag{13.63a–c}$$

$$\gamma_{xy} = \frac{\partial d_x}{\partial y} + \frac{\partial d_y}{\partial x} = v' - \theta \qquad \gamma_{yz} = \frac{\partial d_y}{\partial z} + \frac{\partial d_z}{\partial y} = 0 \qquad \gamma_{xz} = \frac{\partial d_x}{\partial z} + \frac{\partial d_z}{\partial x} = 0 \tag{13.63d–f}$$

where ε_x and γ_{xy} are the non-zero strains for the Timoshenko beam model.

The expression for γ_{xy} highlights how its value remains constant at a cross-section (i.e. it does not vary over the thickness of the member). This representation leads to some inaccuracies that are discussed and addressed in the following sections when considering the calculation of the internal shear force.

13.3.2 Finite element formulation

The weak form of the problem is obtained using the principle of virtual work. For generality, the member is assumed to be subjected to n and w, which represent the vertical and horizontal distributed member loads, respectively, as shown in Figure 13.3. On the basis of the adopted strain field, the principle of virtual work can then be expressed as:

$$\int_L \int_A \sigma_x \hat{\varepsilon}_x \, dA \, dx + \int_L \int_A \tau_{xy} \hat{\gamma}_{xy} \, dA \, dx$$

$$= \int_L (w\hat{v} + n\hat{u}) \, dx + S_L \hat{v}_L + N_L \hat{u}_L + M_L \hat{\theta}_L + S_R \hat{v}_R + N_R \hat{u}_R + M_R \hat{\theta}_R \qquad (13.64)$$

where the variables with the hat '^' again represent virtual variations of displacements or strains. In Equation 13.64, σ_x and τ_{xy} represent those longitudinal and shearing stresses that produce internal work, and the terms on the right-hand side represent the work done by the member forces along the beam length and by the nodal actions at the beam ends.

Equation 13.64 can be re-written ignoring the nodal actions included on its right-hand side, because the use of zero virtual displacements at the segment ends does not influence the calculation of the stiffness matrix and the vector accounting for the member loads n and w. This is carried out by highlighting the terms related to the internal actions (N, M and S) and those describing the strain field (u', v', θ and θ') as follows:

$$\int_L \begin{bmatrix} N \\ M \\ S \end{bmatrix} \cdot \begin{bmatrix} \hat{u}' \\ \hat{\theta}' \\ \hat{v}' - \hat{\theta} \end{bmatrix} dx = \int_L \begin{bmatrix} n \\ w \\ 0 \end{bmatrix} \cdot \begin{bmatrix} \hat{u} \\ \hat{v} \\ \hat{\theta} \end{bmatrix} dx \qquad (13.65a)$$

or in compact form as:

$$\int_L \mathbf{r} \cdot \mathbf{A}\hat{\mathbf{e}} \, dx = \int_L \mathbf{p} \cdot \hat{\mathbf{e}} \, dx \qquad (13.65b)$$

where $\mathbf{r}$ collects the internal actions ($\mathbf{r} = [N \quad M \quad S]^T$) and $\mathbf{p}$ specifies the member loads ($\mathbf{p} = [n \quad w \quad 0]^T$).

Assuming linear–elastic properties, i.e. $\sigma_x = E\varepsilon_x$ and $\tau_{xy} = G\gamma_{xy}$, the internal actions can be expressed as a function of the generalised displacements on the basis of the strain field described in Equations 13.63:

$$N = \int_A \sigma_x \, dA = \int_A E(\varepsilon_r - y\kappa) \, dA = R_A \varepsilon_r - R_B \kappa \qquad (13.66a)$$

$$M = -\int_A y\sigma_x \, dA = -\int_A Ey(\varepsilon_r - y\kappa) \, dA = -R_B \varepsilon_r + R_I \kappa \qquad (13.66b)$$

$$S = \int_A \tau_{xy}\,dA = \int_A G\gamma_{xy}\,dA = R_S\gamma_{xy} \tag{13.66c}$$

where the expressions for N and M are identical to those previously obtained for the Euler–Bernoulli beam model in Equations 13.11, and R_S represents the shear stiffness, i.e. the rigidity associated with the shear deformation, which is given by:

$$R_S = AG \tag{13.67}$$

where G is the shear modulus of the material.

The use of a constant shear strain γ_{xy} does not satisfy equilibrium at the top and bottom fibres of the cross-section (which are stress free). This is a consequence of the fact that the shear stress requires the use of a higher-order polynomial to describe its actual variation. To address this aspect, a shear correction factor is commonly used with the Timoshenko beam model to modify the cross-sectional shear rigidity to:

$$R_S = k_S AG \tag{13.68}$$

where the value for k_S depends on the geometric and material properties, as well as the loading and boundary conditions of the member analysed. It is beyond the scope of this book to provide more details on its calculation. For example, for a rectangular cross-section of area A $(= b \times d)$ and with an applied shear force parallel to d: $k_S = 5/6$. The inclusion of k_S might become unnecessary if the order of the functions describing the out-of-plane displacements of the cross-section (i.e. warping) is increased and, because of this, these formulations are usually referred to as higher-order beam models.

On the basis of the adopted material properties, Equations 13.66 can be rewritten in more compact form as:

$$\mathbf{r} = \mathbf{D}\boldsymbol{\varepsilon} \tag{13.69}$$

where:

$$\mathbf{r} = \begin{bmatrix} N \\ M \\ S \end{bmatrix} \qquad \mathbf{D} = \begin{bmatrix} R_A & -R_B & 0 \\ -R_B & R_I & 0 \\ 0 & 0 & R_S \end{bmatrix} \tag{13.70a,b}$$

$$\boldsymbol{\varepsilon} = \begin{bmatrix} \varepsilon_r \\ \kappa \\ \gamma_{xy} \end{bmatrix} = \begin{bmatrix} u' \\ \theta' \\ v' - \theta \end{bmatrix} = \begin{bmatrix} \partial & 0 & 0 \\ 0 & 0 & \partial \\ 0 & \partial & -1 \end{bmatrix} \begin{bmatrix} u \\ v \\ \theta \end{bmatrix} = \mathbf{Ae} \tag{13.70c}$$

and $\partial \equiv d/dx$.

The finite element formulation can then be derived by approximating the generalised displacements ($\mathbf{e} \cong \mathbf{N}_e\mathbf{d}_e$) and substituting these in Equation 13.65b to rewrite the weak form of the problem in terms of nodal displacements $\mathbf{d}_e$. These steps have already been performed when presenting the Euler–Bernoulli beam model and, because of this, are not reproduced here. Reference for these should be made to Equations 13.21 through 13.41. In particular, the stiffness matrix $\mathbf{k}_e$ and the loading vector $\mathbf{q}_e$ associated with the member loads n and w are obtained based on Equations 13.41.

The derivation of the finite element describing the behaviour of a Timoshenko beam is outlined below considering the 7-dof element depicted in Figure 13.12, where the generalised

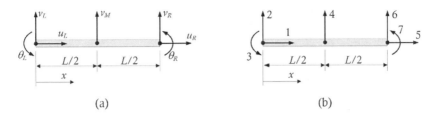

Figure 13.12 The 7-dof finite element (Timoshenko beam). (a) Nodal displacements. (b) Freedom numbering for isolated finite element.

displacements u and θ are approximated by linear functions, while v is described by a parabolic polynomial. Based on these, the expressions for u, v and θ can be re-written as:

$$u = a_0 + a_1 x \qquad (13.71\text{a})$$

$$v = b_0 + b_1 x + b_2 x^2 \qquad (13.71\text{b})$$

$$\theta = c_0 + c_1 x \qquad (13.71\text{c})$$

and re-arranging, the coefficients a_i, b_j and c_i (with $i = 1,2$ and $j = 1,2,3$) may be obtained in terms of the nodal displacements:

$$a_0 = u_{\text{L}} \qquad a_1 = -\frac{u_{\text{L}} - u_{\text{R}}}{L} \qquad (13.72\text{a,b})$$

$$b_0 = v_{\text{L}} \qquad b_1 = -\frac{3v_{\text{L}} - 4v_{\text{M}} + v_{\text{R}}}{L} \qquad b_2 = \frac{2(v_{\text{L}} - 2v_{\text{M}} + v_{\text{R}})}{L^2} \qquad (13.72\text{c–e})$$

$$c_0 = \theta_{\text{L}} \qquad c_1 = -\frac{\theta_{\text{L}} - \theta_{\text{R}}}{L} \qquad (13.72\text{f,g})$$

Based on the adopted approximations, u, v and θ can be expressed as functions of the nodal displacements:

$$
\begin{bmatrix} u \\ v \\ \theta \end{bmatrix} =
\begin{bmatrix}
N_{u1} & 0 & 0 & 0 & N_{u2} & 0 & 0 \\
0 & N_{v1} & 0 & N_{v2} & 0 & N_{v3} & 0 \\
0 & 0 & N_{\theta 1} & 0 & 0 & 0 & N_{\theta 2}
\end{bmatrix}
\begin{bmatrix} u_{\text{L}} \\ v_{\text{L}} \\ \theta_{\text{L}} \\ v_{\text{M}} \\ u_{\text{R}} \\ v_{\text{R}} \\ \theta_{\text{R}} \end{bmatrix}
= \mathbf{N}_e \mathbf{d}_e \qquad (13.73)
$$

where the shape functions are given by:

$$N_{u1} = 1 - \frac{x}{L} \qquad N_{u2} = \frac{x}{L} \qquad (13.74\text{a,b})$$

$$N_{v1} = 1 - \frac{3x}{L} + \frac{2x^2}{L^2} \quad N_{v2} = \frac{4x}{L} - \frac{4x^2}{L^2} \quad N_{v3} = -\frac{x}{L} + \frac{2x^2}{L^2} \tag{13.75a–c}$$

$$N_{\theta1} = 1 - \frac{x}{L} \quad N_{\theta2} = \frac{x}{L} \tag{13.76a,b}$$

The variables describing the strain deformations, collected in vector $\boldsymbol{\varepsilon}$, can be defined in terms of nodal displacements as (Equation 13.70c):

$$\boldsymbol{\varepsilon} = \begin{bmatrix} \varepsilon_r \\ \kappa \\ \gamma_{xy} \end{bmatrix} = \begin{bmatrix} u' \\ \theta' \\ v' - \theta \end{bmatrix} = \mathbf{Ae} = \mathbf{Bd}_e \tag{13.77}$$

in which:

$$\mathbf{B} = \begin{bmatrix} -\dfrac{1}{L} & 0 & 0 & 0 & \dfrac{1}{L} & 0 & 0 \\[2mm] 0 & 0 & -\dfrac{1}{L} & 0 & 0 & 0 & \dfrac{1}{L} \\[2mm] 0 & \dfrac{4x-3L}{L^2} & \dfrac{x-L}{L} & 4\dfrac{L-2x}{L^2} & 0 & \dfrac{4x-L}{L^2} & -\dfrac{x}{L} \end{bmatrix} \tag{13.78}$$

The stiffness matrix $\mathbf{k}_e$ and the loading vector $\mathbf{q}_e$ related to the member loads n and w are calculated with Equations 13.41 as:

$$\mathbf{k}_e = \int_L \mathbf{B}^T \mathbf{D} \mathbf{B} \, dx = \begin{bmatrix} \dfrac{R_A}{L} & 0 & -\dfrac{R_B}{L} & 0 & -\dfrac{R_A}{L} & 0 & \dfrac{R_B}{L} \\[2mm] 0 & \dfrac{7R_S}{3L} & \dfrac{5R_S}{6} & -\dfrac{8R_S}{3L} & 0 & \dfrac{R_S}{3L} & \dfrac{R_S}{6} \\[2mm] -\dfrac{R_B}{L} & \dfrac{5R_S}{6} & \dfrac{LR_S}{3}+\dfrac{R_I}{L} & -\dfrac{2R_S}{3} & \dfrac{R_B}{L} & -\dfrac{R_S}{6} & \dfrac{LR_S}{6}-\dfrac{R_I}{L} \\[2mm] 0 & -\dfrac{8R_S}{3L} & -\dfrac{2R_S}{3} & \dfrac{16R_S}{3L} & 0 & -\dfrac{8R_S}{3L} & \dfrac{2R_S}{3} \\[2mm] -\dfrac{R_A}{L} & 0 & \dfrac{R_B}{L} & 0 & \dfrac{R_A}{L} & 0 & -\dfrac{R_B}{L} \\[2mm] 0 & \dfrac{R_S}{3L} & -\dfrac{R_S}{6} & -\dfrac{8R_S}{3L} & 0 & \dfrac{7R_S}{3L} & -\dfrac{5R_S}{6} \\[2mm] \dfrac{R_B}{L} & \dfrac{R_S}{6} & \dfrac{LR_S}{6}-\dfrac{R_I}{L} & \dfrac{2R_S}{3} & -\dfrac{R_B}{L} & -\dfrac{5R_S}{6} & \dfrac{LR_S}{3}+\dfrac{R_I}{L} \end{bmatrix} \tag{13.79}$$

$$\mathbf{q}_e = \int_L \mathbf{N}_e^T \mathbf{p} \, dx = \begin{bmatrix} \dfrac{nL}{2} & \dfrac{wL}{6} & 0 & \dfrac{2wL}{3} & \dfrac{nL}{2} & \dfrac{wL}{6} & 0 \end{bmatrix}^T \tag{13.80}$$

REFLECTION ACTIVITY 13.2

Derive the stiffness matrix $\mathbf{k}_e$ and the loading vector $\mathbf{q}_e$ for the 6-dof finite element, illustrated in Figure 13.13, describing the behaviour of the Timoshenko beam model.

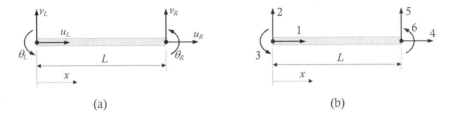

Figure 13.13 The 6-dof finite element (Timoshenko beam). (a) Nodal displacements. (b) Freedom numbering for the isolated finite element.

The generalised displacements of the 6-dof finite element are approximated based on the following linear polynomials:

$$u = a_0 + a_1 x \quad v = b_0 + b_1 x \quad \theta = c_0 + c_1 x$$

Enforcing the following conditions, the coefficients a_i, b_i and c_i (with $i = 0,1$) can be expressed in terms of the nodal displacements shown in Figure 13.13:

$$u(x = 0) = u_L \quad u(x = L) = u_R \quad v(x = 0) = v_L \quad v(x = L) = v_R \quad \theta(x = 0) = \theta_L \quad \theta(x = L) = \theta_R$$

In particular, the expressions for a_i, b_i and c_i become:

$$a_0 = u_L \quad a_1 = -\frac{u_L - u_R}{L} \quad b_0 = v_L \quad b_1 = -\frac{v_L - v_R}{L} \quad c_0 = \theta_L \quad c_1 = -\frac{\theta_L - \theta_R}{L}$$

from which the approximations of the generalised displacements can be re-written as:

$$u = u_L + \left(-\frac{u_L - u_R}{L}\right)x \quad v = v_L + \left(-\frac{v_L - v_R}{L}\right)x \quad \theta = \theta_L + \left(-\frac{\theta_L - \theta_R}{L}\right)x \tag{13.81a–c}$$

These generalised displacements can be re-arranged in a more compact form as:

$$u = N_{u1}u_L + N_{u2}u_R \quad v = N_{v1}v_L + N_{v2}v_R \quad \theta = N_{\theta1}\theta_L + N_{\theta2}\theta_R \tag{13.82a–c}$$

or

$$\begin{bmatrix} u \\ v \\ \theta \end{bmatrix} = \begin{bmatrix} N_{u1} & 0 & 0 & N_{u2} & 0 & 0 \\ 0 & N_{v1} & 0 & 0 & N_{v2} & 0 \\ 0 & 0 & N_{\theta1} & 0 & 0 & N_{\theta2} \end{bmatrix} \begin{bmatrix} u_L \\ v_L \\ \theta_L \\ u_R \\ v_R \\ \theta_R \end{bmatrix} = \mathbf{N}_e \mathbf{d}_e \tag{13.83}$$

where:

$$N_{u1} = 1 - \frac{x}{L} \quad N_{u2} = \frac{x}{L} \quad N_{v1} = 1 - \frac{x}{L} \quad N_{v2} = \frac{x}{L} \quad N_{\theta1} = 1 - \frac{x}{L} \quad N_{\theta2} = \frac{x}{L}$$

The stiffness matrix can be determined using Equation 13.41a as:

$$\mathbf{k}_e = \int_L \mathbf{B}^T \mathbf{D} \mathbf{B}\, dx = \begin{bmatrix} \dfrac{R_A}{L} & 0 & -\dfrac{R_B}{L} & -\dfrac{R_A}{L} & 0 & \dfrac{R_B}{L} \\ 0 & \dfrac{R_S}{L} & \dfrac{R_S}{2} & 0 & -\dfrac{R_S}{L} & \dfrac{R_S}{2} \\ -\dfrac{R_B}{L} & \dfrac{R_S}{2} & \dfrac{R_S L}{3} + \dfrac{R_I}{L} & \dfrac{R_B}{L} & -\dfrac{R_S}{2} & \dfrac{R_S L}{6} - \dfrac{R_I}{L} \\ -\dfrac{R_A}{L} & 0 & \dfrac{R_B}{L} & \dfrac{R_A}{L} & 0 & -\dfrac{R_B}{L} \\ 0 & -\dfrac{R_S}{L} & -\dfrac{R_S}{2} & 0 & \dfrac{R_S}{L} & -\dfrac{R_S}{2} \\ \dfrac{R_B}{L} & \dfrac{R_S}{2} & \dfrac{R_S L}{6} - \dfrac{R_I}{L} & -\dfrac{R_B}{L} & -\dfrac{R_S}{2} & \dfrac{R_S L}{3} + \dfrac{R_I}{L} \end{bmatrix} \tag{13.84}$$

where, in the calculation process, the matrix **B** has been evaluated as follows:

$$\mathbf{B} = \begin{bmatrix} -\dfrac{1}{L} & 0 & 0 & \dfrac{1}{L} & 0 & 0 \\ 0 & 0 & -\dfrac{1}{L} & 0 & 0 & \dfrac{1}{L} \\ 0 & -\dfrac{1}{L} & -1 + \dfrac{x}{L} & 0 & \dfrac{1}{L} & -\dfrac{x}{L} \end{bmatrix} \tag{13.85}$$

The loading vector $\mathbf{q}_e$ required to account for constant member loads n and w is evaluated using Equation 13.41b as:

$$\mathbf{q}_e = \int_L \mathbf{N}_e^T \mathbf{p}\, dx = \begin{bmatrix} \dfrac{nL}{2} & \dfrac{wL}{2} & 0 & \dfrac{nL}{2} & \dfrac{wL}{2} & 0 \end{bmatrix}^T \tag{13.86}$$

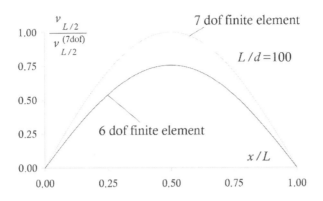

Figure 13.14 Deflection profiles calculated with 6-dof and 7-dof finite elements.

The two finite elements derived for the Timoshenko beam model are applied below for the analysis of a simply-supported beam subjected to a mid-span point load. While the results are very similar for very short spans, their differences tend to increase for longer lengths, with the 6-dof element exhibiting a stiffer response. These differences are shown in Figure 13.14 for a beam with an exaggerated span-to-depth ratio of 100, to illustrate a limit case where the shear deformations are negligible when compared to the flexural deformations. The stiffer behaviour of the 6-dof element is attributed to the fact that the polynomials used for the approximations of the generalised displacements are not consistent. This is the case because their contribution to the strain variables is not of the same order. In particular, the shear strain γ_{xy} $(= v' - \theta)$ is obtained as the sum of the constant term v' (as v is described by a linear polynomial) and the linear term θ. This difference results in the stiffer response produced by the 6-dof element and this is usually referred to as shear locking. This is discussed further in Reflection Activity 13.3.

Other approaches could be used to address the locking problem, such as the *reduced integration* approach, which 'reduces' the order of the numerical integration carried out in the calculation of the stiffness coefficients. The idea at the basis of this approach is that the error introduced in the reduced numerical integration counterbalances the higher stiffness exhibited because of locking. This procedure should be tested very carefully when implemented in a finite element analysis to ensure reliable results.

REFLECTION ACTIVITY 13.3

Consider the results presented in Figure 13.14 for the 6-dof and 7-dof finite elements derived for the Timoshenko beam model. The stiffer response of the 6-dof element is due to shear locking. Demonstrate the occurrence of shear locking for this finite element considering the polynomials adopted to approximate their generalised displacements.

The stiffening effect observed in the results of the 6-dof finite element in Figure 13.14 becomes significant for increasing values of span-to-depth ratios for the beam analysed. In these cases, the shearing deformations have a negligible influence on the structural response and, increasing the slenderness of the member, the solution is well predicted by the Euler–Bernoulli beam model, for which γ_{xy} is negligible. Because of this, we will now try to see what happens to the polynomials used to approximate the generalised displacement for the 6-dof finite element (Equations 13.82 and 13.83) when we consider the case of a very long beam, i.e. when the shear strain γ_{xy} becomes negligible and in the limit when γ_{xy} approaches zero:

$$\gamma_{xy} = v' - \theta = \frac{v_R - v_L}{L} - \theta_L - \frac{\theta_R - \theta_L}{L} x = \left(\frac{v_R - v_L}{L} - \theta_L \right) - \left(\frac{\theta_R - \theta_L}{L} \right) x \to 0 \qquad (13.87)$$

where L is the length of the finite element under consideration.

For Equation 13.87 to approach zero for any value of x, it requires that both its terms equal zero independently of x. This condition is achieved by enforcing both of these terms to equal zero as follows:

$$\frac{v_R - v_L}{L} - \theta_L = 0 \qquad (13.88a)$$

$$\frac{\theta_R - \theta_L}{L} = 0 \qquad (13.88b)$$

Recalling the approximated displacements introduced earlier (see Equations 13.82 and 13.83):

$$u = a_0 + a_1 x \quad v = b_0 + b_1 x \quad \theta = c_0 + c_1 x \qquad (13.89a\text{–}c)$$

and:

$$a_0 = u_L \quad a_1 = -\frac{u_L - u_R}{L} \qquad (13.90a,b)$$

$$b_0 = v_L \quad b_1 = -\frac{v_L - v_R}{L} \qquad (13.90c,d)$$

$$c_0 = \theta_L \quad c_1 = -\frac{\theta_L - \theta_R}{L} \qquad (13.90e,f)$$

At the limit condition, Equation 13.88b forces the coefficient c_1 of Equation 13.90f to equal zero. The main implication of this is that the order of the polynomial describing the rotation reduces to a constant value and, because of this, the first derivative of the rotation θ is now zero. The curvature κ ($= \theta'$) is then forced to remain equal to zero. This behaviour limits (locks) the possible deformations of the element and produces the stiffened behaviour observed in Figure 13.14, hence the name 'shear locking.' Because of its effects, Equation 13.90f can be regarded as a fictitious kinematic restraint imposed on the rotation θ.

One way to avoid this problem is to ensure that the polynomials adopted to describe the generalised displacements contribute with the same order to all non-zero strains appearing in the strain field of the model. In this case, the non-zero variables are ε_x and γ_{xy}. In particular, the contribution of u' and θ included in the expression for ε_x (Equation 13.63a) are both constant and, therefore, of the same order. In the case of γ_{xy} (Equation 13.63d), the contributions of v' and θ

have different orders, v' being constant and θ being linear. In order to ensure compatible contributions from v' and θ to γ_{xy}, we need to increase the order of the polynomial used to describe v by one, so that v is approximated by a parabolic function. In such a way, both v' and θ are then linear functions. Such an element corresponds to the 7-dof element presented in Figure 13.12. Other procedures are available to avoid the occurrence of locking problems, such as the use of reduced integration when calculating the stiffness coefficients of the element numerically.

REFLECTION ACTIVITY 13.4

Consider the two beams shown in Figure 13.15 and discuss the differences in the structural response, in terms of mid-span deflection, calculated based on the Euler–Bernoulli and Timoshenko beam models. Use the 7-dof elements previously derived with the two beam formulations. The beams of Figure 13.15 are prismatic, with a rectangular cross-section of width $b = 100$ mm and depth $d = 400$ mm. Take $E = 20$ GPa.

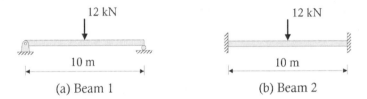

(a) Beam 1   (b) Beam 2

Figure 13.15 Beams for Reflection Activity 13.4.

The steps involved in the numerical calculations have been covered in previous sections of this chapter and are not repeated here. In the following, we will only highlight some of the main differences related to the responses observed using the two beam models.

In order to ensure an adequate discretisation for the analysis, a convergence study is carried out on the simply-supported beam (beam 1) using different number of elements with the Timoshenko beam model. These results are summarised in Figure 13.16 and show the expected convergence, even if at a lower rate than the one observed for 7-dof element of the Euler–Bernoulli

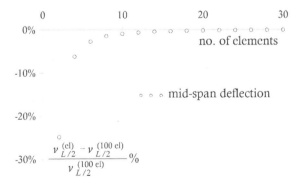

Figure 13.16 Mid-span deflections for different levels of mesh discretisation.

beam in Worked Example 13.1. Because of this, a very fine mesh is specified in the following calculations.

Figure 13.17a presents the differences observed for the mid-span deflections determined using the two beam models, which are particularly significant up to span-to-depth ratios of about 10 or 20 for the simply-supported and fixed-ended beam, respectively. These differences are more pronounced for the fixed-ended beam, as highlighted in Figure 13.17b. The ratio between the mid-span deflections calculated for a fixed-ended and a simply-supported beam for the Euler–Bernoulli beam model remains constant and is equal to 0.25 (Figure 3.17b). Obviously, this ratio could have also been calculated algebraically using the closed-form solutions to the problem presented in earlier chapters. When using the Timoshenko beam model, these differences are more pronounced for low span-to-depth ratios and tend to 0.25 as the beams become more slender, when shear deformations are negligible. For these long-span beams, the deflection predicted by the Timoshenko beam model approaches the value observed for the Euler–Bernoulli beam. In particular, the ratio of 0.25 obtained for the Euler–Bernoulli beam model highlights how the change in support conditions is capable of reducing the deflections induced by flexural deformations. This is the case because the fixity provided by the supports changes the distribution of the bending moment along the beam length.

On the other hand, the shearing deformations, calculated as the difference between the Timoshenko beam deflection results and the Euler–Bernoulli beam deflections, remain equal despite the change in support conditions, as shown in Figure 13.17c. This is a consequence of the fact that the fixity of the supports does not change the shear distribution along the beam length.

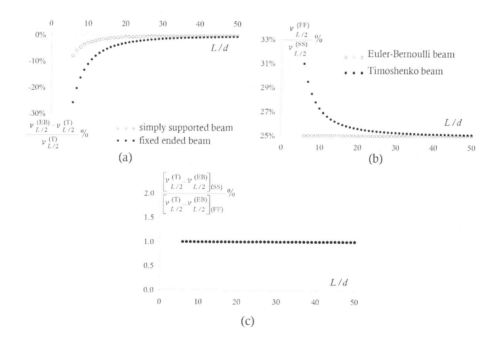

Figure 13.17 Differences between Euler–Bernoulli and Timoshenko beam models.

PROBLEMS

13.1 Consider the finite elements shown which are derived based on the assumptions of the Euler–Bernoulli beam model. For each element, specify the order of the corresponding polynomials required for the approximations of the generalised displacements.

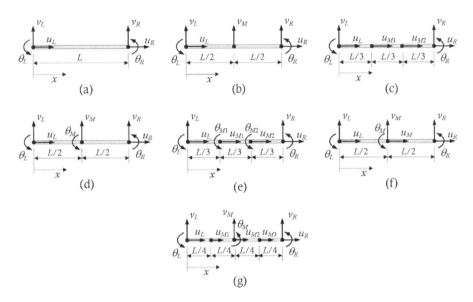

13.2 Under the assumptions of the Euler–Bernoulli beam model, derive the shape functions required to describe the variations along the coordinate x of the generalised displacements for the finite element shown.

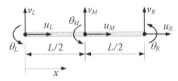

13.3 Derive the stiffness matrix and the loading vector for a uniformly distributed load for the finite element illustrated below. Assume the element to follow the assumptions of the Euler–Bernoulli beam model.

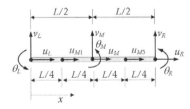

13.4 For the finite elements of Problem 13.1, clarify whether their polynomials satisfy the consistency requirements for finite elements under the assumptions of the Euler–Bernoulli beam model.

13.5 Consider the 7-dof finite element derived in the chapter for the Euler–Bernoulli beam model (Figure 13.4) and derive the loading vector associated with a (i) triangular and (ii) trapezoidal distributed load applied along the element length as illustrated below.

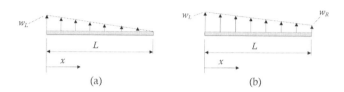

(a)   (b)

13.6 Reconsider the finite elements of Problem 13.1 and evaluate the order of the polynomials required for the approximations of the generalised displacements assuming these line elements to follow the Timoshenko beam model.

13.7 For the finite elements of Problem 13.1, clarify whether their polynomials satisfy the consistency requirements for finite elements under the assumptions of the Timoshenko beam model.

13.8 Derive the loading vector associated with the trapezoidal distributed load shown below and applied along the element length of the 7-dof finite element derived in the chapter for the Timoshenko beam model (Figure 13.12).

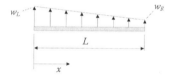

Chapter 14

Introduction to the structural stability of columns

14.1 INTRODUCTION

A structural engineer must ensure that the design of a structure is safe and serviceable, so that the chance of it failing during its design lifetime is sufficiently small. This is achieved by satisfying a number of limit states. This chapter deals with the limit state associated with the possible occurrence of *instability* and presents some fundamental concepts related to basic *stability* theory.

From an intuitive viewpoint, *instability* describes those situations in which a small change in applied forces or deformations produces a large and uncontrolled change in the associated displacements. Instability can occur at different levels in a structure in the form of (i) *local instability* when it affects the response of a local part of a structural element (such as a flange or a web plate), (ii) *member instability* when it is associated with the behaviour of a single member and (iii) *structure* (or *system*) *instability* when it relates to the entire structure. Depending on the layout and arrangement of a structure, the occurrence of local and member instabilities can lead to unstable conditions of the entire structure.

This chapter provides an introduction to the structural stability of columns when subjected to compressive forces. Particular attention is devoted to the identification of the level of load at which the column moves from a stable to an unstable configuration. This load is usually referred to as the *critical load* (or *buckling load*) and it depends on the geometry, support conditions and material properties of the column. The influence of different support conditions on the buckling load is outlined. Concepts, such as effective length and slenderness, are then presented and discussed with numerical examples. The chapter closes with considerations of column imperfections and how these affect the equilibrium conditions and the buckling load.

14.2 ASSUMPTIONS

In previous chapters, we analysed structures based on *small displacement theory* where the following assumptions are made:

1. displacements and deformations are small — in this case, it is acceptable to approximate the cosine and sine of the rotation θ with $\cos \theta \approx 1$ and $\sin \theta = \theta$ and, for small curvatures, to use $\kappa = v''$ (Equation 5.14);
2. equilibrium within the structure is not influenced by its displacements — it follows that equilibrium can be enforced using the geometry of the undeformed structure.

We will now revisit the validity of these assumptions by considering the perfectly straight and infinitely rigid member shown in Figure 14.1a, which is subjected to a transverse load P

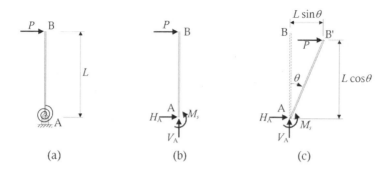

Figure 14.1 Member subjected to a transverse load. (a) Structural layout. (b) Undeformed configuration. (c) Deformed configuration.

at one end and pinned at the other, and can only deform in the plane of the page. An elastic rotational spring is specified at the support, which is activated once the column rotates and its behaviour is described by:

$$M_s = k_s \theta \tag{14.1}$$

where k_s is the rotational stiffness of the spring and M_s represents the restoring moment induced by the spring when deformed by a rotation θ.

The structural response can be described by relating the transverse load P and the rotation of the member θ. This is performed by applying moment equilibrium at the support on the basis of a free-body diagram of the undeformed shape as shown in Figure 14.1b:

$$M_s - PL = 0 \tag{14.2}$$

This expression can be re-written in terms of the rotation θ substituting Equation 14.1 into Equation 14.2 as:

$$\frac{PL}{k_s} = \theta \tag{14.3}$$

The response of the structure defined in Equation 14.3 is plotted in Figure 14.2a.

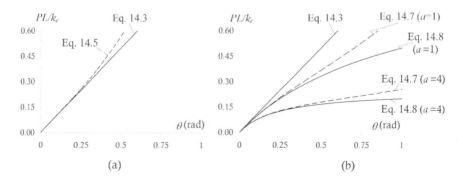

Figure 14.2 Rotations calculated under different loading conditions. (a) Transverse load. (b) Transverse and axial loads.

The previous calculations are now repeated writing the moment equilibrium at the support by considering the free-body diagram of the displaced shape of the structure shown in Figure 14.1c:

$$M_s - PL \cos \theta = 0 \qquad (14.4)$$

which can be re-arranged as:

$$\frac{PL}{k_s} = \frac{\theta}{\cos \theta} \qquad (14.5)$$

In this case, we can see that the relationship between the load P and the rotation θ is nonlinear. Despite this, its overall trend is very similar to the response predicted with Equation 14.3 at least for relatively small angles, as illustrated in Figure 14.2a. Adopting the assumptions of small displacement theory (i.e. $\cos \theta = 1$), Equation 14.5 becomes identical to Equation 14.3.

These results confirm the adequacy of using small displacement theory in predicting the response of a structure when subjected to transverse loads, at least for the range of angles where the curves represented by Equations 14.3 and 14.5 are close together, i.e. for small displacements.

A similar comparison is now performed for the column of Figure 14.3a, which is loaded by an axial force aP that is proportional, through a coefficient a, to the applied transverse force P. Moment equilibrium applied to the free-body diagram of the undeformed structure (Figure 14.3b) produces the same relationship between applied transverse load P and rotation θ as defined in Equation 14.3 and shown in Figure 14.2b. This is because the applied axial force induces no moment with respect to point A in the undeformed structure.

We will now consider the free-body diagram of the deformed shape shown in Figure 14.3c. Moment equilibrium about the node A gives:

$$M_A - aPL \sin \theta - PL \cos \theta = 0 \qquad (14.6)$$

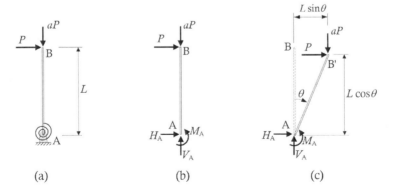

(a)   (b)   (c)

Figure 14.3 Column subjected to axial and transverse loads. (a) Structural layout. (b) Undeformed configuration. (c) Deformed configuration.

which can be re-arranged by isolating the term PL/k_s and including the spring response of Equation 14.1 as:

$$\frac{PL}{k_s} = \frac{\theta}{a\sin\theta + \cos\theta}$$ (14.7)

Equation 14.7 is now modified based on small displacement theory (with $\cos\theta \approx 1$ and $\sin\theta \approx \theta$) as:

$$\frac{PL}{k_s} = \frac{\theta}{a\theta + 1}$$ (14.8)

which is referred to as the linearised solution because it is valid only for small rotations. Unlike the previous case of a column subjected only to a transverse load, this simplification leads to an expression different from the one obtained with the undeformed shape (expressed in Equation 14.3).

Equations 14.7 and 14.8 are plotted in Figure 14.2b for $a = 1$ and $a = 4$, and are compared with the response of the structure in the undeformed shape (as given by Equation 14.3). It can be seen that, for an adequate prediction of the structural response of the column of Figure 14.3, it is necessary to consider the deformed shape of the structure, while it is still acceptable to rely on small displacement theory, at least for relatively small values of θ.

14.3 CRITICAL LOAD FROM EQUILIBRIUM

We will now consider a column subjected to a vertical force P as shown in Figure 14.4 and investigate its buckling and post-buckling response. By enforcing moment equilibrium of the column in its deformed configuration (Figure 14.4b), the following expression is obtained:

$$PL \sin\theta - k_s\theta = 0$$ (14.9)

Under the assumptions of small displacements (with $\sin\theta \approx \theta$), Equation 14.9 can be simplified to:

$$(PL - k_s)\,\theta = 0$$ (14.10)

which is satisfied when $\theta = 0$ or $(PL - k_s) = 0$. Let us now consider these two cases in more detail. When $\theta = 0$, the column remains vertical for any level of load P, while the condition $(PL - k_s) = 0$ leads to the solution:

$$P_{cr} = \frac{k_s}{L}$$ (14.11)

where P_{cr} represents the critical buckling load.

These results are plotted in Figure 14.4c from which it can be observed that for an applied force $P < P_{cr}$, the column remains vertical (with $\theta = 0$) until the load reaches P_{cr} (shown as $P/P_{cr} = 1$ in Figure 14.4c). At this point, equilibrium (defined by Equations 14.10 and 14.11)

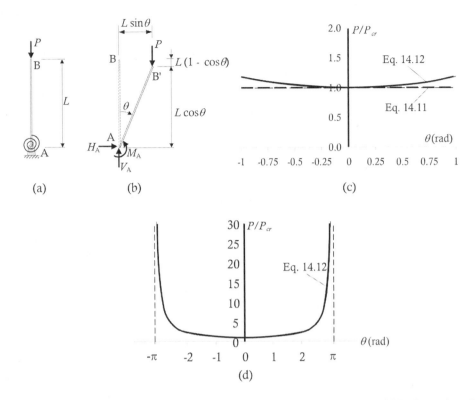

Figure 14.4 Response of a column subjected to a vertical force. (a) Structural layout. (b) Deformed configuration. (c) Structural response (for $-1 \le \theta \le 1$). (d) Structural response (for $-\pi \le \theta \le \pi$).

is satisfied if the column remains vertical (with $\theta = 0$) or for arbitrary values for θ (with $P = P_{cr}$). Because of these different equilibrium paths, this point is also referred to as the bifurcation point. This solution (Equation 14.10) is said to be *linearised*, because on the basis of the approximation $\sin \theta \approx \theta$, it is only valid for small values of θ.

The prediction of the column response for larger values of θ requires the use of Equation 14.9, which accounts for large displacements. This expression can be re-written as a function of the buckling load P_{cr} (Equation 14.11) as:

$$\frac{P}{P_{cr}} = \frac{\theta}{\sin \theta} \tag{14.12}$$

and is also illustrated in Figure 14.4c. The plots of Equation 14.11 (linearised solution in small displacement theory) and Equation 14.12 (with large displacements) are equivalent for small values of θ. This justifies the evaluation of the critical load with small displacement theory based on the deformed shape of the structure.

Once the structure has reached the buckling load, it is necessary to use the nonlinear solution (Equation 14.12) to predict its post-buckling response as highlighted by the growing differences between the values of Equations 14.11 and 14.12 for increasing values of θ. A limit condition is reached when θ approaches π in which case the column is pointing downwards with the applied load P inducing tension along its length. At this value of θ, the load can increase to infinity (or until material fracture occurs in a real column) as shown in Figure 14.4d because the column cannot buckle when subjected to tension.

REFLECTION ACTIVITY 14.1

Determine and comment on the buckling and post-buckling response of the column shown in Figure 14.5a. An elastic spring, with rigidity k_s, is connected to node B, which tends to restrain its horizontal displacement. Answer this question by enforcing moment equilibrium of the deformed shape of the structure (shown in Figure 14.5b), relying on both small and large displacements.

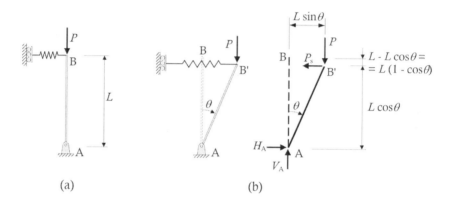

(a)   (b)

Figure 14.5 Column arrangement for Reflection Activity 14.1.

On the basis of the free-body diagram of Figure 14.5b, the moment equilibrium enforced under the assumptions of large displacements can be expressed as:

$P_s L \cos \theta - PL \sin \theta = 0$

Considering that k_s is the elastic rigidity of the spring and with $P_s = k_s L \sin \theta$, we get:

$(k_s L \sin \theta) L \cos \theta - PL \sin \theta = 0$ $\qquad$ (14.13)

Therefore:

$$\frac{P}{k_s L} = \cos \theta \qquad (14.14)$$

The buckling load P_{cr} can be determined by simplifying the expression of Equation 14.14 under the conditions of small displacement theory (linearised solution) as:

$$\frac{P}{k_s L} = 1 \quad \text{from which:}$$

$$P_{cr} = k_s L \qquad (14.15)$$

The curves representing Equations 14.14 and 14.15 are plotted in Figure 14.6 and show that, for values of P/P_{cr} smaller than 1 (before P first reaches P_{cr}), the column remains in its vertical

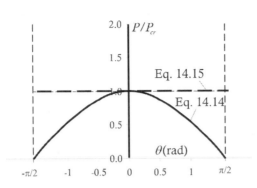

Figure 14.6 Response of the column for Reflection Activity 14.1.

equilibrium with $\theta = 0$. Once the critical load P_{cr} is reached (i.e. $P/P_{cr} = 1$), equilibrium can be satisfied for either $\theta = 0$ or $P/P_{cr} = \cos \theta$ in large displacements ($P/P_{cr} = 1$ in small displacements). Because of these different equilibrium paths, the point where $P/P_{cr} = 1$ is also referred to as the bifurcation point. The post-buckling behaviour is described by Equation 14.14, which shows how the column is able to carry decreasing loads for increasing values of θ. The load becomes zero for $\theta = \pm\pi/2$ because, at this condition, the column buckles under the load induced by the spring.

14.4 CRITICAL LOAD FROM POTENTIAL ENERGY

In the previous section, we have considered simple examples to illustrate the assumptions required in the analysis to evaluate the critical load based on considerations of equilibrium. In this section, we will present an approach based on the principle of minimum total potential energy of the structure to determine the critical load and to distinguish between different types of stability conditions.

We now consider a conservative elastic system in which energy is not dissipated. For this case, the potential energy V is determined as the sum of the elastic strain energy, referred to as U_e (already introduced in Chapter 7), and the potential due to the external actions, denoted as W:

$$V = U_e + W \tag{14.16}$$

In particular, the elastic strain energy U_e equals the work of the internal actions and the potential of the external actions W quantifies the capacity of these actions to produce external work.

Reconsidering the column of Figure 14.4a, the elastic strain energy U_e is defined as the work done by the spring:

$$U_e = \frac{1}{2}k_s\theta^2 \tag{14.17}$$

while the potential of the external actions W can be written as:

$$W = -PL(1 - \cos \theta) \tag{14.18}$$

where the minus sign represents the fact that when the force produces external work, it reduces the potential energy of the system.

Substituting Equations 14.17 and 14.18 into Equation 14.16 leads to the following expression:

$$V = U_e + W = \frac{1}{2}k_s\theta^2 - PL(1 - \cos\theta) \tag{14.19}$$

The principle of minimum total potential energy requires V to be stationary when equilibrium is satisfied and this is evaluated as follows:

$$\frac{dV}{d\theta} = k_s\theta - PL\sin\theta = 0 \tag{14.20}$$

which is equivalent, as expected, to the equilibrium conditions obtained in Equation 14.12 with large displacements. In a similar manner, the buckling loads calculated with the equilibrium and the potential energy methods coincide, as shown by simplifying Equation 14.20 within the framework of small displacements (with $\sin\theta = \theta$), which then equals Equation 14.11.

The second derivative of V provides information on the type of equilibrium exhibited by the structure. In the case of a positive second derivative, the structure is said to be in *stable equilibrium* because, even if subjected to a small perturbation, the system returns to its original configuration. This is illustrated in Figure 14.7a with an example of a ball placed on a surface with a profile that is concave upwards. If we try to move the ball sideways by a small amount, it always returns to its original position after a few oscillations. In the case where the second derivative is negative, the equilibrium is said to be *unstable*. This can be represented by a ball placed on a concave downward surface (Figure 14.7b), where a small sideways movement (small perturbation) given to the ball changes the configuration of the system and the ball continues to move. When the second derivative is nil, the system is in *neutral equilibrium*. In this case, a small perturbation changes the system equilibrium from one configuration to another. Considering the example of the ball, a small movement applied to the ball changes its original position to a new arbitrary one, as there are an infinite number of equilibrium positions for the ball (Figure 14.7c).

We will now evaluate the type of equilibrium exhibited by the column of Figure 14.4a for different combinations of P and θ. In this case, the second derivative of V can be obtained by twice differentiating Equation 14.19 as:

$$\frac{d^2V}{d\theta^2} = k_s - PL\cos\theta \tag{14.21}$$

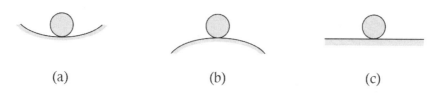

(a)   (b)   (c)

Figure 14.7 Possible equilibrium configurations. (a) Stable equilibrium. (b) Unstable equilibrium. (c) Neutral equilibrium.

Three conditions of equilibrium can now be identified:

$$\frac{d^2 V}{d\theta^2} = k_s - PL\cos\theta > 0 \quad \text{stable equilibrium} \tag{14.22a}$$

$$\frac{d^2 V}{d\theta^2} = k_s - PL\cos\theta = 0 \quad \text{neutral equilibrium} \tag{14.22b}$$

$$\frac{d^2 V}{d\theta^2} = k_s - PL\cos\theta < 0 \quad \text{unstable equilibrium} \tag{14.22c}$$

These can be re-written by substituting the expression for P_{cr} (= k_s/L from Equation 14.11):

$$\frac{P}{P_{cr}} < \frac{1}{\cos\theta} \quad \text{stable equilibrium} \tag{14.23a}$$

$$\frac{P}{P_{cr}} = \frac{1}{\cos\theta} \quad \text{neutral equilibrium} \tag{14.23b}$$

$$\frac{P}{P_{cr}} > \frac{1}{\cos\theta} \quad \text{unstable equilibrium} \tag{14.23c}$$

These conditions of equilibrium are illustrated in Figure 14.8 as a function of θ and P/P_{cr}, on the basis of the classification provided in Equations 14.23. In particular, during the initial loading, the column is in a stable equilibrium until the applied axial force reaches P_{cr}. At this point, the column reaches a bifurcation point at which it can either continue on the curve described by Equation 14.23a in a stable equilibrium or remain undeformed in an unstable equilibrium.

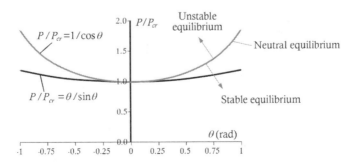

Figure 14.8 Summary of equilibrium conditions.

REFLECTION ACTIVITY 14.2

Consider the column in Figure 14.5 and evaluate the buckling and post-buckling behaviour using the potential energy method. Clarify the type of stability, i.e. stable, neutral or unstable equilibrium, for the different combinations of applied load P and rotation θ.

The potential energy V for this problem is determined as the sum of the elastic strain energy of the spring (with k_s being the spring stiffness):

$$U_e = \frac{1}{2} k_s (L \sin\theta)^2 \tag{14.24}$$

and the potential W of the external force P:

$$W = -PL(1 - \cos\theta) \tag{14.25}$$

Combining the contributions of Equations 14.24 and 14.25, the expression for V can be written as:

$$V = \frac{1}{2} k_s (L \sin\theta)^2 - PL(1 - \cos\theta) \tag{14.26}$$

The equilibrium conditions are determined by enforcing the total potential to be stationary, i.e. $dV/d\theta = 0$, as:

$$\frac{dV}{d\theta} = k_s L^2 \sin\theta \cos\theta - PL \sin\theta = 0 \tag{14.27a}$$

which is satisfied when:

$$P = k_s L \cos\theta \tag{14.27b}$$

as also obtained with the equilibrium approach in Equation 14.14.

Based on the assumption of small displacement theory (with $\cos\theta \approx 1$), the buckling load P_{cr} can then be evaluated as:

$$P_{cr} = k_s L \tag{14.28}$$

As expected, this is identical to Equation 14.15.

The type of equilibrium exhibited by the column is determined based on the values of the second derivative of the potential energy, which is:

$$\frac{d^2 V}{d\theta^2} = -k_s L^2 \sin^2\theta + k_s L^2 \cos^2\theta - PL \cos\theta$$

and simplifying:

$$\frac{d^2 V}{d\theta^2} = 2LP_{cr} \cos^2\theta - P_{cr}L - PL \cos\theta \tag{14.29}$$

Depending on the sign of Equation 14.29, the equilibrium is said to be stable, neutral or unstable as specified below:

$$2\cos^2\theta - 1 - \frac{P}{P_{cr}}\cos\theta > 0 \quad \text{or} \quad \frac{P}{P_{cr}} < \frac{2\cos^2\theta - 1}{\cos\theta} \quad \text{stable equilibrium} \tag{14.30a,b}$$

$$2\cos^2\theta - 1 - \frac{P}{P_{cr}}\cos\theta = 0 \quad \text{or} \quad \frac{P}{P_{cr}} = \frac{2\cos^2\theta - 1}{\cos\theta} \quad \text{neutral equilibrium} \tag{14.31a,b}$$

$$2\cos^2\theta - 1 - \frac{P}{P_{cr}}\cos\theta < 0 \quad \text{or} \quad \frac{P}{P_{cr}} > \frac{2\cos^2\theta - 1}{\cos\theta} \quad \text{unstable equilibrium} \tag{14.32a,b}$$

These results are plotted in Figure 14.9.

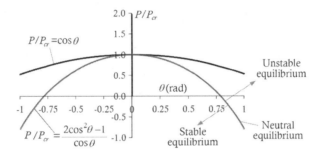

Figure 14.9 Summary of equilibrium conditions for Reflection Activity 14.2.

14.5 BUCKLING OF AN ELASTIC COLUMN

The evaluations of the buckling loads of the columns in Sections 14.2 through 14.4 were based on the assumption that the columns were infinitely rigid. In this section, we extend the analysis to the case of elastic and perfectly straight columns. For this purpose, the relationship between the internal moment and the curvature along the column is determined with the method of double integration (see Section 5.5). This method is particularly useful for structures that are statically determinate, i.e. for which the internal moment M can be determined from equilibrium considerations.

Let us consider the simply-supported column subjected to an axial force P as illustrated in Figure 14.10a. We have seen in previous sections that the buckling load can be determined with small displacement theory, by considering the deformed shape of the column when evaluating the internal moment distribution. The method of double integration is based on Equation 5.36 reproduced here for ease of reference:

$$EI\kappa = EIv'' = M \tag{14.33}$$

The support reactions are calculated from statics as $H_A = P$, $V_A = 0$ and $V_B = 0$. The expression for the internal moment M is obtained by making a cut along the member length in the deformed shape as shown in Figure 14.10c as:

$$M = -H_A v = -Pv \tag{14.34}$$

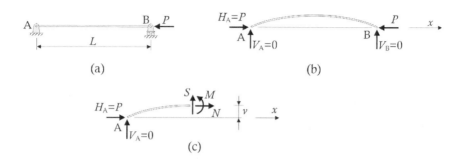

Figure 14.10 Simply-supported column subjected to an axial load. (a) Structural layout. (b) Deformed configuration. (c) Cut along the column length in the deformed configuration.

Substituting Equation 14.34 into Equation 14.33 leads to the governing differential equation of the problem defined by:

$$EIv'' + Pv = 0 \tag{14.35}$$

Before solving Equation 14.35 for the displacement v, it is convenient to collect the known coefficients and to rewrite Equation 14.35 as:

$$v'' + \alpha^2 v = 0 \tag{14.36}$$

where:

$$\alpha^2 = \frac{P}{EI} \tag{14.37}$$

The parameter α^2 has been introduced to highlight that its sign is always positive (or equal to zero for the trivial case of $P = 0$), on the basis of which the general solution of the problem (defined in Equation 14.36) can be written as:

$$v = C_1 \cos \alpha x + C_2 \sin \alpha x \tag{14.38}$$

where the constants of integration C_1 and C_2 are determined from the boundary conditions $v(x = 0) = 0$ and $v(x = L) = 0$:

$$v(x = 0) = C_1 \cos(\alpha \times 0) + C_2 \sin(\alpha \times 0) = C_1 = 0 \tag{14.39a}$$

$$v(x = L) = C_1 \cos \alpha L + C_2 \sin \alpha L = 0 \tag{14.39b}$$

This system of equation has the trivial solution of:

$$C_1 = 0 \qquad C_2 = 0 \tag{14.40a,b}$$

which describes the case in which the column remains undeformed with $v = 0$ throughout its length. Other solutions can be obtained from Equation 14.39b (instead of $C_2 = 0$) when:

$$\sin \alpha L = 0 \tag{14.41}$$

which is satisfied when $\alpha L = n\pi$ (with $n = 1,2,...$), while the case of $n = 0$ is ignored because it is related to an unloaded member (i.e. for which $P = 0$). There are an infinite number of solutions for Equation 14.41, which are valid for any value of C_2. Substituting these results into the general solution of Equation 14.38 produces the following expression for the deflection:

$$v = C_2 \sin \alpha x \qquad (14.42)$$

with:

$$\alpha = \frac{n\pi}{L} \qquad (14.43)$$

Recalling Equation 14.37, Equation 14.43 can be re-written in terms of the applied load P as:

$$P = n^2 \pi^2 \frac{EI}{L^2} \qquad (14.44)$$

We will now evaluate the solution associated with $n = 1$ because it is the lowest value of Equation 14.44 for which the column can buckle. For this particular case, $\alpha = \pi/L$ (Equation 14.43) and the corresponding load is (Equation 14.44):

$$P_E = \pi^2 \frac{EI}{L^2} \qquad (14.45)$$

where P_E is the critical load, usually known as the *Euler buckling load* (or the *elastic buckling load*).

The deflected shape at this load level is shown in Figure 14.11 and is determined from Equation 14.42 by assigning an arbitrary value for C_2 (say for example, $C_2 = 1$):

$$v_1 = C_2 \sin \frac{\pi x}{L} \qquad (14.46)$$

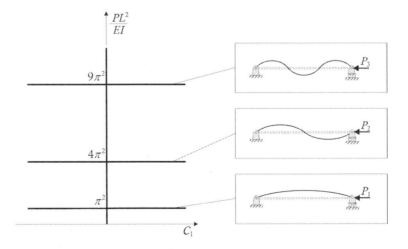

Figure 14.11 First buckling modes for a simply-supported column subjected to an axial load.

In a similar way, the loads and the deflected shapes associated with the solutions calculated using $n = 2$ (with $\alpha = 2\pi/L$) and $n = 3$ (with $\alpha = 3\pi/L$) are:

$$P_2 = 4\pi^2 \frac{EI}{L^2} \qquad v_2 = C_2 \sin \frac{2\pi x}{L} \tag{14.47a,b}$$

$$P_3 = 9\pi^2 \frac{EI}{L^2} \qquad v_3 = C_2 \sin \frac{3\pi x}{L} \tag{14.48a,b}$$

The deflected shapes plotted in Figure 14.11 are also referred to as *buckling modes*, because they describe the deformed shape in which buckling can occur in a column. The subscripts adopted for P and v in Equations 14.46 to 14.48 indicate the numbering of the buckling modes, with the first mode being the one related to the lowest buckling load.

When the applied load reaches one of the critical values (calculated from Equation 14.44), it is possible to satisfy equilibrium either in the undeformed shape or in the deformed configuration (as shown in Figure 14.11). Because of these different equilibrium paths, these branching points are usually referred to as *bifurcation points*. In real structures, a column would always become unstable at the lowest of the critical loads, as it will be shown when discussing the influence produced by initial imperfections (always present in real members) on the structural response.

WORKED EXAMPLE 14.1

Consider the cantilever of Figure 14.12, which is subjected to an axial force P at its tip. Calculate the loads and deformed shapes related to its first three buckling modes.

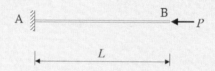

Figure 14.12 Column for Worked Example 14.1.

The first step in the solution is to determine the expression for the internal moment. This is carried out by performing a cut along the member length in its deformed state (Figure 14.13a) and considering the free-body diagram shown in Figure 14.13b.

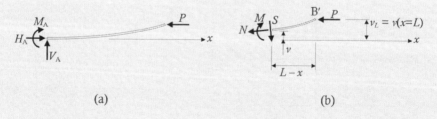

(a)   (b)

Figure 14.13 Free-body diagrams for Worked Example 14.1. (a) Deformed configuration. (b) Cut along the column length in the deformed configuration.

The expression for M is (Figure 14.13b):

$$M = P\,(v_L - v) \tag{14.49}$$

where v_L defines the deflection at $x = L$. The value of v_L is unknown at the beginning of this derivation and will be determined from the boundary conditions of the problem. Equation 14.49 is now substituted into Equation 14.33 to give:

$$EIv'' = Pv_L - Pv \tag{14.50}$$

and introducing α (defined in Equation 14.37):

$$v'' + \alpha^2 v = \alpha^2 v_L \tag{14.51}$$

The solution of the differential equation (Equation 14.51) is:

$$v = C_1 \cos \alpha x + C_2 \sin \alpha x + v_L \tag{14.52}$$

The constants of integration C_1 and C_2, as well as the value for v_L, are evaluated by enforcing the following boundary conditions:

$$v(0) = 0 \qquad v'(0) = 0 \qquad v(L) = v_L \tag{14.53a–c}$$

where the latter condition is consistent with the expression for M in Equation 14.49. Performing the calculations:

$$v(0) = C_1 + v_L = 0 \tag{14.54a}$$

$$v'(0) = C_2 \alpha = 0 \tag{14.54b}$$

$$v(L) = C_1 \cos \alpha L + C_2 \sin \alpha L + v_L = v_L \tag{14.54c}$$

and simplifying:

$$v_L = -C_1 \tag{14.55a}$$

$$C_2 = 0 \tag{14.55b}$$

$$C_1 \cos \alpha L = 0 \tag{14.55c}$$

The system of Equations 14.55 enables the trivial solution of $C_1 = C_2 = v_L = 0$, which is not relevant for the evaluation of the buckling response because it considers the entire beam to remain undeformed, i.e. $v = 0$ throughout the beam length. Other solutions can be obtained by reconsidering Equation 14.55c and seeking values for αL that satisfy:

$$\cos \alpha L = 0 \tag{14.56a}$$

The solutions occur when:

$$\alpha L = \frac{\pi}{2} + n\pi \quad \text{or} \quad \alpha = \frac{\pi(1+2n)}{2L} \tag{14.56b,c}$$

with $n = 0,1,2...$ and so on.

We will now calculate the load and deformed shape associated with the first buckling mode (represented by $n = 0$). Based on Equation 14.56c, the corresponding value for α is:

$$\alpha = \frac{\pi}{2L} \tag{14.57}$$

The buckling load P_1 (where the subscript indicates the first buckling mode corresponding to $n = 0$) is obtained by substituting Equation 14.57 into Equation 14.37 as:

$$\alpha^2 = \frac{P_1}{EI} = \frac{\pi^2}{4L^2} \tag{14.58}$$

from which:

$$P_1 = \pi^2 \frac{EI}{4L^2} \tag{14.59}$$

The deformed shape related to $n = 0$ is described by substituting Equations 14.55 and 14.57 into Equation 14.52:

$$v_1 = v_L \left[1 - \cos\left(\frac{\pi}{2L} x\right) \right] \tag{14.60}$$

The load and deformed shapes associated with the buckling modes corresponding to $n = 1$ (second buckling mode) and $n = 2$ (third buckling mode) are obtained following the procedures illustrated for $n = 0$ and are described by:

$$n = 1 \quad \alpha = \frac{3}{2}\frac{\pi}{L} \quad P_2 = \frac{9}{4}\frac{\pi^2 EI}{L^2} \quad v_2 = v_L \left[1 - \cos\left(\frac{3\pi}{2L} x\right) \right]$$

$$n = 2 \quad \alpha = \frac{5}{2}\frac{\pi}{L} \quad P_3 = \frac{25}{4}\frac{\pi^2 EI}{L^2} \quad v_3 = v_L \left[1 - \cos\left(\frac{5\pi}{2L} x\right) \right]$$

These results are summarised in Figure 14.14 showing the deformed shapes associated with the different buckling modes, plotted for an arbitrary value for v_L. In a real structure, the column would buckle once its applied load reaches P_1 (Equation 14.59) with the relevant shape shown in Figure 14.14.

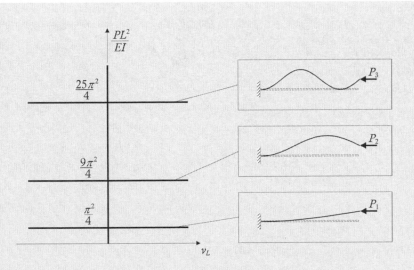

Figure 14.14 First buckling modes for Worked Example 14.1.

WORKED EXAMPLE 14.2

The propped cantilever beam shown in Figure 14.15 is subjected to an axial force *P* applied at node B. Determine the loads and deformed shapes related to its first three buckling modes.

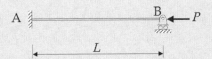

Figure 14.15 Column for Worked Example 14.2.

The propped cantilever is a statically indeterminate structure, because there are four unknown reactions (Figure 14.16a). For this reason, the three equilibrium equations available from statics are used to express three unknown reactions (H_A, V_A and V_B) in terms of the fourth one (M_A).

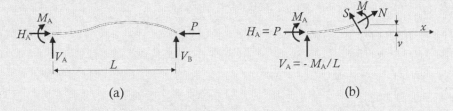

(a)   (b)

Figure 14.16 Free-body diagrams for Worked Example 14.2. (a) Deformed configuration. (b) Cut along the column length in the deformed configuration.

From horizontal equilibrium, it is possible to determine that $H_A = P$, while moment equilibrium about A leads to $V_B = M_A/L$. Finally, from vertical equilibrium, $V_A = -V_B = -M_A/L$. The expression for the internal moment M is then obtained by applying moment equilibrium at the cut of the free-body diagram shown in Figure 14.16b.

The variation for M can be written as:

$$M = -Pv + M_A + \left(-\frac{M_A}{L}\right)x \qquad (14.61)$$

and the unknown moment reaction at A will be determined when applying the boundary conditions of the problem. Based on the method of double integration (Equation 14.33):

$$EIv'' + Pv = M_A\left(1 - \frac{x}{L}\right) \qquad (14.62)$$

or rewriting in terms of α (Equation 14.37):

$$v'' + \alpha^2 v = \frac{M_A}{EI}\left(1 - \frac{x}{L}\right) \qquad (14.63)$$

The solution of the differential equation (Equation 14.63) is:

$$v = C_1 \cos \alpha x + C_2 \sin \alpha x + \frac{M_A}{\alpha^2 EI}\left(1 - \frac{x}{L}\right) \qquad (14.64)$$

The constants of integration C_1 and C_2 and the unknown reaction M_A are calculated by applying the following boundary conditions:

$$v(0) = 0 \qquad v'(0) = 0 \qquad v(L) = 0 \qquad (14.65a–c)$$

which are expressed as:

$$v(0) = C_1 + \frac{M_A}{\alpha^2 EI} = 0 \qquad (14.66a)$$

$$v'(0) = C_2\alpha - \frac{M_A}{\alpha^2 EI\, L} = 0 \qquad (14.66b)$$

$$v(L) = C_1 \cos \alpha L + C_2 \sin \alpha L = 0 \qquad (14.66c)$$

and simplifying gives:

$$C_1 = -\frac{M_A}{\alpha^2 EI} \qquad (14.67a)$$

$$C_2 = \frac{M_A}{\alpha^3 EIL} \qquad (14.67b)$$

$$\frac{M_A}{\alpha^2 EI}\left(\frac{\sin \alpha L}{\alpha L} - \cos \alpha L\right) = 0 \qquad (14.67c)$$

The trivial solution $C_1 = C_2 = M_A = 0$ does not identify the occurrence of buckling because it describes the column in its undeformed configuration. Possible non-trivial solutions of the system of Equations 14.66 are obtained by seeking values for αL that satisfy Equation 14.67c and in particular:

$$\left(\frac{\sin\alpha L}{\alpha L} - \cos\alpha L\right) = 0 \quad\text{or}\quad \tan\alpha L = \alpha L \tag{14.68a,b}$$

Solutions of Equation 14.68b cannot be obtained in closed form and need to be determined numerically. The lowest three values of αL that satisfy it (excluding the trivial solution $\alpha L = 0$ corresponding to no applied load) are $\alpha L = 1.430\pi$, $\alpha L = 2.459\pi$, and $\alpha L = 3.471\pi$.

The load and deformed shapes associated with these values for αL represent the first three buckling modes and are given by:

Mode 1:

$$\alpha = \frac{1.430\pi}{L} \quad P_1 = \frac{(1.430)^2\pi^2 EI}{L^2}$$

$$v_1 = \frac{M_A}{\alpha^2 EI}\left[-\cos\left(\frac{1.430\pi}{L}x\right) + \frac{1}{1.430\pi}\sin\left(\frac{1.430\pi}{L}x\right) + 1 - \frac{x}{L}\right]$$

Mode 2:

$$\alpha = \frac{2.459\pi}{L} \quad P_2 = \frac{(2.459)^2\pi^2 EI}{L^2}$$

$$v_2 = \frac{M_A}{\alpha^2 EI}\left[-\cos\left(\frac{2.459\pi}{L}x\right) + \frac{1}{2.459\pi}\sin\left(\frac{2.459\pi}{L}x\right) + 1 - \frac{x}{L}\right]$$

Mode 3:

$$\alpha = \frac{3.471\pi}{L} \quad P_3 = \frac{(3.471)^2\pi^2 EI}{L^2}$$

$$v_3 = \frac{M_A}{\alpha^2 EI}\left[-\cos\left(\frac{3.471\pi}{L}x\right) + \frac{1}{3.471\pi}\sin\left(\frac{3.471\pi}{L}x\right) + 1 - \frac{x}{L}\right]$$

The deformed shapes associated with the different buckling modes are plotted in Figure 14.17.

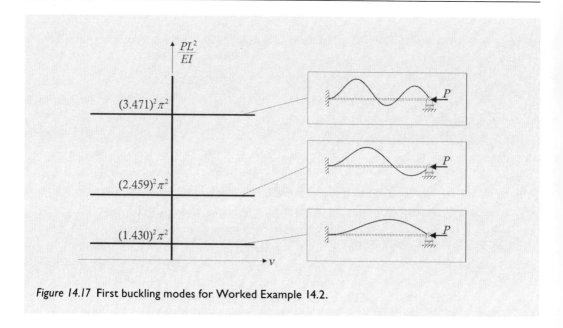

Figure 14.17 First buckling modes for Worked Example 14.2.

The non-trivial solutions obtained to date have been based on a careful evaluation of the boundary conditions provided by the supports of the column. Similar results could have been obtained by seeking values of αL for which the determinant of the system of equation defining the boundary conditions equals zero. The calculations related to the three column cases considered are now briefly outlined to clarify the procedure involved.

The boundary conditions of the simply-supported column were specified in Equations 14.39 and are written in matrix form as:

$$\begin{bmatrix} 1 & 0 \\ \cos \alpha L & \sin \alpha L \end{bmatrix} \begin{bmatrix} C_1 \\ C_2 \end{bmatrix} = \begin{bmatrix} 0 \\ 0 \end{bmatrix} \tag{14.69}$$

The non-trivial solutions for the simply-supported column can then be obtained by solving the following determinant for αL:

$$\det \begin{bmatrix} 1 & 0 \\ \cos \alpha L & \sin \alpha L \end{bmatrix} = \sin \alpha L = 0 \tag{14.70}$$

which is identical to Equation 14.41, previously used in the solution.

In the case of the cantilever beam of Worked Example 14.1, the boundary conditions can be expressed in matrix form as:

$$\begin{bmatrix} 1 & 0 & 1 \\ 0 & \alpha & 0 \\ \cos \alpha L & \sin \alpha L & 0 \end{bmatrix} \begin{bmatrix} C_1 \\ C_2 \\ v_L \end{bmatrix} = \begin{bmatrix} 0 \\ 0 \\ 0 \end{bmatrix} \tag{14.71}$$

Setting the determinant of the matrix of coefficients to be zero leads to an expression equivalent to Equation 14.56a:

$$\det \begin{bmatrix} 1 & 0 & 1 \\ 0 & \alpha & 0 \\ \cos\alpha L & \sin\alpha L & 0 \end{bmatrix} = -\alpha\cos\alpha L = 0 \tag{14.72}$$

For the propped cantilever (Worked Example 14.2), the boundary conditions are expressed by the following system of equations:

$$\begin{bmatrix} 1 & 0 & 1 \\ 0 & \alpha & -\dfrac{1}{L} \\ \cos\alpha L & \cos\alpha L & 0 \end{bmatrix} \begin{bmatrix} C_1 \\ C_2 \\ \dfrac{M_A}{\alpha^2 EI} \end{bmatrix} = \begin{bmatrix} 0 \\ 0 \\ 0 \end{bmatrix} \tag{14.73}$$

and the non-trivial solutions are obtained from its corresponding determinant as:

$$\det \begin{bmatrix} 1 & 0 & 1 \\ 0 & \alpha & -\dfrac{1}{L} \\ \cos\alpha L & \sin\alpha L & 0 \end{bmatrix} = -\alpha\cos\alpha L + \dfrac{\sin\alpha L}{L} = 0 \tag{14.74}$$

which is identical to Equations 14.68.

14.6 EFFECTIVE BUCKLING LENGTH

In the previous section, we have derived the expressions for the buckling loads exhibited by an elastic column for different boundary conditions, expressed in the form of:

$$P_E = \psi\pi^2 \dfrac{EI}{L^2} \tag{14.75}$$

where the values for ψ varies depending on the support conditions of the column and have been summarised in Table 14.1 for the three cases considered.

Table 14.1 Summary of effective length factors

Structural static configuration		ψ	$k_e = 1/\sqrt{\psi}$
A ⌒————————————— B ⌒ P		1	1
A ▌————————————— B ← P		0.25	2
A ▌————————————— B ⌒ P		2.046	0.7

From a design viewpoint, it is convenient to introduce the concept of the effective length of the column L_e to account for the different boundary conditions, where L_e is calculated from:

$$L_e = \frac{L}{\sqrt{\psi}} = k_e L \qquad (14.76)$$

Equation 14.75 can now be re-written as:

$$P_E = \pi^2 \frac{EI}{L_e^2} \qquad (14.77)$$

In this manner, the buckling load of a column with arbitrary support conditions is related to an equivalent simply-supported column with the length replaced by L_e. This can be observed by considering that Equation 14.77 is similar to Equation 14.45 (i.e. Euler buckling load for the simply-supported column subjected to an axial force) with the only difference being that the effective length rather than the actual length is used.

Values for k_e describing the buckling response under different support conditions are also provided in Table 14.1 and are calculated from Equation 14.76 as:

$$k_e = \frac{1}{\sqrt{\psi}} \qquad (14.78)$$

14.7 BUCKLING STRESSES

It is often convenient to express the buckling load in terms of stresses:

$$\sigma_E = \frac{P_E}{A} = \pi^2 \frac{EI}{AL_e^2} \qquad (14.79)$$

where σ_E is the buckling stress and A is the cross-sectional area of the column.

Equation 14.79 can be re-arranged to highlight the key parameters influencing the buckling behaviour, which are the elastic modulus E and the slenderness λ of the column:

$$\sigma_E = \pi^2 \frac{E}{\lambda^2} \qquad (14.80)$$

where λ is given by:

$$\lambda = k_e L \sqrt{\frac{A}{I}} = \frac{L_e}{r} \qquad (14.81)$$

where r is the radius of gyration of the column cross-section equal to $\sqrt{I/A}$. Equation 14.80 is usually known as *Euler's formula*.

The slenderness value λ is a useful parameter for predicting the likelihood of a column buckling. It accounts for the geometric and material properties of the column, as well as its

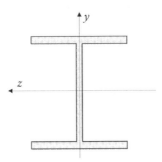

Figure 14.18 Typical I-shaped steel section.

support conditions. For example, a column with a low value for λ is less likely to buckle than a column with a large value of λ, because its buckling stress σ_E is higher (Equation 14.80).

Let us consider a typical doubly symmetric I-section shown in Figure 14.18. Such a section is commonly used in steel structures. Let us assign the z- and y-axes to be the strong and weak axes of the cross-section, respectively. This implies that the second moment of area about the z-axis, I_z, is greater than that calculated about the y-axis, I_y. We will now consider the slenderness values with respect to the z- and y-axes to evaluate the possibility of the column buckling in the xy or xz planes, respectively. In reality, a column will always buckle in the plane associated with its lowest slenderness value.

Based on this, if the column has identical support conditions with respect to both z- and y-axes, it will always buckle about the y-axis, i.e. in the xz plane. This is due to the fact that the slenderness calculated about the z-axis, λ_z, is smaller than the one for the y-axis, λ_y, i.e. $\lambda_z < \lambda_y$. A different response might occur if different support conditions are present about the two axes, which is often the case in practice.

Equation 14.80 is plotted in Figure 14.19a, where the relationship between the buckling stress σ_E and the column slenderness λ for a linear–elastic material is illustrated. In reality, a material is not able to carry an infinite level of stress and its carrying capacity is limited by its strength. For example, let us consider a steel section, whose material properties can be described by an elasto-perfectly plastic constitutive model, i.e. its stress–strain relationship is linear–elastic up to the yield stress of the steel σ_y, after which the steel continues to deform at constant stress as shown in Figure 14.19b (see also Chapter 15 for more details on nonlinear behaviour). On the basis of these assumptions, it is not possible for a steel section

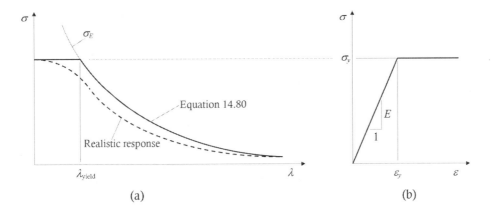

Figure 14.19 Buckling stress for different levels of column slenderness.

to resist a stress greater than its yield stress. Because of this, there is a particular value of slenderness, here referred to as λ_{yield}, for which the buckling stress equals the yield stress. If a column possesses a λ value lower than λ_{yield}, it is expected to yield before buckling. In the case $\lambda > \lambda_{\text{yield}}$, the column would buckle before the steel yields. These different responses can be described as follows:

$$\sigma_{\text{max}} = \sigma_y \qquad\qquad \text{when } \lambda \leq \lambda_{\text{yield}} \qquad\qquad (14.82\text{a})$$

$$\sigma_{\text{max}} = \sigma_E = \pi^2 \frac{E}{\lambda^2} (<\sigma_y) \quad \text{when } \lambda > \lambda_{\text{yield}} \qquad\qquad (14.82\text{b})$$

The relationship described by Equations 14.82 is only theoretical as, in reality, because of the presence of imperfections (such as out-of-straightness, and material and cross-sectional variations), the buckling stress–slenderness curve follows the trend illustrated in Figure 14.19a by the dashed curve.

The procedure required to predict the occurrence of buckling in columns with different boundary conditions is outlined in Worked Examples 14.3 and 14.4.

WORKED EXAMPLE 14.3

Determine the buckling load for the 5 m long column shown in Figure 14.20. The I-shaped section is shown in Figure 14.20a and has the following geometry: $A = 8000$ mm^2, $I_z = 60 \times 10^6$ mm^4 and $I_y = 20 \times 10^6$ mm^4. The column can buckle in either xy or xz planes. The relevant support conditions are shown for clarity for the two planes separately in Figures 14.20b and c. Assume the material to be linear–elastic with $E = 200$ GPa. Comment on how the results would change if the column length reduces to 3 m.

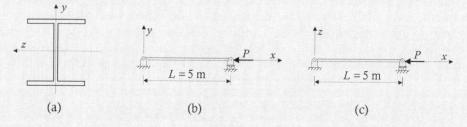

(a)   (b)   (c)

Figure 14.20 Column arrangement for Worked Example 14.3. (a) Cross-section. (b) Boundary conditions in the xy plane about the z-axis. (c) Boundary conditions in the xz plane about the y-axis.

The calculation of the buckling load P_E is carried out by calculating first the slenderness about the y- and z-axes and then the relevant buckling stress.

The radius of gyration $r \left(= \sqrt{I/A}\right)$ is calculated about the y- and z-axes as:

$$r_z = \sqrt{\frac{I_z}{A}} = \sqrt{\frac{60 \times 10^6}{8000}} = 86.6 \text{ mm} \qquad r_y = \sqrt{\frac{I_y}{A}} = \sqrt{\frac{20 \times 10^6}{8000}} = 50 \text{ mm}$$

The support conditions are simply-supported in both planes, i.e. $k_e = 1$ (from Table 14.1):

$$k_{ez} = 1 \quad k_{ey} = 1$$

and the effective lengths become:

$$L_{ez} = k_{ez} \times L = 1 \times 5000 = 5000 \text{ mm} \qquad L_{ey} = k_{ey} \times L = 1 \times 5000 = 5000 \text{ mm}$$

The slenderness can then be calculated with Equation 14.81:

$$\lambda_z = \frac{L_{ez}}{r_z} = \frac{5000}{86.6} = 57.7 \quad \lambda_y = \frac{L_{ey}}{r_y} = \frac{5000}{50} = 100$$

Buckling will occur in the xz plane because $\lambda_z < \lambda_y$. Its corresponding buckling stress and load are:

Equation 14.80: $\sigma_E = \pi^2 \dfrac{E}{\lambda_y^2} = \pi^2 \dfrac{200,000}{100^2} = 197.4 \text{ MPa}$

Equation 14.79: $P_E = \sigma_E A = 197.4 \times 8000 = 1579 \text{ kN}$

Considering that the column has identical support conditions in the xy and xz planes, it should be expected that the column buckles about its weakest axis, i.e. in the xz plane with respect to the y-axis.

Calculations are presented below for a column length equal to 3 m (only critical values about the y-axis are provided):

$$L_{ey} = k_{ey} \times L = 1 \times 3000 = 3000 \text{ mm}$$

$$\lambda_y = \frac{L_{ey}}{r_y} = \frac{3000}{50} = 60$$

$$\sigma_E = \pi^2 \frac{E}{\lambda_y^2} = \pi^2 \frac{200,000}{60^2} = 548.3 \text{ MPa}$$

$$P_E = \sigma_E A = 548.3 \times 8000 = 4386 \text{ kN}$$

In a real steel structure, if the yield stress of the steel was, say, 350 MPa, the steel section would have yielded before buckling and it is not possible to reach the stress of 548.3 MPa. After yielding, the linear-elastic material response that forms the basis of the derivation of Equation 14.80 is no longer valid.

WORKED EXAMPLE 14.4

Consider the I-shaped column of Figure 14.21 and calculate its buckling load. The column can buckle in either xy or xz planes. Different support conditions are specified in these two planes as shown in Figures 14.21b and c. The column is 4 m long and its cross-sectional properties are $A = 8000$ mm^2, $I_z = 100 \times 10^6$ mm^4 and $I_y = 10 \times 10^6$ mm^4. The material is linear–elastic with $E = 200$ GPa. Comment on how the results would change if I_y doubles (i.e. $I_y = 20 \times 10^6$ mm^4).

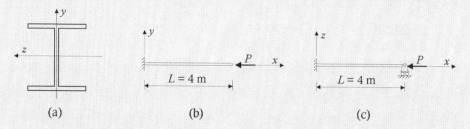

(a)   (b)   (c)

Figure 14.21 Column arrangement for Worked Example 14.4. (a) Cross-section. (b) Boundary conditions in the xy plane about the z-axis. (c) Boundary conditions in the xz plane about the y-axis.

P_E is calculated by considering the slenderness about each axis.

The radius of gyration $r \left(= \sqrt{I/A} \right)$ is calculated about the y- and z-axes as:

$$r_z = \sqrt{\frac{I_z}{A}} = \sqrt{\frac{100 \times 10^6}{8000}} = 111.8 \text{ mm} \qquad r_y = \sqrt{\frac{I_y}{A}} = \sqrt{\frac{10 \times 10^6}{8000}} = 35.35 \text{ mm}$$

The support conditions are those of a cantilever beam in the xy plane (with respect to the z-axis) and those of a propped cantilever in the xz plane (with respect to the y-axis). From Table 14.1:

$$k_{ez} = 2 \quad k_{ey} = 0.7$$

and the effective lengths become:

$$L_{ez} = k_{ez} \times L = 2 \times 4000 = 8000 \text{ mm} \qquad L_{ey} = k_{ey} \times L = 0.7 \times 4000 = 2800 \text{ mm}$$

The slenderness can then be calculated with Equation 14.81:

$$\lambda_z = \frac{L_{ez}}{r_z} = \frac{8000}{111.8} = 71.55 \qquad \lambda_y = \frac{L_{ey}}{r_y} = \frac{2800}{35.35} = 79.20$$

Buckling will occur in the xz plane because $\lambda_z < \lambda_y$. Its corresponding buckling stress and load are:

Equation 14.80: $\sigma_E = \pi^2 \dfrac{E}{\lambda_y^2} = \pi^2 \dfrac{200,000}{79.20^2} = 314.7$ MPa

Equation 14.79: $P_E = \sigma_E A = 314.7 \times 8000 = 2518$ kN

The previous calculations are now repeated in the case $I_y = 20 \times 10^6$ mm⁴ (only for values related to the y-axis because properties with respect to z remained unchanged):

$$r_y = \sqrt{\frac{I_y}{A}} = \sqrt{\frac{20 \times 10^6}{8000}} = 50 \text{ mm}$$

$L_{ey} = 2800$ mm (unchanged)

$$\lambda_y = \frac{L_{ey}}{r_y} = \frac{2800}{50} = 56$$

In this case, buckling will occur in the xy plane because $\lambda_z (= 71.55) > \lambda_y (= 56)$.

$$\sigma_E = \pi^2 \frac{E}{\lambda_z^2} = \pi^2 \frac{200,000}{71.55^2} = 385.5 \text{ MPa}$$

$$P_E = \sigma_E A = 385.5 \times 8000 = 3084 \text{ kN}$$

Note that if the yield stress of the steel was 350 MPa, the steel would have yielded before buckling.

14.8 IMPERFECTIONS IN COLUMNS

In the previous sections, we have analysed perfectly straight columns. We will now consider the influence on the calculated buckling loads of imperfections, such as the out-of-straightness present in real columns. This is carried out following the procedure already introduced in Section 14.5, where the buckling response has been studied with the method of double integration.

The out-of-straightness is here introduced by assuming that the column possesses an initial deformed profile before the application of the load. The selection of this initial deformed geometry is very important and, in this section, we will assume it to follow the shape of the first buckling mode calculated for the same column without imperfections. In this manner, we are considering the worst possible imperfection profile for the overall column buckling.

For example, reconsider the simply-supported column of Figure 14.10, whose first buckling mode is defined by Equation 14.46 (i.e. $v_1 = C_2 \sin \pi x/L$). The initial column geometry is described by:

$$v_i = \bar{v}_i \sin \frac{\pi x}{L} \tag{14.83}$$

where $\bar{v}_i$ represents the magnitude of imperfection at mid-span, as shown in Figure 14.22a.

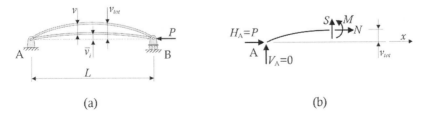

(a)   (b)

Figure 14.22 Column with initial imperfections. (a) Deformed configuration. (b) Cut along the column length in the deformed configuration.

In the adopted notation, v is the deflection induced by the applied load, while v_{tot} represents the total deflection:

$$v_{tot} = v_i + v \tag{14.84}$$

The internal moment is calculated from the free-body diagram of Figure 14.22b as:

$$M = -H_A v_{tot} = -P(v_i + v) \tag{14.85}$$

Substituting Equation 14.85 into Equation 14.33, we get:

$$EIv'' + Pv = -Pv_i \tag{14.86}$$

and including α (from Equation 14.37):

$$v'' + \alpha^2 v = -\alpha^2 \bar{v}_i \sin\left(\frac{\pi x}{L}\right) \tag{14.87}$$

The general solution of the differential Equation 14.87 is:

$$v = C_1 \cos\alpha x + C_2 \sin\alpha x + \frac{\alpha^2}{(\pi/L)^2 - \alpha^2} \bar{v}_i \sin\left(\frac{\pi x}{L}\right) \tag{14.88}$$

and with the boundary conditions:

$$v(x = 0) = 0 \qquad v(x = L) = 0 \tag{14.89a,b}$$

we get:

$$v(x = 0) = C_1 = 0 \tag{14.90a}$$

$$v(x = L) = C_1 \cos\alpha L + C_2 \sin\alpha L = 0 \tag{14.90b}$$

Unlike the case of the perfectly straight column, the solution $C_1 = C_2 = 0$ leads to the following deformed shape:

$$v = \frac{\dfrac{P}{P_E}}{1 - \dfrac{P}{P_E}} \bar{v}_i \sin\left(\frac{\pi x}{L}\right) \tag{14.91}$$

where P_E is the Euler buckling load ($= \pi^2 EI/L^2$) given by Equation 14.45.

The total deflection v_{tot} is then determined combining the initial column geometry v_i (Equation 14.83) with the calculated deflection v (Equation 14.91) as:

$$v_{tot} = v_i + v = \bar{v}_i \sin\frac{\pi x}{L} + \frac{\dfrac{P}{P_E}}{1 - \dfrac{P}{P_E}} \bar{v}_i \sin\left(\frac{\pi x}{L}\right) = \frac{1}{1 - \dfrac{P}{P_E}} \bar{v}_i \sin\left(\frac{\pi x}{L}\right) \tag{14.92}$$

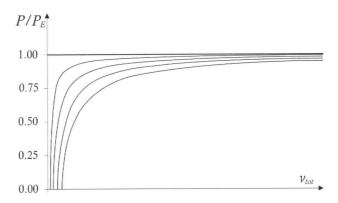

Figure 14.23 Total deflections for different levels of load and for different levels of initial out-of-straightness (based on the linearised solution of the problem).

The deflected shape obtained for different levels of initial out-of-straightness $\bar{v}_i$ is illustrated in Figure 14.23. In the limit case where $\bar{v}_i = 0$, the solution of Equation 14.92 degenerates to $v = 0$ (the undeformed shape).

From Figure 14.23, it is clear that real columns, which always possess some degree of imperfections, cannot follow the equilibrium path described by $v = 0$ as observed for the perfectly straight columns in the previous sections. Depending on the magnitude of the imperfections, the deflection exhibited by the column can be significant even for relatively low levels of load. Despite this, the curves plotted in Figure 14.23, based on the linearised solution of the problem, tend towards the first buckling load for increasing values for the total deflection v_{tot}.

PROBLEMS

14.1 Determine the buckling and post-buckling response of the rigid column shown. An elastic spring, with rigidity k_s, is connected to node B and tends to restrain its horizontal displacement. Answer this question enforcing moment equilibrium on the deformed shape of the structure, relying on both small and large displacements.

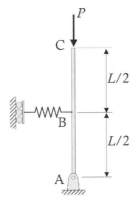

14.2   Reconsider the column of Problem 14.1 and evaluate the buckling and post-buckling behaviour using the potential energy method. Clarify the type of stability, i.e. unstable, neutral or stable equilibrium, for the different combinations of applied load P and rotations θ.

14.3   Determine the buckling load for the column shown. Assume E to be equal to 20 GPa. The column is free to buckle in either xy or xz planes. The relevant support conditions are shown for the two planes separately.

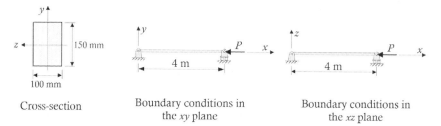

Cross-section   Boundary conditions in the xy plane   Boundary conditions in the xz plane

14.4   Evaluate the buckling load for the column shown which can buckle in either xy or xz planes. Assume the cross-sectional and material properties are $A = 10 \times 10^3$ mm^2, $I_z = 80 \times 10^6$ mm^4, $I_y = 20 \times 10^6$ mm^4 and $E = 200$ GPa.

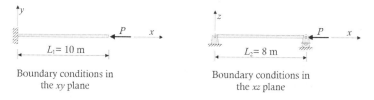

Boundary conditions in the xy plane   Boundary conditions in the xz plane

14.5   Reconsider the column of Problem P14.4 and calculate the buckling load if the cross-sectional and material properties are changed to: $A = 12 \times 10^3$ mm^2, $I_z = 100 \times 10^6$ mm^4, $I_y = 80 \times 10^6$ mm^4 and $E = 200$ GPa.

14.6   For the column illustrated below, calculate the buckling load with the following cross-sectional and material properties: $A = 8 \times 10^3$ mm^2, $I_z = 60 \times 10^6$ mm^4, $I_y = 20 \times 10^6$ mm^4 and $E = 200$ GPa. Assume the column can buckle in either xy or xz planes.

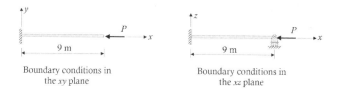

Boundary conditions in the xy plane   Boundary conditions in the xz plane

14.7   The lengths of the column analysed in Problem 14.6 are modified to the dimensions specified below. Evaluate the buckling load based on the new geometry.

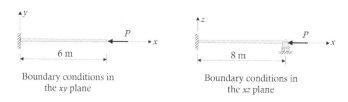

Boundary conditions in the xy plane   Boundary conditions in the xz plane

Chapter 15

Introduction to nonlinear analysis

15.1 INTRODUCTION

This chapter provides an introduction to the nonlinear analysis of structures. The most common causes of nonlinearity in structures include (i) *material nonlinearity* — in this case, the relationship between strains and stresses is nonlinear and cannot be simply expressed in terms by Hooke's law ($\sigma = E\varepsilon$); (ii) *geometric nonlinearity* — this occurs when the displacements and rotations of a structural system are large and the geometry of the structure changes significantly; and (iii) *contact nonlinearity* — this category relates to the situations in which the boundary conditions of a structure change during the analysis.

The use of *nonlinear analysis* is essential when nonlinearities significantly affect the structural behaviour. For example, a nonlinear analysis is required if we want to evaluate the load-carrying capacity of a member or a structure just before it fails. A nonlinear analysis might also be required to predict the day-to-day behaviour of a structure when, for example, deformations or material damage might become significant at service loads.

This chapter intends to provide introductory remarks only on material nonlinearity, providing some insight into the solution techniques required to account for it in an analysis. Some aspects related to geometric nonlinearity were already covered in Chapter 14 when dealing with instability problems. A detailed treatment of nonlinear modeling is beyond the scope of this book and reference should be made to specialised literature in the area.

In the first part of the chapter, we will provide a brief introduction to different nonlinear material properties. These will then be used in a number of illustrative examples to outline some key features of the nonlinear modelling of trusses and beams. This will then be followed by a description of nonlinear numerical solutions based on the Newton–Raphson method applied first to cross-sectional analyses, to familiarise ourselves with the overall procedure, and then to member analyses implemented with the finite element method.

15.2 NONLINEAR MATERIAL PROPERTIES

Different materials respond in different ways to applied loads and induced deformations. In structural analysis, the material response is described by a mathematical equation or set of equations usually referred to as the *constitutive models* or *material stress–strain relationships*. The complexity of these mathematical representations varies depending on the number of material features that need to be captured and are usually expressed in terms of algebraic, differential or integral equations.

For the purpose of this chapter, we will only focus on uniaxial constitutive models that relate axial deformations and stresses in one direction and, because of this, are usually referred to as *uniaxial stress–strain relationships*. These are applicable when dealing with line elements, such as those considered in previous chapters for the modelling of trusses, beams and frames. An example of a linear uniaxial constitutive model, already encountered in previous chapters, is Hooke's law, for which the relationship between stress σ and strain ε is expressed by $\sigma = E\varepsilon$.

Stress–strain relationships of materials are usually obtained either by performing standard tests specified in national design guidelines or from formulations based on some theoretical assumptions. The actual details and requirements of a standard test vary depending on the failure modes expected for the material under consideration. In this section, we will consider the response of typical metals whose material properties can be evaluated by means of tensile tests. This experiment is usually carried out on a representative piece of material taken out of a structural component or member. For the sake of simplicity, we will consider the round bar shown in its unloaded condition in Figure 15.1a, with length L_0 and cross-sectional area A_0 (with diameter D_0). Once we start to apply a tensile load P, as shown in Figure 15.1b, the bar begins to elongate to a new length L and to decrease its area to A (and corresponding diameter D).

The overall response of the bar is described by plotting the deformations achieved at different levels of applied load or, alternatively, the loads exhibited for different levels of applied deformations. The measurements recorded from a tensile test can be converted to a stress–strain relationship on the basis of the following calculations:

$$\varepsilon = \frac{L - L_0}{L_0} \qquad \sigma = \frac{P}{A_0} \tag{15.1a,b}$$

where P represents the load at which the sample reaches length L, and ε and σ are referred to as the *nominal* or *engineering strain and stress*, respectively, because they are calculated on the basis of the initial geometry of the sample. These calculations assume that the test is carried out under ideal conditions so that the sample is uniformly deformed over its entire cross-section and the sample is monitored over a sufficient length to ignore edge effects.

Typical stress–strain curves that can be obtained from a brittle and a ductile material are shown in Figure 15.2. In both cases, we have an initial *linear–elastic branch* (line OA in Figure 15.2). The *elastic* condition reflects the ability of a deformed material in this range to return to its initial conditions upon removal of the applied load, while the *linear* property

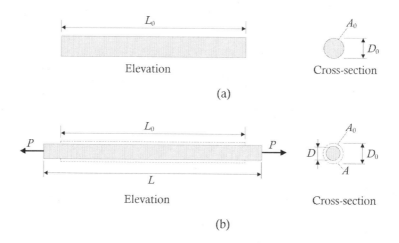

Figure 15.1 Tensile test of a round bar. (a) Unloaded sample. (b) Loaded sample.

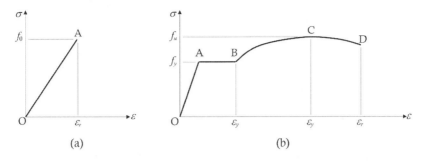

Figure 15.2 Typical stress–strain curves. (a) Brittle material. (b) Ductile material.

highlights the linear proportionality between stresses and strains, defined by means of the elastic modulus E. For a brittle material, it is common that the material fractures at a *limit stress* f_0 at the end of the linear–elastic region, i.e. at Point A in Figure 15.2a. In the case of ductile materials, the linear–elastic response is followed by an inelastic behaviour characterised, for example, in Figure 15.2b by a yielding plateau AB, where the material continues to deform at its *yield stress* f_y, and to strain-harden up to its *ultimate* or *peak stress* f_u (at Point C in Figure 15.2b). After this point, the engineering stress–strain curve starts to decrease until failure of the material occurs at Point D in Figure 15.2b. This decrease is due to the fact that the local stress and strain are calculated with Equations 15.1, using the initial geometry of the sample. In reality, the local stress continues to increase and this could be calculated by evaluating the stress based on the actual cross-sectional area A. This is usually referred to as the *true stress*. During the test, the sample elongates and, as it elongates, its cross-sectional area decreases, as illustrated in Figure 15.1b. When the peak stress f_u is reached (at Point C in Figure 15.2b), there is usually some localised *necking* over a very short length of the test sample and the cross-sectional area A in this region reduces significantly as strain increases until rupture occurs at ε_r (Point D in Figure 15.2b).

Depending on the material considered, the compressive and tensile stress–strain relationships may exhibit significant differences, such as shown in Figure 15.3 for a typical stress–strain curve of concrete. In this case, the behaviour of the material subjected to tensile (positive) stresses and strains is brittle, while the material behaves in a nonlinear but generally more ductile manner under compression (for typical low- to medium-strength concrete), when stresses and strains are negative.

For the purpose of this chapter, we will use only simple nonlinear material properties, such as the idealised elastic–perfectly plastic behaviour shown in Figure 15.4. We will also assume that behaviour is the same in both compression and tension. This is usually

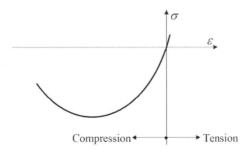

Figure 15.3 Typical stress–strain curve for concrete.

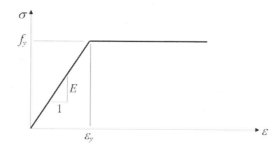

Figure 15.4 Idealised stress–strain curve for an elastic–perfectly plastic material.

suitable for metals and, in particular, for mild steel. The constitutive stress–strain equations describing this behavior are defined as (see Figure 15.4):

$$\sigma = E\varepsilon \quad \text{for } |\varepsilon| \le \varepsilon_y \tag{15.2a}$$

$$\sigma = f_y \quad \text{for } \varepsilon > \varepsilon_y \quad \text{and} \quad \sigma = -f_y \quad \text{for } \varepsilon < -\varepsilon_y \tag{15.2b,c}$$

where $\varepsilon_y = (f_y/E)$ is the strain at which the material starts to yield, commonly referred to as the *yield strain,* and the absolute value of the strain $|\varepsilon|$ has been introduced to account for both tensile (positive) and compressive (negative) strains.

15.3 ILLUSTRATIVE EXAMPLES

15.3.1 Axially loaded members

The nonlinear analysis of a member subjected to an axial force is illustrated by means of a simple example that consists of one element fixed at one end and subjected to an applied load P at the other end, as shown in Figure 15.5a. The cross-section is made up of three rectangular layers bonded together as illustrated in Figure 15.5b. The material of each layer is assumed to have an elastic–perfectly plastic stress–strain relationship as given by

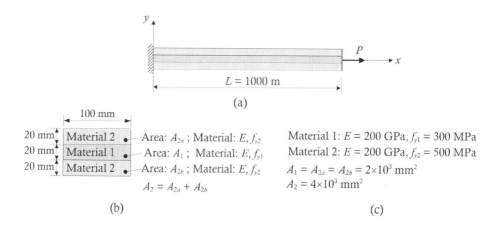

Figure 15.5 Example of an axially loaded composite member. (a) Elevation. (b) Cross-section. (c) Material and cross-sectional properties.

Equations 15.2. The doubly-symmetric cross-section ensures that the external force induces only axial deformations and axial forces in each material.

For low levels of P, the applied load is shared among the layers proportionally to their axial rigidities (since both materials are loaded in the linear–elastic range):

$$N_1 = \frac{EA_1}{E(A_1 + A_2)} P = \frac{A_1}{A_1 + A_2} P = \frac{1}{3} P \qquad (15.3a)$$

$$N_2 = \frac{EA_2}{E(A_1 + A_2)} P = \frac{A_2}{A_1 + A_2} P = \frac{2}{3} P \qquad (15.3b)$$

where N_1 and N_2 represent the axial forces resisted by materials 1 and 2, respectively. The consequent deformation that must be the same in each material can be calculated recalling Equations 4.25 and 5.8 as:

$$e = \frac{N_1 L}{EA_1} = \frac{N_2 L}{EA_2} = \frac{PL}{E(A_1 + A_2)} \qquad \varepsilon = \frac{e}{L} = \frac{P}{E(A_1 + A_2)} \qquad (15.4a,b)$$

Equations 15.3 and 15.4 are valid provided the two materials remain in the linear–elastic range. This is satisfied when the strain in each material determined by Equation 15.4b is less than its yield strain or, alternatively, as long as the stress in each material is less than its yield stress.

Before investigating the nonlinear response of the member, it is useful to identify the load P at which one of the materials starts to yield. We do this by considering the yield strains of each material. From Figure 15.4:

$$\varepsilon_{y1} = \frac{f_{y1}}{E} = \frac{300}{200 \times 10^3} = 0.0015 \qquad (15.5a)$$

$$\varepsilon_{y2} = \frac{f_{y2}}{E} = \frac{500}{200 \times 10^3} = 0.0025 \qquad (15.5b)$$

Material 1 yields at a lower strain than material 2 (because f_{y1} is smaller than f_{y2}). The load P, which produces a strain of $\varepsilon_{y1} = 0.0015$, is determined by re-arranging Equation 15.4b:

$$P_{y1} = \varepsilon_{y1} E(A_1 + A_2) = 0.0015 \times (200 \times 10^3) \times (2 \times 10^3 + 4 \times 10^3) = 1800 \text{ kN} \qquad (15.6)$$

which causes an extension e_{y1} (from Equations 15.4):

$$e_{y1} = \varepsilon_{y1} L = 0.0015 \times 1000 = 1.5 \text{ mm} \qquad (15.7)$$

The yield force N_{y1} resisted by material 1 is obtained from Equation 15.3a:

$$N_{y1} = \frac{1}{3} P_{y1} = 600 \text{ kN} \qquad (15.8)$$

After material 1 has yielded, under the assumption of the elastic–perfectly plastic behaviour, it cannot carry any additional load. Therefore, if the applied load P increases above P_{y1}, the load carried by material 1 remains at N_{y1} while all additional force is taken by material 2:

$$N_1 = N_{y1} = 600 \text{ kN} \quad N_2 = P - N_{y1} = P - 600 \text{ kN} \tag{15.9a,b}$$

The structure will be able to carry higher levels of P until material 2 yields. This occurs at:

$$P_{y2} = N_1 + N_2 = N_{y1} + \varepsilon_{y2}EA_2 = 600 \times 10^3 + 0.0025 \times (200 \times 10^3) \times 4 \times 10^3 = 2600 \text{ kN} \tag{15.10}$$

which takes place at an elongation equal to:

$$e_{y2} = \varepsilon_{y2}L = 0.0025 \times 1000 = 2.5 \text{ mm} \tag{15.11}$$

The force P_{y2} is usually referred to as the collapse load because unlimited deformation can take place in the structure at this level of load.

The response of the structure, expressed in terms of load and elongation, is summarised in Figure 15.6, which highlights how the changes in the overall rigidity are affected by the development of yielding in the two materials.

The incremental solution process followed in Equations 15.3 through 15.11 is useful in outlining the key aspects involved in a nonlinear solution. When dealing with nonlinear analyses, it is usually preferable to implement numerical solutions that are more flexible in dealing with complex structures and material properties. This will be outlined in Section 15.4 presenting a widely used solution procedure, i.e. the Newton–Raphson method.

15.3.2 Beams in bending

In Chapter 5, we introduced the moment–curvature relationship to describe the flexural response of beams made up of linear–elastic materials (Equation 5.36):

$$\kappa = \frac{M}{EI} \tag{5.36}$$

The moment M is related to the curvature κ by means of the flexural rigidity EI. All terms are calculated with respect to the z-axis (based on a centroidal coordinate system) and assuming that the cross-section possesses an axis of symmetry coincident with the y-axis. If

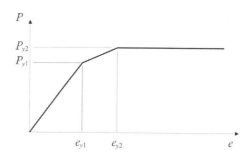

Figure 15.6 Nonlinear response of the axially loaded composite member.

the curvature along a beam is known, this expression can be used to determine the rotations and deflections along the member length, as for example carried out using the method of double integration (see Section 5.5).

When dealing with nonlinear material properties, the relationship between the moment and the curvature becomes *nonlinear* as well.

To determine the nonlinear moment–curvature relationship, we will consider the simple case of a rectangular cross-section (with width b and depth d) fabricated from an elastic–perfectly plastic material (with a stress–strain relationship given by Equations 15.2) and subjected to an external moment only. This example has been selected to minimise the complexity of the derivation, but other cross-sections and other nonlinear material properties can also be considered with the same approach.

Taking the y- and z-axes as the centroidal axes, typical stress and strain distributions induced at the cross-section for low levels of moments are shown in Figure 15.7. The materials are assumed to remain within their linear–elastic range because their maximum strains at the top and bottom fibres of the cross-section fall below the yield strains ε_y. Under these conditions, stresses σ remain proportional to strains ε, with the proportionality constant being the elastic modulus E, as in Equation 15.2a.

For the cross-section of Figure 15.7a, the expressions relating ε and σ can be written as:

$$\varepsilon = -y\kappa \quad \sigma = E\varepsilon = -Ey\kappa \tag{15.12a,b}$$

which represent simplified versions of Equations 5.18 and 5.20 because, in this case, the axial force is zero. Recalling the definition of the internal moment, the relationship between the moment and curvature can be derived as (Equation 5.22):

$$M = -\int_A y\sigma\,dA = -\int_A Ey(-y\kappa)\,dA = E\kappa\int_A y^2\,dA = EI\kappa \tag{15.13}$$

which, re-arranged, produces Equation 5.36. This expression is valid up to the point at which the material starts to yield. This occurs when the extreme fibers of the cross-section reach the material yield stress as shown in Figure 15.8, i.e. when:

$$|\varepsilon| = \varepsilon_y = \frac{f_y}{E} \tag{15.14}$$

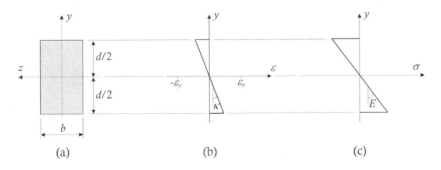

Figure 15.7 Strain and stress diagrams in the linear–elastic range. (a) Cross-section. (b) Strain diagram. (c) Stress diagram.

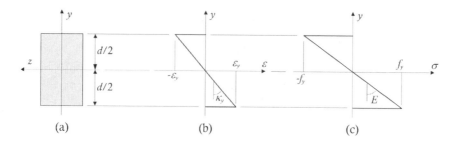

Figure 15.8 Strain and stress diagrams at first yield. (a) Cross-section. (b) Strain diagram. (c) Stress diagram.

The corresponding *curvature at first yield* is referred to as κ_y and can be calculated from Equations 15.12a and 15.14 as:

$$\kappa_y = \frac{\varepsilon_y}{d/2} = \frac{2f_y}{Ed} \tag{15.15}$$

where $d/2$ represents the distance between the z-axis and the extreme fibre of the cross-section (Figure 15.8). The corresponding *moment at first yield* M_y can then be calculated using Equations 15.13 substituting $\kappa = \kappa_y$:

$$M_y = EI\kappa_y \tag{15.16}$$

Recalling that $I = bd^3/12$ for a rectangular section, Equation 15.16 can be re-written using Equation 5.15 as:

$$M_y = f_y\frac{bd^2}{6} = f_yZ \tag{15.17}$$

where Z is the elastic section modulus and is equal to $bd^2/6$ for a rectangular section.

For levels of moment greater than M_y, yielding extends into the cross-section as shown in Figure 15.9. In this case, the yielded part of the cross-section exhibits a strain greater than ε_y (Figure 15.9b) and is subjected to a stress equal to the yield stress f_y (Figure 15.9c), while the remaining part of the cross-section remains in the linear–elastic range. For ease of notation, the extent of yielding is defined by the parameter α, which is equal to 1 for a section at first yield and 0 for a fully yielded section, i.e. $0 \le \alpha \le 1$. The moment corresponding

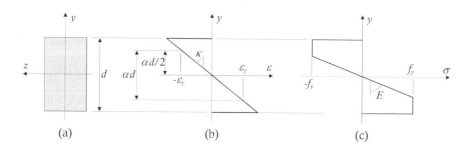

Figure 15.9 Strain and stress diagrams in the nonlinear range. (a) Cross-section. (b) Strain diagram. (c) Stress diagram.

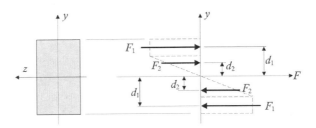

Figure 15.10 Summary of cross-sectional resultants in the nonlinear range.

to strain diagrams for which $\alpha < 1$ can be calculated by summing the moments induced by the stress blocks shown in Figure 15.9c whose resultant and lever arms are specified in Figure 15.10:

$$M = 2d_1 F_1 + 2d_2 F_2 \tag{15.18}$$

where:

$$d_1 = \frac{d}{2}\frac{1+\alpha}{2} \quad d_2 = \alpha\frac{d}{3} \quad F_1 = f_y bd\frac{1-\alpha}{2} \quad F_2 = f_y bd\frac{\alpha}{4} \tag{15.19a–d}$$

Simplifying Equations 15.18 and 15.19, the nonlinear moment–curvature relationship can be expressed as:

$$M = M_p\left(1 - \frac{\alpha^2}{3}\right) \tag{15.20}$$

where M_p is the plastic moment of the cross-section and is defined as:

$$M_p = f_y S \tag{15.21}$$

with S being the plastic section modulus, calculated as $S = bd^2/4$ for rectangular cross-sections. M_p is the moment corresponding to $\alpha = 0$, with the entire cross-section above the neutral axis yielding in compression and the entire cross-section below the neutral axis yielding in tension.

The values and differences between the first yield and plastic moments (M_y and M_p), and equivalently between the elastic and plastic section moduli (Z and S), depend on the shape of the cross-section and are captured in the *shape factor* λ defined as:

$$\lambda = \frac{M_p}{M_y} = \frac{S}{Z} \tag{15.22}$$

with λ equal to 1.5 for rectangular cross-sections, about 1.15 for I-shaped sections and 1.7 for circular sections.

It is useful to express α in terms of the curvature considering that, from its definition, it specifies the location $y = \pm\alpha d/2$ at which the strain equals $\pm\varepsilon_y$ and, recalling Equation 15.12a:

$$\varepsilon_y = \frac{f_y}{E} = \kappa\left(\alpha\frac{d}{2}\right) \tag{15.23}$$

Re-arranging and substituting the expression for κ_y in Equation 15.15, we get:

$$\alpha = \frac{\kappa_y}{\kappa} \tag{15.24}$$

In summary, the moment–curvature relationship for a rectangular cross-section of an elastic–perfectly plastic material is described by Equations 15.13, 15.20 and 15.24, reproduced here for ease of reference:

$$M = EI\kappa \qquad \text{for } |\kappa| \le \kappa_y \tag{15.25a}$$

$$M = M_p\left(1 - \frac{\kappa_y^2}{3\kappa^2}\right) \quad \text{for } |\kappa| > \kappa_y \tag{15.25b}$$

The graphical representation of Equations 15.25 is illustrated in Figure 15.11, which shows how the cross-section remains in its linear–elastic range until the external moment M reaches M_y, after which yielding starts to develop and expands to the entire cross-section at $M = M_p$ when the curvature is infinite, that is, when $\alpha = \kappa_y/\kappa = 0$. The limiting capacity for the cross-section is the plastic moment M_p.

In the case of a statically determinate beam, we can determine the structural response by applying the method of double integration to Equations 15.25. This procedure is now outlined for a simply-supported beam with a rectangular cross-section and subjected to a point load applied at mid-span, as illustrated in Figure 15.12. Because of the symmetry of both the loading and boundary conditions, only half of the beam will be considered in the analysis. Considering the two different expressions derived to describe the moment–curvature

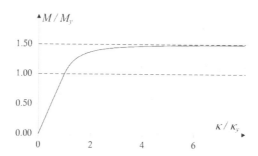

Figure 15.11 Normalised moment–curvature diagram for a linear–perfectly plastic material.

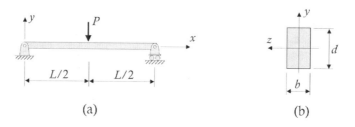

(a)   (b)

Figure 15.12 Beam and cross-section. (a) Elevation. (b) Cross-section.

relationship (Equations 15.25), the method of double integration needs to be applied twice: the first for the case where the beam remains in its linear–elastic range for its entire length and the second when part of the beam is yielding.

The determination of the rotations and deflections when the beam is linear–elastic follows the procedure already outlined in Sections 5.5 and 6.2 where we introduced the method of double integration.

For the beam of Figure 15.12, the expression for the internal moment is calculated using the free-body diagram shown in Figure 15.13b (reactions were obtained from statics: $H_A = 0$, $V_A = P/2$, $V_B = P/2$):

$$M = \frac{P}{2}x \quad \text{for } 0 \le x \le \frac{L}{2} \tag{15.26}$$

Applying the method of double integration to the curvature of Equation 15.25a, the general expressions for the rotation and deflection can be written as:

$$\kappa = \frac{M}{EI} = \frac{Px}{2EI} \tag{15.27a}$$

$$\theta = \int \kappa \, dx + C_a = \int \frac{M}{EI} dx + C_a = \frac{Px^2}{4EI} + C_a \tag{15.27b}$$

$$v = \int \theta \, dx + C_b = \int \int \frac{M}{EI} dx \, dx + C_b = \frac{Px^3}{12EI} + C_a x + C_b \tag{15.27c}$$

whose constants of integration C_a and C_b are evaluated by enforcing the boundary conditions $v(x = 0) = 0$ and $\theta(x = L/2) = 0$ as:

$$C_a = -\frac{PL^2}{16EI} \quad C_b = 0 \tag{15.28a,b}$$

Substituting Equations 15.28 into Equations 15.27b and c, the expressions for the rotation and deflection are:

$$\theta = \frac{P}{4EI}\left(x^2 - \frac{L^2}{4} \right) \quad v = \frac{Px}{4EI}\left(\frac{x^2}{3} - \frac{L^2}{4} \right) \tag{15.29a,b}$$

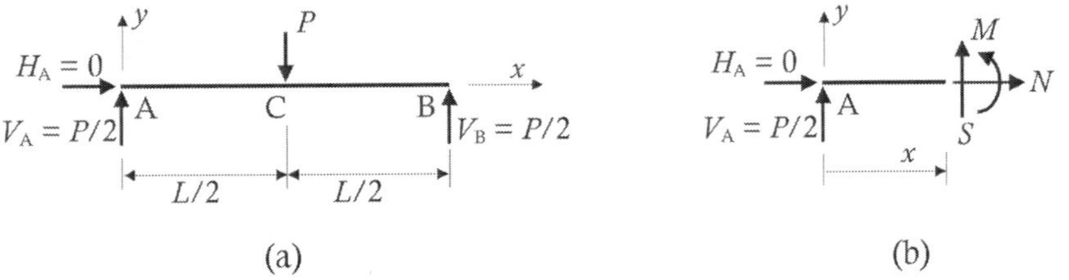

(a)

(b)

Figure 15.13 Free-body diagrams.

and the mid-span deflection is:

$$v\left(x=\frac{L}{2}\right) = -\frac{PL^3}{48EI} \tag{15.29c}$$

When the mid-span moment is greater than M_y, i.e. when P is greater than $4M_y/L$, part of the cross-section at mid-span starts to yield. As load increases, the region of the beam where M exceeds M_y expands. The part of the beam of length a adjacent to each supports where $M \le M_y$ exhibits linear–elastic behaviour (regions 1 in Figure 15.14) and its response can be described by the moment–curvature relationship of Equation 15.25a. In the remaining central length of the beam, referred to as region 2 in Figure 15.14, the material is partly yielded and its response is represented by the moment–curvature relationship of Equation 15.25b.

For clarity, we will use subscripts '1' or '2' on the symbols for curvature, rotation and deflection to indicate when they relate to regions 1 or 2, respectively.

The expressions for the curvature in the two regions in the first half span of the beam are expressed as:

$$\kappa_1 = \frac{M}{EI} = \frac{Px}{2EI} \qquad \text{for } 0 \le x \le a \tag{15.30a}$$

$$\kappa_2 = \frac{\sqrt{6M_p}\,\kappa_y}{3}(2M_p - Px)^{-\frac{1}{2}} \quad \text{for } a < x \le \frac{L}{2} \tag{15.30b}$$

where κ_2 is obtained by re-arranging Equation 15.25b and substituting the expression for the internal moment (Equation 15.26). Applying the method of double integration to the two regions, we obtain the general expressions for the rotation and deflection:

$$\theta_1 = \int \kappa_1 \, dx + C_{a1} = \frac{Px^2}{4EI} + C_{a1} \tag{15.31a}$$

$$v_1 = \int \theta_1 \, dx + C_{b1} = \frac{Px^3}{12EI} + C_{a1}x + C_{b1} \tag{15.31b}$$

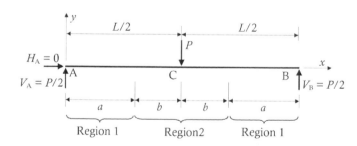

Note: Region 1: linear-elastic region $(M \le M_y)$
       Region 2: nonlinear region $(M > M_y)$

Figure 15.14 Linear–elastic and nonlinear regions along the beam length.

$$\theta_2 = \int \kappa_2 \, dx + C_{a2} = -\frac{2}{3} \frac{\sqrt{6M_p}\kappa_y}{P} (2M_p - Px)^{\frac{1}{2}} + C_{a2} \tag{15.32a}$$

$$v_2 = \int \theta_2 \, dx + C_{b2} = \frac{4}{9} \frac{\sqrt{6M_p}\kappa_y}{P^2} (2M_p - Px)^{\frac{3}{2}} + C_{a2}x + C_{b2} \tag{15.32b}$$

The constants of integration are evaluated by enforcing the following boundary conditions:

$$v_1(x = 0) = 0 \quad \theta_2\left(x = \frac{L}{2}\right) = 0 \tag{15.33a,b}$$

$$v_1(x = a) = v_2(x = a) \quad \theta_1(x = a) = \theta_2(x = a) \tag{15.34a,b}$$

which specify zero deflection at the pinned support, zero rotation at mid-span owing to symmetry, and continuity of deflection and rotation at $x = a$. These are determined as:

$$C_{a1} = \frac{3M_y^2 \beta_1\beta_3 + \beta_2\beta_3\left(4M_p^2 - 4M_pM_y\right) + 2\beta_1\beta_2 M_p(PL - 4M_p)}{-3P\beta_1\beta_3 EI} \tag{15.35a}$$

$$C_{b1} = 0 \tag{15.35b}$$

$$C_{a2} = \frac{2\sqrt{3}\kappa_y\beta_3}{3P} \tag{15.36a}$$

$$C_{b2} = \frac{-4\left[M_y^2\left(3M_y\beta_1 - 2M_p\beta_2\right) + 2M_p^2\beta_2(2M_p - M_y)\right]}{9\beta_1 P^2 EI} \tag{15.36b}$$

with:

$$\beta_1 = \sqrt{(M_p - M_y)M_p} \quad \beta_2 = \sqrt{3}\kappa_y EI \quad \beta_3 = \sqrt{(4M_p - PL)M_p} \tag{15.37a–c}$$

By substituting Equations 15.35 and 15.36 into Equations 15.31 and 15.32, we obtain the expressions for the rotations and deflections in regions 1 and 2. In particular, the mid-span deflection can be calculated as:

$$v_2\left(x = \frac{L}{2}\right) = \frac{\beta_1\beta_2\beta_3(PL + 8M_p) - 4\left(3M_y^3\beta_1 + 4M_p^3\beta_2\right) + 8\beta_2 M_y M_p(M_y + M_p)}{9P^2\beta_1 EI} \tag{15.38}$$

These equations are applicable for load levels that induce mid-span moments between M_y and M_p. The limit case of a beam with a mid-span moment M_p leads to the formation of a hinge, because the curvature tends to infinity as shown by the asymptote of Figure 15.11.

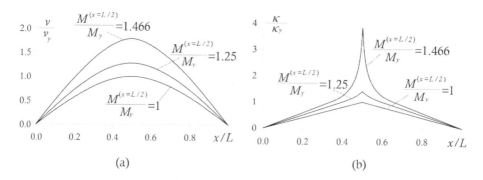

Figure 15.15 Deflection and curvature distributions for different levels of load.

An overview of the structural response of the beam, in terms of deflection and curvature, is illustrated in Figure 15.15 for different load levels, expressed as a function of mid-span moment over moment at first yield. In particular, three cases have been considered:

  i. when the mid-span moment just reaches M_y, i.e. shown as $M^{(x=L/2)}/M_y = 1$ in the plot (this case shows behaviour when the entire beam is linear–elastic)
 ii. at a level of load inducing a mid-span moment equal to the average of the yield and plastic moment (i.e. equal to $(M_y + M_p)/2$)
iii. at a moment approaching M_p, i.e. with $M^{(x=L/2)}/M_y = 1.466$ which corresponds to about 98% of M_p

In the elastic–plastic region, the curvature tends to increase significantly as yielding propagates in the cross-section and in central region of the beam. As a consequence, also the deflection increases significantly.

It is not always possible to derive analytical solutions to describe the nonlinear response of a structural member and, for this purpose, a more general numerical approach will be outlined in the following section based on the Newton–Raphson method.

15.4 NONLINEAR ANALYSIS USING THE NEWTON–RAPHSON METHOD

15.4.1 Overview of the Newton–Raphson method

The Newton–Raphson method is an iterative procedure suitable for the solution of nonlinear problems. In this section, we outline the key steps of the procedure considering that a structural problem can be expressed as a system of nonlinear equations enforcing equilibrium at particular locations or along specified freedoms of the structure. When dealing with material nonlinearities, these equations can be written in general form as:

$$\mathbf{K}(\mathbf{D}) = \mathbf{Q} \qquad (15.39)$$

where $\mathbf{Q}$ represents the vector of applied loads and vector $\mathbf{K}(\mathbf{D})$ collects the nonlinear functions describing the internal actions, expressed in terms of displacements $\mathbf{D}$. In this context, the components of vector $\mathbf{D}$ depend on the formulation of the problem being defined by the nonlinear equations. For clarity, no separation is specified at this stage between known and unknown variables collected in $\mathbf{D}$.

The iterative procedure starts from an assumed set of results, adopted in the first iteration (referred to as $i = 1$, with i being the counter for the iterations). The assumed initial set of results is usually based on an informed guess. The solution is then improved in subsequent iterations until a selected convergence criterion is satisfied. For a structure subjected to a particular level of applied load, if we denote the solution at the i-th iteration as $\mathbf{D}^{(i)}$, the solution in the next iteration $(i + 1)$ is obtained by approximating Equation 15.39 with the first terms of its Taylor expansion as:

$$\mathbf{K}(\mathbf{D}^{(i+1)}) \approx \mathbf{K}(\mathbf{D}^{(i)}) + \mathbf{K}_t(\mathbf{D}^{(i)})\Delta\mathbf{D}^{(i)} = \mathbf{Q} \tag{15.40}$$

where $\Delta\mathbf{D}^{(i)}$ represents the vector of increments of the structure displacements and $\mathbf{K}_t(\mathbf{D}^{(i)})$ describes the tangent behaviour of the structure based on the stiffness calculated at the displaced conditions set by $\mathbf{D}^{(i)}$. Equation 15.40 can be re-arranged to separate the terms associated with the increment displacements $\Delta\mathbf{D}^{(i)}$ from the remaining terms as:

$$\mathbf{K}_t(\mathbf{D}^{(i)})\Delta\mathbf{D}^{(i)} = \mathbf{Q}_R^{(i)} \tag{15.41}$$

where $\mathbf{Q}_R^{(i)}$ is usually referred to as the unbalanced load vector (or residual load) and is defined as:

$$\mathbf{Q}_R^{(i)} = \mathbf{Q} - \mathbf{K}(\mathbf{D}^{(i)}) \tag{15.42}$$

In this expression, the vector $\mathbf{K}(\mathbf{D}^{(i)})$ describes the internal actions corresponding to displacements $\mathbf{D}^{(i)}$.

At this point, the displacement increments $\Delta\mathbf{D}^{(i)}$ are calculated following the standard solutions procedures presented in Appendix C and used in previous chapters when dealing with linear–elastic material properties. For this purpose, in the detailed solution, the known and unknown terms of displacement increments and loads can be separated by partitioning.

At the end of each iteration, the possible convergence of the solution is evaluated based on a specified criteria and the current displacement vector $\mathbf{D}^{(i)}$ is updated by adding the displacement increments $\Delta\mathbf{D}^{(i)}$ calculated at the i-th iteration. There are different possible termination criteria that can be used in the analysis. The convergence criteria are usually based on the calculations of normalised norms that involve either displacements or actions, such as those specified in the following:

$$\text{norm}_{\text{tol1}} = \frac{\left|\Delta\mathbf{D}^{(i)}\right|}{\left|\mathbf{D}^{(i+1)}\right|} \qquad \text{norm}_{\text{tol2}} = \frac{\left|\mathbf{Q}_R^{(i+1)}\right|}{\left|\mathbf{Q}\right|} \tag{15.43a,b}$$

If the values obtained with the adopted normalised norm fall below a certain tolerance value norm_{tol}, then the solution is said to converge; otherwise, another iteration is required and performed. Adequate values to be used for norm_{tol} depend on the problem being considered and the accuracy sought. Usually, these are in the order of 10^{-2} to 10^{-6}, but its validity can be easily checked for the particular problem being considered by evaluating the magnitude of the residual load vector for different norm_{tol} values.

The overall strategy of the Newton–Raphson method is illustrated in Figure 15.16 considering a scalar problem for clarity.

The case of one load level is considered in Figure 15.16a to better highlight the details of the solution process. In particular, the first iteration starts by adopting an initial tangent stiffness, referred to as $K_t(D^{(1)})$, calculated based on the 'guessed' displacement $D^{(1)}$. This

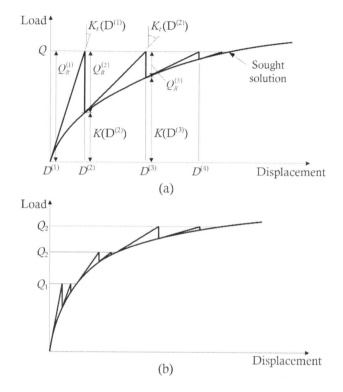

Figure 15.16 Newton–Raphson method.

enables the calculation of the displacement increment $\Delta D^{(1)}$ from which we get displacement $D^{(2)}$ $(= D^{(1)} + \Delta D^{(1)})$. Convergence is then verified for this solution and, if satisfied, the analysis is terminated. If convergence is not satisfied, another iteration is carried out with a revised tangent stiffness $K_t(D^{(2)})$ and the previous process is repeated as many times as necessary.

The ability of the Newton–Raphson method to follow the nonlinear response of a structure relies on the use of relatively small increments to avoid the solution diverging or oscillations in the results. For this purpose, the loads are applied in increments and each of these load levels is usually referred to as a load step. The solution procedure applied to consider different load increments is identical to the one previously outlined for a load Q and illustrated in Figure 15.16b.

Other solution strategies that build on the Newton–Raphson method are available in the literature. For example, it is possible to avoid building the tangent stiffness between adjacent iterations and to keep on using the tangent stiffness calculated in the first iteration. This approach is usually referred to as the Modified Newton–Raphson method and becomes useful with problems of large size, where the calculation of a new tangent stiffness can be very demanding computationally. When running an analysis for increasing levels of applied load, it is usually possible to approach the peak load, while different solution strategies need to be implemented to follow the post-peak response, such as the arc-length method.

15.4.2 Cross-sectional analysis using the Newton–Raphson method

The cross-sectional analysis considered in this section is based on the assumptions of the Euler–Bernoulli beam model. In this context, the variables included in the vector **D** are

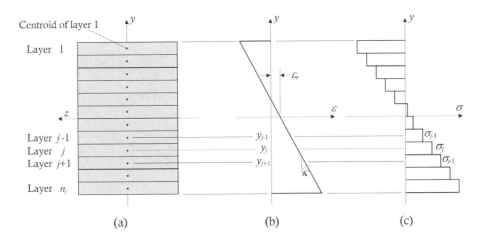

Figure 15.17 Cross-sectional discretisation. (a) Cross-section. (b) Strain diagram. (c) Stress diagram.

deformations, namely, the strain measured at the level of the reference axis ε_r and the curvature κ, as illustrated in Figure 15.17b. The variation of the strain ε over the cross-section can then be written as:

$$\varepsilon = \varepsilon_r - y\kappa \tag{15.44}$$

The terms included in the vector $\mathbf{K(D)}$ are the internal actions that, in this case, are the internal axial force N and the internal moment M with respect to the reference z-axis. External loads considered are expressed in terms of the external actions applied to the cross-section, consisting of the external axial force N_e and external moment M_e calculated with respect to the reference axis.

To remain consistent with the notation introduced in Chapter 13, the variables $\mathbf{D}$, $\mathbf{K(D)}$ and $\mathbf{Q}$ presented in the previous section are replaced with $\boldsymbol{\varepsilon}$, $\mathbf{r}(\boldsymbol{\varepsilon})$ and $\mathbf{r}_e$, respectively. These terms are defined as:

$$\boldsymbol{\varepsilon} = \begin{bmatrix} \varepsilon_r \\ \kappa \end{bmatrix} \quad \mathbf{r}(\boldsymbol{\varepsilon}) = \begin{bmatrix} N \\ M \end{bmatrix} \quad \mathbf{r}_e = \begin{bmatrix} N_e \\ M_e \end{bmatrix} \tag{15.45a–c}$$

and Equation 15.39 can be re-written as:

$$\mathbf{r}(\boldsymbol{\varepsilon}) = \mathbf{r}_e \tag{15.46}$$

We will now revisit the nonlinear procedure presented in Equations 15.39 through 15.43. In particular, the internal actions to be calculated at the $(i + 1)$-th iteration are approximated with the first terms of the Taylor expansion of the internal actions calculated at the i-th iteration as:

$$\mathbf{r}(\boldsymbol{\varepsilon}^{(i+1)}) \approx \mathbf{r}(\boldsymbol{\varepsilon}^{(i)}) + \mathbf{r}_t(\boldsymbol{\varepsilon}^{(i)})\Delta\boldsymbol{\varepsilon}^{(i)} = \mathbf{r}_e \tag{15.47}$$

Equation 15.47 is equivalent to Equation 15.40 and can be re-arranged to separate the vector of residual loads $\mathbf{r}_R^{(i)}$ as (equivalent to Equation 15.41):

$$\mathbf{r}_t(\boldsymbol{\varepsilon}^{(i)})\Delta\boldsymbol{\varepsilon}^{(i)} = \mathbf{r}_R^{(i)} \tag{15.48}$$

with:

$$\mathbf{r}_R^{(i)} = \mathbf{r}_e - \mathbf{r}(\boldsymbol{\varepsilon}^{(i)}) \tag{15.49}$$

To better illustrate the solution procedure, we separate the two equations included in the nonlinear system of Equations 15.48:

$$\frac{\partial N\left(\varepsilon_r^{(i)}, \kappa^{(i)}\right)}{\partial \varepsilon_r} \Delta \varepsilon_r^{(i)} + \frac{\partial N\left(\varepsilon_r^{(i)}, \kappa^{(i)}\right)}{\partial \kappa} \Delta \kappa^{(i)} = N_R^{(i)} \tag{15.50a}$$

$$\frac{\partial M\left(\varepsilon_r^{(i)}, \kappa^{(i)}\right)}{\partial \varepsilon_r} \Delta \varepsilon_r^{(i)} + \frac{\partial M\left(\varepsilon_r^{(i)}, \kappa^{(i)}\right)}{\partial \kappa} \Delta \kappa^{(i)} = M_R^{(i)} \tag{15.50b}$$

in which the residual load vectors $N_R^{(i)}$ and $M_R^{(i)}$ are calculated as:

$$N_R^{(i)} = N_e - N\left(\varepsilon_r^{(i)}, \kappa^{(i)}\right) \tag{15.51a}$$

$$M_R^{(i)} = M_e - M\left(\varepsilon_r^{(i)}, \kappa^{(i)}\right) \tag{15.51b}$$

with $N\left(\varepsilon_r^{(i)}, \kappa^{(i)}\right)$ and $M\left(\varepsilon_r^{(i)}, \kappa^{(i)}\right)$ being the internal axial force and moment determined at the i-th iteration.

Based on the notation adopted in Equations 15.50 and 15.51, it is possible to define all terms included in Equation 15.48 at the i-th iteration as follows:

$$\mathbf{r}_t(\boldsymbol{\varepsilon}^{(i)}) = \begin{bmatrix} \dfrac{\partial N\left(\varepsilon_r^{(i)}, \kappa^{(i)}\right)}{\partial \varepsilon_r} & \dfrac{\partial N\left(\varepsilon_r^{(i)}, \kappa^{(i)}\right)}{\partial \kappa} \\ \dfrac{\partial M\left(\varepsilon_r^{(i)}, \kappa^{(i)}\right)}{\partial \varepsilon_r} & \dfrac{\partial M\left(\varepsilon_r^{(i)}, \kappa^{(i)}\right)}{\partial \kappa} \end{bmatrix} \tag{15.52a}$$

$$\Delta \boldsymbol{\varepsilon}^{(i)} = \begin{bmatrix} \Delta \varepsilon_r^{(i)} \\ \Delta \kappa^{(i)} \end{bmatrix} \qquad \mathbf{r}_R^{(i)} = \begin{bmatrix} N_R^{(i)} \\ M_R^{(i)} \end{bmatrix} \tag{15.52b,c}$$

The partial derivatives of N and M with respect to ε_r and κ included in Equations 15.50 (and collected in Equation 15.52a) can be re-arranged in a more practical form, recalling the definitions of internal actions (Equation 13.7), as:

$$\frac{\partial N\left(\varepsilon_r^{(i)}, \kappa^{(i)}\right)}{\partial \varepsilon_r} = \int_A \frac{\partial \sigma}{\partial \varepsilon_r} dA \qquad \frac{\partial N\left(\varepsilon_r^{(i)}, \kappa^{(i)}\right)}{\partial \kappa} = \int_A \frac{\partial \sigma}{\partial \kappa} dA \tag{15.53a,b}$$

$$\frac{\partial M\left(\varepsilon_r^{(i)}, \kappa^{(i)}\right)}{\partial \varepsilon_r} = -\int_A y \frac{\partial \sigma}{\partial \varepsilon_r} dA \qquad \frac{\partial M\left(\varepsilon_r^{(i)}, \kappa^{(i)}\right)}{\partial \kappa} = -\int_A y \frac{\partial \sigma}{\partial \kappa} dA \tag{15.53c,d}$$

where the value of the stress depends on the constitutive models adopted for the materials and on the magnitude of the strain, which is defined by the strain variables ε_r and κ as well as by the location y of the point being considered within the cross-section.

For the implementation of a nonlinear analysis, it is possible to rewrite the terms of Equations 15.53 making use of the chain rule for the calculation of the partial derivatives. This is carried out for Equation 15.53a recalling the expression for the strain of Equation 15.44:

$$\frac{\partial N\left(\varepsilon_r^{(i)},\kappa^{(i)}\right)}{\partial \varepsilon_r} = \int_A \frac{\partial \sigma}{\partial \varepsilon_r}\,dA = \int_A \frac{\partial \sigma}{\partial \varepsilon}\frac{\partial \varepsilon}{\partial \varepsilon_r}\,dA = \int_A \frac{\partial \sigma}{\partial \varepsilon}\frac{\partial (\varepsilon_r - y\kappa)}{\partial \varepsilon_r}\,dA = \int_A \frac{\partial \sigma}{\partial \varepsilon}\,dA \qquad (15.54a)$$

and applying the same procedure to Equations 15.53b through d:

$$\frac{\partial N\left(\varepsilon_r^{(i)},\kappa^{(i)}\right)}{\partial \kappa} = \int_A \frac{\partial \sigma}{\partial \kappa}\,dA = \int_A \frac{\partial \sigma}{\partial \varepsilon}\frac{\partial (\varepsilon_r - y\kappa)}{\partial \kappa}\,dA = -\int_A y\frac{\partial \sigma}{\partial \varepsilon}\,dA \qquad (15.54b)$$

$$\frac{\partial M\left(\varepsilon_r^{(i)},\kappa^{(i)}\right)}{\partial \varepsilon_r} = -\int_A y\frac{\partial \sigma}{\partial \varepsilon_r}\,dA = -\int_A y\frac{\partial \sigma}{\partial \varepsilon}\frac{\partial (\varepsilon_r - y\kappa)}{\partial \varepsilon_r}\,dA = -\int_A y\frac{\partial \sigma}{\partial \varepsilon}\,dA \qquad (15.54c)$$

$$\frac{\partial M\left(\varepsilon_r^{(i)},\kappa^{(i)}\right)}{\partial \kappa} = -\int_A y\frac{\partial \sigma}{\partial \kappa}\,dA = -\int_A y\frac{\partial \sigma}{\partial \varepsilon}\frac{\partial (\varepsilon_r - y\kappa)}{\partial \kappa}\,dA = +\int_A y^2\frac{\partial \sigma}{\partial \varepsilon}\,dA \qquad (15.54d)$$

To gain a better understanding of the use of these equations, let us reconsider the elastic–perfectly plastic model illustrated in Figure 15.4 and defined in Equations 15.2, recalling that $\varepsilon = \varepsilon_r - y\kappa$:

$$\sigma = E\varepsilon \quad \text{for } |\varepsilon| \le \varepsilon_y \qquad (15.55a)$$

$$\sigma = f_y \quad \text{for } \varepsilon > \varepsilon_y \quad \text{and} \quad \sigma = -f_y \quad \text{for } \varepsilon < -\varepsilon_y \qquad (15.55b,c)$$

The partial derivatives of the stress σ with respect to ε are:

$$\frac{\partial \sigma}{\partial \varepsilon} = \frac{\partial (E\varepsilon)}{\partial \varepsilon} = E \quad \text{for } |\varepsilon| \le \varepsilon_y \qquad (15.56a)$$

$$\frac{\partial \sigma}{\partial \varepsilon} = \frac{\partial (f_y)}{\partial \varepsilon} = 0 \quad \text{for } |\varepsilon| > \varepsilon_y \qquad (15.56b)$$

The integrals of Equations 15.52 and 15.54 are usually performed numerically and, for this purpose, the cross-section is subdivided into a number of layers as shown in Figure 15.17a. In this discretisation, we assume that the stress resisted by each layer is constant and determined by the strain calculated at the centroid of the layer, as shown in Figure 15.17 for the j-th layer (with $j = 1,...,n_j$). Other assumptions could be introduced in regards to the stress distribution within each layer. For example, it is possible to adopt a linearly varying profile defined by the strain values calculated at the top and bottom of the layer.

With the adopted discretisation and under the assumption of constant stress over each layer, the integrals defining the internal actions $N\left(\varepsilon_r^{(i)},\kappa^{(i)}\right)$ and $M\left(\varepsilon_r^{(i)},\kappa^{(i)}\right)$ can be approximated by means of the rectangular rule as follows:

$$N\left(\varepsilon_r^{(i)},\kappa^{(i)}\right)=\int_A \sigma\,dA=\sum_{j=1}^{n_j}\sigma\left(y_j,\varepsilon_r^{(i)},\kappa^{(i)}\right)A_j \tag{15.57a}$$

$$M\left(\varepsilon_r^{(i)},\kappa^{(i)}\right)=-\int_A y\sigma\,dA=-\sum_{j=1}^{n_j}y_j\sigma\left(y_j,\varepsilon_r^{(i)},\kappa^{(i)}\right)A_j \tag{15.57b}$$

In a similar manner, the integrals of Equations 15.54 are approximated as:

$$\frac{\partial N\left(\varepsilon_r^{(i)},\kappa^{(i)}\right)}{\partial\varepsilon_r}=\int_A\frac{\partial\sigma}{\partial\varepsilon}\,dA=\sum_{j=1}^{n_j}\frac{\partial\sigma\left(y_j,\varepsilon_r^{(i)},\kappa^{(i)}\right)}{\partial\varepsilon}A_j \tag{15.58a}$$

$$\frac{\partial N\left(\varepsilon_r^{(i)},\kappa^{(i)}\right)}{\partial\kappa}=-\int_A y\frac{\partial\sigma}{\partial\varepsilon}\,dA=-\sum_{j=1}^{n_j}y_j\frac{\partial\sigma\left(y_j,\varepsilon_r^{(i)},\kappa^{(i)}\right)}{\partial\varepsilon}A_j \tag{15.58b}$$

$$\frac{\partial M\left(\varepsilon_r^{(i)},\kappa^{(i)}\right)}{\partial\varepsilon_r}=-\int_A y\frac{\partial\sigma}{\partial\varepsilon}\,dA=-\sum_{j=1}^{n_j}y_j\frac{\partial\sigma\left(y_j,\varepsilon_r^{(i)},\kappa^{(i)}\right)}{\partial\varepsilon}A_j \tag{15.58c}$$

$$\frac{\partial M\left(\varepsilon_r^{(i)},\kappa^{(i)}\right)}{\partial\kappa}=+\int_A y^2\frac{\partial\sigma}{\partial\varepsilon}\,dA=\sum_{j=1}^{n_j}y_j^2\frac{\partial\sigma\left(y_j,\varepsilon_r^{(i)},\kappa^{(i)}\right)}{\partial\varepsilon}A_j \tag{15.58d}$$

When performing a nonlinear cross-sectional analysis, the load is usually applied in successive load increments, usually referred to as load steps. At each level of load (i.e. at each load step), the iterative procedure is applied until convergence is achieved. This is carried out by solving Equations 15.48 for $\Delta\varepsilon_r^{(i)}$ and $\Delta\kappa^{(i)}$ in subsequent iterations until the selected convergence criterion is satisfied. For the cross-sectional analysis, we will use the following normalised norms as termination criteria:

$$\text{norm}_{tol1}=\frac{\left|\Delta\boldsymbol{\varepsilon}^{(i)}\right|}{\left|\boldsymbol{\varepsilon}^{(i+1)}\right|}\qquad \text{norm}_{tol2}=\frac{\left|\mathbf{r}_R^{(i+1)}\right|}{\left|\mathbf{r}_e\right|} \tag{15.59a,b}$$

The use of the nonlinear procedure is now outlined in Worked Examples 15.1 and 15.2.

WORKED EXAMPLE 15.1

Consider the member fixed at one end and subjected to an axial force P, previously analysed in Section 15.3.1 and illustrated in Figure 15.5. Assume the material properties to be elastic–perfectly plastic with values detailed in Figure 15.5c. Determine the deformations, expressed in terms of strain at the level of the reference axis (taken at mid-height of the section) and curvature, with a cross-sectional analysis implemented with the Newton–Raphson method for the following load steps: (1) N_e = 1000 kN, M_e = 0 kNm; (2) N_e = 2200 kN, M_e = 0 kNm; and

(3) N_e = 2600 kN, M_e = 0 kNm. For the convergence criteria, use norm$_{tol2}$ defined in Equation 15.59b equal to 0.001.

We will discretise the cross-section into three layers to account for the presence of different materials, as shown in Figure 15.18. A higher number of layers is usually adopted in the modelling, but for this simple problem, three layers are acceptable and have been specified here to keep the complexity of the solution to a minimum.

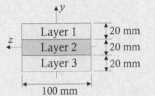

Layer	y_j (mm)	A_j (mm^2)	E_j (MPa)	f_{yj} (MPa)
j=1	20	2×10^5	2×10^5	500
j=2	0	2×10^5	2×10^5	300
j=3	-20	2×10^5	2×10^5	500

Figure 15.18 Summary of the cross-sectional properties.

On the basis of the elastic–perfectly plastic material assumptions, the partial derivatives of the stress–strain relationship can be re-written based on Equations 15.56 as:

$$\frac{\partial \sigma}{\partial \varepsilon} = \frac{\partial(E\varepsilon)}{\partial \varepsilon} = E \quad \text{for } |\varepsilon| \le \varepsilon_y \tag{15.60a}$$

$$\frac{\partial \sigma}{\partial \varepsilon} = \frac{\partial(f_y)}{\partial \varepsilon} = 0 \quad \text{for } |\varepsilon| > \varepsilon_y \tag{15.60b}$$

Load step 1: N_e = 1000 kN, M_e = 0 kNm
Iteration 1 (i = 1)
The initial tangent rigidity $r_t(\varepsilon^{(1)})$ to be used in the analysis is calculated assuming the member to be unloaded with $\varepsilon_r^{(1)} = 0$ and $\kappa^{(1)} = 0$, which is based on an initial 'guessed' condition of the structure being undeformed. The strains are all zero and, consequently, the internal actions and their derivatives are calculated as (Equations 15.57 and 15.58):

$$N\left(\varepsilon_r^{(1)},\kappa^{(1)}\right) = \sum_{j=1}^{3}\sigma\left(y_j,\varepsilon_r^{(1)},\kappa^{(1)}\right)A_j = E_1\left(\varepsilon_r^{(1)}-y_1\kappa_r^{(1)}\right)A_1 + E_2\left(\varepsilon_r^{(1)}-y_2\kappa_r^{(1)}\right)A_2 + E_3\left(\varepsilon_r^{(1)}-y_3\kappa_r^{(1)}\right)A_3 = 0 \text{ N}$$

$$M\left(\varepsilon_r^{(1)},\kappa^{(1)}\right) = -\sum_{j=1}^{3}y_j\sigma\left(y_j,\varepsilon_r^{(1)},\kappa^{(1)}\right)A_j = -y_1E_1\left(\varepsilon_r^{(1)}-y_1\kappa_r^{(1)}\right)A_1 - y_2E_2\left(\varepsilon_r^{(1)}-y_2\kappa_r^{(1)}\right)A_2 - y_3E_3\left(\varepsilon_r^{(1)}-y_3\kappa_r^{(1)}\right)A_3$$

$$= 0 \text{ Nmm}$$

$$\frac{\partial N\left(\varepsilon_r^{(1)},\kappa^{(1)}\right)}{\partial \varepsilon_r} = \sum_{j=1}^{3}\frac{\partial\sigma(y_j)}{\partial \varepsilon}A_j = E_1A_1 + E_2A_2 + E_3A_3$$

$$= (2\times10^5)\times(2\times10^3)+(2\times10^5)\times(2\times10^3)+(2\times10^5)\times(2\times10^3)$$

$$= 12\times10^8 \text{ N}$$

$$\frac{\partial N\left(\varepsilon_r^{(1)}, \kappa^{(1)}\right)}{\partial \kappa} = -\sum_{j=1}^{3}\frac{\partial \sigma(y_j)}{\partial \varepsilon} y_j A_j = -E_1 y_1 A_1 - E_2 y_2 A_2 - E_3 y_3 A_3$$
$$= -(2 \times 10^5) \times 20 \times (2 \times 10^3) - (2 \times 10^5) \times 0 \times (2 \times 10^3) - (2 \times 10^5) \times (-20) \times (2 \times 10^3)$$
$$= 0 \text{ Nmm}$$

$$\frac{\partial M\left(\varepsilon_r^{(1)}, \kappa^{(1)}\right)}{\partial \varepsilon_r} = -\sum_{j=1}^{3}\frac{\partial \sigma(y_j)}{\partial \varepsilon} y_j A_j = -E_1 y_1 A_1 - E_2 y_2 A_2 + E_3 y_3 A_3$$
$$= -(2 \times 10^5) \times 20 \times (2 \times 10^3) - (2 \times 10^5) \times 0 \times (2 \times 10^3) - (2 \times 10^5) \times (-20) \times (2 \times 10^3)$$
$$= 0 \text{ Nmm}$$

$$\frac{\partial M\left(\varepsilon_r^{(1)}, \kappa^{(1)}\right)}{\partial \kappa} = \sum_{j=1}^{3}\frac{\partial \sigma(y_j)}{\partial \varepsilon} y_j^2 A_j = +E_1 y_1^2 A_1 + E_2 y_2^2 A_2 + E_3 y_3^2 A_3$$
$$= +(2 \times 10^5) \times 20^2 \times (2 \times 10^3) + (2 \times 10^5) \times 0^2 \times (2 \times 10^3) + (2 \times 10^5) \times (-20)^2 \times (2 \times 10^3)$$
$$= 3.2 \times 10^{11} \text{ Nmm}^2$$

The residual loads to be used in the first iteration are (Equation 15.51):

$$N_R^{(1)} = N_e - N\left(\varepsilon_r^{(1)}, \kappa^{(1)}\right) = 1000 \times 10^3 - 0 = 1000 \times 10^3 \text{ N}$$

$$M_R^{(1)} = M_e - M\left(\varepsilon_r^{(1)}, \kappa^{(1)}\right) = 0 - 0 = 0 \text{ Nmm}$$

We can now write the system of Equations 15.50 to solve for the unknown $\Delta\varepsilon_r^{(1)}$ and $\Delta\kappa^{(1)}$:

$$12 \times 10^8 \times \Delta\varepsilon_r^{(1)} + 0 \times \Delta\kappa^{(1)} = 1000 \times 10^3$$

$$0 \times \Delta\varepsilon_r^{(1)} + 3.2 \times 10^{11} \times \Delta\kappa^{(1)} = 0$$

from which: $\Delta\varepsilon_r^{(1)} = 0.8333 \times 10^{-3}$ and $\Delta\kappa^{(1)} = 0 \text{ mm}^{-1}$.

The convergence of the solution is calculated based on norm_{tol2} (Equation 15.59b), which requires the evaluation of the residual internal actions $N_R^{(2)}$ and $M_R^{(2)}$ (Equations 15.51). To achieve this, we calculate the strain values for the second iteration:

$$\varepsilon_r^{(2)} = \varepsilon_r^{(1)} + \Delta\varepsilon_r^{(1)} = 0 + 0.8333 \times 10^{-3} = 0.8333 \times 10^{-3}$$

$$\kappa^{(2)} = \kappa^{(1)} + \Delta\kappa^{(1)} = 0 \text{ mm}^{-1}$$

and the internal actions based on $\varepsilon_r^{(2)}$ and $\kappa^{(2)}$:

$$N\left(\varepsilon_r^{(2)}, \kappa^{(2)}\right) = \sum_{j=1}^{3}\sigma\left(y_j, \varepsilon_r^{(2)}, \kappa^{(2)}\right) A_j = 1000 \times 10^3 \text{ N}$$

$$M\left(\varepsilon_r^{(2)}, \kappa^{(2)}\right) = -\sum_{j=1}^{3} y_j \sigma\left(y_j, \varepsilon_r^{(2)}, \kappa^{(2)}\right) A_j = 0 \text{ kNm}$$

The residual loads are then evaluated as:

$$N_R^{(2)} = N_e - N\left(\varepsilon_r^{(2)}, \kappa^{(2)}\right) = 1000 \times 10^3 - 1000 \times 10^3 = 0 \text{ N}$$

$$M_R^{(2)} = M_e - M\left(\varepsilon_r^{(2)}, \kappa^{(2)}\right) = 0 - 0 = 0 \text{ Nmm}$$

on the basis of which:

$$\text{norm}_{\text{tol2}} = \frac{\left|\mathbf{r}_R^{(i+1)}\right|}{\left|\mathbf{r}_e\right|} = \frac{\sqrt{N_R^{(2)2} + M_R^{(2)2}}}{\sqrt{N_e^2 + M_e^2}} = \frac{\sqrt{0^2 + 0^2}}{\sqrt{(1000 \times 10^3)^2 + 0^2}} = 0$$

which satisfies the convergence criterion being $\text{norm}_{\text{tol2}}$ below 0.001. A nil value for $\text{norm}_{\text{tol2}}$ is expected when dealing with linear–elastic materials. We can now move to the next load step. If we would have used $\text{norm}_{\text{tol1}}$ (Equation 15.59a) as the convergence criterion, its value at the end of iteration 1 would be:

$$\text{norm}_{\text{tol1}} = \frac{\left|\Delta \boldsymbol{\varepsilon}^{(1)}\right|}{\left|\boldsymbol{\varepsilon}^{(2)}\right|} = \frac{\sqrt{\Delta \varepsilon_r^{(1)2} + \Delta \kappa_r^{(1)2}}}{\sqrt{\varepsilon_r^{(2)2} + \kappa^{(2)2}}} = \frac{\sqrt{(0.8333 \times 10^{-3})^2 + 0^2}}{\sqrt{(0.8333 \times 10^{-3})^2 + 0^2}} = 1$$

and using this criterion, we would have required a second iteration to be performed in this case. This is the case because the materials are still in the linear–elastic range.

Load step 2: $N_e = 2200$ kN, $M_e = 0$ kNm
We will now continue the analysis from the previous load step adopting the strain values obtained at the end of the previous load step for our first iteration with $i = 1$ (at load step 2): $\varepsilon_r^{(1)} = 0.8333 \times 10^{-3}$ and $\kappa^{(1)} = 0$ mm^{-1}. We have already calculated the corresponding internal actions as $N\left(\varepsilon_r^{(1)}, \kappa^{(1)}\right) = 1000 \times 10^3$ N and $M\left(\varepsilon_r^{(1)}, \kappa^{(1)}\right) = 0$ kNm. The residual actions to be used in the first iteration of load step 2 are:

$$N_R^{(1)} = N_e - N\left(\varepsilon_r^{(1)}, \kappa^{(1)}\right) = 2200 \times 10^3 - 1000 \times 10^3 = 1200 \times 10^3 \text{ N}$$

$$M_R^{(1)} = M_e - M\left(\varepsilon_r^{(1)}, \kappa^{(1)}\right) = 0 \text{ Nmm}$$

Considering the fact that at $\varepsilon_r^{(1)} = 0.8333 \times 10^{-3}$ and $\kappa^{(1)} = 0$ mm^{-1}, the material properties of the cross-sections are still linear–elastic:

$$\frac{\partial N\left(\varepsilon_r^{(1)}, \kappa^{(1)}\right)}{\partial \varepsilon_r} = \sum_{j=1}^{3} \frac{\partial \sigma(y_j)}{\partial \varepsilon} A_j = E_1 A_1 + E_2 A_2 + E_3 A_3 = 12 \times 10^8 \text{ N}$$

$$\frac{\partial N\left(\varepsilon_r^{(1)}, \kappa^{(1)}\right)}{\partial \kappa} = -\sum_{j=1}^{3} \frac{\partial \sigma(y_j)}{\partial \varepsilon} y_j A_j = -E_1 y_1 A_1 - E_2 y_2 A_2 - E_3 y_3 A_3 = 0 \text{ Nmm}$$

$$\frac{\partial M\left(\varepsilon_r^{(1)}, \kappa^{(1)}\right)}{\partial \varepsilon_r} = -\sum_{j=1}^{3} \frac{\partial \sigma(y_j)}{\partial \varepsilon} y_j A_j = -E_1 y_1 A_1 - E_2 y_2 A_2 - E_3 y_3 A_3 = 0 \text{ Nmm}$$

$$\frac{\partial M\left(\varepsilon_r^{(1)}, \kappa^{(1)}\right)}{\partial \kappa} = \sum_{j=1}^{3} \frac{\partial \sigma(y_j)}{\partial \varepsilon} y_j^2 A_j = E_1 y_1^2 A_1 + E_2 y_2^2 A_2 + E_3 y_3^2 A_3 = 3.2 \times 10^{11} \text{ Nmm}^2$$

The system of Equations 15.50 can then be written as:

$$12 \times 10^8 \times \Delta \varepsilon_r^{(1)} + 0 \times \Delta \kappa^{(1)} = 1200 \times 10^3$$

$$0 \times \Delta \varepsilon_r^{(1)} + 3.2 \times 10^{11} \times \Delta \kappa^{(1)} = 0$$

and solving gives $\Delta \varepsilon_r^{(1)} = 1 \times 10^{-3}$ and $\Delta \kappa^{(1)} = 0$ mm^{-1}.
Based on this:

$$N\left(\varepsilon_r^{(2)}, \kappa^{(2)}\right) = 2066.6 \times 10^3 \text{ N} \quad \text{and} \quad M\left(\varepsilon_r^{(2)}, \kappa^{(2)}\right) = 0 \text{ kNm}$$

and

$$N_R^{(2)} = N_e - N\left(\varepsilon_r^{(2)}, \kappa^{(2)}\right) = 2200 \times 10^3 - 2066.6 \times 10^3 = 133.3 \times 10^3 \text{ N}$$

$$M_R^{(2)} = M_e - M\left(\varepsilon_r^{(2)}, \kappa^{(2)}\right) = 0 - 0 = 0 \text{ Nmm}$$

Convergence is then evaluated with norm$_{tol2}$:

$$\text{norm}_{tol2} = \frac{\sqrt{N_R^{(2)^2} + M_R^{(2)^2}}}{\sqrt{N_e^2 + M_e^2}} = \frac{\sqrt{(133.3 \times 10^3)^2 + 0^2}}{\sqrt{(2200 \times 10^3)^2 + 0^2}} = 0.0606$$

which is greater than the limit value 0.001 and the analysis continues to the next iteration $i = 2$. The results of the iterations carried out for load step 2 are summarised in the tables below.

i	$\varepsilon_r^{(i)} \times 10^{-3}$	$\kappa^{(i)}$ mm^{-1}	$\Delta \varepsilon_r^{(i)} \times 10^{-3}$	$\Delta \kappa^{(i)}$ mm^{-1}	norm$_{tol2}$
1	0.8333	0	1	0	0.0606
2	1.8333	0	166.7×10^{-3}	0	0
3	2	0	–	–	–

i	$\dfrac{\partial N\left(\varepsilon_r^{(i)}, \kappa^{(i)}\right)}{\partial \varepsilon_r}$ N	$\dfrac{\partial N\left(\varepsilon_r^{(i)}, \kappa^{(i)}\right)}{\partial \kappa}$ Nmm	$\dfrac{\partial M\left(\varepsilon_r^{(i)}, \kappa^{(i)}\right)}{\partial \varepsilon_r}$ Nmm	$\dfrac{\partial M\left(\varepsilon_r^{(i)}, \kappa^{(i)}\right)}{\partial \kappa}$ Nmm^2
1	1200×10^6	0	0	320×10^9
2	800×10^6	0	0	320×10^9

i	$N_e \times 10^3$ N	M_e Nmm	$N\left(\varepsilon_r^{(i)}, \kappa^{(i)}\right) \times 10^3$ N	$M\left(\varepsilon_r^{(i)}, \kappa^{(i)}\right)$ Nmm	$N_R^{(i)} \times 10^3$ N	$M_R^{(i)}$ Nmm
1	2200	0	1000	0	1200	0
2	2200	0	2066.7	0	133.3	0
3	2200	0	2200	0	0	0

Load step 3: $N_e = 2600$ kN, $M_e = 0$ kNm

The calculations carried out for load step 3 follow the procedure used in the previous load steps and the results are summarised in the following tables:

i	$\varepsilon_r^{(i)} \times 10^{-3}$	$\kappa^{(i)}$ mm^{-1}	$\Delta\varepsilon_r^{(i)} \times 10^{-3}$	$\Delta\kappa^{(i)}$ mm^{-1}	$norm_{tol2}$
1	2	0	500×10^{-3}	0	0
2	2.5	0	–	–	–

i	$\dfrac{\partial N\left(\varepsilon_r^{(i)},\kappa^{(i)}\right)}{\partial\varepsilon_r}$ N	$\dfrac{\partial N\left(\varepsilon_r^{(i)},\kappa^{(i)}\right)}{\partial\kappa}$ Nmm	$\dfrac{\partial N\left(\varepsilon_r^{(i)},\kappa^{(i)}\right)}{\partial\varepsilon_r}$ Nmm	$\dfrac{\partial N\left(\varepsilon_r^{(i)},\kappa^{(i)}\right)}{\partial\kappa}$ Nmm^2
1	800×10^6	0	0	320×10^9

i	$\dfrac{N_e \times 10^3}{N}$	$\dfrac{M_e}{Nmm}$	$\dfrac{N\left(\varepsilon_r^{(i)},\kappa^{(i)}\right)\times 10^3}{N}$	$\dfrac{M\left(\varepsilon_r^{(i)},\kappa^{(i)}\right)}{Nmm}$	$\dfrac{N_R^{(i)}\times 10^3}{N}$	$\dfrac{M_R^{(i)}}{Nmm}$
1	2600	0	2200	0	400	0
	2600	0	2600	0	0	0

It is worth pointing out that the ability to converge to a solution when material 2 also starts to yield is possible only because the initial 'guessed' solution used at the beginning of load step 3 did not cause yielding in material 2. In fact, if we were to repeat the same calculation starting with higher values for $\varepsilon_r^{(1)}$ and $\kappa^{(1)}$, we would not have been able to find a solution because $\mathbf{r}_t(\varepsilon^{(i)})$ would become a zero matrix, therefore not enabling the calculation of the unknowns $\Delta\varepsilon_r^{(1)}$ and $\Delta\kappa^{(1)}$. This consideration highlights the importance of carefully interpreting numerical results, especially when obtained from nonlinear analyses.

It is noted that this solution is identical to that calculated in the illustrative example in Section 15.3.1.

WORKED EXAMPLE 15.2

Consider a rectangular cross-section with a width of 100 mm and a height of 360 mm bending about its strong axis. Assume the material to be elastic–perfectly plastic with $E = 200$ GPa and $f_y = 300$ MPa. Determine the deformations, expressed in terms of the strain at the level of the reference axis (taken at mid-height of the section) and the curvature, with a cross-sectional analysis implemented using the Newton–Raphson method for the following load steps:

(1) $N_e = 0$ kN, $M_e = 648$ kNm

(2) $N_e = 0$ kN, $M_e = 810$ kNm

(3) $N_e = 0$ kN, $M_e = 950$ kNm

For the convergence criteria, use $norm_{tol2}$ defined in Equation 15.59b equal to 0.001. Compare the results with those obtained from the moment–curvature expressions of Equations 15.25.

The first step in the solution is to discretise the cross-section. We will consider a very simple approach to select the number of layers. It involves the evaluation of the flexural rigidity based on different number of layers, assuming linear–elastic material properties. The preferred number of layers will be the number for which the error of the flexural rigidity is less than 0.1% of the value of EI calculated for the cross-section without layering. For the cross-section considered in this example:

$$EI = E\frac{bd^3}{12} = 200\times10^3\frac{100\times360^3}{12} = 77.76\times10^{12}\ Nmm^2$$

A discretisation with 40 layers is preferred on the basis of the comparisons shown in the table below, where different numbers of layers have been considered.

Number of layers	$\sum E_j y_j^2 A_j\ Nmm^2$	$\left\{\left[\left(\sum E_j y_j^2 A_j\right)-\left(Ebd^3/12\right)\right]/\left(Ebd^3/12\right)\right\}\times100$ Error (%)
10	76.98×10^{12}	-1
20	77.56×10^{12}	-0.25
30	77.67×10^{12}	-0.11
40	77.71×10^{12}	-0.06

The elastic–perfectly plastic material properties are calculated based on Equations 15.60 with $E = 200$ GPa and $f_y = 300$ MPa.

For each load step, Equations 15.45 through 15.59 are applied until the convergence criterion based on $norm_{tol2}$ (Equation 15.59b) is satisfied, following the procedure adopted in the previous worked example.

The various results calculated for the iterations of each load step are summarised in the tables below.

Load step	i	$\varepsilon_r^{(i)}\times10^{-3}$	$\kappa^{(i)}\times10^{-6}$ mm^{-1}	$\Delta\varepsilon_r^{(i)}\times10^{-3}$	$\Delta\kappa^{(i)}\times10^{-6}$ mm^{-1}	$norm_{tol2}$
1	1	0	0	0	8.3385	0
1	2	0	8.3385	—	—	—
2	1	0	8.3385	0	2.0846	0.0562
2	2	0	10.4231	0	1.1441	0.0076
2	3	0	11.5673	0	0.2310	0
2	4	0	11.7983	—	—	—
3	1	0	11.7983	0	5.2558	0.0586
3	2	0	17.0541	0	5.7396	0.0226
3	3	0	22.7936	0	6.4712	0.0047
3	4	0	29.2648	0	2.1413	0.0011
3	5	0	31.4062	0	0.8090	0
3	6	0	32.2151	—	—	—

Load step	i	$\dfrac{\partial N\left(\varepsilon_r^{(i)},\kappa^{(i)}\right)}{\partial \varepsilon_r}$ N	$\dfrac{\partial N\left(\varepsilon_r^{(i)},\kappa^{(i)}\right)}{\partial \kappa}$ Nmm	$\dfrac{\partial M\left(\varepsilon_r^{(i)},\kappa^{(i)}\right)}{\partial \varepsilon_r}$ Nmm	$\dfrac{\partial M\left(\varepsilon_r^{(i)},\kappa^{(i)}\right)}{\partial \kappa}$ Nmm²
1	1	7.2×10^9	0	0	77.7114×10^{12}
2	1	7.2×10^9	0	0	77.7114×10^{12}
2	2	5.76×10^9	0	0	39.7742×10^{12}
2	3	5.04×10^9	0	0	26.6377×10^{12}
3	1	5.04×10^9	0	0	26.6377×10^{12}
3	2	3.6×10^9	0	0	9.6957×10^{12}
3	3	2.52×10^9	0	0	3.3170×10^{12}
3	4	2.16×10^9	0	0	2.0849×10^{12}
3	5	1.8×10^9	0	0	1.2029×10^{12}

Load step	i	N_e N	$M_e \times 10^6$ Nmm	$N\left(\varepsilon_r^{(i)},\kappa^{(i)}\right)\times 10^3$ N	$M\left(\varepsilon_r^{(i)},\kappa^{(i)}\right)\times 10^6$ Nmm	$N_R^{(i)}\times 10^3$ N	$M_R^{(i)}\times 10^6$ Nmm
1	1	0	648	0	0	0	648
1	2	0	648	0	648	0	0
2	1	0	810	0	648	0	162
2	2	0	810	0	764.5	0	45.5
2	3	0	810	0	803.8	0	6.2
2	4	0	810	0	810	0	0
3	1	0	950	0	810	0	140
3	2	0	950	0	894.4	0	55.6
3	3	0	950	0	928.5	0	21.5
3	4	0	950	0	945.5	0	4.5
3	5	0	950	0	949.0	0	1.0
3	6	0	950	0	950	0	0

The results are plotted in Figure 15.19, together with the moment–curvature curve obtained from the analytical solutions of Equations 15.25 (with moment values non-dimensionalised against $M_y = 648$ kNm calculated for the cross-section under consideration). As expected, these results perfectly match.

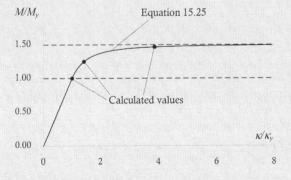

Figure 15.19 Nonlinear moment–curvature response.

15.5 FINITE ELEMENT ANALYSIS USING THE NEWTON–RAPHSON METHOD

In this section, the Newton–Raphson method applied to the finite element approach is presented. In particular, the procedure considering the displacement-based finite element formulation presented in Chapter 13 for the Euler–Bernoulli beam model is outlined. We will only focus on the aspects related to the nonlinear solution and reference should be made to Chapter 13 for issues related to the general finite element modelling, such as the transformation from local to global coordinate for an isolated element, its assembling, the definition of the support conditions, the determination of the unknown displacements and the post-processing.

The Newton–Raphson iterative procedure required for the finite element solution is based on the steps provided in Equations 15.39 through 15.43 where the vector of displacements $\mathbf{D}^{(i)}$ is replaced with the finite element nodal freedoms $\mathbf{D}_{e}^{(i)}$ of the assembled structure. Equations 15.41 and 15.42 can be re-written as:

$$\mathbf{K}_{t}\left(\mathbf{D}_{e}^{(i)}\right)\Delta\mathbf{D}_{e}^{(i)} = \mathbf{Q}_{R}^{(i)} \tag{15.61}$$

and

$$\mathbf{Q}_{R}^{(i)} = \mathbf{Q} - \mathbf{K}\left(\mathbf{D}_{e}^{(i)}\right) \tag{15.62}$$

where vectors $\mathbf{Q}$ and $\mathbf{Q}_{R}^{(i)}$ are expressed in terms of nodal freedoms. Convergence is then evaluated by calculating the norms of Equations 15.43 based on the finite element nodal displacements or the residual actions:

$$\text{norm}_{\text{tol1}} = \frac{\left|\Delta\mathbf{D}_{e}^{(i)}\right|}{\left|\mathbf{D}_{e}^{(i+1)}\right|} \quad \text{norm}_{\text{tol2}} = \frac{\left|\mathbf{Q}_{R}^{(i+1)}\right|}{\left|\mathbf{Q}\right|} \tag{15.63a,b}$$

Equations 15.61 and 15.62 describe the behaviour of the entire structure, obtained by assembling the contributions of the various elements. We will now derive the stiffness matrix and the tangent stiffness matrix of an isolated finite element, by following the formulation already described in Chapter 13 and extending it to account for material nonlinearity. We start from the weak form for the Euler–Bernoulli beam model (Equation 13.18):

$$\int_{L}\mathbf{r}\cdot\mathbf{A}\hat{\mathbf{e}}\,dx = \int_{L}\mathbf{p}\cdot\hat{\mathbf{e}}\,dx \tag{15.64}$$

where, for ease of notation, we omit the iteration index i previously introduced in the Newton–Raphson iterative procedure (this will be included again when describing the details of the iterative scheme). As specified in Chapter 13, $\mathbf{r}$ is the vector of internal actions, $\mathbf{A}$ is a differential operator, $\mathbf{e}$ is the vector of the generalised displacements and $\mathbf{p}$ is the vector of member loads:

$$\mathbf{r} = \begin{bmatrix} N \\ M \end{bmatrix} \quad \mathbf{A} = \begin{bmatrix} \partial & 0 \\ 0 & \partial^{2} \end{bmatrix} \quad \mathbf{e} = \begin{bmatrix} u \\ v \end{bmatrix} \quad \mathbf{p} = \begin{bmatrix} n \\ w \end{bmatrix} \tag{15.65a–d}$$

The key step in the derivation of a displacement-based finite element is to approximate the generalised displacements e by means of polynomial functions, which can be described in compact form as:

$$e \approx N_e d_e \tag{15.66}$$

where matrix N_e collects the shape functions adopted for the displacements and d_e is the vector of nodal displacements.

On the basis of this approximation, it is possible to rewrite the weak form of the problem (Equation 15.64) as:

$$\int_L B^T r \cdot \hat{d}_e \, dx = \int_L N_e^T p \cdot \hat{d}_e \, dx \tag{15.67}$$

where matrix B is introduced to describe the strain field ε expressed in terms of the nodal displacements (Equation 13.36):

$$\varepsilon = A N_e d_e = B d_e \tag{15.68}$$

The stiffness relationship for an isolated element can then be written as:

$$k_e(d_e) = q_e \tag{15.69}$$

in which $k_e(d_e)$ defines the internal actions and q_e is the vector of nodal actions describing member loads (Equation 15.65d). These terms may be expressed as:

$$k_e(d_e) = \int_L B^T r \, dx \tag{15.70a}$$

$$q_e = \int_L N_e^T p \, dx \tag{15.70b}$$

In Equation 15.70a, we have highlighted the dependency of k_e on the nodal displacements d_e. In fact, unlike the linear–elastic case outlined in Chapter 13, the rigidity of the structure depends on its material properties and its deformed shape.

For the Euler–Bernoulli beam model, we need to specify a uniaxial constitutive model (see Section 15.2) relating stresses and strains:

$$\sigma = f(\varepsilon) \tag{15.71}$$

where the function $f(\varepsilon)$ is assigned for the actual material under consideration. In finite element modelling, the strains are calculated in terms of the nodal displacements d_e as:

$$\varepsilon = [1 \quad -y]\varepsilon = [1 \quad -y]A N_e d_e = [1 \quad -y] \, B d_e \tag{15.72}$$

and Equation 15.71 can be re-written, highlighting the independent variables, as:

$$\sigma = f(x, y, d_e) \tag{15.73}$$

We can then reconsider the expressions for the internal actions highlighting their dependency on the member axis x and nodal displacements $\mathbf{d}_e$. This is carried out by discretising the cross-section into n_j layers (with $j = 1,...,n_j$) to account for the nonlinear material behaviour as previously performed for the cross-sectional analysis in Figure 15.17 and Equations 15.57:

$$\mathbf{r}(x,\mathbf{d}_e) = \begin{bmatrix} N(x,\mathbf{d}_e) \\ M(x,\mathbf{d}_e) \end{bmatrix} = \begin{bmatrix} \int_A \sigma(x,y,\mathbf{d}_e)\,dA \\ -\int_A y\sigma(x,y,\mathbf{d}_e)\,dA \end{bmatrix} = \begin{bmatrix} \sum_{j=1}^{n_j} \sigma(x,y_j,\mathbf{d}_e)A_j \\ -\sum_{j=1}^{n_j} y_j\sigma(x,y_j,\mathbf{d}_e)A_j \end{bmatrix} \tag{15.74}$$

When performing nonlinear analyses, the integrals of Equations 15.70 and 15.74 are usually evaluated numerically. For the purpose of this chapter, the Gauss–Legendre formulae are used to enable the calculation of an integral I between a and b of a function $f(x)$ based on the following weighted summation:

$$I = \int_a^b f(x)\,dx = \frac{b-a}{2}\int_{-1}^1 f\left(\frac{a+b}{2} + \frac{b-a}{2}\bar{x}\right)d\bar{x} \approx \frac{b-a}{2}\sum_{k=1}^{n_G} w_k f\left(\frac{a+b}{2} + \frac{b-a}{2}\bar{x}_k\right) \tag{15.75}$$

Some of the possible values for the weighting functions w_k and function arguments x_k are provided in Table 15.1 for different numbers of Gauss integration points n_G. The level of accuracy achieved by the numerical integration depends on n_G. When specifying n_G integration points, it is possible to integrate exactly a polynomial of degree $(2n_G + 1)$.

Based on Equation 15.75, the numerical integrals of Equations 15.70 can be carried out as:

$$\mathbf{k}_e(x,\mathbf{d}_e) = \int_L \mathbf{B}^T(x)\,\mathbf{r}(x,\mathbf{d}_e)\,dx \approx \frac{L}{2}\sum_{k=1}^{n_G} w_k \mathbf{B}^T(x_k)\,\mathbf{r}(x_k,\mathbf{d}_e) \tag{15.76a}$$

$$\mathbf{q}_e = \int_L \mathbf{N}_e^T(x)\,\mathbf{p}(x)\,dx \approx \frac{L}{2}\sum_{k=1}^{n_G} w_k \mathbf{N}_e^T(x_k)\,\mathbf{p}(x_k) \tag{15.76b}$$

where x_k is calculated as a function of $\bar{x}_k$ (Table 15.1) based on:

$$x_k = \frac{L}{2}(\bar{x}_k + 1) \tag{15.76c}$$

assuming the limits of the integral to vary between 0 and L, with L being the length of the finite element.

Considering the terms included in Equations 15.76, matrices $\mathbf{B}$ and $\mathbf{N}_e$ are known, once n_G is specified, because they are defined in terms of x_k. Vector $\mathbf{p}(x_k)$ is also known because it describes the known applied loads for the specific problem, while vector $\mathbf{r}(x_k,\mathbf{d}_e)$ represents the internal actions resisted by the member and these depend on the current deformations and material properties (see Equation 15.74).

Table 15.1 Function arguments and weighting factors for the Gauss–Legendre formulae

n_G	$\overline{x}_k$	w_k
1	$x_1 = 0$	$w_1 = 2$
2	$x_1 = -0.577350269$	$w_1 = 1$
	$x_2 = 0.577350269$	$w_2 = 1$
3	$x_1 = -0.774596669$	$w_1 = 0.5555556$
	$x_2 = 0$	$w_2 = 0.8888888$
	$x_3 = 0.774596669$	$w_3 = 0.5555556$
4	$x_1 = -0.861136312$	$w_1 = 0.3478548$
	$x_2 = -0.339981044$	$w_2 = 0.6521452$
	$x_3 = 0.339981044$	$w_3 = 0.6521452$
	$x_4 = 0.861136312$	$w_4 = 0.3478548$
5	$x_1 = -0.906179846$	$w_1 = 0.2369269$
	$x_2 = -0.538469310$	$w_2 = 0.4786287$
	$x_3 = 0$	$w_3 = 0.5688888$
	$x_4 = 0.538469310$	$w_4 = 0.4786287$
	$x_5 = 0.906179846$	$w_5 = 0.2369269$
6	$x_1 = -0.932469514$	$w_1 = 0.1713245$
	$x_2 = -0.661209386$	$w_2 = 0.3607616$
	$x_3 = -0.238619186$	$w_3 = 0.4679139$
	$x_4 = 0.238619186$	$w_4 = 0.4679139$
	$x_5 = 0.661209386$	$w_5 = 0.3607616$
	$x_6 = 0.932469514$	$w_6 = 0.1713245$

We will now describe in more detail the proposed finite element derivation considering the 7-dof finite element of Figure 13.4 (reproduced here in Figure 15.20). This is preferred to the 6-dof finite element described in Figure 13.7 because the latter element produces inaccurate results when the reference system used in the derivation of its stiffness coefficients is not centroidal (see Figure 13.8), as can occur when dealing with material nonlinearities.

To better outline the following calculations, it is more convenient to rewrite Equation 15.66 highlighting the terms included in $\mathbf{N}_e$ and $\mathbf{d}_e$ (Equation 13.34), as well as their dependency on x:

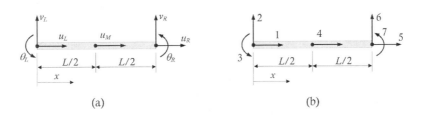

Figure 15.20 The 7-dof finite element (Euler–Bernoulli beam). (a) Nodal displacements. (b) Freedom numbering for the isolated finite element.

$$\begin{bmatrix} u(x) \\ v(x) \end{bmatrix} = \begin{bmatrix} N_{u1}(x) & 0 & 0 & N_{u2}(x) & N_{u3}(x) & 0 & 0 \\ 0 & N_{v1}(x) & N_{v2}(x) & 0 & 0 & N_{v3}(x) & N_{v4}(x) \end{bmatrix} \begin{bmatrix} u_L \\ v_L \\ \theta_L \\ u_M \\ u_R \\ v_R \\ \theta_R \end{bmatrix} = \mathbf{N}_e(x)\mathbf{d}_e \qquad (15.77)$$

where the terms of $\mathbf{N}_e(x)$ are (Equations 13.27 and 13.33):

$$N_{u1}(x) = 1 - \frac{3x}{L} + \frac{2x^2}{L^2} \quad N_{u2}(x) = \frac{4x}{L} - \frac{4x^2}{L^2} \quad N_{u3}(x) = -\frac{x}{L} + \frac{2x^2}{L^2} \qquad (15.78a\text{–}c)$$

$$N_{v1}(x) = 1 - \frac{3x^2}{L^2} + \frac{2x^3}{L^3} \quad N_{v2}(x) = x - \frac{2x^2}{L} + \frac{x^3}{L^2} \qquad (15.78d,e)$$

$$N_{v3}(x) = \frac{3x^2}{L^2} - \frac{2x^3}{L^3} \quad N_{v4}(x) = -\frac{x^2}{L} + \frac{x^3}{L^2} \qquad (15.78f,g)$$

Separating the expressions for u and v, we can re-arrange Equation 15.77 as:

$$u(x) = N_{u1}(x)u_L + N_{u2}(x)u_M + N_{u3}(x)u_R \qquad (15.79a)$$

$$v(x) = N_{v1}(x)v_L + N_{v2}(x)\theta_L + N_{v3}(x)v_R + N_{v4}(x)\theta_R \qquad (15.79b)$$

Based on Equation 15.68, matrix $\mathbf{B}(x)$ is calculated as:

$$\mathbf{B}(x) = \begin{bmatrix} N'_{u1}(x) & 0 & 0 & N'_{u2}(x) & N'_{u3}(x) & 0 & 0 \\ 0 & N''_{v1}(x) & N''_{v2}(x) & 0 & 0 & N''_{v3}(x) & N''_{v4}(x) \end{bmatrix} \qquad (15.80)$$

with:

$$N'_{u1}(x) = -\frac{3}{L} + \frac{4x}{L^2} \quad N'_{u2}(x) = \frac{4}{L} - \frac{8x}{L^2} \quad N'_{u3}(x) = -\frac{1}{L} + \frac{4x}{L^2} \qquad (15.81a\text{–}c)$$

$$N''_{v1}(x) = \frac{12x}{L^3} - \frac{6}{L^2}; \quad N''_{v2}(x) = \frac{6x}{L^2} - \frac{4}{L}; \quad N''_{v3}(x) = \frac{6}{L^2} - \frac{12x}{L^3}; \quad N''_{v4}(x) = \frac{6x}{L^2} - \frac{2}{L} \qquad (15.81d\text{–}g)$$

Recalling Equations 15.76 and 15.77, the expressions describing $\mathbf{k}_e\,(x, \mathbf{d}_e)$ and $\mathbf{q}_e$ can now be written for the 7-dof finite element as follows (calculated at the i-th iteration of the solution process):

$$\mathbf{k}_\mathrm{e}\left(x,\mathbf{d}_\mathrm{e}^{(i)}\right)=\frac{L}{2}\sum_{k=1}^{n_G}w_k\ \mathbf{B}^\mathrm{T}(x_k)\ \mathbf{r}^{(i)}\left(x_k,\mathbf{d}_\mathrm{e}^{(i)}\right)=\frac{L}{2}\sum_{k=1}^{n_G}w_k\begin{bmatrix}N'_{u1}(x_k)&0\\0&N''_{v1}(x_k)\\0&N''_{v2}(x_k)\\N'_{u2}(x_k)&0\\N'_{u3}(x_k)&0\\0&N''_{v3}(x_k)\\0&N''_{v4}(x_k)\end{bmatrix}\begin{bmatrix}N\left(x_k,\mathbf{d}_\mathrm{e}^{(i)}\right)\\M\left(x_k,\mathbf{d}_\mathrm{e}^{(i)}\right)\end{bmatrix} \tag{15.82a}$$

$$\mathbf{q}_\mathrm{e}=\frac{L}{2}\sum_{k=1}^{n_G}w_k\begin{bmatrix}N_{u1}(x_k)&0\\0&N_{v1}(x_k)\\0&N_{v2}(x_k)\\N_{u2}(x_k)&0\\N_{u3}(x_k)&0\\0&N_{v3}(x_k)\\0&N_{v4}(x_k)\end{bmatrix}\begin{bmatrix}n(x_k)\\w(x_k)\end{bmatrix} \tag{15.82b}$$

Considering the cross-sectional discretisations of Figure 15.17 (Equation 15.57), the internal actions are:

$$N\left(x_k,\mathbf{d}_\mathrm{e}^{(i)}\right)=\sum_{j=1}^{n_j}\sigma\left(x_k,y_j,\mathbf{d}_\mathrm{e}^{(i)}\right)A_j \tag{15.83a}$$

$$M\left(x_k,\mathbf{d}_\mathrm{e}^{(i)}\right)=-\sum_{j=1}^{n_j}y_j\sigma\left(x_k,y_j,\mathbf{d}_\mathrm{e}^{(i)}\right)A_j \tag{15.83b}$$

The Newton–Raphson method requires the calculation of the tangent stiffness matrix of an isolated element (Equation 15.61):

$$\mathbf{k}_\mathrm{et}(x,\mathbf{d}_\mathrm{e})=\int_L\mathbf{B}^\mathrm{T}(x)\ \mathbf{r}_\mathrm{t}(x,\mathbf{d}_\mathrm{e})\,\mathrm{d}x \tag{15.84}$$

where $\mathbf{r}_\mathrm{t}(x,\mathbf{d}_\mathrm{e})$ is similar to the tangent properties evaluated in the previous section for the cross-sectional analysis (see Equations 15.48 and 15.52). In this case, we are expressing the problem in terms of the nodal displacements and, therefore, the partial derivatives need to be carried out with respect to $\mathbf{d}_\mathrm{e}$. At the i-th iteration of the solution process, $\mathbf{k}_\mathrm{et}(x,\mathbf{d}_\mathrm{e})$ can be calculated as follows:

$$\mathbf{k}_\mathrm{et}\left(x,\mathbf{d}_\mathrm{e}^{(i)}\right)=\frac{\partial}{\partial\mathbf{d}_\mathrm{e}}\int_L\mathbf{B}^\mathrm{T}(x)\ \mathbf{r}\left(x,\mathbf{d}_\mathrm{e}^{(i)}\right)\mathrm{d}x=\int_L\mathbf{B}^\mathrm{T}(x)\frac{\partial\mathbf{r}\left(x,\mathbf{d}_\mathrm{e}^{(i)}\right)}{\partial\mathbf{d}_\mathrm{e}}\mathrm{d}x \tag{15.85}$$

Based on the numerical approximations introduced for the integrals carried out over the cross-section (with n_j layers) and over the member length (at the n_G Gauss integration points), the tangent vector $\mathbf{k}_\mathrm{et}\left(x,\mathbf{d}_\mathrm{e}^{(i)}\right)$ can be approximated by:

$$\mathbf{k}_{et}\left(x,\mathbf{d}_e^{(i)}\right)=\int_L \mathbf{B}^{\mathrm{T}}(x)\frac{\partial \mathbf{r}\left(x,\mathbf{d}_e^{(i)}\right)}{\partial \mathbf{d}_e}dx \approx \frac{L}{2}\sum_{k=1}^{n_G}w_k\begin{bmatrix} N'_{u1}(x_k) & 0 \\ 0 & N''_{v1}(x_k) \\ 0 & N''_{v2}(x_k) \\ N'_{u2}(x_k) & 0 \\ N'_{u3}(x_k) & 0 \\ 0 & N''_{v3}(x_k) \\ 0 & N''_{v4}(x_k) \end{bmatrix}\begin{bmatrix} \dfrac{\partial N\left(x_k,\mathbf{d}_e^{(i)}\right)}{\partial \mathbf{d}_e} \\ \\ \dfrac{\partial M\left(x_k,\mathbf{d}_e^{(i)}\right)}{\partial \mathbf{d}_e} \end{bmatrix} \quad (15.86)$$

where the partial derivatives can be simplified by applying the chain rule, as already performed for the nonlinear cross-sectional analysis in Equations 15.54:

$$\frac{\partial N\left(x_k,\mathbf{d}_e^{(i)}\right)}{\partial \mathbf{d}_e}=\int_A \frac{\partial \sigma\left(x_k,y,\mathbf{d}_e^{(i)}\right)}{\partial \mathbf{d}_e}dA=\int_A \frac{\partial \sigma\left(x_k,y,\mathbf{d}_e^{(i)}\right)}{\partial \varepsilon}\frac{\partial \varepsilon\left(x_k,y,\mathbf{d}_e^{(i)}\right)}{\partial \mathbf{d}_e}dA \quad (15.87a)$$

$$\frac{\partial M\left(x_k,\mathbf{d}_e^{(i)}\right)}{\partial \mathbf{d}_e}=-\int_A y\frac{\partial \sigma\left(x_k,y,\mathbf{d}_e^{(i)}\right)}{\partial \mathbf{d}_e}dA=-\int_A y\frac{\partial \sigma\left(x_k,y,\mathbf{d}_e^{(i)}\right)}{\partial \varepsilon}\frac{\partial \varepsilon\left(x_k,y,\mathbf{d}_e^{(i)}\right)}{\partial \mathbf{d}_e}dA \quad (15.87b)$$

and considering the cross-sectional discretisation of Figure 15.17:

$$\frac{\partial N\left(x_k,\mathbf{d}_e^{(i)}\right)}{\partial \mathbf{d}_e}=\int_A \frac{\partial \sigma\left(x_k,y,\mathbf{d}_e^{(i)}\right)}{\partial \varepsilon}\frac{\partial \varepsilon\left(x_k,y,\mathbf{d}_e^{(i)}\right)}{\partial \mathbf{d}_e}dA \approx \sum_{j=1}^{n_j}A_j\frac{\partial \sigma\left(x_k,y_j,\mathbf{d}_e^{(i)}\right)}{\partial \varepsilon}\frac{\partial \varepsilon\left(x_k,y_j,\mathbf{d}_e^{(i)}\right)}{\partial \mathbf{d}_e}$$

$$(15.88a)$$

$$\frac{\partial M\left(x_k,\mathbf{d}_e^{(i)}\right)}{\partial \mathbf{d}_e}=-\int_A y\frac{\partial \sigma\left(x_k,y,\mathbf{d}_e^{(i)}\right)}{\partial \varepsilon}\frac{\partial \varepsilon\left(x_k,y,\mathbf{d}_e^{(i)}\right)}{\partial \mathbf{d}_e}dA \approx -\sum_{j=1}^{n_j}y_j A_j\frac{\partial \sigma\left(x_k,y_j,\mathbf{d}_e^{(i)}\right)}{\partial \varepsilon}\frac{\partial \varepsilon\left(x_k,y_j,\mathbf{d}_e^{(i)}\right)}{\partial \mathbf{d}_e}$$

$$(15.88b)$$

We will now calculate $\mathbf{k}_{et}\left(x,\mathbf{d}_e^{(i)}\right)$ of Equation 15.86 for the 7-dof finite element of Figure 15.20 as follows:

$$\mathbf{k}_{et}\left(x,\mathbf{d}_e^{(i)}\right)=\frac{L}{2}\sum_{k=1}^{n_G}w_k\begin{bmatrix} N'_{u1}(x_k) & 0 \\ 0 & N''_{v1}(x_k) \\ 0 & N''_{v2}(x_k) \\ N'_{u2}(x_k) & 0 \\ N'_{u3}(x_k) & 0 \\ 0 & N''_{v3}(x_k) \\ 0 & N''_{v4}(x_k) \end{bmatrix}$$

$$\times \begin{bmatrix} \dfrac{\partial N_k^{(i)}}{\partial d_{e1}} & \dfrac{\partial N_k^{(i)}}{\partial d_{e2}} & \dfrac{\partial N_k^{(i)}}{\partial d_{e3}} & \dfrac{\partial N_k^{(i)}}{\partial d_{e4}} & \dfrac{\partial N_k^{(i)}}{\partial d_{e5}} & \dfrac{\partial N_k^{(i)}}{\partial d_{e6}} & \dfrac{\partial N_k^{(i)}}{\partial d_{e7}} \\ \\ \dfrac{\partial M_k^{(i)}}{\partial d_{e1}} & \dfrac{\partial M_k^{(i)}}{\partial d_{e2}} & \dfrac{\partial M_k^{(i)}}{\partial d_{e3}} & \dfrac{\partial M_k^{(i)}}{\partial d_{e4}} & \dfrac{\partial M_k^{(i)}}{\partial d_{e5}} & \dfrac{\partial M_k^{(i)}}{\partial d_{e6}} & \dfrac{\partial M_k^{(i)}}{\partial d_{e7}} \end{bmatrix}$$

$$(15.89)$$

where, for ease of notation, $N\left(x_k,\mathbf{d}_e^{(i)}\right)$, $M\left(x_k,\mathbf{d}_e^{(i)}\right)$ and $\varepsilon\left(x_k,y,\mathbf{d}_e^{(i)}\right)$ are replaced with $N_k^{(i)}$, $M_k^{(i)}$ and $\varepsilon_k^{(i)}$. Considering that:

$$\frac{\partial \varepsilon_k^{(i)}}{\partial d_{e1}} = \frac{\partial\left\{\dfrac{\left[1 \quad -y_j\right]\mathbf{B}(x_k)\mathbf{d}_e^{(i)}}{}\right\}}{\partial d_{e1}} = -\frac{3}{L} + \frac{4x_k}{L^2} \tag{15.90a}$$

$$\frac{\partial \varepsilon_k^{(i)}}{\partial d_{e2}} = -y_j\left(\frac{12x_k}{L^3} - \frac{6}{L^2}\right) \qquad \frac{\partial \varepsilon_k^{(i)}}{\partial d_{e3}} = -y_j\left(\frac{6x_k}{L^2} - \frac{4}{L}\right) \tag{15.90b,c}$$

$$\frac{\partial \varepsilon_k^{(i)}}{\partial d_{e4}} = \frac{4}{L} - \frac{8x_k}{L^2} \qquad \frac{\partial \varepsilon_k^{(i)}}{\partial d_{e5}} = -\frac{1}{L} + \frac{4x_k}{L^2} \tag{15.90d,e}$$

$$\frac{\partial \varepsilon_k^{(i)}}{\partial d_{e6}} = -y_j\left(\frac{6}{L^2} - \frac{12x_k}{L^3}\right) \qquad \frac{\partial \varepsilon_k^{(i)}}{\partial d_{e7}} = -y_j\left(\frac{6x_k}{L^2} - \frac{2}{L}\right) \tag{15.90f,g}$$

the different terms included in the second matrix of Equation 15.89 are calculated as:

$$\frac{\partial N_k^{(i)}}{\partial d_{e1}} \approx \sum_{j=1}^{n_j} A_j\, \frac{\partial \sigma\left(x_k,y_j,\mathbf{d}_e^{(i)}\right)}{\partial \varepsilon}\frac{\partial \varepsilon_k^{(i)}}{\partial d_{e1}} = \sum_{j=1}^{n_j} A_j\, \frac{\partial \sigma\left(x_k,y_j,\mathbf{d}_e^{(i)}\right)}{\partial \varepsilon}\frac{\partial\left\{\left[1 \quad -y_j\right]\mathbf{B}(x_k)\mathbf{d}_e^{(i)}\right\}}{\partial d_{e1}}$$
$$= \sum_{j=1}^{n_j} A_j\, \frac{\partial \sigma\left(x_k,y_j,\mathbf{d}_e^{(i)}\right)}{\partial \varepsilon}\left(-\frac{3}{L} + \frac{4x_k}{L^2}\right) \tag{15.91a}$$

$$\frac{\partial N_k^{(i)}}{\partial d_{e2}} \approx \sum_{j=1}^{n_j} A_j\, \frac{\partial \sigma\left(x_k,y_j,\mathbf{d}_e^{(i)}\right)}{\partial \varepsilon}\frac{\partial \varepsilon_k^{(i)}}{\partial d_{e2}} = \sum_{j=1}^{n_j} A_j\, \frac{\partial \sigma\left(x_k,y_j,\mathbf{d}_e^{(i)}\right)}{\partial \varepsilon}\left[-y_j\left(\frac{12x_k}{L^3} - \frac{6}{L^2}\right)\right] \tag{15.91b}$$

$$\frac{\partial N_k^{(i)}}{\partial d_{e3}} \approx \sum_{j=1}^{n_j} A_j\, \frac{\partial \sigma\left(x_k,y_j,\mathbf{d}_e^{(i)}\right)}{\partial \varepsilon}\left[-y_j\left(\frac{6x_k}{L^2} - \frac{4}{L}\right)\right] \tag{15.91c}$$

$$\frac{\partial N_k^{(i)}}{\partial d_{e4}} \approx \sum_{j=1}^{n_j} A_j\, \frac{\partial \sigma\left(x_k,y_j,\mathbf{d}_e^{(i)}\right)}{\partial \varepsilon}\left(\frac{4}{L} - \frac{8x_k}{L^2}\right) \tag{15.91d}$$

$$\frac{\partial N_k^{(i)}}{\partial d_{e5}} \approx \sum_{j=1}^{n_j} A_j\, \frac{\partial \sigma\left(x_k,y_j,\mathbf{d}_e^{(i)}\right)}{\partial \varepsilon}\left(-\frac{1}{L} + \frac{4x_k}{L^2}\right) \tag{15.91e}$$

$$\frac{\partial N_k^{(i)}}{\partial d_{e6}} \approx \sum_{j=1}^{n_j} A_j\, \frac{\partial \sigma\left(x_k,y_j,\mathbf{d}_e^{(i)}\right)}{\partial \varepsilon}\left[-y_j\left(\frac{6}{L^2} - \frac{12x_k}{L^3}\right)\right] \tag{15.91f}$$

$$\frac{\partial N_k^{(i)}}{\partial d_{e7}} \approx \sum_{j=1}^{n_j} A_j \frac{\partial \sigma\left(x_k, y_j, \mathbf{d}_e^{(i)}\right)}{\partial \varepsilon}\left[-y_j\left(\frac{6x_k}{L^2} - \frac{2}{L}\right)\right] \tag{15.91g}$$

$$\frac{\partial M_k^{(i)}}{\partial d_{e1}} \approx -\sum_{j=1}^{n_j} y_j A_j \frac{\partial \sigma\left(x_k, y_j, \mathbf{d}_e^{(i)}\right)}{\partial \varepsilon}\left(-\frac{3}{L} + \frac{4x_k}{L^2}\right) \tag{15.92a}$$

$$\frac{\partial M_k^{(i)}}{\partial d_{e2}} \approx -\sum_{j=1}^{n_j} y_j A_j \frac{\partial \sigma\left(x_k, y_j, \mathbf{d}_e^{(i)}\right)}{\partial \varepsilon}\left[-y_j\left(\frac{12x_k}{L^3} - \frac{6}{L^2}\right)\right]$$
$$= \sum_{j=1}^{n_j} y_j^2 A_j \frac{\partial \sigma\left(x_k, y_j, \mathbf{d}_e^{(i)}\right)}{\partial \varepsilon}\left(\frac{12x_k}{L^3} - \frac{6}{L^2}\right) \tag{15.92b}$$

$$\frac{\partial M_k^{(i)}}{\partial d_{e3}} \approx \sum_{j=1}^{n_j} y_j^2 A_j \frac{\partial \sigma\left(x_k, y_j, \mathbf{d}_e^{(i)}\right)}{\partial \varepsilon}\left(\frac{6x_k}{L^2} - \frac{4}{L}\right) \tag{15.92c}$$

$$\frac{\partial M_k^{(i)}}{\partial d_{e4}} \approx -\sum_{j=1}^{n_j} y_j A_j \frac{\partial \sigma\left(x_k, y_j, \mathbf{d}_e^{(i)}\right)}{\partial \varepsilon}\left(\frac{4}{L} - \frac{8x_k}{L^2}\right) \tag{15.92d}$$

$$\frac{\partial M_k^{(i)}}{\partial d_{e5}} \approx -\sum_{j=1}^{n_j} y_j A_j \frac{\partial \sigma\left(x_k, y_j, \mathbf{d}_e^{(i)}\right)}{\partial \varepsilon}\left(-\frac{1}{L} + \frac{4x_k}{L^2}\right) \tag{15.92e}$$

$$\frac{\partial M_k^{(i)}}{\partial d_{e6}} \approx \sum_{j=1}^{n_j} y_j^2 A_j \frac{\partial \sigma\left(x_k, y_j, \mathbf{d}_e^{(i)}\right)}{\partial \varepsilon}\left(\frac{6}{L^2} - \frac{12x_k}{L^3}\right) \tag{15.92f}$$

$$\frac{\partial M_k^{(i)}}{\partial d_{e7}} \approx \sum_{j=1}^{n_j} y_j^2 A_j \frac{\partial \sigma\left(x_k, y_j, \mathbf{d}_e^{(i)}\right)}{\partial \varepsilon}\left(\frac{6x_k}{L^2} - \frac{2}{L}\right) \tag{15.92g}$$

Reconsidering the elastic–perfectly plastic constitutive model of Equations 15.2, the partial derivatives of the stress σ with respect to ε have already been calculated in Equations 15.60.

The stiffness matrix, the tangent stiffness components and the loading vector of the entire structure, required in the Newton–Raphson approach in Equations 15.61 and 15.62, can be obtained by assembling the contributions of the individual elements produced by Equations 15.82, 15.85 and 15.89 following standard finite element procedures (see Section 13.2).

The use of the Newton–Raphson procedure for the 7-dof finite element of Figure 15.20 is outlined in Worked Example 15.3.

WORKED EXAMPLE 15.3

Consider a simply-supported beam subjected to a mid-span point load and with span of 8 m. Calculate the deflection and curvature along the member length with the finite element approach for the following two applied mid-span loads: (1) $P = 405$ kN and (2) $P = 475$ kN. Use the Newton–Raphson method and, for the convergence criteria, take $norm_{tol2}$ equal to 10^{-5} (Equation 15.59b). Consider the rectangular cross-section (width of 100 mm and height of 360 mm) and the elastic–perfectly plastic material properties ($E = 200$ GPa and $f_y = 300$ MPa) of Worked Example 15.2. Compare the calculated results with those obtained using the analytical solutions in Section 15.3.2 and plotted in Figure 15.15. For the numerical integrations, use three Gauss integration points.

---

(1) The load $P = 405$ kN is applied in one load step and is included in the load vector $\mathbf{Q}$.

In the first iteration, we assume that the beam is undeformed, therefore described by a nil vector of displacements. Based on this, we can calculate the internal actions resisted by the beam $\mathbf{K}\left(\mathbf{D}_e^{(1)}\right)$ and substitute it in Equation 15.62 for the calculation of the residual loads to be used in the analysis: $\mathbf{Q}_R^{(1)} = \mathbf{Q} - \mathbf{K}\left(\mathbf{D}_e^{(1)}\right)$. In particular, $\mathbf{K}\left(\mathbf{D}_e^{(1)}\right)$ is obtained by collecting the contribution of each element $\mathbf{k}_e\left(x, \mathbf{d}_e^{(1)}\right)$ from Equation 15.82a.

We then calculate the tangent stiffness matrices for each element from Equations 15.89 to 15.92 and assemble these in $\mathbf{K}_t\left(\mathbf{D}_e^{(1)}\right)$. At this point, we can evaluate the unknown displacement increment by solving Equation 15.61 for $\Delta\mathbf{D}_e^{(1)}$.

We then assign $\mathbf{D}_e^{(2)} = \mathbf{D}_e^{(1)} + \Delta\mathbf{D}_e^{(1)} = \Delta\mathbf{D}_e^{(1)}$ and recalculate the internal actions $\mathbf{K}\left(\mathbf{D}_e^{(1)}\right)$ so that we can verify the acceptance criteria of the convergence on the basis of $norm_{tol2} = \left|\mathbf{Q}_R^{(2)}\right| / |\mathbf{Q}|$, where $\mathbf{Q}_R^{(2)} = \mathbf{Q} - \mathbf{K}\left(\mathbf{D}_e^{(2)}\right)$. If $norm_{tol2}$ is less than 10^{-5}, then convergence is reached and the solution can be post-processed; otherwise, an additional iteration is required. The actual number of iterations depends on the number of layers and mesh discretisation specified in the solution process.

Comparisons between the results obtained at mid-span with the analytical solutions derived in Section 15.3.2 (referred to in the figure as 'CFS' for closed-form solution) and the finite element results (referred to in the figure as 'FEA') are plotted in Figure 15.21 for different levels of discretisation to provide better insight into the convergence behaviour. The number of layers used at the cross-section has been varied between 40, 80 and 200 to highlight how the results obtained with the discretisation of 40 layers (selected in Worked Example 15.2 for this particular cross-section) provide an acceptable prediction of the structural response. As expected, the convergence of the curvature is slower than convergence of the deflection because the former is based on the second derivative of the latter.

(2) A similar procedure is followed for the calculation of the solution when the beam is subjected to the load $P = 475$ kN. Comparisons between analytical (CFS) and numerical (FEA) values for the mid-span results of deflection and curvature are illustrated in Figure 15.22, which show similar trends to those observed at point 1. Because of the higher level of nonlinearity developed in the beam, a larger number of elements are required, when compared to point 1, to approach the values calculated with the analytical solution.

The variations along the member length are presented in Figure 15.23 for both deflection and curvature using the values plotted in Figure 15.15 as reference. Good agreement between the analytical and finite element results (calculated with a highly refined mesh) is evident.

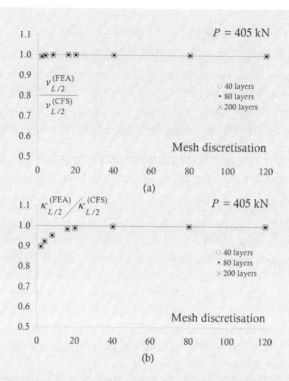

Figure 15.21 Convergence of the solution for load 1. (a) Mid-span deflection. (b) Mid-span curvature.

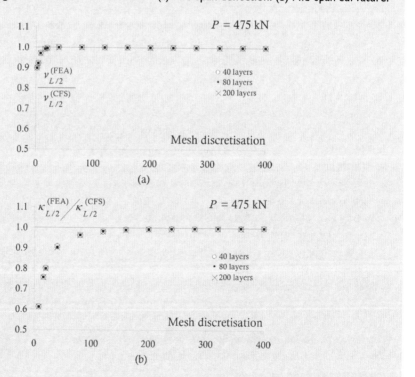

Figure 15.22 Convergence of the solution for load 2. (a) Mid-span deflection. (b) Mid-span curvature.

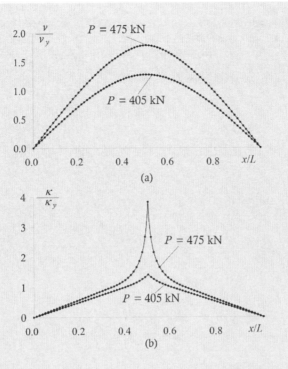

Figure 15.23 Variation of the solution along the member length. (a) Deflection. (b) Curvature.

PROBLEMS

15.1 Consider the following constitutive model (known as Ramberg–Osgood model):

$$\varepsilon = \frac{\sigma}{E} + 0.002\left(\frac{\sigma}{\sigma_{0.2}}\right)^{n}$$

where E is the elastic modulus, $\sigma_{0.2}$ represents the 0.2% proof stress and n defines the strain-hardening parameter. Plot this expression considering the following properties: $E = 200$ GPa, $\sigma_{0.2} = 400$ MPa and $n = 5$, and evaluate the derivatives of the stress σ with respect to the strain ε.

15.2 Consider a rectangular cross-section with width 100 mm and height 460 mm. Based on a cross-sectional analysis implemented with the Newton–Raphson method, calculate the curvature for the following levels of applied moments: (1) $M_e = 1000$ kNm and (2) $M_e = 1500$ kNm. Assume the material to follow the Ramberg–Osgood constitutive model specified in Problem 15.1, with $E = 200$ GPa, $\sigma_{0.2} = 400$ MPa and $n = 5$. In the solution, determine and adopt a suitable number of layers for the discretisation of the cross-section. Evaluate the convergence based on $\mathrm{norm}_{tol2} = 10^{-5}$.

15.3 Consider a simply-supported beam 10 m long with the rectangular cross-section of Problem 15.2. Calculate the variations of the deflection and curvature along the member length induced by the following point loads applied at mid-span:

(1) $P = 400$ kN and (2) $P = 600$ kN. Assume the material to follow the Ramberg–Osgood constitutive model, with $E = 200$ GPa, $\sigma_{0.2} = 400$ MPa and $n = 5$ (as considered in Problem 15.1). In the solution, determine and adopt a suitable number of layers for the discretisation of the cross-section. Evaluate the convergence based on $\text{norm}_{tol2} = 10^{-5}$. For the numerical integrations use 3 Gauss integration points.

Appendix A: Properties of plane sections

In structural analysis, the geometrical properties of the cross-sections of structural members are required to determine the structural deformation and the distribution of internal actions. In addition, the geometrical properties of shapes are needed to find the magnitude and position of the resultant of a distributed load acting on a plane surface and in many other types of problems.

A.1 CENTROID

Consider a plane surface of arbitrary shape subjected to a uniform pressure of intensity p as shown in Figure A.1. The resultant of the pressure distribution is P and it acts at a point on the arbitrary shape called the *centroid*.

The pressure acting on an infinitesimal area dA is $dP = p\,dA$ and the resultant force P acting on the shape is obtained by integration:

$$P = \int p\,dA = p \int dA = pA \tag{A.1}$$

where A is the area of the surface. The area A for regular-shaped surfaces (such as often occur in structural engineering) is easily calculated and integration is only necessary for unusual shapes. The resultant force P is equal to the volume of the pressure block of area A and thickness p (shown in Figure A.1).

Consider the plane surface of area A and the arbitrary axes Oz and Oy shown in Figure A.2a. To find the position of the centroid of the area (z_c, y_c), we simply have to find the position of the resultant force P if the area is subjected to a uniform pressure p (assumed to be acting into the page). To find the ordinate y_c, consider the infinitesimal force dP acting on the infinitesimal area dA, shown in Figure A.2b, where $dP = p\,dA$. The moment of this infinitesimal force about the axis Oz is $dM = y\,dP = yp\,dA$. The sum of the moments of all the infinitesimal forces about the axis Oz is therefore:

$$M = p \int_A y\,dA = pB_z \tag{A.2}$$

where the integral $\int_A y\,dA$ is known as the first moment of area about the axis Oz. It is a geometrical property of the surface of area A and is denoted here as B_z.

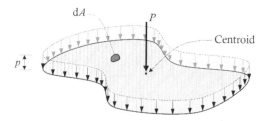

Figure A.1 Plane surface of arbitrary shape subjected to a uniform pressure.

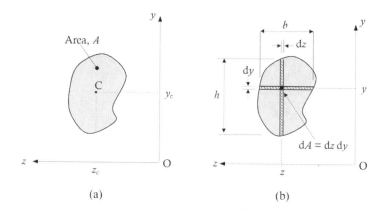

(a)   (b)

Figure A.2 Plane surface of a cross-section.

The sum of the moments of all the infinitesimal forces about the axis Oz (given by Equation A.2) must equal the moment of the resultant force P about the axis Oz:

$$M = pB_z = Py_c = pAy_c$$

and therefore:

$$y_c = \frac{B_z}{A} \quad \text{with} \quad B_z = \int_A y\,dA \tag{A.3a,b}$$

Similarly, the z coordinate of the centroid z_c is found by taking moments about the Oy axis and is given by:

$$z_c = \frac{B_y}{A} \quad \text{with} \quad B_y = \int_A z\,dA \tag{A.3c,d}$$

In the determination of the first moments of area about the y- and z-axes, B_z and B_y, respectively, the calculations of the integrations can be simplified by considering the infinitesimal strips of varying widths shown in Figure A.2b rather than the infinitesimal area dA. Consider the horizontal strip of width b and thickness dy. The infinitesimal force on this strip is d$P = pb$ dy and its moment about the Oz axis is d$M = y$ d$P = pby$ dy. The total moment about Oz is:

$$M_z = p\int by\,dy = pB_z$$

Similarly, the moment about Oy of the force on the vertical strip of height h and thickness dz shown in Figure A.2b is:

$$M_y = p\int hz\,dz = pB_y$$

The first moments of area about the z- and y-axes can then be calculated considering an infinitesimal strip of varying width as:

$$B_z = \int_A y\,dA = \int by\,dy \quad \text{and} \quad B_y = \int_A z\,dA = \int hz\,dz \qquad \text{(A.4a,b)}$$

where b and h need to be expressed (if necessary) in terms of y and z, respectively.

WORKED EXAMPLE A.1

For the shaded triangular area shown in Figure A.3, find the y coordinate of its centroid. The dimensions of the triangle are in mm.

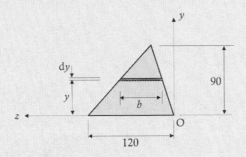

Figure A.3 Cross-section for Worked Example A.1.

The width b of the infinitesimal strip shown in Figure A.3 is first expressed in terms of y. From similar triangles:

$$\frac{90}{120} = \frac{90-y}{b} \qquad \therefore b = 120\left(1 - \frac{y}{90}\right)$$

From Equation A.4a:

$$B_z = \int_0^{90} by\,dy = \int_0^{90} 120\left(1 - \frac{y}{90}\right)y\,dy = \left[60y^2 - \frac{120}{270}y^3\right]_0^{90} = 162 \times 10^3 \text{ mm}^3$$

The area of the triangle is $A = 0.5 \times 90 \times 120 = 5400 \text{ mm}^2$, and from Equation A.3a:

$$y_c = \frac{B_z}{A} = \frac{162,000}{5400} = 30.0 \text{ mm}$$

The centroid of the triangle lies 30 mm above the base (i.e. one-third of the height of the triangle above the base).

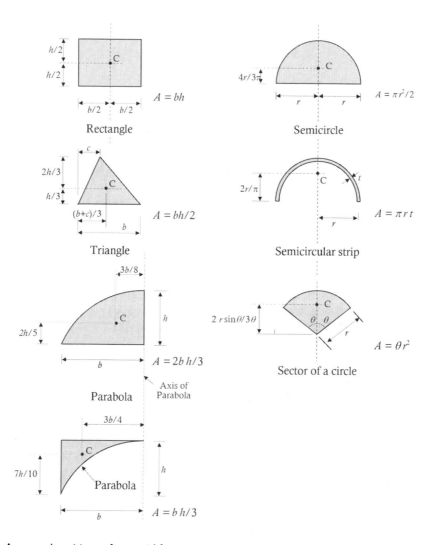

Figure A.4 Areas and positions of centroid for common geometrical shapes.

In Figure A.4, the area and the position of the centroid for some common geometrical shapes are provided.

Some particular features related to the centroid are noted below.

1. If a plane area has an axis of symmetry, the centroid will lie on that axis of symmetry.
2. If a plane area has two axes of symmetry, the centroid will lie on the intersection of the two axes of symmetry.
3. Axes that pass through the centroid are called *centroidal axes*. Since the resultant force P acts at the centroid, the first moment of area B about any axis passing through the centroid must be zero.
4. The first moment of an area about any axis is the product of the area A and the perpendicular distance of the axis from the centroid.

In most problems in structural engineering, the cross-sectional shapes for which the geometrical properties are required can be sub-divided into simple shapes whose areas and centroids are known. In such cases, integration is replaced by summation. This is illustrated in Worked Example A.2.

WORKED EXAMPLE A.2

Find the position of the centroid C of the cross-section shown in Figure A.5. All dimensions are in millimetres. The vertical y-axis drawn in the figure is an axis of symmetry.

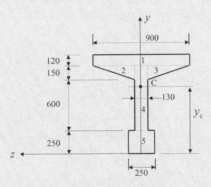

Figure A.5 Cross-section for Worked Example A.2.

The cross-section is here divided into five regular-shaped elements as shown in Figure A.5: elements 1, 4, and 5 are rectangular with dimensions 120 × 900 mm, 750 × 130 mm, and 250 × 250 mm, respectively, while elements 2 and 3 are triangular both with dimensions 150 × 385 mm. The position of the centroid of the cross-section is conveniently found using the following tabulation. In the second column, A_i is the area of the i-th element, and in the third column, y_{ci} is the y coordinate of the centroid of element i. The product $A_i\, y_{ci}$ in the fourth column is the first moment of the i-th element about the Oz axis.

Element	Area, A_i (mm^2)	y_{ci} (mm)	$A_i y_{ci}$ (mm^3)
I	108,000	1060	114.48 × 10^6
2	28,875	950	27.43 × 10^6
3	28,875	950	27.43 × 10^6
4	97,500	625	60.94 × 10^6
5	62,500	125	7.81 × 10^6
Sum	325,750		238.09 × 10^6

The height of the centroid above the bottom of the cross-section (i.e. above the Oz axis) is therefore:

$$y_c = \frac{\sum A_i y_{ci}}{\sum A_i} = 730.9 \text{ mm}$$

The centroid lies on the axis of symmetry at 730.9 mm above the Oz axis.

A.2 SECOND MOMENT AND PRODUCT MOMENT OF AREA

A.2.1 Second moment of area

We will now consider the cross-section in Figure A.6 subjected to a linearly varying stress from top to bottom (i.e. in the direction of the Oy axis). The stress is uniform in the direction of Oz. The position of the Oz axis corresponds to the level at which the linearly varying stress would be zero, as shown. The stress at any value of y in the range $y_{\text{btm}} \le y \le y_{\text{top}}$ is obtained from simple geometry as:

$$\sigma(y) = \frac{\sigma(y_{\text{top}})}{y_{\text{top}}} y = \frac{\sigma(y_{\text{btm}})}{y_{\text{btm}}} y = \frac{\sigma(y_A)}{y_A} y \tag{A.5}$$

where y_A defines the vertical position of an arbitrary point A within the cross-section and $\sigma(y_A)$ denotes the stress at A. The infinitesimal force on the horizontal strip shown in Figure A.6 is $dF = \sigma(y)dA$ and the resultant force F on the cross-section is:

$$F = \int_A \sigma(y)\,dA = \int_A \frac{\sigma(y_A)}{y_A} y\,dA = \frac{\sigma(y_A)}{y_A} \int_A y\,dA = \frac{\sigma(y_A)}{y_A} B_z \tag{A.6}$$

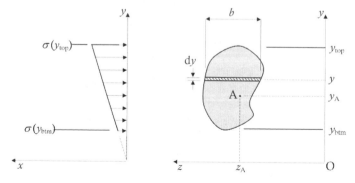

Figure A.6 Cross-section subjected to a linearly varying stress distribution.

recalling that B_z is the first moment of area about Oz and is given by Equation A.4a.

If the position of the centroid above Oz is y_c, then we know that:

$$B_z = y_c A \quad \text{and therefore} \quad F = \frac{\sigma(y_A)}{y_A} y_c A = \sigma(y_c)A \tag{A.7a,b}$$

To find the position y_F of the resultant force, we take moments about the Oz axis:

$$M_z = \int_A y\sigma(y)\,\mathrm{d}A = \int_A \frac{\sigma(y_A)}{y_A} y^2\,\mathrm{d}A = \frac{\sigma(y_A)}{y_A}\int_A y^2\,\mathrm{d}A = \frac{\sigma(y_A)}{y_A} I_{zz} \tag{A.8}$$

where I_{zz} is known as the *second moment of area* about Oz and is given by:

$$I_{zz} = \int_A y^2\,\mathrm{d}A \tag{A.9a}$$

or performing the integral using an infinitesimal strip of varying width:

$$I_{zz} = \int by^2\,\mathrm{d}y \tag{A.9b}$$

The moment M_z given by Equation A.8 is equal to the resultant force F multiplied by its distance above Oz (i.e. $M_z = Fy_F$), and therefore, from Equations A.7b and A.8:

$$M_z = \frac{\sigma(y_A)}{y_A} I_{zz} = \frac{\sigma(y_A)}{y_A} y_c A y_F$$

and rearranging gives:

$$y_F = \frac{I_{zz}}{y_c A} = \frac{I_{zz}}{B_z} \tag{A.10}$$

WORKED EXAMPLE A.3

For the shaded triangular area shown in Figure A.3 (and reproduced here as A.7), find the second moment of area about the Oz axis.

As in Worked Example A.1, the width b of the infinitesimal strip shown in Figure A.7 is first expressed in terms of y. From similar triangles:

$$b = 120\left(1 - \frac{y}{90}\right)$$

From Equation A.9b:

$$I_{zz} = \int_0^{90} by^2\,\mathrm{d}y = \int_0^{90} 120\left(1 - \frac{y}{90}\right)y^2\,\mathrm{d}y = \left[40y^3 - \frac{120}{360}y^4\right]_0^{90} = 7.29\times10^6 \text{ mm}^3$$

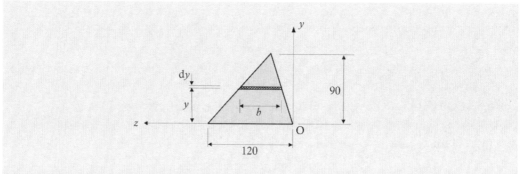

Figure A.7 Cross-section for Worked Example A.3.

Similar considerations could have been done for a stress distribution varying along the z-axis, based on which the second moment of area calculated with respect to y-axis is defined as:

$$I_{yy} = \int_A z^2 \, dA$$

or performing the integral using an infinitesimal strip of varying with h:

$$I_{yy} = \int h z^2 \, dz$$

A.2.2 Product moment of area

To determine the z coordinate of the position of the resultant force F, we will consider the elemental area dA shown in Figure A.8a. The force acting on dA is $dF = [\sigma(y_A)/y_A] y dA$ and the moment of that infinitesimal force about the Oy axis is:

$$dM_y = dFz = \frac{\sigma(y_A)}{y_A} \, y \, dA z \tag{A.11}$$

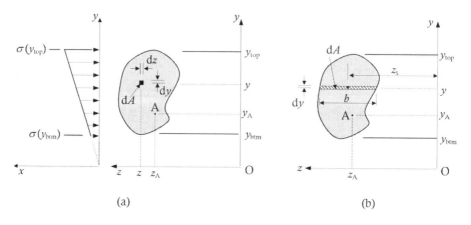

(a)   (b)

Figure A.8 Cross-sections with linearly varying stress distribution.

The moment of the stress block about the Oy axis is obtained by integrating over the area:

$$M_y = \int_A \frac{\sigma(y_A)}{y_A} yz \, dA = \frac{\sigma(y_A)}{y_A} \int_A yz \, dA = \frac{\sigma(y_A)}{y_A} I_{zy} \qquad (A.12)$$

where I_{zy} is called the product moment of area and is given by:

$$I_{zy} = \int_A yz \, dA \qquad (A.13)$$

The moment M_y given by Equation A.12 is equal to the resultant force F multiplied by its distance above Oy (i.e. $M_y = Fz_F = [\sigma(y_A)/y_A]y_c A z_F$), and therefore, from Equations A.7b and A.12:

$$M_y = \frac{\sigma(y_A)}{y_A} I_{zy} = \frac{\sigma(y_A)}{y_A} y_c A z_F$$

and rearranging gives:

$$x_F = \frac{I_{zy}}{y_c A} = \frac{I_{zy}}{B_z} \qquad (A.14)$$

If we now consider the infinitesimal strip of area $dA = b \, dy$ shown in Figure A.8b, subjected to a stress of $\sigma(y) = [\sigma(y_A)/y_A]y$, the moment of the force on dA about Oy is:

$$dM_y = \frac{\sigma(y_A)}{y_A} yb \, dy \, z_S \qquad (A.15)$$

where z_S is the distance from the Oy axis to the centroid of the infinitesimal strip dA. Equation A.15 can be expressed as:

$$M_y = \frac{\sigma(y_A)}{y_A} \int ybz_S \, dy = \frac{\sigma(y_A)}{y_A} I_{zy} \qquad (A.16)$$

where the product moment of area is calculated by performing the integral using an infinitesimal strip of width b as follows:

$$I_{zy} = \int_A yz \, dA = \int ybz_S \, dy \qquad (A.17)$$

where b needs to be expressed (if necessary) in terms of the axis coordinates.

Similar considerations could have been carried out considering a stress distribution varying along the z-axis.

WORKED EXAMPLE A.4

For the triangular area shown in Figure A.9, find the product moment of area, I_{zy}.

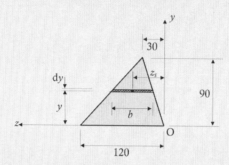

Figure A.9 Cross-section for Worked Example A.4.

As in Worked Example A.1, the width b of the elemental strip shown in Figure A.9 is:

$$b = 120\left(1 - \frac{y}{90}\right) = 120 - \frac{4y}{3}$$

and the z coordinate of its centroid is:

$$z_s = y/3 + b/2 = \frac{y}{3} + 60 - \frac{2y}{3} = 60 - \frac{y}{3}$$

Using Equation A.17 and considering the infinitesimal strip shown in Figure A.9:

$$I_{zy} = \int ybz_s\,dy$$

$$= \int_0^{90} y\left[120 - \frac{4y}{3}\right]\left[60 - \frac{y}{3}\right]dy = 7.29 \times 10^6 \text{ mm}^4.$$

A.2.3 Parallel axis theorems

When the value of the second moment of area is calculated about a particular axis, such as I_{zz} about axis Oz (I_{yy} about axis Oy), it is a relatively simple matter to determine the second moment of area about any parallel axis. Reconsidering Figure A.6 and Equation A.9, we can see that I_{zz} is always a positive quantity, irrespective of the position of the Oz axis (since y^2 must always be positive). It is noted that a change in position of the Oz axis corresponds to a change in position of the axis of zero stress (see Figure A.6).

Figure A.10 shows an infinitesimal area dA on a cross-section a distance y above the Oz axis. Also shown is the position of the centroidal axis Cz' parallel to Oz and y_c above it.

The second moment of area about the centrodial axis and about the Oz axis are, respectively:

$$I_{z'z'} = \int y'^2\,dA \quad \text{and} \quad I_{zz} = \int y^2\,dA$$

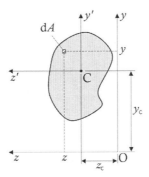

Figure A.10 Typical cross-section.

With $y = y' + y_c$, I_{zz} becomes:

$$I_{zz} = \int (y' + y_c)^2 \, dA = \int y'^2 \, dA + 2y_c \int y' \, dA + y_c^2 \int dA \qquad (A.18)$$

We saw in Section A.1 that the first moment of area about the centroidal axis is zero, i.e. $\int y' \, dA = 0$. Therefore, Equation A.18 becomes:

$$I_{zz} = \int y'^2 \, dA + y_c^2 \int dA = I_{z'z'} + y_c^2 A \qquad (A.19)$$

Both terms on the right-hand side of Equation A.19 are always positive, so it follows that the second moment of area about a centroidal axis is always less than the second moment of area about any other parallel axis.

The value of the product moment of inertia also changes with a change in the axis system. If the coordinates of the centroid of a cross-section with respect to the orthogonal axes Oz and Oy are z_c and y_c and if the product moment of inertia about the orthogonal centroidal axes Cz' and Cy' is $I_{z'y'}$, it can be readily shown that:

$$I_{zy} = I_{z'y'} + z_c y_c A \qquad (A.20)$$

Equations A.19 and A.20 are often called the *parallel axis theorem* and they are readily used to determine the second moment of area and product moment of area of irregular cross-sectional shapes that can be sub-divided into simple shapes, whose areas, centroids and second moments of area (I_{zz}) are known. This is illustrated in Worked Example A.5. The second moments of area and the product moment of area of some simple shapes are provided in Figure A.11.

The parallel theorem can also be derived for the calculation of the second moments of area with respect to the y-axis and this can be expressed as:

$$I_{yy} = I_{y'y'} + z_c^2 A$$

where $I_{y'y'}$ represents the second moment of area about the centroidal axis y' and z_c defines the location of the centroid of area A in the z direction.

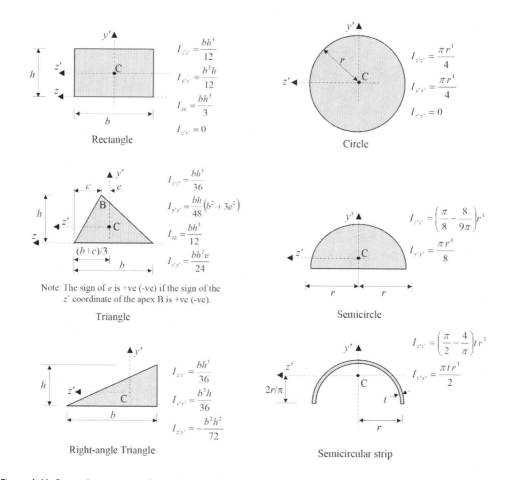

Figure A.11 Second moments of area and product moments of area for common geometrical shapes.

WORKED EXAMPLE A.5

For the Z-shaped cross-section shown in Figure A.12, determine the position of the centroid and the values of $I_{\bar{z}\bar{z}}, I_{\bar{y}\bar{y}}$ and $I_{\bar{z}\bar{y}}$ calculated with respect to the centroidal axes of the cross-section. All dimensions are in millimetres.

For convenience, we sub-divide the cross-section into three rectangular elements as shown. The top flange plate (element 1) is 30 × 120, the vertical web plate (element 2) is 300 × 40, and the bottom flange plate (element 3) is 40 × 240.

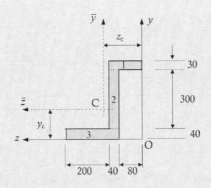

Figure A.12 Cross-section for Worked Example A.5.

We first calculate the location of the centroid.

Element i	A_i (mm²)	z_i (mm)	$B_{yi} = A_i z_i$ (mm³)	y_i (mm)	$B_{zi} = A_i y_i$ (mm³)
1	3600	60	216 × 10³	355	1278 × 10³
2	12,000	100	1200 × 10³	190	2280 × 10³
3	9600	200	1920 × 10³	20	192 × 10³
Sum	25,200		3336 × 10³		3750 × 10³

The coordinates of the centroid of the whole section are:

$$z_c = \frac{B_y}{A} = \frac{3336 \times 10^3}{25,200} = 132.4 \text{ mm} \quad \text{and} \quad y_c = \frac{B_z}{A} = \frac{3750 \times 10^3}{25,200} = 148.8 \text{ mm}$$

The coordinates of the centroids of the three elements with respect to the centroidal axes are:

Element 1:	$\bar{z}_1 = 60 - 132.4 = -72.4$	$\bar{y}_1 = 355 - 148.8 = +206.2$
Element 2:	$\bar{z}_2 = 100 - 132.4 = -32.4$	$\bar{y}_2 = 190 - 148.8 = +41.2$
Element 3:	$\bar{z}_3 = 200 - 132.4 = +67.6$	$\bar{y}_3 = 20 - 148.8 = -128.8$

and the values of $I_{\bar{z}\bar{z}}$, $I_{\bar{y}\bar{y}}$, and $I_{\bar{z}\bar{y}}$ for each element about its own centroidal axis are:

Element 1: $I_{z'z'1} = \dfrac{b_1 h_1^3}{12} = \dfrac{120 \times 30^3}{12} = 0.27 \times 10^6 \text{ mm}^4$

$I_{y'y'1} = \dfrac{b_1^3 h_1}{12} = \dfrac{120^3 \times 30}{12} = 4.32 \times 10^6 \text{ mm}^4 \quad \text{and} \quad I_{z'y'1} = 0$

Element 2: $I_{z'z'2} = \dfrac{b_2 h_2^3}{12} = \dfrac{40 \times 300^3}{12} = 90.0 \times 10^6 \text{ mm}^4$

$I_{y'y'2} = \dfrac{b_2^3 h_2}{12} = \dfrac{40^3 \times 300}{12} = 1.60 \times 10^6 \text{ mm}^4 \quad \text{and} \quad I_{z'y'2} = 0$

Element 3: $I_{z'z'3} = \dfrac{b_3 h_3^3}{12} = \dfrac{240 \times 40^3}{12} = 1.28 \times 10^6 \ \text{mm}^4$

$I_{y'y'3} = \dfrac{b_3^3 h_3}{12} = \dfrac{240^3 \times 40}{12} = 46.08 \times 10^6 \ \text{mm}^4 \quad \text{and} \quad I_{z'y'3} = 0$

We now calculate the values of $I_{\bar{z}\bar{z}}$, $I_{\bar{y}\bar{y}}$ and $I_{\bar{z}\bar{y}}$ for the whole section making use of Equations A.19 and A.20:

Element i	A_i (mm^2)	$I_{z'z'i}$	$A_i \bar{y}_i^2$	$I_{y'y'i}$ $(\times 10^6 \ mm^4)$	$A_i \bar{z}_i^2$	$I_{z'y'i}$	$A_i \bar{z}_i \bar{y}_i$
1	3600	0.27	153.07	4.32	18.87	0	−53.74
2	12,000	90.00	20.37	1.60	12.60	0	−16.02
3	9600	1.28	159.26	46.08	43.87	0	−83.59
Sum		91.55	332.70	52.00	75.34	0	−153.35

From Equation A.19:

$I_{\bar{z}\bar{z}} = \sum I_{z'z'i} + \sum A \bar{y}_i^2 = 424.25 \times 10^6 \ \text{mm}^4$

$I_{\bar{y}\bar{y}} = \sum I_{y'y'i} + \sum A \bar{z}_i^2 = 127.34 \times 10^6 \ \text{mm}^4$

$I_{\bar{z}\bar{y}} = \sum I_{z'y'i} + \sum A \bar{z}_i \bar{y}_i = -153.35 \times 10^6 \ \text{mm}^4$

Appendix B: Fixed-end moments

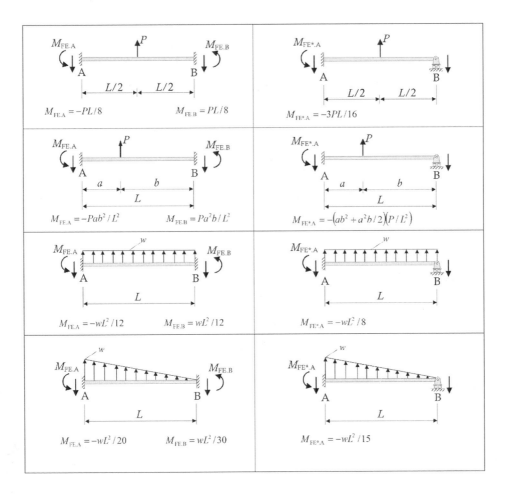

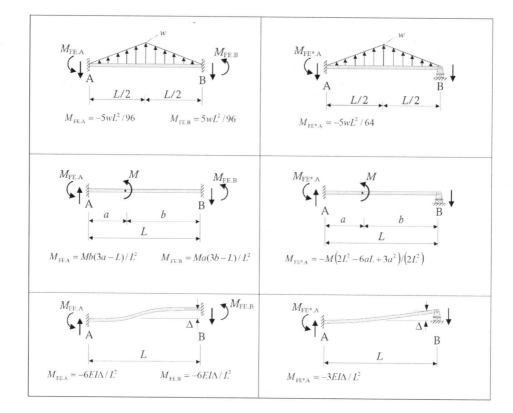

$M_{FE.A} = -5wL^2/96$ $M_{FE.B} = 5wL^2/96$

$M_{FE^*.A} = -5wL^2/64$

$M_{FE.A} = Mb(3a-L)/L^2$ $M_{FE.B} = Ma(3b-L)/L^2$

$M_{FE^*.A} = -M(2L^2 - 6aL + 3a^2)/(2L^2)$

$M_{FE.A} = -6EI\Delta/L^2$ $M_{FE.B} = -6EI\Delta/L^2$

$M_{FE^*.A} = -3EI\Delta/L^2$

Appendix C: Matrix algebra

C.I MATRICES

A *matrix* is a rectangular array of numbers, variables or functions. A matrix is said to have an order of $m \times n$ when it possesses m rows and n columns. For example, an $m \times n$ matrix **A** can be arranged as:

$$\mathbf{A} = \begin{bmatrix} A_{11} & A_{12} & \cdots & A_{1n} \\ A_{21} & A_{22} & & A_{2n} \\ \vdots & & \ddots & \vdots \\ A_{m1} & A_{m2} & \cdots & A_{mn} \end{bmatrix} \tag{C.1}$$

where A_{ij} represents the term located in the i-th row and j-th column.

A *row vector* is a matrix with a single row, while a *column vector* is a matrix with a single column.

A matrix that possesses the same number of columns and rows is referred to as a *square matrix*. A square matrix with non-zero terms only on the principal diagonal is said to be a *diagonal matrix*. When all diagonal terms (in a diagonal matrix) are equal to unity, it is defined as an *identity matrix*. Examples of a diagonal matrix **A** and an identity matrix **I** are:

$$\mathbf{A} = \begin{bmatrix} A_{11} & 0 & \cdots & 0 \\ 0 & A_{22} & & 0 \\ \vdots & & \ddots & \vdots \\ 0 & 0 & \cdots & A_{nm} \end{bmatrix} \qquad \mathbf{I} = \begin{bmatrix} 1 & 0 & \cdots & 0 \\ 0 & 1 & & 0 \\ \vdots & & \ddots & \vdots \\ 0 & 0 & \cdots & 1 \end{bmatrix} \tag{C.2,3}$$

A matrix is *symmetric* when $A_{ij} = A_{ji}$. An example of a symmetric matrix is provided below:

$$\mathbf{A} = \begin{bmatrix} 1 & 2 & 12 & -8 \\ 2 & 14 & 7 & 0 \\ 12 & 7 & 11 & 6 \\ -8 & 0 & 6 & 3 \end{bmatrix} \tag{C.4}$$

Two matrices are said to be equal only if all their terms coincide. For example, matrices **A** and **B** are equal only if $A_{ij} = B_{ij}$ for all values of i and j.

C.2 OPERATIONS WITH MATRICES

Two matrices can be added or subtracted by simply adding and subtracting their terms included in corresponding rows and columns. Additions and subtractions are possible only between matrices that possess the same numbers of rows and columns. For example, the addition and subtraction of two matrices $\mathbf{A}$ and $\mathbf{B}$:

$$\mathbf{A} = \begin{bmatrix} 3 & 2 & 5 \\ 7 & 7 & 7 \\ 4 & 6 & 2 \end{bmatrix} \qquad \mathbf{B} = \begin{bmatrix} 1 & 2 & 3 \\ 4 & 5 & 6 \\ 7 & 8 & 9 \end{bmatrix} \qquad \text{(C.5a,b)}$$

can be carried out based on:

$$\mathbf{A} + \mathbf{B} = \begin{bmatrix} 3 & 2 & 5 \\ 7 & 7 & 7 \\ 4 & 6 & 2 \end{bmatrix} + \begin{bmatrix} 1 & 2 & 3 \\ 4 & 5 & 6 \\ 7 & 8 & 9 \end{bmatrix} = \begin{bmatrix} 3+1 & 2+2 & 5+3 \\ 7+4 & 7+5 & 7+6 \\ 4+7 & 6+8 & 2+9 \end{bmatrix} = \begin{bmatrix} 4 & 4 & 8 \\ 11 & 12 & 13 \\ 11 & 14 & 11 \end{bmatrix} \quad \text{(C.6a)}$$

$$\mathbf{A} - \mathbf{B} = \begin{bmatrix} 3 & 2 & 5 \\ 7 & 7 & 7 \\ 4 & 6 & 2 \end{bmatrix} - \begin{bmatrix} 1 & 2 & 3 \\ 4 & 5 & 6 \\ 7 & 8 & 9 \end{bmatrix} = \begin{bmatrix} 3-1 & 2-2 & 5-3 \\ 7-4 & 7-5 & 7-6 \\ 4-7 & 6-8 & 2-9 \end{bmatrix} = \begin{bmatrix} 2 & 0 & 2 \\ 3 & 2 & 1 \\ -3 & -2 & -7 \end{bmatrix} \quad \text{(C.6b)}$$

The result of a matrix $\mathbf{A}$ multiplied by a scalar number k is another matrix whose components are equal to the components of matrix $\mathbf{A}$ multiplied by k. For example, reconsidering $\mathbf{A}$ from Equation C.5a, its product with k can be written as:

$$k\mathbf{A} = k \begin{bmatrix} 3 & 2 & 5 \\ 7 & 7 & 7 \\ 4 & 6 & 2 \end{bmatrix} = \begin{bmatrix} 3k & 2k & 5k \\ 7k & 7k & 7k \\ 4k & 6k & 2k \end{bmatrix} \qquad \text{(C.7)}$$

The product of two matrices $\mathbf{A}$ and $\mathbf{B}$ is equal to a matrix $\mathbf{C}$ whose coefficients are calculated based on:

$$C_{ij} = \sum_{k=1}^{n} A_{ik} B_{kj} \qquad \text{(C.8)}$$

which highlights the need for matrix $\mathbf{A}$ to have the number of columns equal to the number of rows of matrix $\mathbf{B}$. If this latter condition is not satisfied, it is not possible to carry out the multiplication. Based on this, the product of a matrix $\mathbf{A}$ $(m \times n)$ with $\mathbf{B}$ $(n \times q)$ is equal to a matrix $\mathbf{C}$ $(m \times q)$. For example, the product of matrices $\mathbf{A}$ and $\mathbf{B}$

$$\mathbf{A} = \begin{bmatrix} A_{11} & A_{12} \\ A_{21} & A_{22} \\ A_{31} & A_{32} \end{bmatrix} \qquad \mathbf{B} = \begin{bmatrix} B_{11} & B_{12} \\ B_{21} & B_{22} \end{bmatrix} \qquad \text{(C.9a,b)}$$

is equal to:

$$C = AB = \begin{bmatrix} A_{11} & A_{12} \\ A_{21} & A_{22} \\ A_{31} & A_{32} \end{bmatrix} \begin{bmatrix} B_{11} & B_{12} \\ B_{21} & B_{22} \end{bmatrix} = \begin{bmatrix} A_{11}B_{11}+A_{12}B_{21} & A_{11}B_{12}+A_{12}B_{22} \\ A_{21}B_{11}+A_{22}B_{21} & A_{21}B_{12}+A_{22}B_{22} \\ A_{31}B_{11}+A_{32}B_{21} & A_{31}B_{12}+A_{32}B_{22} \end{bmatrix} \qquad (C.10)$$

The multiplication of two matrices is:
 i. not commutative, i.e. $AB \neq AB$;
 ii. distributive, i.e. $A(B+C) = AB + AC$;
 iii. associative, i.e. $A(BC) = (AB)C$.

The transpose of a matrix A is referred to as A^T and is determined by interchanging its rows and columns. For example, the transpose of the matrix A defined in Equation C.9a is:

$$A^T = \begin{bmatrix} A_{11} & A_{21} & A_{31} \\ A_{12} & A_{22} & A_{32} \end{bmatrix} \qquad (C.11)$$

The transpose of the sum of two matrices is given in Equation C.12a. The transpose of the product of a matrix and a scalar is given in Equation C.12b, while the transpose of the product of two matrices may be expressed by Equation C.12c:

$$(A + B)^T = A^T + B^T \qquad (C.12a)$$

$$(kA)^T = kA^T \qquad (C.12b)$$

$$(AB)^T = B^T A^T \qquad (C.12c)$$

At times, it is useful to subdivide a matrix into sub-matrices. This process is usually referred to as partitioning. For example, the following matrix A is partitioned into submatrices A_{11}, A_{12}, A_{21} and A_{22}:

$$A = \begin{bmatrix} A_{11} & A_{12} & A_{13} & A_{14} \\ A_{21} & A_{22} & A_{23} & A_{24} \\ A_{31} & A_{32} & A_{33} & A_{34} \end{bmatrix} = \begin{bmatrix} A_{11} & A_{12} \\ A_{21} & A_{22} \end{bmatrix} \qquad (C.13)$$

where:

$$A_{11} = [A_{11}]; \; A_{12} = [A_{12} \quad A_{13} \quad A_{14}]; \; A_{21} = \begin{bmatrix} A_{21} \\ A_{31} \end{bmatrix}; \; A_{22} = \begin{bmatrix} A_{22} & A_{23} & A_{24} \\ A_{32} & A_{33} & A_{34} \end{bmatrix} \qquad (C.14a)$$

C.3 DETERMINANT OF A MATRIX

The determinant of a square matrix A is usually represented by either $\det(A)$ or $|A|$. A matrix whose determinant is equal to zero is denoted as *singular*.

In the case of a 2 × 2 matrix:

$$\mathbf{A} = \begin{bmatrix} A_{11} & A_{12} \\ A_{21} & A_{22} \end{bmatrix} \tag{C.15a}$$

the determinant is calculated as:

$$\det(\mathbf{A}) = |\mathbf{A}| = A_{11}A_{22} - A_{12}A_{21} \tag{C.15b}$$

For example, for $\mathbf{A} = \begin{bmatrix} 1 & 2 \\ 3 & 4 \end{bmatrix}$, we have $|\mathbf{A}| = (1)(4) - (2)(3) = -2$.

When dealing with a 3 × 3 matrix:

$$\mathbf{A} = \begin{bmatrix} A_{11} & A_{12} & A_{13} \\ A_{21} & A_{22} & A_{23} \\ A_{31} & A_{32} & A_{33} \end{bmatrix} \tag{C.16}$$

its determinant is evaluated as:

$$\begin{aligned} \det(\mathbf{A}) = |\mathbf{A}| &= A_{11}A_{22}A_{33} + A_{12}A_{23}A_{31} + A_{21}A_{32}A_{13} \\ &\quad - (A_{13}A_{22}A_{31} + A_{21}A_{12}A_{33} + A_{11}A_{32}A_{23}) \end{aligned} \tag{C.17}$$

For example, $\mathbf{A} = \begin{bmatrix} 2 & 3 & 4 \\ 3 & 5 & 7 \\ 11 & 4 & 9 \end{bmatrix}$, and its determinant is $\det(\mathbf{A}) = 12$.

The calculation of the determinant of a larger matrix $\mathbf{A}$ $(n \times n)$ is more tedious and is carried out as follows:

$$\det(\mathbf{A}) = A_{j1}C_{j1} + A_{j2}C_{j2} + \ldots + A_{jn}C_{jn} \tag{C.18}$$

where:

$$\mathbf{A} = \begin{bmatrix} A_{11} & A_{12} & \cdots & A_{1n} \\ A_{21} & A_{22} & & A_{2n} \\ \vdots & & \ddots & \vdots \\ A_{n1} & A_{n2} & \cdots & A_{nn} \end{bmatrix} \tag{C.19}$$

and A_{ji} (with i = 1,...,n) represent the terms in the j-th row of matrix, while C_{jk} is the cofactor defined as:

$$C_{jk} = (-1)^{j+k}M_{jk} \tag{C.20}$$

M_{jk} is referred to as the minor of matrix $\mathbf{A}$ and is the determinant of the matrix obtained by eliminating row j and column k from $\mathbf{A}$.

The same results could have been obtained by considering the components of $\mathbf{A}$ in its k-th column (unlike the j-th row as described in Equation C.18):

$$\det(\mathbf{A}) = |\mathbf{A}| = A_{1k}C_{1k} + A_{2k}C_{2k} + \ldots + A_{nk}C_{nk} \tag{C.21}$$

For example, considering the following matrix $\mathbf{A}$:

$$\mathbf{A} = \begin{bmatrix} 1 & 2 & 5 \\ 5 & 3 & 4 \\ 2 & 8 & 9 \end{bmatrix} \tag{C.22}$$

it is possible to determine the determinant following Equation C.18:

$$\det(\mathbf{A}) = |\mathbf{A}| = A_{11}C_{11} + A_{12}C_{12} + A_{13}C_{13} \tag{C.23}$$

where its cofactors are obtained as:

$$C_{11} = (-1)^{(1+1)}(-5) = -5 \quad C_{12} = (-1)^{(1+2)}(37) = -37 \tag{C.24a,b}$$

$$C_{13} = (-1)^{(1+3)}(34) = 34 \tag{C.24c}$$

based on the following minors:

$$M_{11} = \begin{vmatrix} 3 & 4 \\ 8 & 9 \end{vmatrix} = 3 \times 9 - 4 \times 8 = -5 \tag{C.25a}$$

$$M_{12} = \begin{vmatrix} 5 & 4 \\ 2 & 9 \end{vmatrix} = 5 \times 9 - 4 \times 2 = 37 \tag{C.25b}$$

$$M_{13} = \begin{vmatrix} 5 & 3 \\ 2 & 8 \end{vmatrix} = 5 \times 8 - 3 \times 2 = 34 \tag{C.25c}$$

The determinant can then be evaluated by substituting the calculated values in Equation C.23 as:

$$\det(\mathbf{A}) = |\mathbf{A}| = 1 \times (-5) + 2 \times (-37) + 5 \times 34 = 91 \tag{C.26}$$

C.4 INVERSE OF A MATRIX

The inverse of a square matrix $\mathbf{A}$ is denoted as $\mathbf{A}^{-1}$. Multiplying a matrix by its inverse (and vice-versa) produces the identity matrix $\mathbf{I}$ (see Equation C.3).

For the particular case of a 2 × 2 matrix, the inverse can be calculated as

$$\mathbf{A}^{-1} = \frac{1}{|\mathbf{A}|} \begin{bmatrix} A_{22} & -A_{12} \\ -A_{21} & A_{11} \end{bmatrix}$$
(C.27)

where $\mathbf{A} = \begin{bmatrix} A_{11} & A_{12} \\ A_{21} & A_{22} \end{bmatrix}$ and det $(\mathbf{A}) = |\mathbf{A}| = A_{11}A_{22} - A_{12}A_{21}$ (see Equations C.15).

For a larger matrix $\mathbf{A}$, its inverse is evaluated using:

$$\mathbf{A}^{-1} = \frac{1}{\det(\mathbf{A})} \mathbf{C}^{\mathrm{T}}$$
(C.28)

in which $\mathbf{C}$ is the matrix collecting all cofactors C_{ij} (already defined in Equation C.20).

From Equations C.27 and C.28, it becomes apparent that if det($\mathbf{A}$) = 0, the inverse would tend to infinity. Based on this, a matrix cannot be inverted if its determinant is zero, i.e. if it is singular.

Reconsidering matrix $\mathbf{A}$ of Equation C.22, its inverse is evaluated as (Equation C.28):

$$\mathbf{A}^{-1} = \frac{1}{\det(\mathbf{A})} \mathbf{C}^{\mathrm{T}} = \frac{1}{91} \begin{bmatrix} -5 & 22 & -7 \\ -37 & -1 & 21 \\ 34 & -4 & -7 \end{bmatrix}$$
(C.29)

where det($\mathbf{A}$) was calculated in Equation C.26, and $\mathbf{C}$ and its transpose are:

$$\mathbf{C} = \begin{bmatrix} -5 & -37 & 34 \\ 22 & -1 & -4 \\ -7 & 21 & -7 \end{bmatrix}$$
(C.30a)

and

$$\mathbf{C}^{\mathrm{T}} = \begin{bmatrix} -5 & 22 & -7 \\ -37 & -1 & 21 \\ 34 & -4 & -7 \end{bmatrix}$$
(C.30b)

All the cofactors are obtained as follows (Equation C.20):

$$C_{11} = (-1)^{1+1} M_{11} = \begin{vmatrix} 3 & 4 \\ 8 & 9 \end{vmatrix} = -5 \quad ; \quad C_{12} = (-1)^{1+2} M_{12} = -1 \times \begin{vmatrix} 5 & 4 \\ 2 & 9 \end{vmatrix} = -37 \quad \text{(C.31a,b)}$$

$$C_{13} = (-1)^{1+3} M_{13} = \begin{vmatrix} 5 & 3 \\ 2 & 8 \end{vmatrix} = 34 \quad ; \quad C_{21} = (-1)^{2+1} M_{21} = -1 \times \begin{vmatrix} 2 & 5 \\ 8 & 9 \end{vmatrix} = 22 \quad \text{(C.31c,d)}$$

$$C_{22} = (-1)^{2+2} M_{22} = \begin{vmatrix} 1 & 5 \\ 2 & 9 \end{vmatrix} = -1 \quad ; \quad C_{23} = (-1)^{2+3} M_{23} = -1 \times \begin{vmatrix} 1 & 2 \\ 2 & 8 \end{vmatrix} = -4 \quad \text{(C.31e,f)}$$

$$C_{31} = (-1)^{3+1} M_{31} = \begin{vmatrix} 2 & 5 \\ 3 & 4 \end{vmatrix} = -7 \quad ; \quad C_{32} = (-1)^{3+2} M_{32} = -1 \times \begin{vmatrix} 1 & 5 \\ 5 & 4 \end{vmatrix} = 21 \qquad \text{(C.31g,h)}$$

$$C_{33} = (-1)^{3+3} M_{33} = \begin{vmatrix} 1 & 2 \\ 5 & 3 \end{vmatrix} = -7 \qquad \text{(C.31i)}$$

C.5 SOLVING A SYSTEM OF LINEAR EQUATIONS

In this section, three procedures are introduced that can be used for the solution of a system of linear equations. In particular, we will consider:
- a method based on matrix algebra
- Cramer's rule
- the triangulation method, also known as Gauss elimination

These will be illustrated by considering the following system of three linear equations:

$$\begin{aligned}
A_{11}x_1 + A_{12}x_2 + A_{13}x_3 &= b_1 \\
A_{21}x_1 + A_{22}x_2 + A_{23}x_3 &= b_2 \\
A_{31}x_1 + A_{32}x_2 + A_{33}x_3 &= b_3
\end{aligned} \qquad \text{(C.32)}$$

which can be written in more compact form as:

$$\begin{bmatrix} A_{11} & A_{12} & A_{13} \\ A_{21} & A_{22} & A_{23} \\ A_{31} & A_{32} & A_{33} \end{bmatrix} \begin{bmatrix} x_1 \\ x_2 \\ x_3 \end{bmatrix} = \begin{bmatrix} b_1 \\ b_2 \\ b_3 \end{bmatrix} \qquad \text{(C.33)}$$

or

$$\mathbf{Ax} = \mathbf{b} \qquad \text{(C.34)}$$

Obviously, the proposed procedures are also applicable to larger and smaller systems of linear equations.

C.5.1 Method based on matrix algebra

The solution method based on matrix algebra involves the calculation of the inverse of matrix **A**. In particular, the vector of unknowns **x** is evaluated based on the following steps starting from Equation C.34.

1. Pre-multiplying both sides of Equation C.34 by the inverse of **A**:

$$\mathbf{A}^{-1}\mathbf{Ax} = \mathbf{A}^{-1}\mathbf{b} \qquad \text{(C.35)}$$

2. Recalling that $I = A^{-1}A$:

$$Ix = A^{-1}b \tag{C.36}$$

3. From which x can be calculated as:

$$x = A^{-1}b \tag{C.37}$$

C.5.2 Cramer's rule

This method is very useful for solving a system of linear equations when only a few unknowns need to be evaluated. For example, reconsidering the system of equations outlined in Equation C.32, the unknown value for x_1 can be determined using:

$$x_1 = \frac{\begin{vmatrix} b_1 & A_{12} & A_{13} \\ b_2 & A_{22} & A_{23} \\ b_3 & A_{32} & A_{33} \end{vmatrix}}{|A|} \tag{C.38}$$

where the matrix considered in the numerator is obtained by replacing the first column of matrix A with the vector of known coefficients b. In a similar manner, the value for x_2 is calculated as:

$$x_2 = \frac{\begin{vmatrix} A_{11} & b_1 & A_{13} \\ A_{21} & b_2 & A_{23} \\ A_{31} & b_3 & A_{33} \end{vmatrix}}{|A|} \tag{C.39}$$

In general, the unknown x_j can be calculated as the ratio of the determinants of two matrices, where the matrix considered in the numerator is obtained replacing the j-th column of matrix A with the column of known coefficients b and the matrix considered in the denominator is A.

C.5.3 Triangulation method (Gauss elimination)

The procedure involved with the triangulation method, also known as the Gauss elimination, is outlined in this section by means of an example. For this purpose, we will consider the following system of linear equations:

$$x_1 + 4x_2 - 2x_3 = 2 \tag{C.40a}$$

$$4x_1 + 4x_2 + 3x_3 = 1 \tag{C.40b}$$

$$2x_1 + 7x_2 + 5x_3 = 5 \tag{C.40c}$$

1. We first determine the expression for x_1 from the first equation (Equation C.40a)

$$x_1 = -4x_2 + 2x_3 + 2 \tag{C.41}$$

and substitute it in all remaining equations (Equations C.40b and c):

$$4(2 - 4x_2 + 2x_3) + 4x_2 + 3x_3 = 1 \tag{C.42a}$$

$$2(2 - 4x_2 + 2x_3) + 7x_2 + 5x_3 = 5 \tag{C.42b}$$

Simplifying:

$$12x_2 - 11x_3 = 7 \tag{C.43a}$$

$$x_2 - 9x_3 = -1 \tag{C.43b}$$

2. We then evaluate x_2 from Equation C.43a:

$$x_2 = \frac{11}{12}x_3 + \frac{7}{12} \tag{C.44}$$

and substitute in into the remaining equation (Equation C.43b):

$$\left(\frac{11}{12}x_3 + \frac{7}{12}\right) - 9x_3 = -1 \quad \text{and simplifying:} \quad \frac{97}{12}x_3 = \frac{19}{12} \tag{C.45a,b}$$

3. Finally, we calculate x_3 from Equation C.45b:

$$x_3 = \frac{19}{97} \tag{C.46}$$

4. At this point, we can determine the expressions for all unknowns, i.e. x_1 and x_2, by back-substituting x_3 (Equation C.46) into Equations C.41 and C.44:

$$x_2 = \frac{11}{12}\frac{19}{97} + \frac{7}{12} = \frac{74}{97} \tag{C.47a}$$

$$x_1 = -4\frac{74}{97} + 2\frac{19}{97} + 2 = -\frac{64}{97} \tag{C.47b}$$

C.5.4 Matrix form of the triangulation method (Gauss elimination)

The triangulation method (Gauss elimination) previously outlined can also be implemented in matrix form. This is outlined below using the previous example (Equations C.40).

The idea at the basis of the matrix form of the triangulation method is to convert the system of equations from its initial format:

$$\begin{bmatrix} 1 & 4 & -2 \\ 4 & 4 & 3 \\ 2 & 7 & 5 \end{bmatrix} \begin{bmatrix} x_1 \\ x_2 \\ x_3 \end{bmatrix} = \begin{bmatrix} 2 \\ 1 \\ 5 \end{bmatrix} \tag{C.48}$$

to:

$$\begin{bmatrix} 1 & 0 & 0 \\ 0 & 1 & 0 \\ 0 & 0 & 1 \end{bmatrix} \begin{bmatrix} x_1 \\ x_2 \\ x_3 \end{bmatrix} = \begin{bmatrix} b_1 \\ b_2 \\ b_3 \end{bmatrix} \tag{C.49}$$

so that the unknowns x_1, x_2, and x_3 equal the terms included in the vector of known terms. To achieve this, we need to perform a number of operations on the equations of the system to transform the matrix $\mathbf{A}$ into an identity matrix $\mathbf{I}$. The required steps are outlined in the following:

1. We start by verifying whether the term A_{11} is equal to unity. In this case, it is (see Equation C.48), and we do not need to change the first equation.

2. We then manipulate the matrix so that all terms below A_{11} are nil. For example, the new second line is equal to the old one minus 4 times line 1. In this way, we make sure that A_{21} will be zero in the revised line:

$$A_{21}(\text{new}) = A_{21}(\text{old}) - A_{21}(\text{old})A_{11} = 4 - 4 \times 1 = 0 \tag{C.50a}$$

$$A_{22}(\text{new}) = A_{22}(\text{old}) - A_{21}(\text{old})A_{12} = 4 - 4 \times 4 = -12 \tag{C.50b}$$

$$A_{23}(\text{new}) = A_{23}(\text{old}) - A_{21}(\text{old})A_{13} = 3 - 4 \times (-2) = 11 \tag{C.50c}$$

$$b_2(\text{new}) = b_2(\text{old}) - A_{21}(\text{old})b_1 = 1 - 4 \times 2 = -7 \tag{C.50d}$$

from which:

$$\begin{bmatrix} 1 & 4 & -2 \\ 0 & -12 & 11 \\ 2 & 7 & 5 \end{bmatrix} \begin{bmatrix} x_1 \\ x_2 \\ x_3 \end{bmatrix} = \begin{bmatrix} 2 \\ -7 \\ 5 \end{bmatrix} \tag{C.51}$$

In a similar way, the third line of the system of Equation C.51 is modified as follows: Line 3 (new) = Line 3 (old) − A_{31}(old) × Line 1, which leads to A_{31}(new) = 0 and:

$$\begin{bmatrix} 1 & 4 & -2 \\ 0 & -12 & 11 \\ 0 & -1 & 9 \end{bmatrix} \begin{bmatrix} x_1 \\ x_2 \\ x_3 \end{bmatrix} = \begin{bmatrix} 2 \\ -7 \\ 1 \end{bmatrix} \tag{C.52}$$

3. We move to the second column and make sure to equate A_{22} to unity. This is achieved by dividing the second equation (specified in Equation C.52) by A_{22}. In this case, we divide the second equation by −12:

$$\begin{bmatrix} 1 & 4 & -2 \\ 0 & 1 & -11/12 \\ 0 & -1 & 9 \end{bmatrix} \begin{bmatrix} x_1 \\ x_2 \\ x_3 \end{bmatrix} = \begin{bmatrix} 2 \\ 7/12 \\ 1 \end{bmatrix} \tag{C.53}$$

4. We can then operate on the terms below A_{22} and ensure that these become zero following a similar procedure to the one adopted at step 2:

$$\begin{bmatrix} 1 & 4 & -2 \\ 0 & 1 & -11/12 \\ 0 & 0 & 97/12 \end{bmatrix} \begin{bmatrix} x_1 \\ x_2 \\ x_3 \end{bmatrix} = \begin{bmatrix} 2 \\ 7/12 \\ 19/12 \end{bmatrix} \tag{C.54}$$

5. Finally, we ensure that A_{33} is 1 (by dividing the third equation by 97/12 as specified in Equation C.54 for A_{33}):

$$\begin{bmatrix} 1 & 4 & -2 \\ 0 & 1 & -11/12 \\ 0 & 0 & 1 \end{bmatrix} \begin{bmatrix} x_1 \\ x_2 \\ x_3 \end{bmatrix} = \begin{bmatrix} 2 \\ 7/12 \\ 19/97 \end{bmatrix} \tag{C.55}$$

6. The system of equation is further manipulated to ensure that all terms above A_{33}, and consequently above A_{22}, become nil, as shown in the following:

$$\begin{bmatrix} 1 & 4 & -2 \\ 0 & 1 & 0 \\ 0 & 0 & 1 \end{bmatrix} \begin{bmatrix} x_1 \\ x_2 \\ x_3 \end{bmatrix} = \begin{bmatrix} 2 \\ 74/97 \\ 19/97 \end{bmatrix} \tag{C.56a}$$

$$\begin{bmatrix} 1 & 4 & 0 \\ 0 & 1 & 0 \\ 0 & 0 & 1 \end{bmatrix} \begin{bmatrix} x_1 \\ x_2 \\ x_3 \end{bmatrix} = \begin{bmatrix} 232/97 \\ 74/97 \\ 19/97 \end{bmatrix} \tag{C.56b}$$

$$\begin{bmatrix} 1 & 0 & 0 \\ 0 & 1 & 0 \\ 0 & 0 & 1 \end{bmatrix} \begin{bmatrix} x_1 \\ x_2 \\ x_3 \end{bmatrix} = \begin{bmatrix} -64/97 \\ 74/97 \\ 19/97 \end{bmatrix} \tag{C.56c}$$

7. The values of the unknown variables are obtained directly from Equation C.56c as:

$$\begin{bmatrix} x_1 \\ x_2 \\ x_3 \end{bmatrix} = \begin{bmatrix} -64/97 \\ 74/97 \\ 19/97 \end{bmatrix} \tag{C.57}$$

C.6 DETERMINATION OF THE RANK OF A MATRIX

The rank of a matrix $\mathbf{A}$ is usually referred to as rank($\mathbf{A}$) and it represents the number of linearly independent rows or columns of $\mathbf{A}$. In the case of an $m \times m$ square matrix, it is said to be non-singular only if its rank is equal to m.

In the following, we will consider a possible approach suitable for hand calculation that requires the matrix to be reduced into a simpler form following steps 1 to 5 previously adopted in the matrix manipulation for the triangulation method (see Equations C.50 through C.55). This approach is acceptable as these operations on row or columns do not influence the matrix rank. At the completion of the matrix manipulation, the rank of the matrix is equal to the number of rows (or columns) with non-zero values.

For example, reconsidering the revised matrix of $\mathbf{A}$ included in Equation C.55:

$$\mathbf{A} = \begin{bmatrix} 1 & 4 & -2 \\ 0 & 1 & -11/12 \\ 0 & 0 & 1 \end{bmatrix} \tag{C.58}$$

All three rows have non-zero values and, because of this, the rank of the matrix is 3.

Let us now consider the following matrix $\mathbf{B}$:

$$\mathbf{B} = \begin{bmatrix} 3 & 6 & 15 \\ 10 & 6 & 8 \\ 6 & 5 & 9 \end{bmatrix} \tag{C.59}$$

After applying steps 1 to 5 (see Equations C.50 through C.55), we obtain the following revised version of $\mathbf{B}$:

$$\mathbf{B} = \begin{bmatrix} 1 & 2 & 5 \\ 0 & 1 & 3 \\ 0 & 0 & 0 \end{bmatrix} \tag{C.60}$$

in which case the rank is equal to 2 because there are only 2 rows with non-zero values (as all terms in the last row are equal to zero).

Index